Electronics Theory and Practice

Electronics Theory and Practice

Gerardo Mesias, CEng, MIEE

Newnes
An imprint of Butterworth-Heinemann Ltd
Linacre House, Jordan Hill, Oxford OX2 8DP

A member of the Reed Elsevier group

OXFORD LONDON BOSTON
MUNICH NEW DELHI SINGAPORE SYDNEY
TOKYO TORONTO WELLINGTON

First published 1993

British Library Cataloguing in Publication Data

Mesias, Gerardo
Electronics Theory and Practice
I. Title
621.38

ISBN 0-7506-1679-2

Library of Congress Cataloging in Publication Data

Mesias, Gerardo.
Electronics theory and practice/Gerardo Mesias.
p. cm.
Includes index.
ISBN 0-7506-1679-2
1. Electronics. I. Title.
TK7816.M47 1993
621.381—dc20 93-25888
CIP

Typeset by Scribe Design, Gillingham, Kent
Printed and bound in Great Britain

Contents

Preface

This book has been written to help you learn how to solve problems in electronics. Whether studying for BEng, HND or Graduate Diploma, you need to be able to cope with assignments and face examinations with confidence.

The basis of this confidence is practice in tackling problems. In solving a problem in electronics you are trying to express the circuit in mathematical terms: you are building a mathematical model of the circuit. The problems in this book, which are the result of long experience of students' needs in tutorial and remedial work, show how this is done.

All the problems are supplied with answers and complete worked solutions. This is useful because the answer obtained varies according to the method followed and the approximations made: two different results, such as 6.94 V and 7.06 V, may both be acceptable answers to a problem. Some simpler problems can be solved in a number of different ways. This can be a way of checking your result – by comparing the results by different methods. You should always arrive at the same result unless approximations have been made somewhere along the line. If you have gone wrong, you can check against the solution given in the book: try to identify exactly where you went wrong and how you can put it right.

The first chapter covers all the main laws and theorems needed to solve the problems in the following chapters. Each chapter starts with a concise explanation of the theory, which is followed by graded problems, starting with simple examples and progressing to the more complicated problems. The chapters are self-contained, and can be tackled independently in any order, referring to the first chapter as required for the basic theorems.

This book is intended for your own study. Once you are familiar with it you will find the way of using it that suits you best: in mastering the fundamental theorems and the different electronics topics, and in preparing for your examinations.

To
Camilo,
Thomas Oliver,
Ana-Claudia and
Gaston

1

Fundamental theorems

Ohm's law

Ohm's law is given by

$$V = IR$$

The voltage in a passive element is given by the product of the current multiplied by the resistance.

We can also easily deduce that:

$$I = \frac{V}{R} \qquad R = \frac{V}{I}$$

Some people prefer to use the magic triangle. This is of unknown origin, but is apparently widely used in secondary education. The magic triangle can be seen in Figure 1.1. In order to find V in the triangle, you cover V and you are left with IR. If you want to select another value, you cover the one you want and get the answer in the uncovered part.

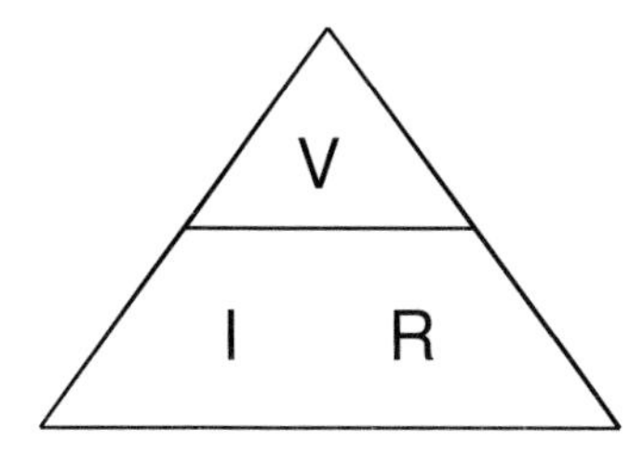

Figure 1.1

There are two points that need to be made at this stage, about Ohm's law. First, in Ohm's law the three elements of the formula, V , I and R must be related, i.e., the current must be going through the resistor and the voltage will appear as a consequence of that current across the resistor. Remember that the voltage refers to the difference in voltage between two points. If a point is said to have 12 V, we imply that there are 12 V from that point to a reference point. The reference point is usually ground, at 0 V. If the reference point is other than ground, the value is clearly stated.

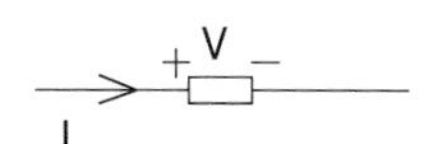

Figure 1.2

Secondly, it is important to realise from the very beginning, that the voltage will have a polarity, depending on the direction of the current that goes through the resistor. The relationship is shown in Figure 1.2. The side of the resistor where the current goes in is the positive side of the voltage. The side where the current goes out is the negative side of the voltage.

Another point that we would like to make is that Ohm's law does not apply to a voltage source, nor to a current source. By definition, a 12 V source will provide any current at 12 V. We can have 12 V and zero current or 12 V and 10 A, but we would not try to find a resistor in this case. Similarly for a current source. The current source will produce a current at any voltage. It is important to take this into account when solving problems with current and voltage sources.

Kirchhoff's current law (KCL)

This law refers to the currents in a junction or node. It is illustrated in Figure 1.3. The algebraic sum of the currents going into a node is equal to zero. Currents going in are positive and those going out of the node are negative.

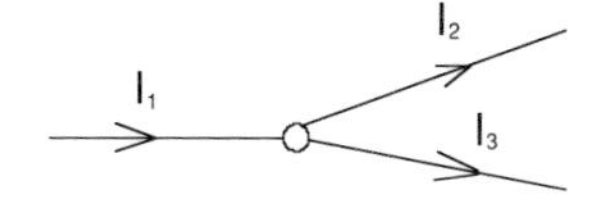

Figure 1.3

Another easier way to understand is to say the sum of the currents going in are equal to the sum of the currents going out of the node:

$$\Sigma I_{in} = \Sigma I_{out}$$

Kirchhoff's voltage law (KVL)

The sum of the voltages around a closed loop are equal to zero. Another way of saying the same thing, that is perhaps easier to understand is: the sum of the voltage sources is equal to the voltage drops around a closed loop. This can be seen in Figure 1.4.

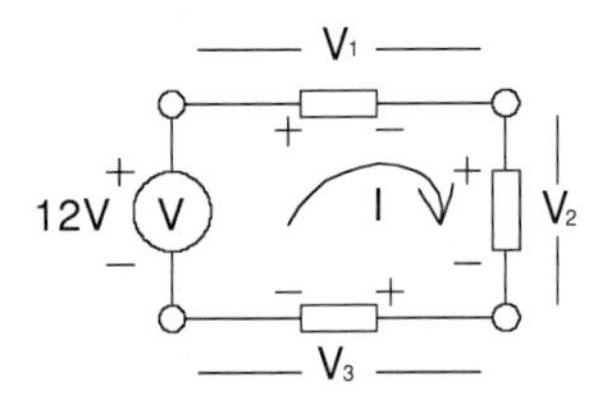

Figure 1.4

The equations are:

For Kirchhoff's law:

$$-12V + V_1 + V_2 + V_3 = 0$$

For the second alternative:

$$12V = V_1 + V_2 + V_3$$

PROBLEM 1.1 KIRCHHOFF

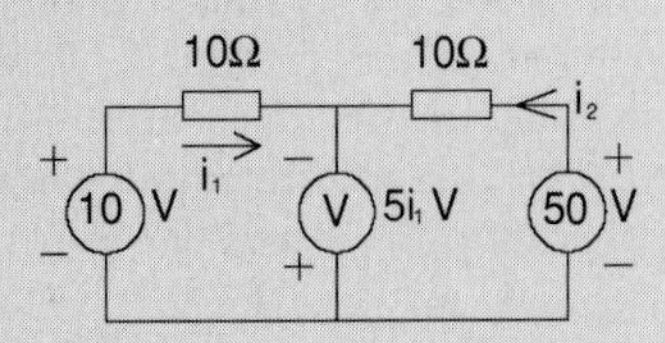

Figure 1.5

The two-mesh circuit has a current-dependent voltage source V. The polarities are as indicated. Find the value of i_2.

Answer: $i_2 = 6$ A

The sum of the voltages is equal to the sum of the voltage drops in the left mesh.

$$10 + 5i_1 = 10i_1 \quad (1)$$

Similarly, in the second loop

$$50 + 5i_1 = 10i_2 \quad (2)$$

From equation (1)

$$10 = 5i_1$$

$$i_1 = 2 \text{ A}$$

Replacing this value in equation (2)

$$50 + 5i_1 = 10i_2$$

$$50 + 10 = 10i_2$$

$$i_2 = 6 \text{ A}$$

PROBLEM 1.2 KIRCHHOFF

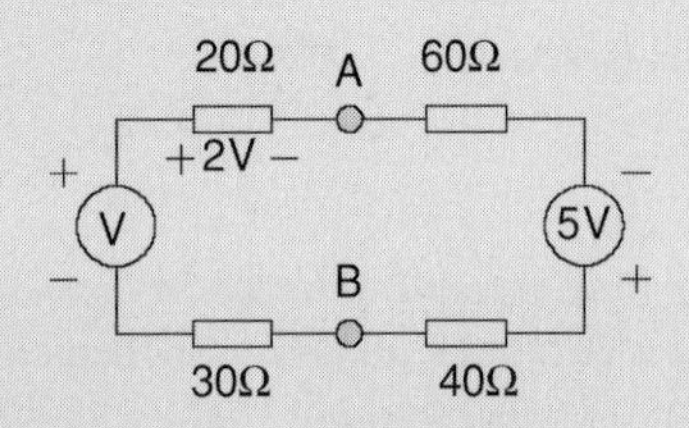

Figure 1.6

Find the value of V and V_{AB}.

Answer: 10 V and 5 V

There is only one current and to produce 2 V in the 20 Ω resistor it must be

$$i = \frac{2}{20} = 0.1 \text{ A}$$

From KVL

$$V + 5 = iR_T$$

$$= i(20 + 30 + 40 + 60)$$

$$= 0.1 \times 150 = 15$$

$$V = 10 \text{ V}$$

$$V_{AB} = V - i(20 + 30)$$

$$= 10 - 0.1(50)$$

$$= 10 - 5 = 5 \text{ V}$$

Also

$$V_{AB} = -5 + i(60 + 40)$$

$$= -5 + 0.1(100)$$

$$= -5 + 10 = 5 \text{ V}$$

$$= 10 - 5 = 5 \text{ V}$$

PROBLEM 1.3 KIRCHHOFF

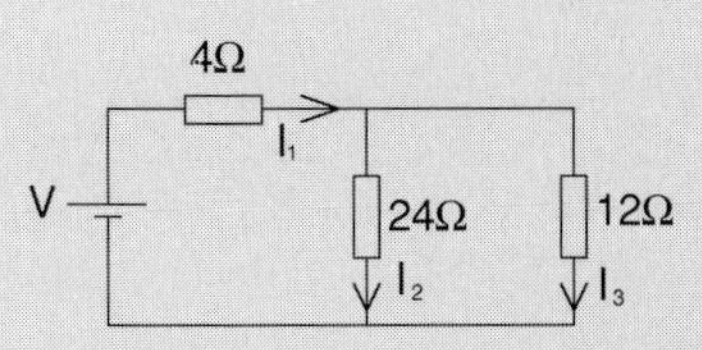

Figure 1.7

Calculate I_1, I_2 and V if $I_3 = 2$ A.

Answer: $I_1 = 3$ A, $I_2 = 1$ A and $V = 36$ V

On the 12 Ω branch we can apply Ohm's law

$$V = IR = 2 \times 12 = 24 \text{ V}$$

The 24 Ω branch has the same voltage as the 12 Ω branch. We apply Ohm's law again

$$I_2 = \frac{V}{R} = \frac{24}{24} = 1\text{A}$$

According to KCL

$$I_1 = I_2 + I_3$$
$$= 1 + 2 = 3 \text{ A}$$

V is given, according to KVL, by

$$V = 24 + I_1R = 24 + 3 \times 4$$
$$= 24 + 12 = 36 \text{ V}$$

PROBLEM 1.4 KIRCHHOFF

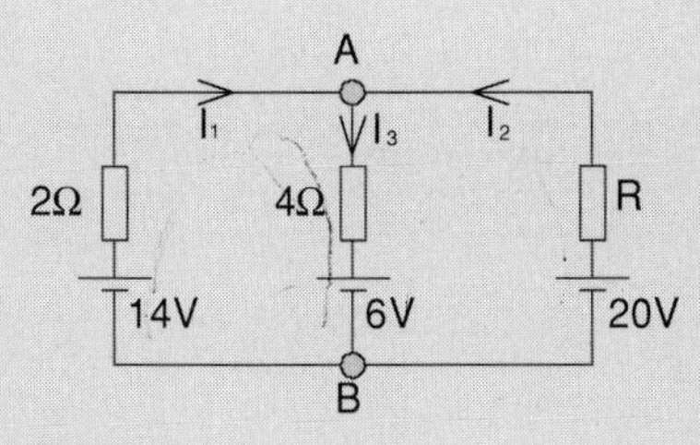

Figure 1.8

Find the value of I_1, I_2 and R, if $I_3 = 1.5$ A.

Answer: $I_1 = 1$ A, $I_2 = 0.5$ A, $R = 16$ Ω

$$V_{AB} = 6 + I_3R = 6 + 1.5 \times 4 = 12 \text{ V}$$

On the left branch

$$V_{AB} = 14 - 2I_1$$
$$12 = 14 - 2I_1$$
$$-2 = -2I_1$$
$$I_1 = 1 \text{ A}$$

On the right branch

$$V_{AB} = 20 - I_2R$$

Note the sign of the voltages, given by the rule of polarity. I_2 can be found from KCL

$$I_2 = I_3 - I_1$$
$$= 1.5 - 1 = 0.5 \text{ A}$$
$$V_{AB} = 20 - I_2R$$
$$12 = 20 - 0.5R$$
$$0.5R = 8$$
$$R = 16 \ \Omega$$

PROBLEM 1.5 KIRCHHOFF

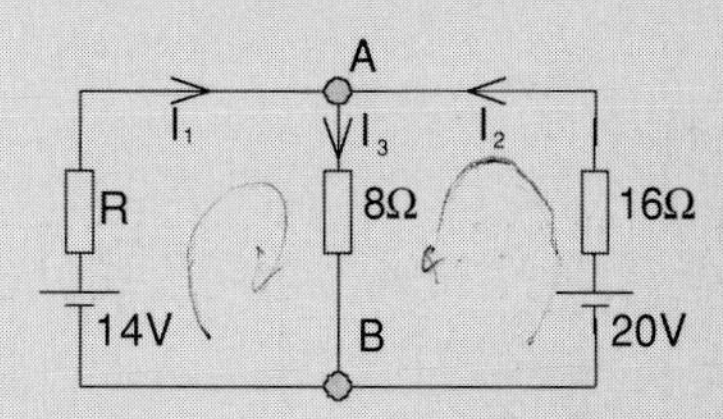

Figure 1.9

Find I_1, I_3 and R if I_2 is equal to 0.5 A.

Answer: $I_1 = 1$ A, $I_3 = 1.5$ A and $R = 2$ Ω

Following the sense of the currents shown, the polarity of the 16 Ω resistor will oppose the 20 V supply.

The voltage AB will be

$$V_{AB} = 20 - 16 \times 0.5 = 20 - 8 = 12 \text{ V}$$

Then

$$I_3 = \frac{V}{R} = \frac{12}{8} = 1.5\text{A}$$

From KCL

$$I_1 = I_3 - I_2 = 1.5 - 0.5 = 1 \text{ A}$$

R produces a voltage drop of

$$V_{AB} = 14 - I_1R$$
$$12 = 14 - I_1R = 14 - R$$
$$R = 2 \ \Omega$$

PROBLEM 1.6 KIRCHHOFF

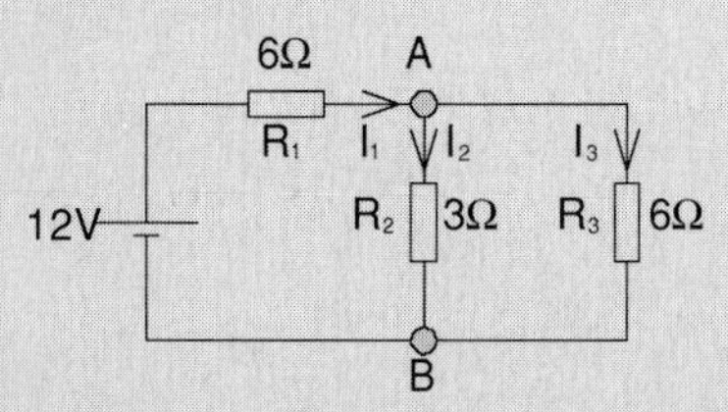

Figure 1.10

Find I_1, I_2 and I_3.

Answer: 1.5 A, 1 A and 0.5 A

Total resistance

$$R_T = R_1 + R_2 || R_3$$

The parallel bar signifies that R_2 and R_3 are in parallel.

$$R_T = R_1 + \frac{R_2 R_3}{R_2 + R_3}$$

$$= 6 + \frac{6 \times 3}{6+3} = 8\Omega$$

Ohm's law

$$I_1 = \frac{V}{R_T} = \frac{12}{8} = 1.5\text{A}$$

Voltage AB

$$VA_{AB} = 12 - R_1 I_1 = 12 - 6 \times 1.5 = 3\text{V}$$

Ohm's law

$$I_2 = \frac{V_{AB}}{R_2} = \frac{3}{3} = 1\text{A}$$

$$I_3 = \frac{V_{AB}}{6} = \frac{3}{6} = 0.5\text{A}$$

Check

$$I_2 + I_3 = I_1$$

$$1 + 0.5 = 1.5$$

PROBLEM 1.7 KIRCHHOFF

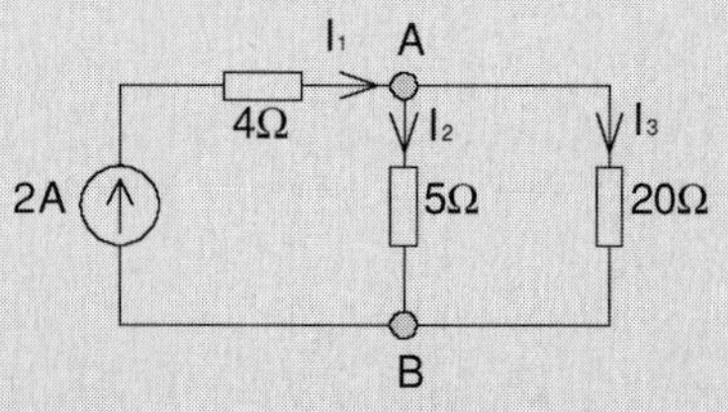

Figure 1.11

Find I_1, I_2 and I_3.

Answer: 2 A, 1.6 A and 0.4 A

I_1 = 2 A, as this is the same as the current source.

$$V_{AB} = 5I_2 = 20I_3$$

Therefore

$$I_2 = \frac{20}{5} I_3 = 4I_3$$

KCL

$$I_1 = I_2 + I_3$$

$$2 = I_2 + I_3$$

$$2 = 4I_3 + I_3$$

$$2 = 5I_3 \qquad I_3 = \frac{2}{5} = 0.4\text{A}$$

Now the value of I_2

$$2 = I_2 + 0.4$$

$$I_2 = 2 - 0.4 = 1.6\text{A}$$

PROBLEM 1.8 KIRCHHOFF

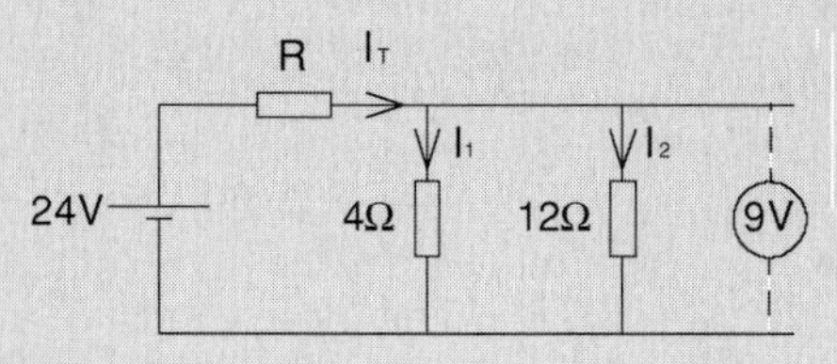

Figure 1.12

Find the value of R.

Answer: $R = 5\Omega$

The value of R is given by Ohm' s law

$$R = \frac{V_R}{I_R}$$

The voltage of R is

$$V_R = 24 - 9 = 15\,\text{V}$$

The current is given by KCL

$$I_T = I_1 + I_2$$

$$I_1 = \frac{9}{4} = 2.25\,\text{A}$$

$$I_2 = \frac{9}{12} = 0.75\,\text{A} \text{ and } I_T = 3\,\text{A}$$

$$R = \frac{15}{3} = 5\Omega$$

PROBLEM 1.9 KIRCHHOFF

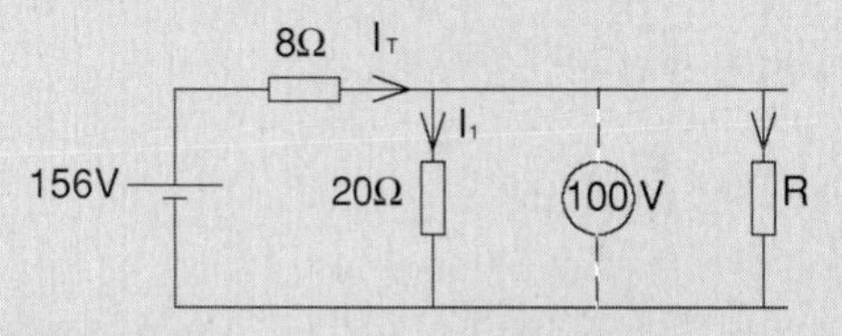

Figure 1.13

Find the value of R.

Answer: $R = 50\ \Omega$

We can find the total current using Ohm's law. We know the resistance, we can calculate the voltage and then the current.

$$I_T = \frac{156 - 100}{8} = \frac{56}{8} = 7\,\text{A}$$

$$I_1 = \frac{100}{20} = 5\,\text{A}$$

KCL

$$I_T = I_i + I_2$$

$$I_2 = I_T - I_1 = 7 - 5 = 2\,\text{A}$$

We now apply Ohm' s law

$$R = \frac{V}{I} = \frac{100}{2} = 50\Omega$$

PROBLEM 1.10 KIRCHHOFF

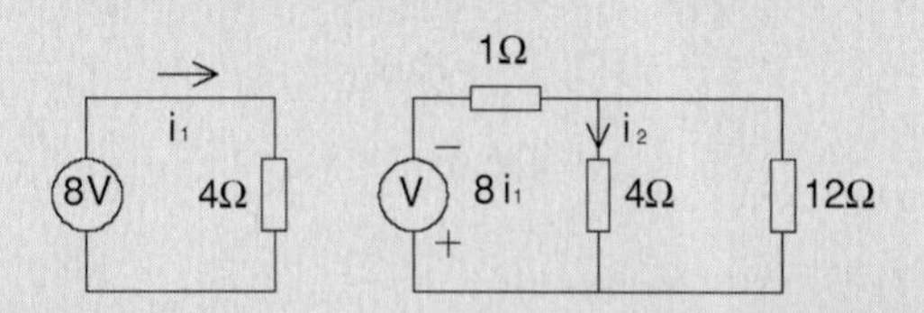

Figure 1.14

The voltage in the centre mesh is a current dependent voltage source V.

Find the value of i_2 as indicated.

Answer: -3 A

The circuit on the left mesh is independent.

$$i_1 = \frac{8}{4} = 2\,\text{A} \qquad V = 16\,\text{V}$$

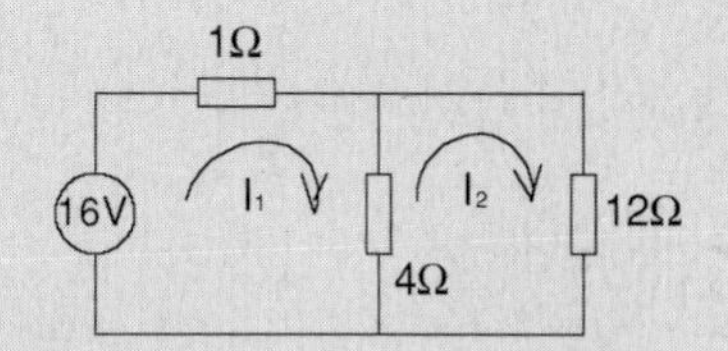

Figure 1.15

First mesh

$$-16 = I_1(1+4) - I_2 4 \tag{1}$$

Second mesh

$$0 = -I_1 4 + I_2(4+12) \tag{2}$$

$$4I_1 = 16I_2$$

$$I_1 = 4I_2$$

Substitute in equation (1)

$$4I_2 5 - 4I_2 = -16$$

$$20I_2 - 4I_2 = -16$$

$$16I_2 = -16$$

$$I_2 = -1$$

$$i_2 = I_1 - I_2 = -4 - (-1) = -3\,\text{A}$$

PROBLEM 1.11 KIRCHHOFF

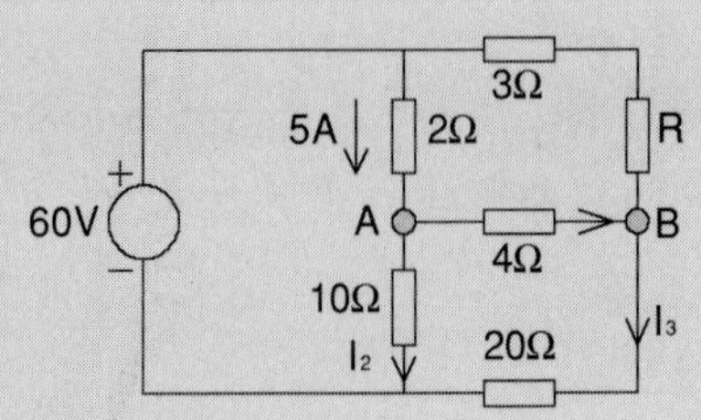

Figure 1.16

Find R, I_1, I_2 and I_3.

Answer: $R = 1\Omega$, $I_1 = 0\text{A}$, $I_2 = 5\text{A}$, and $I_3 = 2.5\text{A}$

The voltage on the 2Ω resistor is

$$V = IR = 5 \times 2 = 10\text{V}$$

Using KVL, the voltage on the 10Ω resistor is

$$60 - 10 = 50\ \text{V}$$

(Work out the polarities to confirm this.)

Ohm's law

$$I_2 = \frac{50}{10} = 5\text{A}$$

In node A, 5 A go in and 5 A go out.
Therefore, I_1 must be zero (KCL).

$$I_3 = \frac{50}{20} = 2.5\text{A}$$

Resistor R

$$(3 + R)2.5 = 10\text{V}$$

$$3 + R = 4$$

$$R = 4 - 3 = 1\Omega$$

Dividing voltage and current

When solving electronic problems you will often find that you need to use either the current division rule, or the voltage division rule. As long as the resistors are known you will be able to find the total voltage from a portion of the voltage or vice versa. And similarly with the current. We are going to examine them now starting with voltage division.

Voltage division rule

The voltage is directly proportional to the resistance in a potential divider. In the potential divider, shown in Figure 1.17, the current is:

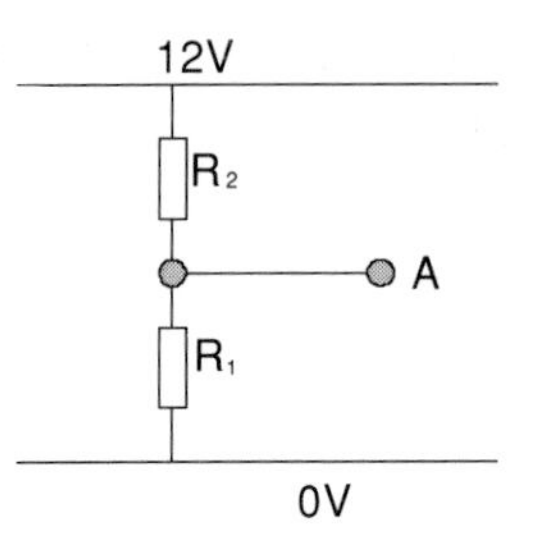

Figure 1.17

$$I = \frac{V}{R} = \frac{V}{R_1 + R_2}$$

The current is common to both resistors. The voltage at R_1 is given by

$$V_{R_1} = I \times R_1 = \frac{V}{R_1 + R_2} \times R_1$$

$$= V\frac{R_1}{R_1 + R_2}$$

The voltage at R_2 is given by

$$V_{R_2} = V\frac{R_2}{R_1 + R_2}$$

We can make a double check by adding both voltages and we should get the total voltage V, as expected.

$$V_{R_1} + V_{R_2} = V\frac{R_1}{R_1 + R_2} + V\frac{R_2}{R_1 + R_2} = V$$

The voltage division, although the same problem as before, could appear in a more

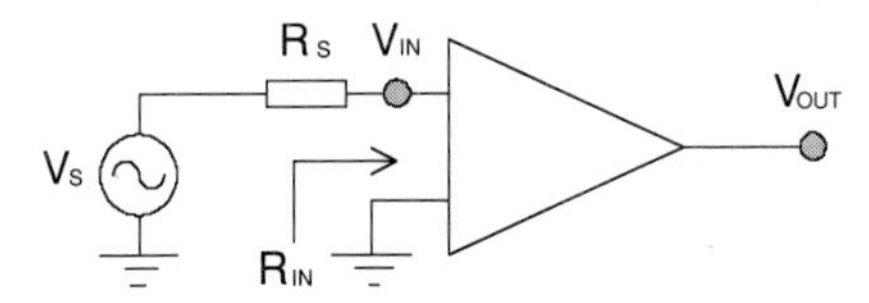

Figure 1.18

complex situation. This is the case in Figure 1.18, where a voltage source feeds a voltage amplifier. In this example the output voltage v_{OUT} depends on the input voltage v_{IN} and the amplification A of the amplifier. The input voltage depends on the value of the voltage source V_S, the resistance of the voltage source R_S and the internal resistance of the amplifier R_{IN}.

R_S and R_{IN} form a potential divider and the same rule applies, i.e.

$$V_{IN} = V_S \frac{R_{IN}}{R_S + R_{IN}}$$

Current division rule

To illustrate this rule we refer to Figure 1.19. We know the total current and we know R_1 and R_2.

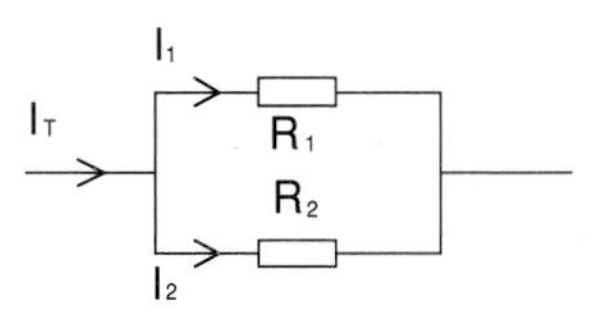

Figure 1.19

It is reasonable to think that the branch with more resistance will carry less current. To demonstrate this we start with the fact that R_1 and R_2 have the same voltage as they are in parallel.

$$I_1 R_1 = I_2 R_2$$

$$\frac{I_1}{I_2} = \frac{R_2}{R_1}$$

Using the principle of proportions:

$$\frac{I_1 + I_2}{I_2} = \frac{R_1 + R_2}{R_1}$$

But

$$I_1 + I_2 = I_T$$

so

$$I_2 = I_T \frac{R_1}{R_1 + R_2}$$

Similarly,

$$I_1 = I_T \frac{R_2}{R_1 + R_2}$$

Try to remember that when calculating I_1, that R_2 goes on top in the formula, that is to say for a current in one branch, we use the resistor of the other branch in the formula. This makes sense when you think that the more resistance a path has, the less current it will carry. In mathematical terms we say that resistance and current, for a given circuit, are inversely proportional.

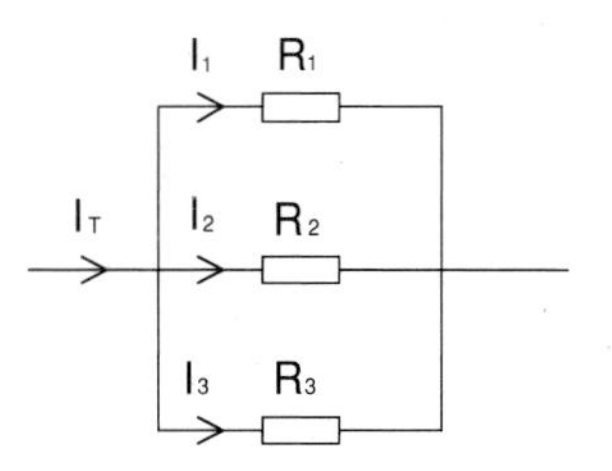

Figure 1.20

The problem of three resistors, shown in Figure 1.20, can be reduced to the previous one of two resistors in parallel, if we join two of them into one using the parallel resistor formula.

Normally, we are interested in one current on any current divider. As this may not always be the case, it is worth considering the more general case, as we do next.

General case

Alternatively, we can look at the general case, that is to say N resistors in parallel, where we know the total resistance and we want to know the current going through one particular branch. The circuit in Figure 1.21, is equivalent to the circuit of Figure 1.22, where the parallel resistors have been replaced by R_P.

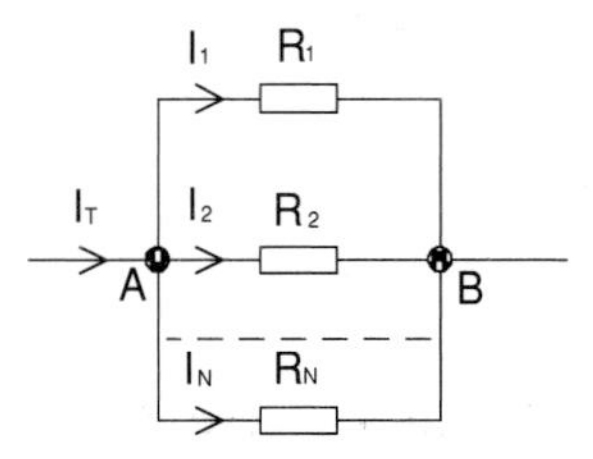

Figure 1.21

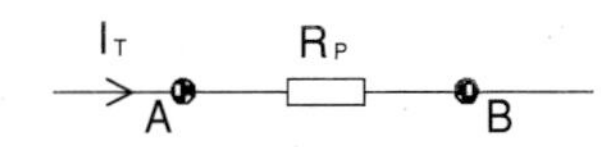

Figure 1.22

The voltage between A and B will be

$$V_{AB} = I_T \text{ x } R_P$$

Similar expressions are valid for any of the branches:

$V_{AB} = I_1 \times R1$

$V_{AB} = I_2 \times R2$

$V_{AB} = I_x \times Rx$

When we are looking for a particular path, or current divider, we can say

$$I_x R_x = I_T R_P$$

and

$$I_X = I_T \frac{R_P}{R_X}$$

This is the general formula where:
- I_x is the current in any of the branches
- I_T is the total current coming into A
- R_P is the parallel equivalent of all the resistances in parallel
- R_x is the resistance associated with current I_x

This general case, has to apply to the particular case of two resistors R_1 and R_2 in parallel. Let us see if this is true.

If we are looking for I_2

$$I_2 = I_T \frac{R_P}{R_2}$$

R_P for two resistors is

$$R_P = \frac{R_1 R_2}{R_1 + R_2}$$

and replacing R_P we obtain

$$I_2 = I_T \frac{R_1}{R_1 + R_2}$$

Which takes us back to the basic current division rule, for the case of two resistors in parallel.

We will be using this particular form of current division (two resistors in parallel), in problems throughout the book. As you will see later, the current amplification of a transistor is calculated with the aid of the current division rule, in the input circuit and the output circuit.

PROBLEM 1.12 VOLTAGE AND CURRENT DIVISION

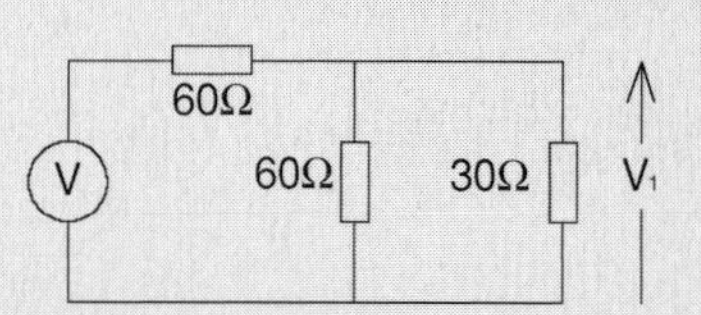

Figure 1.23

Find the value of V_1 in terms of V.

Answer: $V_1 = \frac{V}{4}$

60Ω and 30Ω are in parallel

$$60 \| 30 = \frac{60 \times 30}{60 + 30} = 20\Omega$$

Voltage division

$$V_1 = \frac{V\,20}{20 + 60} = \frac{V}{4}$$

PROBLEM 1.13 VOLTAGE AND CURRENT DIVISION

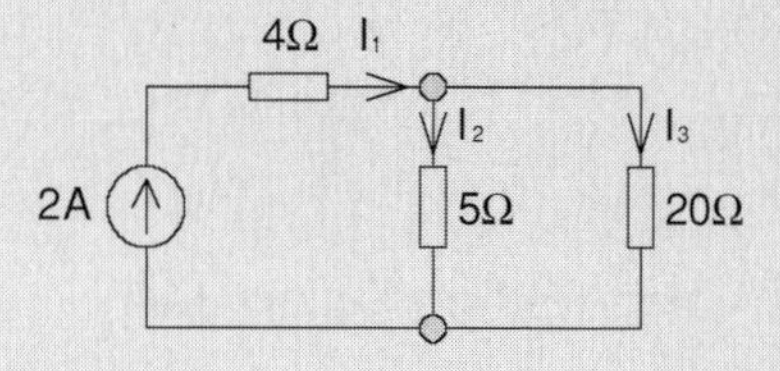

Figure 1.24

This problem was done (problem 1.7) using only Ohm's law and KCL. It is repeated here using the current division rule.

Answer: 2A, 1.6A and 0.4A

I is the same as the source

$I_1 = 2$ A

Current division rule

$$I_2 = I_1 \frac{20}{20 + 5}$$

$$= 2\frac{20}{25} = 1.6\,\text{A}$$

I is the same as the source

$$I_1 = 2\text{ A}$$

Current division rule

$$I_2 = I_1 \frac{20}{20+5}$$

$$= 2\frac{20}{25} = 1.6\text{ A}$$

Current division rule

PROBLEM 1.14 VOLTAGE AND CURRENT DIVISION

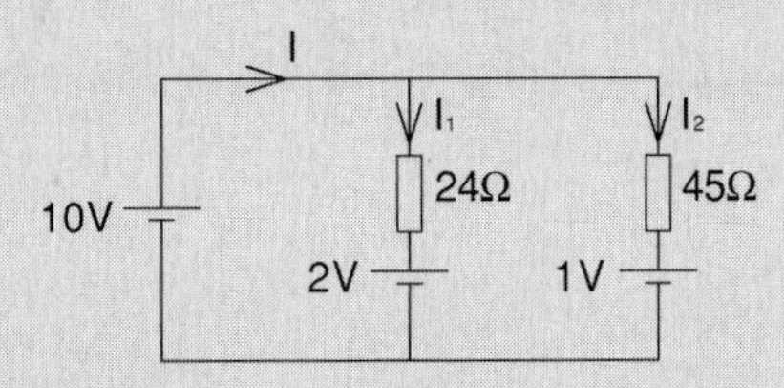

Figure 1.25

Find I_1, I_2 and the total current I.

Answer: $I_1 = \frac{1}{3}\text{A}$, $I_2 = \frac{1}{5}\text{A}$ and $I = \frac{8}{15}\text{A}$

The 24Ω resistor has 8 V applied to it.

$$I_1 = \frac{V}{R} = \frac{8}{24} = \frac{1}{3}$$

The 45Ω resistor has 9 V applied to it.

$$I_2 = \frac{9}{45} = \frac{1}{5}$$

Total current

$$\frac{1}{3} + \frac{1}{5} = \frac{5+3}{15} = \frac{8}{15}\text{A}$$

PROBLEM 1.15 VOLTAGE AND CURRENT DIVISION

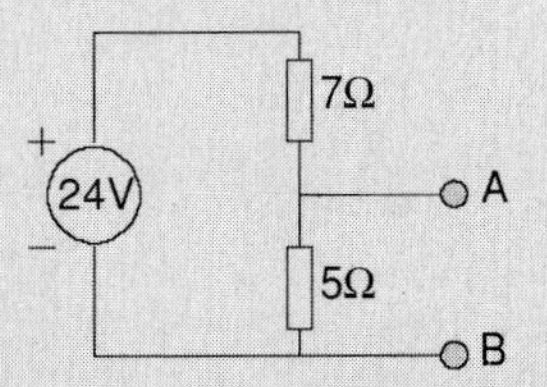

Figure 1.26

What is the change of the voltage at A, if we connected a 5Ω load in AB?

Answer: From 10V to 6.32V

Without the 5Ω resistor

$$V_A = 24\frac{5}{7+5} = 10\text{ V}$$

Connecting the resistor, we have

$$5||5 = \frac{5\times 5}{5+5} = \frac{25}{10}$$

$$= 2.5\Omega$$

Total resistance this time

$$7 + 2.5 = 9.5\Omega$$

And the voltage is

$$V_A = 24\frac{2.5}{7+2.5} = \frac{2.5\times 24}{9.5} = 6.32\text{ V}$$

PROBLEM 1.16 VOLTAGE AND CURRENT DIVISION

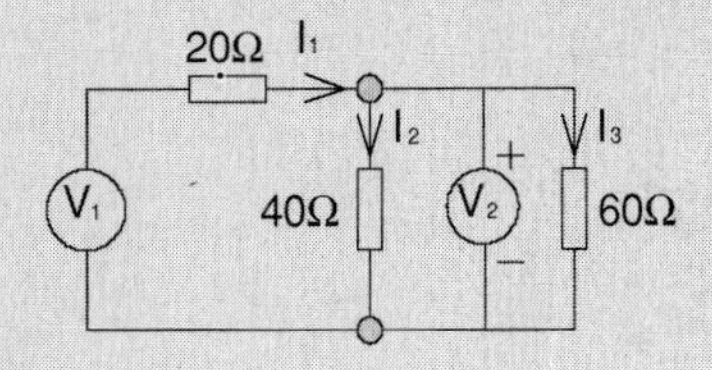

Figure 1.27

Find I_1, I_2 and V_1.

Answer: 5A, 3A and 200V

$$I_2 = \frac{V}{R} = \frac{120}{40} = 3\,\text{A}$$

I_3 is not required by the question, so we will not calculate it. Instead we can go straight to I_1 using the current division rule.

$$I_2 = I_1 \frac{60}{40+60}$$

$$3 = I_1 \frac{60}{100}$$

$$I_1 = \frac{3 \times 100}{60} = 5\,\text{A}$$

Voltage

$$V_1 = V_2 + 20I_1$$
$$= 120 + 20 \times 5$$
$$= 120 + 100 = 220\,\text{V}$$

PROBLEM 1.17 VOLTAGE AND CURRENT DIVISION

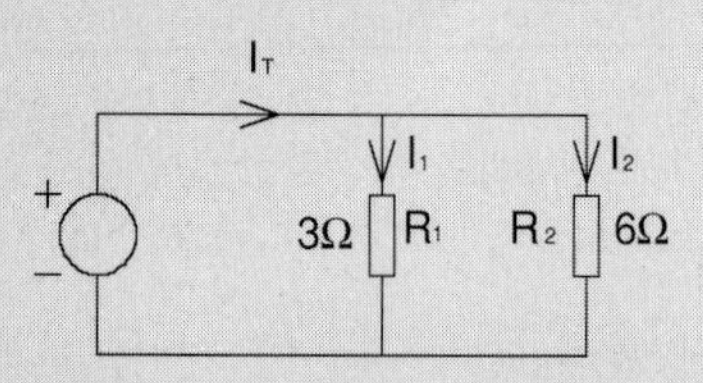

Figure 1.28

Find I_1 and I_T as a function of I_2.

Answer: $I_1 = 2I_2$ and $I_T = 3I_2$

$$V = I_1 R_1 = I_2 R_2$$

$$\frac{I_1}{I_2} = \frac{R_2}{R_1} = \frac{6}{3}$$

$$I_1 = 2I_2$$

Total current

$$I_T = I_1 + I_2 = 2I_2 + I_2$$
$$= 3I_2$$

PROBLEM 1.18 VOLTAGE AND CURRENT DIVISION

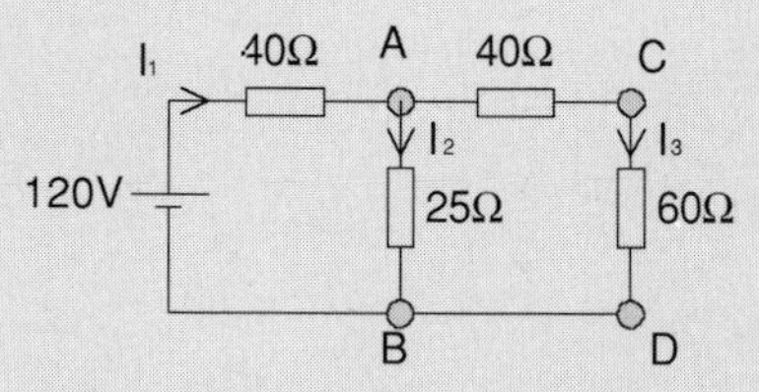

Figure 1.29

$I_1 = 2$A. Find I_2, I_3 and the voltages AB and CD.

Answer: $I_2 = 1.6$A, $I_3 = 0.4$A, $V_{AB} = 40$V and $V_{CD} = 24$V

Voltage AB

$$V_{AB} = 120 - I_1 \times 40$$
$$= 120 - 80 = 40\,\text{V}$$

Voltage division

$$V_{CD} = V_{AB} \frac{60}{40+60} = 40 \frac{60}{100} = 24\,\text{V}$$

Current division

$$I_2 = I_1 \frac{60+40}{25+60+40}$$
$$= 2\frac{100}{125} = 1.6\,\text{A}$$

Current division

$$I_3 = I_1 \frac{25}{25+40+60}$$
$$= 2\frac{25}{125} = 0.4\,\text{A}$$

PROBLEM 1.19 VOLTAGE AND CURRENT DIVISION

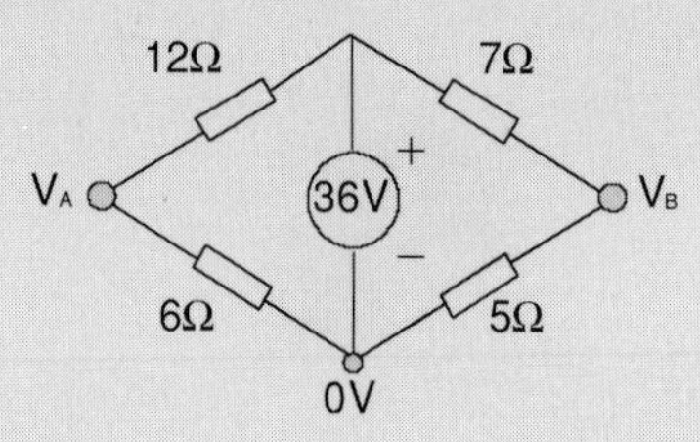

Figure 1.30

Find the value of V_A and V_B.

Answer: $V_A = 12$V and $V_B = 15$V

Voltage division

$$V_A = 36\frac{6}{6+12} = 36\frac{6}{18} = 12\,V$$

Voltage division

$$V_B = 36\frac{5}{7+5} = 36\frac{5}{12} = 15\,V$$

PROBLEM 1.20 VOLTAGE AND CURRENT DIVISION

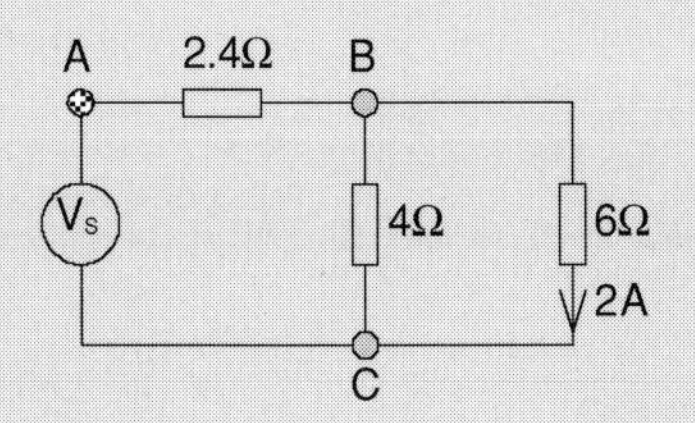

Figure 1.31

Find the voltages AB, BC and V_s.

Answer: 12V, 12V and 24V

Voltage BC

$$V_{BC} = IR = 2\times 6 = 12\,V$$

Current BC

$$I_{BC} = \frac{V}{R} = \frac{12}{4} = 3\,A$$

KCL

$$I_{AB} = 2+3 = 5\,A$$

Voltage AB

$$V_{AB} = I_{AB}R = 5\times 2.4 = 12\,V$$

Source voltage

$$V_S = V_{BC} + V_{AB} = 12+12 = 24\,V$$

PROBLEM 1.21 VOLTAGE AND CURRENT DIVISION

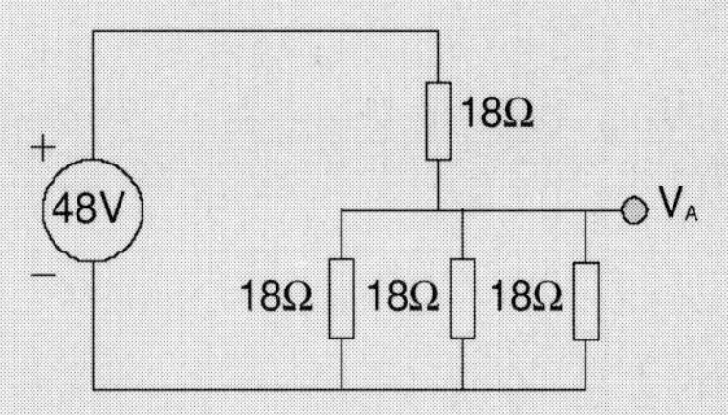

Figure 1.32

Find the voltage V_A, the total current and the branch currents.

Answer: 12 V, 2 A and $\frac{2}{3}$ A for branches

Three resistors in parallel

$$\frac{1}{R_T} = \frac{1}{R_1} + \frac{1}{R_2} + \frac{1}{R_3}$$

$$= \frac{1}{18} + \frac{1}{18} + \frac{1}{18} = \frac{3}{18}$$

$$R_T = \frac{18}{3} = 6\Omega$$

V_A by voltage division

$$V_A = 48\frac{6}{18+6} = \frac{48\times 6}{24} = 12\,V$$

Total resistance of circuit

$$6+18 = 24\Omega$$

Total current

$$I_T = \frac{48}{24} = 2\,A$$

Branch current: All equal.

$$I = \frac{V_A}{R} = \frac{12}{18} = \frac{2}{3}\,A$$

Check current sum

$$\frac{2}{3}+\frac{2}{3}+\frac{2}{3} = \frac{6}{3} = 2\,A \text{ OK.}$$

PROBLEM 1.22 VOLTAGE AND CURRENT DIVISION

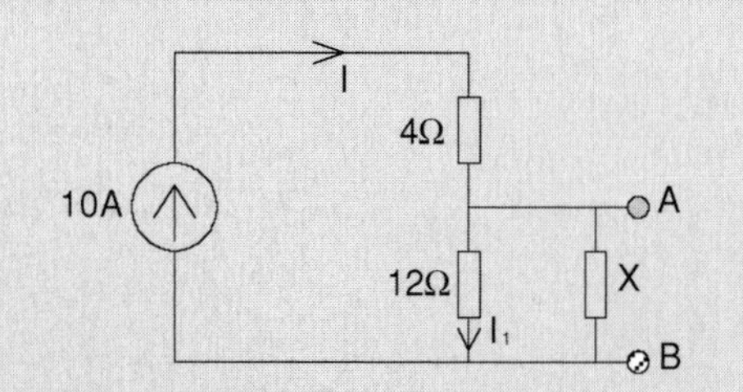

Figure 1.33

Find the value of the resistor to be added to AB, in order to reduce the current in the 12Ω resistor to 7A.

Answer: $R = 28\Omega$

We can use the current division rule with the following information:
total current 10A
partial current 7A
known resistor 12Ω
Unknown resistor

$$I_1 = I\frac{R}{R+12}$$

$$7 = 10\frac{R}{R+12}$$

$$7R + 84 = 10R$$

$$84 = 3R$$

$$R = \frac{84}{3} = 28\Omega$$

Thevenin's theorem

Any circuit or part of a circuit can be replaced by the simple equivalent of a voltage source with a resistor in series. Figure 1.34 shows a general circuit to be converted completely to a Thevenin equivalent. Sometimes it is more practical to convert only part of the circuit into a Thevenin equivalent. We see this situation in Figure 1.35. All we need to find is how to obtain V_{TH} and R_{TH}. Before we do that we must identify points A and B and isolate them from the part of the circuit which will not be included in the equivalent circuit. In Figure 1.34, all the circuit is included in the equivalent circuit, so there is nothing to separate. In Figure 1.35, however, the load resistor will not be included in the equivalent circuit

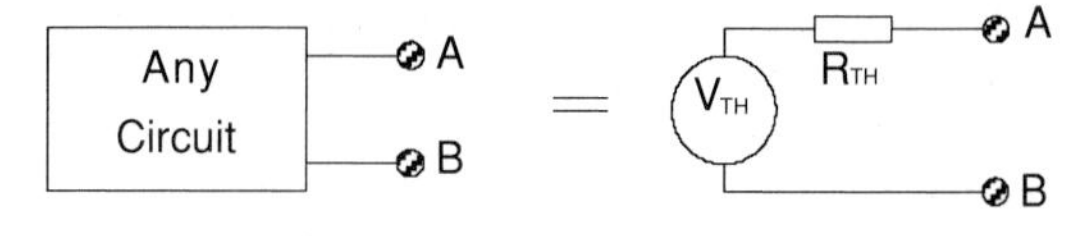

Figure 1.34

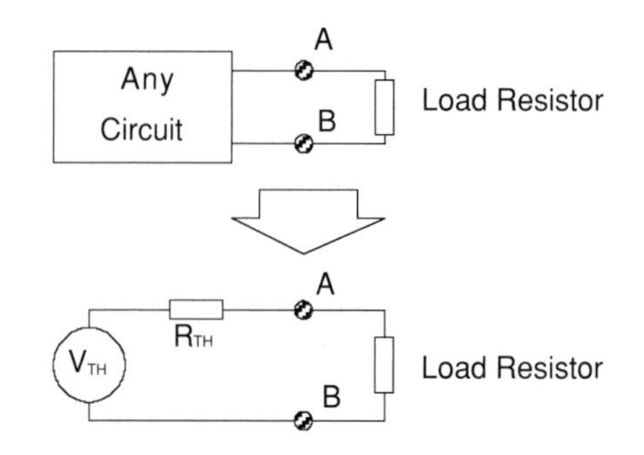

Figure 1.35

and we separate it from the rest of the circuit. This is how we obtain our points A and B.

Once we have separated the circuit that we want to transform, we can proceed to find the value of V_{TH}.

Value of V_{TH}

There are two possibilities:

(a) We are in the lab and we have the actual circuit in front of us. All we then need to do is to measure the voltage AB with a suitable meter. The voltage indicated by the meter is V_{TH}.

(b) More often than not, we will be working on a piece of paper, with a drawing of the circuit. In this case we use our knowledge of circuit analysis to calculate the voltage AB of the separated system.

Value of R_{TH}

Having calculated the value of V_{TH}, we need to move on to the value of R_{TH}. This is more complicated, because we now have to find out the equivalent resistance of the circuit, excluding the influence of the voltage or current sources that are part of the system. What we want is the resistance of the voltage source, without the voltage source, and similarly for a current source.

Once you have mastered the theorem and you understand what it is doing, you will find that this make sense, but it will be difficult to accept it at first. Even if you don't fully understand it, by following step by step the procedure below, you will be quite safe (just like trying a cookery recipe).

(1) Disable all the sources in the circuit. In order to do this you:
- Short circuit the voltage supplies
- Open circuit the current supplies

(2) Measure (if you are in the lab) or calculate (if you are working on paper), the equivalent resistance between points A and B.

PROBLEM 1.23 THEVENIN

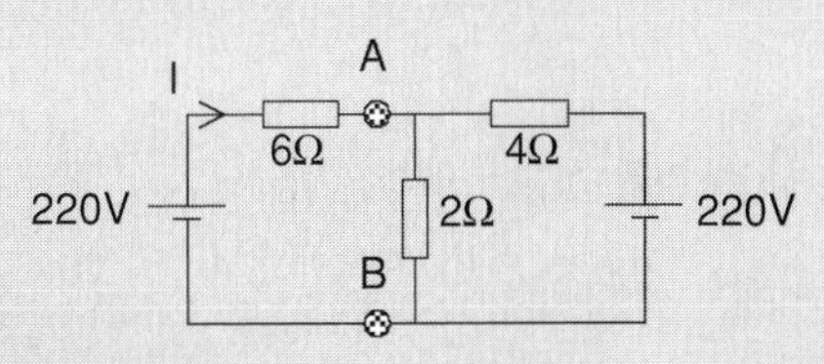

Figure 1.36

Change the right-hand side of AB to Thevenin and then calculate the value of I.

Answer: $I = 20\text{A}$

Thevenin voltage

$$V_{TH} = 220\frac{2}{2+4} = \frac{220}{3}$$

Thevenin resistance

$$R_{TH} = 2 \| 4 = \frac{2\times 4}{2+4} = \frac{8}{6} = \frac{4}{3}$$

Now we can easily identify our next equation

$$I = \frac{220 - \frac{220}{3}}{6+\frac{4}{3}} = \frac{220\times\frac{2}{3}}{\frac{22}{3}}$$

$$= \frac{2\times 220}{22} = 20\,\text{A}$$

PROBLEM 1.24 THEVENIN

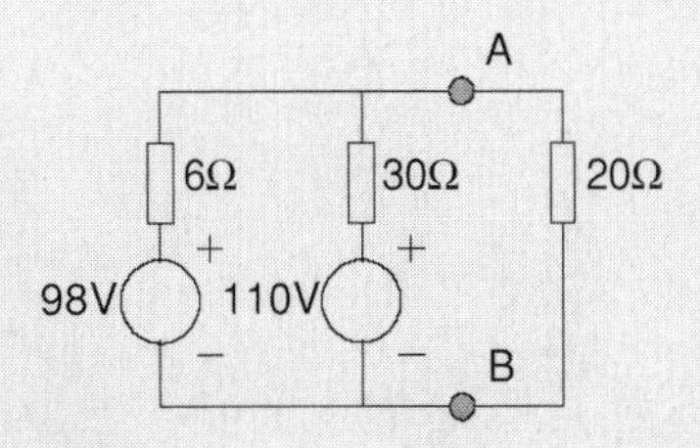

Figure 1.38

Find the value of the current through the 20Ω resistor, by converting the circuit at the left of AB, into a Thevenin equivalent.

Answer: $I = 4\text{A}$

When the circuit is separated at A and B, only the left loop remains. We will calculate the current first and then the voltage at AB which is the Thevenin voltage.

$$I = \frac{110-98}{36} = \frac{12}{36} = \frac{1}{3}\,\text{A}$$

$$V_{AB} = V_{TH} = 110 - \frac{1}{3}\times 30 = 100\,\text{V}$$

Also

$$V_{TH} = 98 + \frac{1}{3}\times 6 = 100\,\text{V}$$

Thevenin resistance

$$R_{TH} = 6 \| 30 = \frac{6\times 30}{6+30} = 5\Omega$$

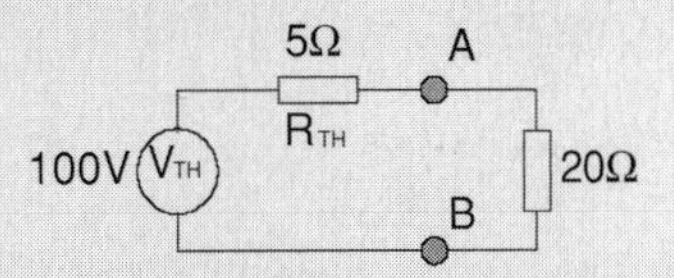

Figure 1.39

With the circuit redrawn, it is easy to find the value of the current

$$I = \frac{100}{20+5} = 4\,\text{A}$$

PROBLEM 1.25 THEVENIN

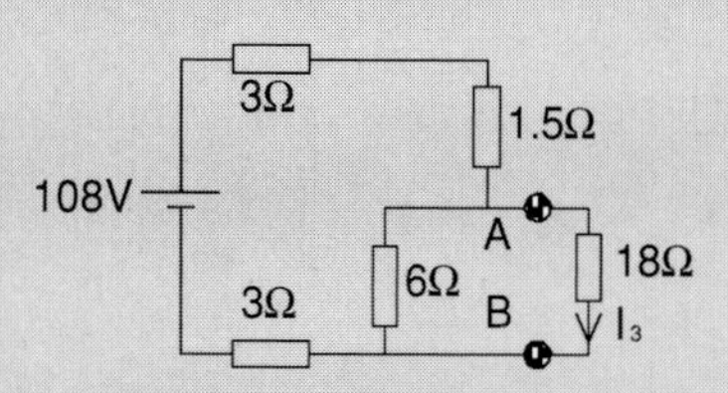

Figure 1.40

Convert to Thevenin equivalent the circuit at the left of AB and then calculate the value of I_3.

Answer: $I_3 = 2.25\text{A}$

Thevenin voltage

$$V_{TH} = 108\frac{6}{3+3+1.5+6} = 108\frac{6}{13.5}$$
$$= 48\,\text{V}$$

Thevenin resistance

$$R_{TH} = 6\|(3+3+1.5) = 6\|7.5$$
$$= \frac{6\times 7.5}{13.5} = 3.333$$

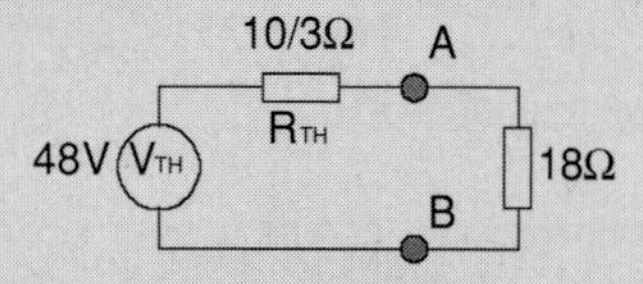

Figure 1.41

It is now easier to calculate the current

$$I = \frac{48}{\frac{10}{3}+18} = \frac{48}{\frac{10}{3}+\frac{54}{3}}$$
$$= \frac{48\times 3}{64} = 2.25\,\text{A}$$

PROBLEM 1.26 THEVENIN

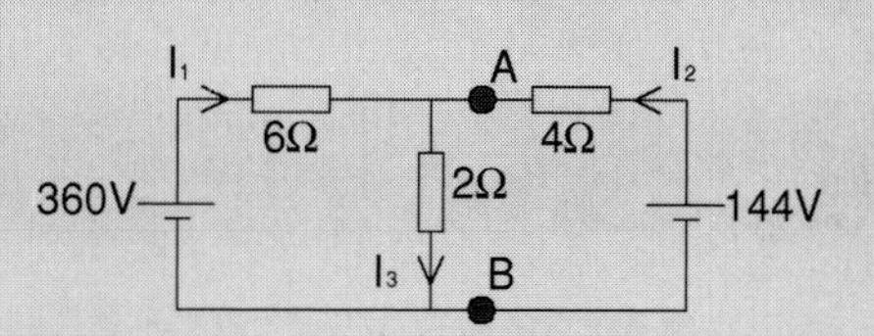

Figure 1.42

The approximate values for the currents are

$I_1 = 42.55\text{A}$, $I_2 = 9.8\text{A}$ and $I_3 = 52.35\text{A}$.

Verify I_2 by converting to Thevenin the circuit at the left of AB.

Answer: Yes, it is 9.8A

Thevenin voltage

$$V_{TH} = \frac{360\times 2}{6+2}$$
$$= 90\,\text{V}$$

Thevenin resistance

$$R_{TH} = \frac{6\times 2}{6+2}$$
$$= 1.5\Omega$$

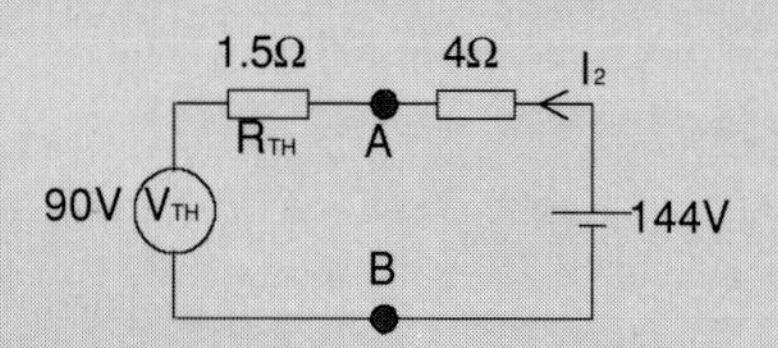

Figure 1.43

The expression for the current now follows

$$I_2 = \frac{144-90}{4+1.5} = \frac{54}{5.5}$$
$$= 9.8\,\text{A}$$

PROBLEM 1.27 THEVENIN

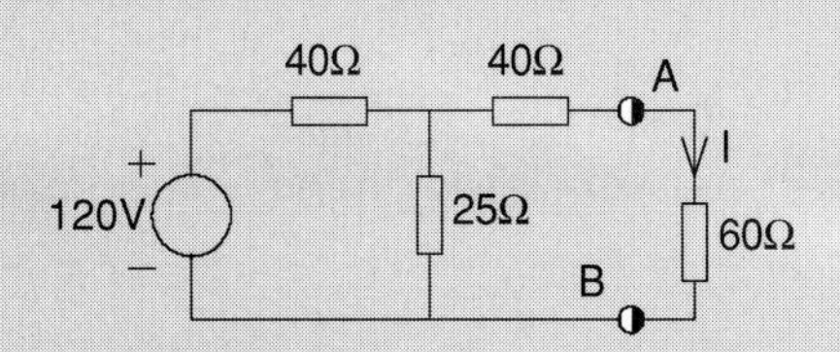

Figure 1.44

Find the value of I by converting the circuit at the left of AB to Thevenin.

Answer: $I = 0.4\text{A}$

Thevenin voltage

$$V_{TH} = \frac{120 \times 25}{40 + 25} = 46.154\text{V}$$

Thevenin resistance

$$R_{TH} = 40 \| 25 + 40 = 55.38\Omega$$

Now the value of the current

$$I = \frac{V_{TH}}{R} = \frac{46.154}{60 + 55.38} = 0.4\text{A}$$

PROBLEM 1.28 THEVENIN

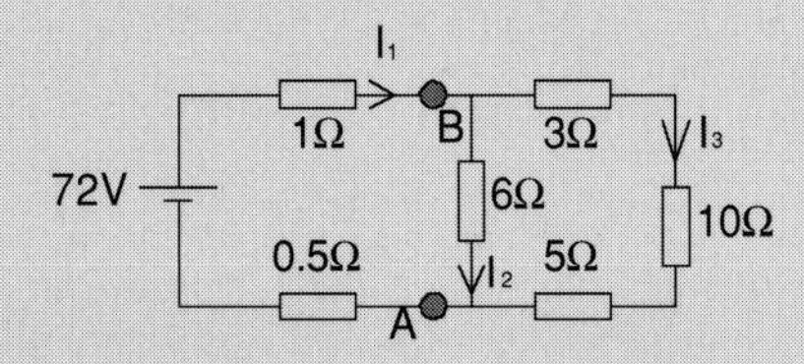

Figure 1.45

In this circuit $I_1 = 12\text{A}$. Verify this by converting to Thevenin the circuit at the right of AB.

Answer: Yes, it is 12A

Thevenin voltage
This is a special case of Thevenin. There is no source of any kind in the circuit. Therefore it is Zero.

$$V_{TH} = 0$$

Thevenin resistance

$$R_{TH} = \frac{6 \times 18}{6 + 18} = 4.5\Omega$$

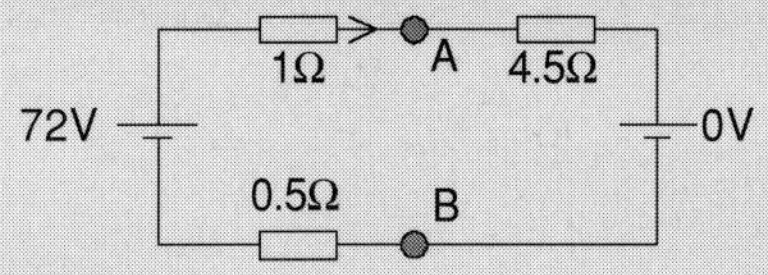

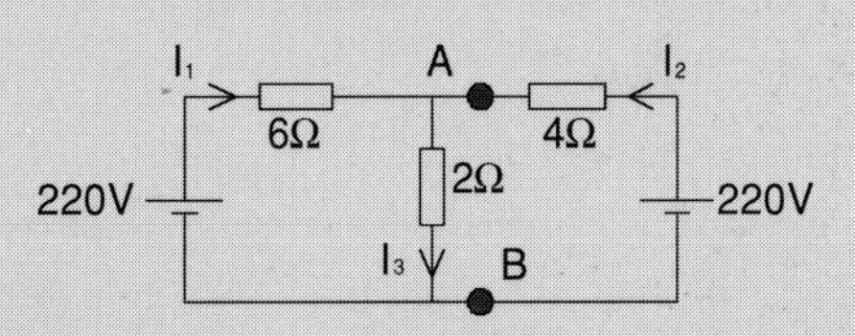

Figure 1.46

The current is therefore

$$I_1 = \frac{72 + 0}{4.5 + 1 + 0.5} = \frac{72}{6} = 12\text{A}$$

PROBLEM 1.29 THEVENIN

Figure 1.47

It has been found that $I_1 = 20\text{A}$, $I_2 = 30\text{A}$ and I_3 is 50A. Verify the value of I_2, by converting the circuit at the left of AB to Thevenin.

Answer: Yes, $I_2 = 30\text{A}$

Thevenin voltage

$$V_{TH} = \frac{220 \times 2}{6 + 2} = 55\text{V}$$

Thevenin resistance

$$R_{TH} = 6 \| 2 = \frac{6 \times 2}{6 + 2} = 1.5\Omega$$

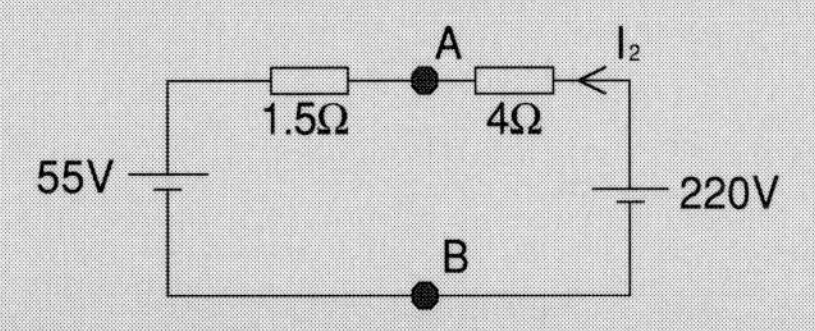

Figure 1.48

We can now produce the expression for I_2

$$I_2 = \frac{220 - 55}{1.5 + 4} = \frac{165}{5.5} = 30\text{A}$$

PROBLEM 1.30 THEVENIN

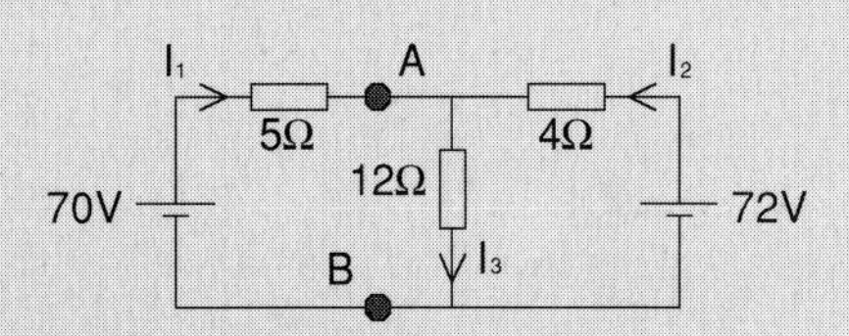

Figure 1.49

In the circuit we have

$I_1 = 2\text{A}$, $I_2 = 3\text{A}$ and $I_3 = 5\text{A}$.

Verify the value of I_1 by converting the circuit at the right of AB to Thevenin.

Answer: Yes, $I_1 = 2\text{A}$

Thevenin voltage

$$V_{TH} = 72\frac{12}{12+4} = 54\,\text{V}$$

Thevenin resistance

$$R_{TH} = 12 \| 4 = \frac{12 \times 4}{12+4} - \frac{48}{16} = 3\Omega$$

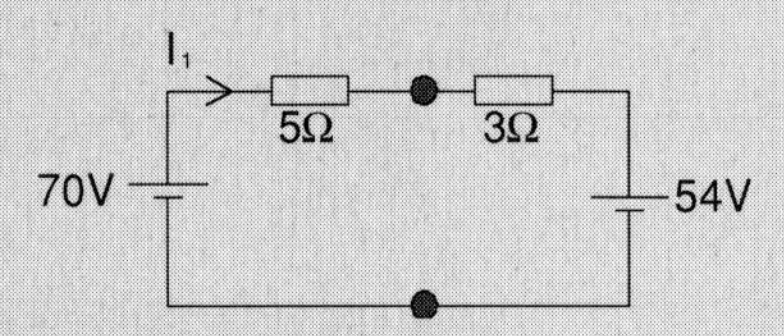

Figure 1.50

The expression for I_1 is now

$$I_1 = \frac{70-54}{5+3} = \frac{16}{8} = 2\,\text{A}$$

PROBLEM 1.31 THEVENIN

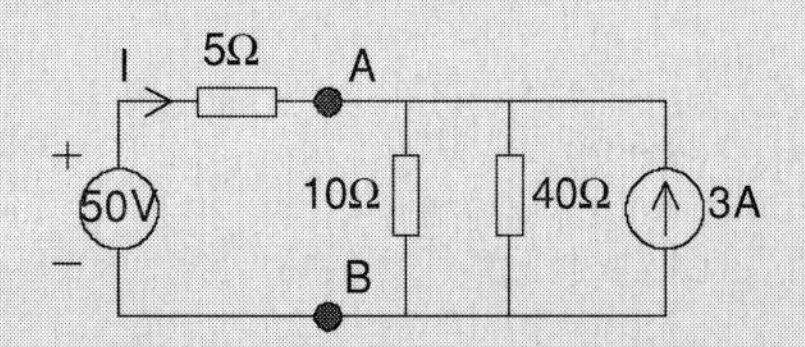

Figure 1.51

Convert the right of AB to the Thevenin equivalent, in order to find the value of I.

Answer: $I = 2\text{A}$

Thevenin voltage

$$V_{TH} = 3 \times (40 \| 10) = 3\frac{40 \times 10}{40+10} = 24\,\text{V}$$

Thevenin resistance

$$R_{TH} = 10 \| 40 = 8\Omega$$

(Current supply open circuit to calculate R_{TH})

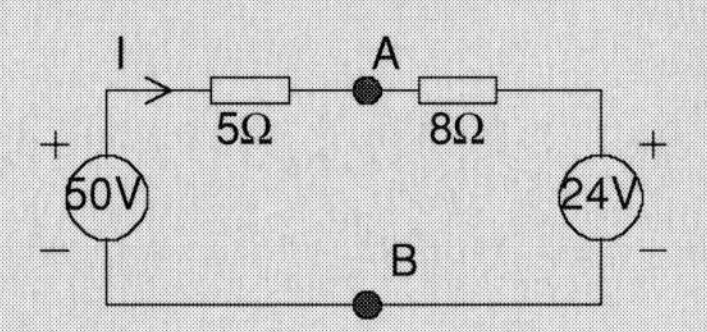

Figure 1.52

The value of I is now

$$I = \frac{50-24}{5+8} = \frac{26}{13} = 2\,\text{A}$$

PROBLEM 1.32 THEVENIN

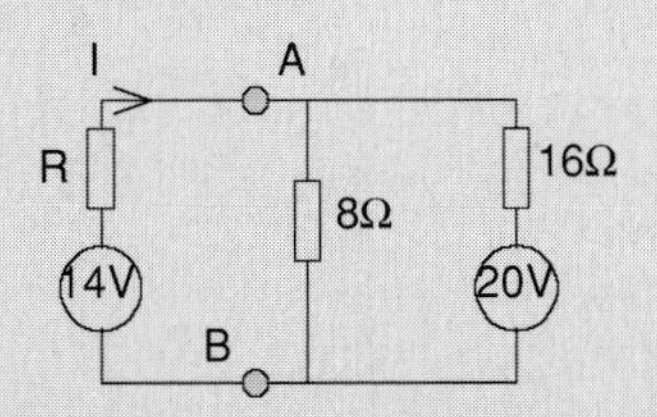

Figure 1.53

Find the value of R to produce $I = -1$A.
Use conversion to Thevenin at the right of AB.

Answer: $R = 2\Omega$

Thevenin voltage

$$V_{TH} = 20\frac{8}{8+16} = 6\frac{2}{3}\text{V}$$

Thevenin resistance

$$R_{TH} = 8 \| 16 = 5\frac{1}{3}\Omega$$

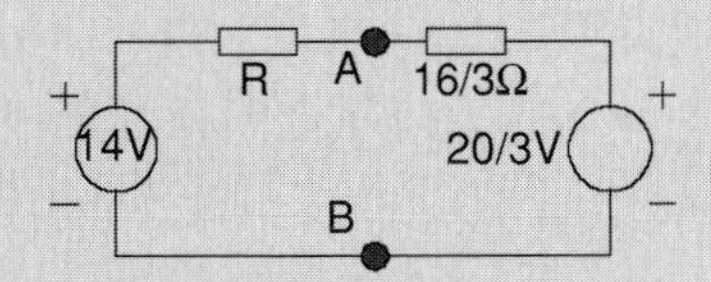

Figure 1.54

The value of I, which we know, is given in terms of R:

$$I = 1\text{A} = \frac{14 - 6\frac{2}{3}}{R + 5\frac{1}{3}}$$

$$R + 5\frac{1}{3} = 14 - 6\frac{2}{3}$$

$$R = 14 - 6\frac{2}{3} - 5\frac{1}{3}$$

$$= 14 - 6 - 5 - \frac{2}{3} - \frac{1}{3}$$

$$R = 2\Omega$$

PROBLEM 1.33 THEVENIN

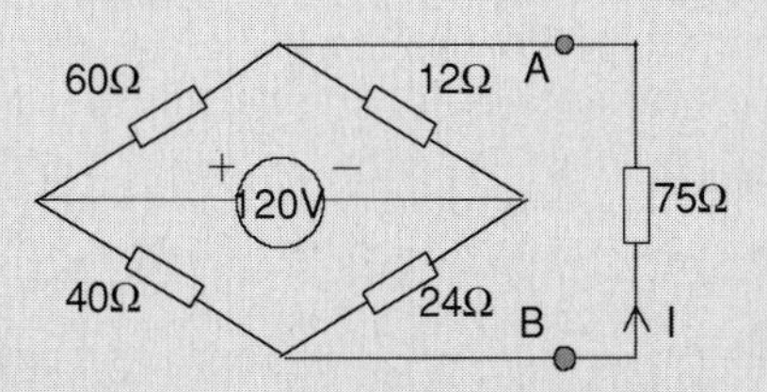

Figure 1.55

Find the value of I using the Thevenin equivalent on the left of AB.

Answer: $I = 0.25$A

When we open A and B, V_A and V_B can be obtained as potential dividers

$$V_A = 120\frac{12}{12+60} = 20\text{V}$$

$$V_B = 120\frac{24}{40+24} = 45\text{V}$$

Thevenin voltage

$$V_{TH} = 45 - 20 = 25\text{V}$$

($V_B > V_A$, so V_{TH} is upside-down, and the current is pointing in the right direction).
Thevenin resistance. (120V supply shorted out to calculate R_{TH})

$$R_{TH} = 60 \| 12 + 40 \| 24 = 10 + 15 = 25\Omega$$

The current is now given by

$$I = \frac{25}{25+75} = \frac{25}{100} = 0.25\text{A}$$

Norton's theorem

This theorem is very similar to Thevenin's theorem and we will follow the same approach, so that you can compare the two theorems. Any circuit or part of a circuit can be replaced by a very simple circuit composed of a current source, with a resistor in parallel. Figure 1.56 shows a

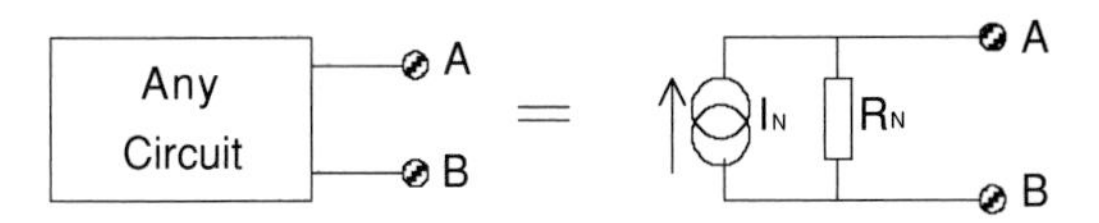

Figure 1.56

general circuit, to be converted completely to a Norton equivalent. In order to solve a problem it might be more practical to convert only part of the circuit into a Norton equivalent, as we see in Figure 1.57.

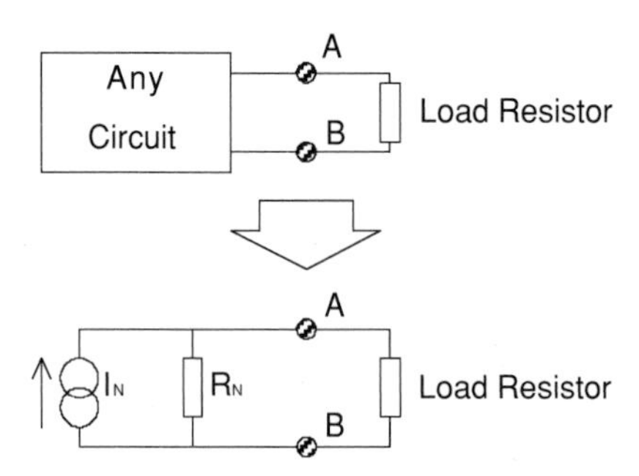

Figure 1.57

All we need to do is to find out how we can obtain I_N and R_N. Before we try to find I_N and R_N, we must be very clear which circuit we are converting to a Norton equivalent. We must clearly identify points A and B of the circuit. Once we have done this, we can move onto the next step which is to find the value of I_N.

Value of I_N

There are two possibilities:

(a) We are in the lab and we have the actual circuit in front of us. In this case we short circuit points A and B and we measure the short circuit current with a suitable meter. The reading of the meter is the current I_N that we are looking for.

(b) More often than not, we will be working on a piece of paper with a drawing of the circuit. In this case we use our knowledge of circuit analysis to calculate the short circuit current through AB, now joined together in short circuit.

Value of R_N

The method of calculating R_N is identical to the calculation of R_{TH}. That will allow us to simplify the explanation and to state the method, as follows:

(1) Disable all sources in the circuit. To do this, you
- Short circuit the voltage supplies
- Open circuit the current supplies

(2) Measure (if you are in the lab) or calculate (if you are working on paper) the equivalent resistance between points A and B.

PROBLEM 1.34 NORTON

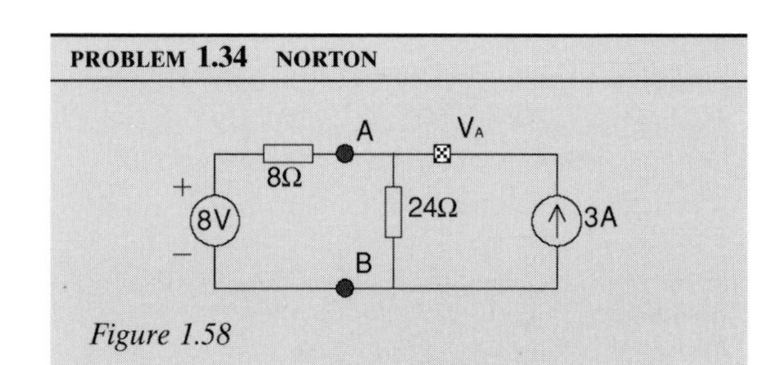

Figure 1.58

Find the value of V_A by converting the circuit at the left of AB, to a Norton equivalent.

Answer: 24V

Norton current

$$I_N = \frac{8}{8} = 1\text{A}$$

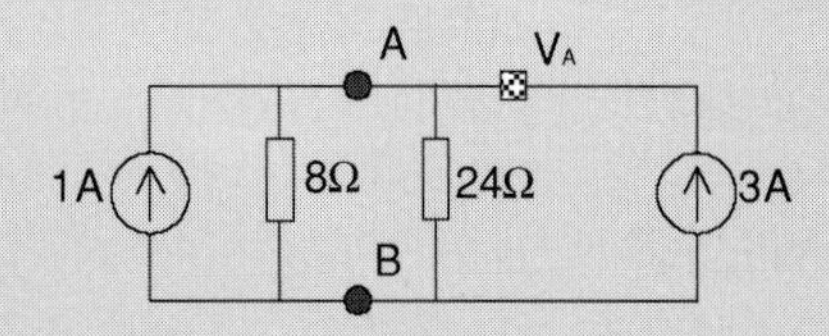

Figure 1.59

Now we have a circuit where both currents feed in the same direction, therefore they add, and there are two resistors in parallel that can be considered as one.

$$V_A = IR = (3+1)\frac{8\times 24}{8+24}$$

$$= \frac{4\times 8\times 24}{32} = 24\text{V}$$

PROBLEM 1.35 NORTON

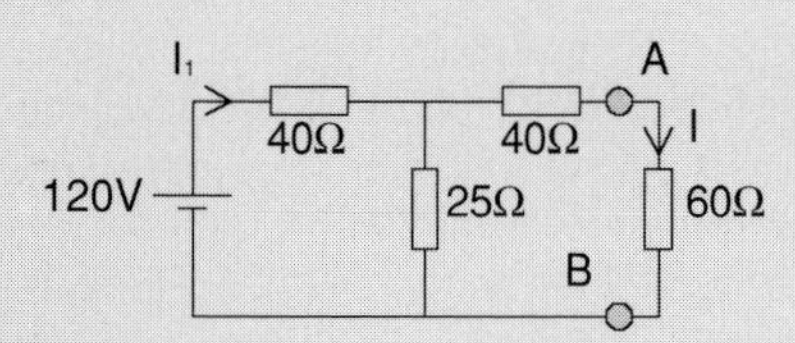

Figure 1.60

Find the value of I by converting to Norton the circuit on the left of AB.

Answer: $I = 0.4\text{A}$

The Norton current is found by calculating the total current of the circuit (with A and B shorted) and then using the current division rule as we will see.

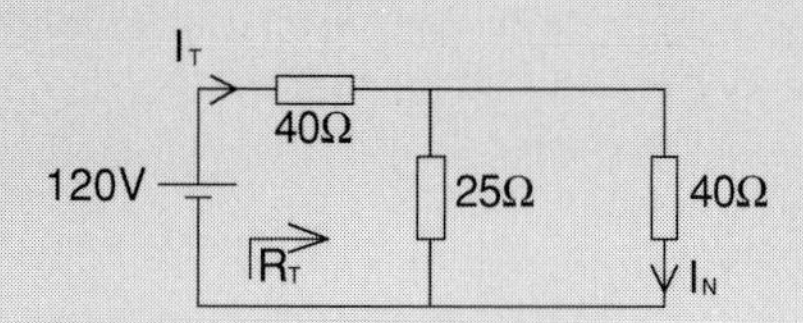

Figure 1.61

Total resistance

$$R_T = 40 + 25 \| 40 = 55.3846\,\Omega$$

Total current

$$I_T = \frac{120}{55.3846} = 2.1666\,\text{A}$$

Norton current (by current division)

$$I_N = I_T \frac{25}{25+40}$$

$$= 2.1666 \frac{25}{65} = 0.8333\,\text{A}$$

Norton resistance

$$R_N = 40 + 25 \| 40 = 55.3846\,\Omega$$

(This is not the same as R_T above, but the values are the same as the circuit is symmetrical.)

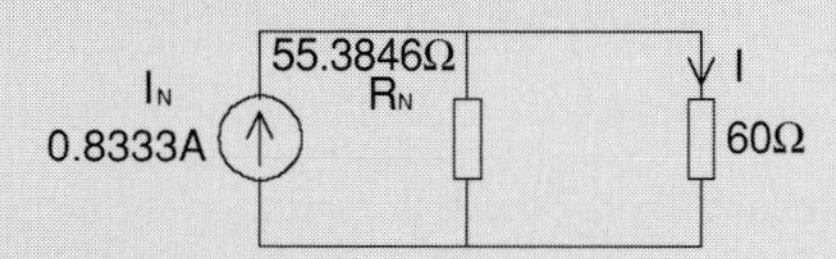

Figure 1.62

The value of I can be obtained from the simpler circuit using the current divison rule

$$I = I_N \frac{R_N}{R_N + 60}$$

$$I = 0.8333 \frac{55.3846}{115.3846}$$

$$I = 0.4\,\text{A}$$

PROBLEM 1.36 NORTON

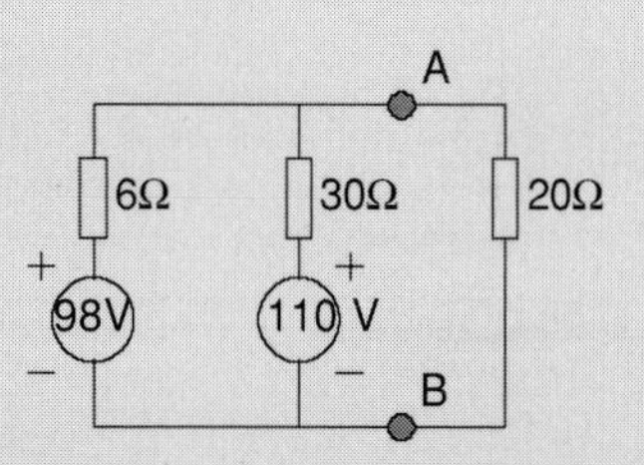

Figure 1.63

Find the value of the current through the 20Ω resistor by converting the circuit at the left of AB into a Norton equivalent.

Answer: $I = 4\text{A}$

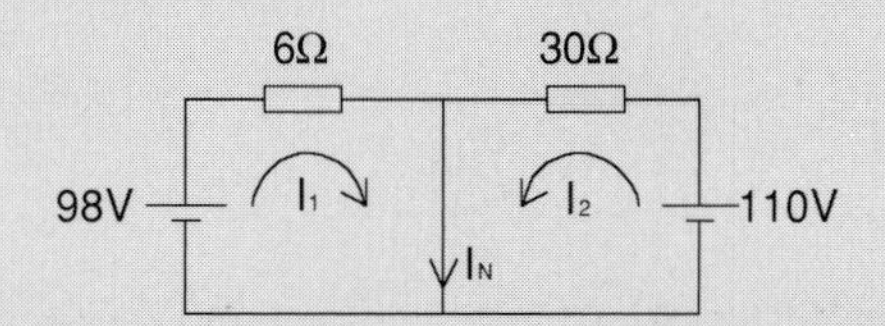

Figure 1.64

We have two independent meshes.

Left loop

$$98 = 6 I_1$$

$$I_1 = 16\frac{1}{3}$$

Right loop

$$110 = 30 I_2$$

$$I_2 = 3\frac{2}{3}$$

Norton current

$$I_N = I_1 + I_2 = 20\,\text{A}$$

Norton resistance

$$R_N = 6 \| 30 = 5\,\Omega$$

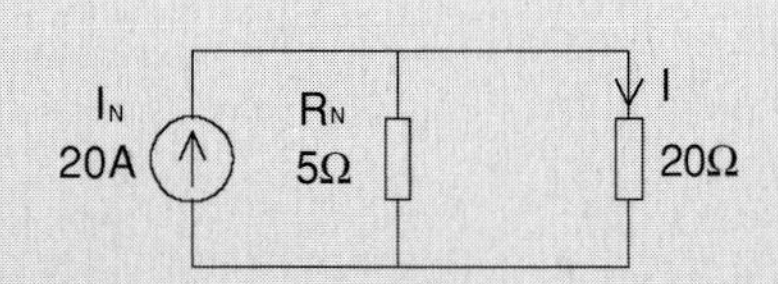

Figure 1.65

The current in question is given by the current division rule

$$I = 20\frac{5}{20+5} = 20\frac{5}{25}$$

$$I = 4\,\text{A}$$

Voltage at AB should also be

$$V_{AB} = (55+20)\times 3\frac{4}{} = 75\frac{4}{3} = 100\,\text{V}$$

This proves that $I_1 = 20\,\text{A}$ is the correct solution.

PROBLEM 1.37 NORTON

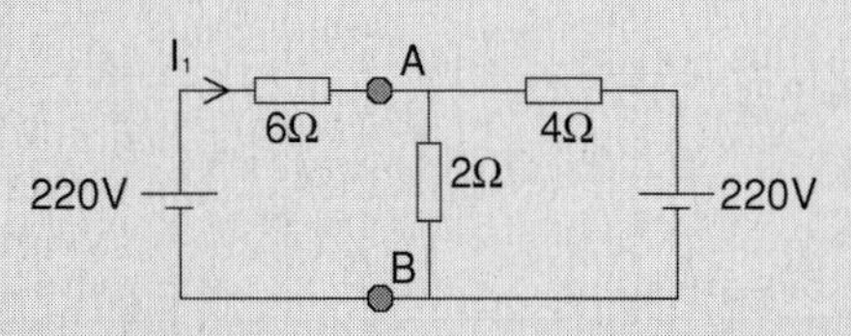

Figure 1.66

I_1 = 20A. Verify this by converting the circuit at the right of AB to Norton.

Answer: Yes, I_1 = 20A

Norton current

$$I_N = \frac{220}{4} = 55\,\text{A}$$

Norton resistance

$$R_N = 2\|4 = \frac{2\times 4}{2+4} = \frac{8}{6} = \frac{4}{3}$$

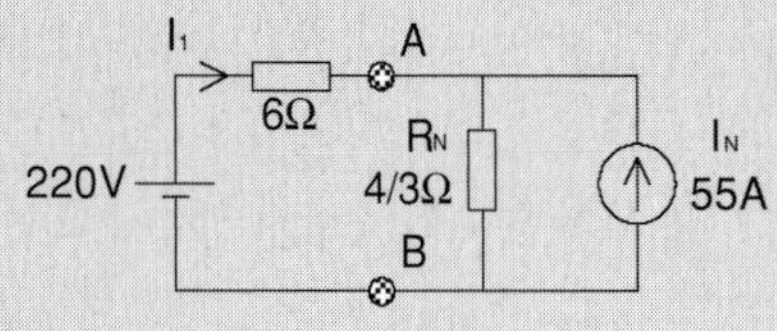

Figure 1.67

We now verify the current I by calculating the voltages at AB from both sides

Voltage at AB

$$V_{AB} = 220 - 20\times 6 = 220 - 120 = 100\,\text{V}$$

PROBLEM 1.38 NORTON

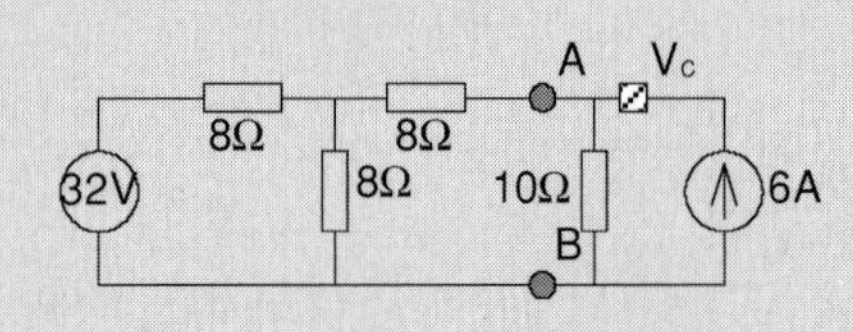

Figure 1.68

Find the voltage at V_C by converting the circuit on the left of AB to a Norton equivalent.

Answer: 40V

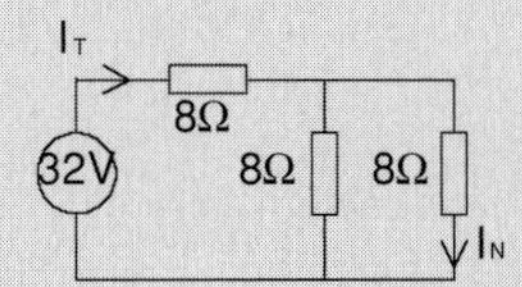

Figure 1.69

The Norton current is obtained by first calculating the total current of the circuit at the left of AB with AB shorted.

$$I_T = \frac{32}{12} = \frac{8}{3}\,\text{A} \qquad I_N = \frac{4}{3}\,\text{A (half)}$$

Norton resistance

$$R_N = 8\|8 + 8 = 12\,\Omega$$

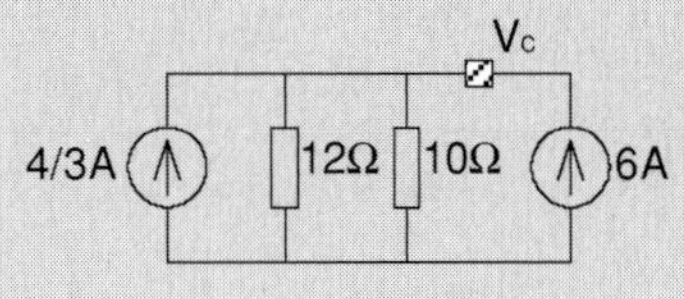

Figure 1.70

Both generators are feeding current in the same direction

Total current

$$6 + \frac{4}{3} = \frac{22}{3}\,\text{A}$$

Voltage V_C

$$V_C = IR = \frac{22}{3}\frac{12\times 10}{12+10}$$

$$= \frac{22}{3}\frac{120}{22} = 40\,\text{V}$$

PROBLEM 1.39 NORTON

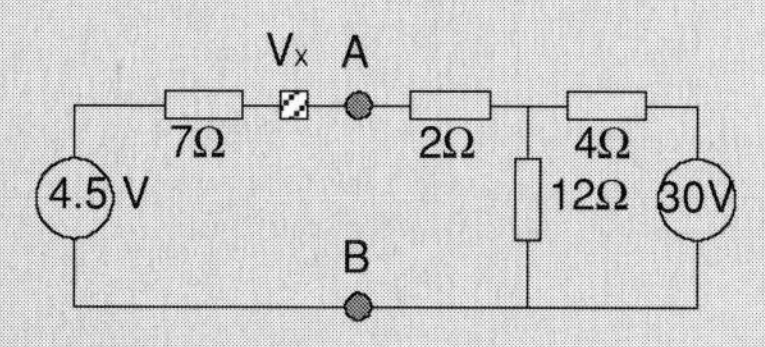

Figure 1.71

Find the voltage V_x by transforming to a Norton equivalent, the circuit on the right of AB.

Answer: $V_x = 15\text{V}$

Norton current
The Norton current is calculated by current division, after finding the total current of the circuit with A and B shorted together.

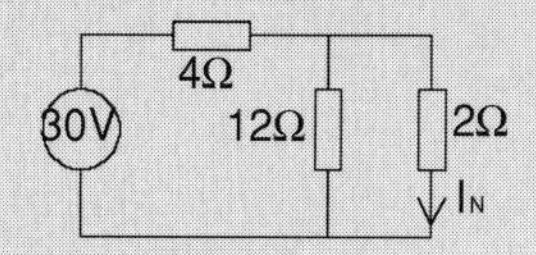

Figure 1.72

$$R_T = 12\|2 + 4 = 5.7143\,\Omega$$

$$I_T = \frac{30}{5.7143} = 5.25\,\text{A}$$

$$I_N = 5.25\frac{12}{12+2} = 4.5\,\text{A}$$

$$R_N = 12\|4 + 2 = 5\,\Omega$$

The circuit now is

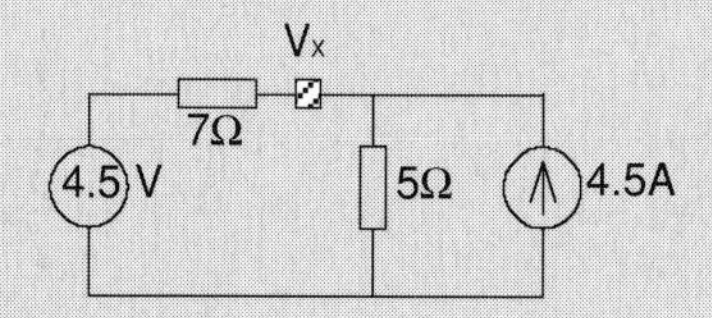

Figure 1.73

Applying KCL, we get

$$4.5 - \frac{V_X}{5} - \frac{V_X - 4.5}{7} = 0 \qquad |\times 35$$

$$157.5 - 7V_X - 5V_X + 22.5 = 0$$

$$-12V_X + 180 = 0$$

$$V_X = 15\,\text{V}$$

Note: A further explanation of this method, a form of nodal analysis, can be found on page 206.

PROBLEM 1.40 NORTON

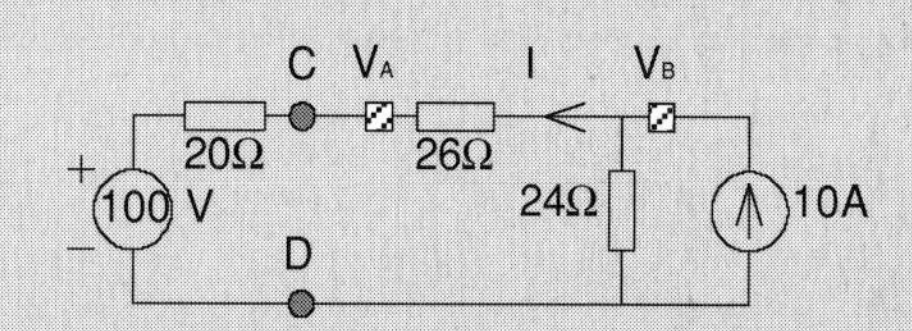

Figure 1.74

Convert to Norton the circuit on the left of CD. Find the value of I by first calculating V_A and V_B.

Answer: $I = 2\text{A}$

Norton current

$$I_N = \frac{100}{20} = 5\,\text{A}$$

Norton resistance

$$R_N = 20\,\Omega$$

The circuit can be redrawn

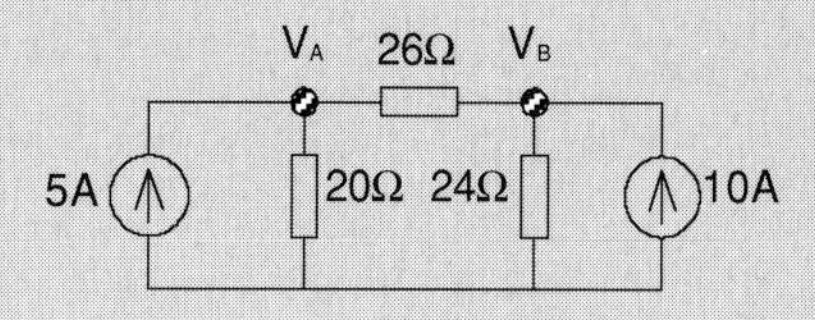

Figure 1.75

Applying KCL(a form of nodal analysis)

$$5-\frac{V_A}{20}-\frac{V_A-V_B}{26}=0 \qquad | \times 520$$

$$2600-26V_A-20V_A+20V_B=0$$

$$2600-46V_A+20V_B=0 \qquad (1)$$

Applying KCL (a form of nodal analysis)

$$10-\frac{V_B}{24}-\frac{V_B-V_A}{26}=0 \qquad | \times 312$$

$$3120-13V_B-12V_B+12V_A=0$$

$$3120-25V_B+12V_A=0 \qquad (2)$$

We multiply equation (1) by 5 and equation (2) by 4

$$1300-230V_A+100V_B=0$$

$$12\,480-100V_B+48V_A=0$$

$$25\,480-182V_A=0 \qquad V_A=140\,\text{V}$$

Replacing V_A in equation (1) we get

$$2600-6440+20V_B=0$$

$$-3840+20V_B=0 \qquad V_B=192\,\text{V}$$

Finally we can calculate the current

$$I=\frac{192-140}{26}=\frac{52}{26}=2\,\text{A}$$

PROBLEM 1.41 NORTON

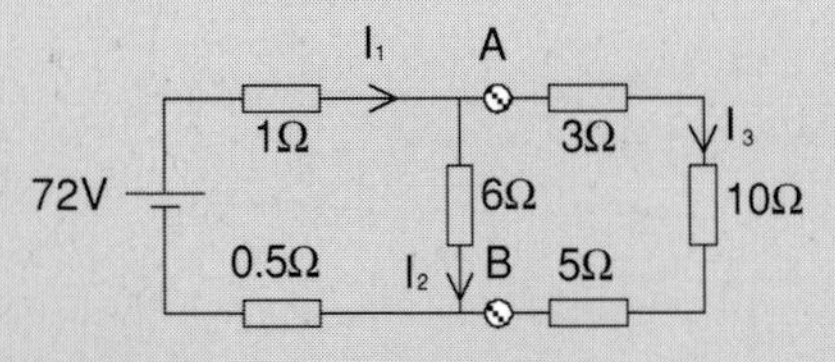

Figure 1.76

In this circuit I_3 = 3A. Verify this by converting the circuit at the left of AB to Norton.

Answer: I_3 = 3A

Norton current

$$I_N=\frac{72}{1.5}=48\,\text{A}$$

Norton resistance

$$R_N=6||1.5=1.2\Omega$$

The circuit can be redrawn

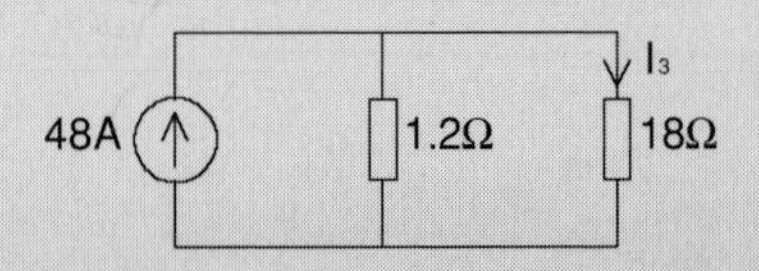

Figure 1.77

The answer is found after applying the current division rule

$$I_3=48\frac{1.2}{18+1.2}=3\,\text{A}$$

PROBLEM 1.42 NORTON

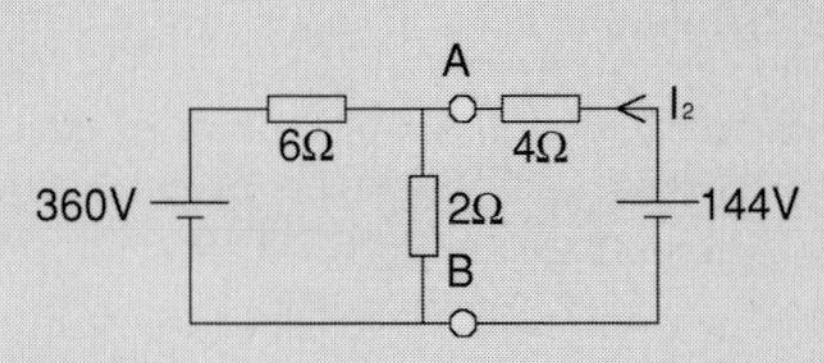

Figure 1.78

The approximate value of I_2 is 9.8A. Verify this by converting the circuit at the left of AB to Norton.

Answer: Yes

$$I_N=\frac{360}{6}=60\,\text{A}$$

Norton resistance

$$R_N=6||2=\frac{2\times 6}{2+6}=1.5\Omega$$

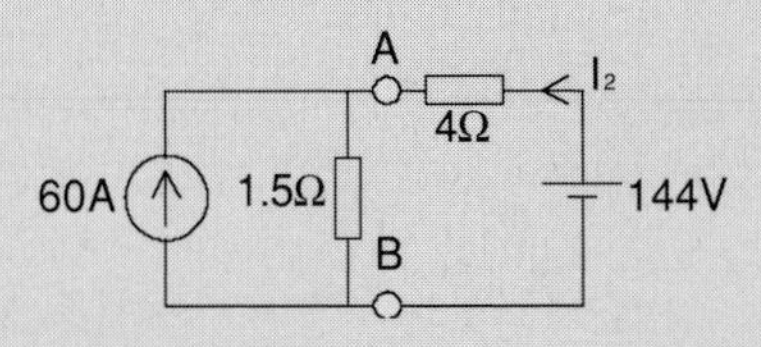

Figure 1.79

Voltage drop on 4Ω resistor

$I_2 \times 4 = 9.8 \times 4 = 39.2\text{V}$

The polarity is such that it is opposing the voltage source.

$V_{AB} = 144 - 39.2 = 104.8\text{V}$

On the 1.5Ω resistor the current is

$60 + 9.8$

Therefore

$V_{AB} = (60 + 9.8)1.5 = 104.7\text{V}$

The answer is yes.

PROBLEM 1.43 NORTON

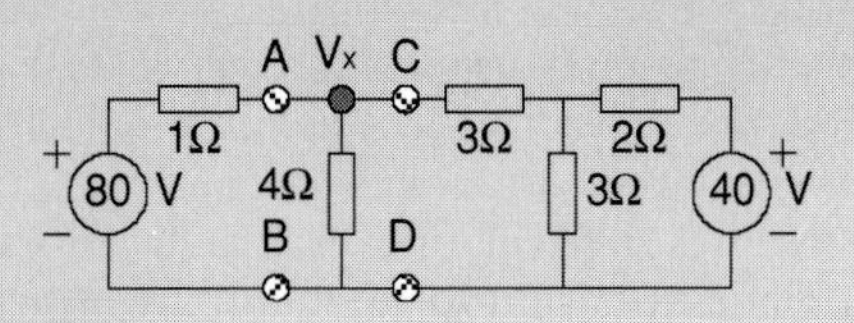

Figure 1.80

Find the value of V_x by converting to Norton at AB and CD.

Answer: $V_x = 57.6\text{V}$

At AB

$$I_N = \frac{80}{1} = 80\,\text{A}$$

$$R_N = 1\Omega$$

At CD

$$I_N = \frac{40}{2+3\|3} \times \frac{1}{2} = \frac{20}{3.5}\,\text{A}$$

$$R_N = 3 + 2\|3 = 3 + \frac{2\times 3}{5} = 3 + \frac{6}{5} = 4.2\Omega$$

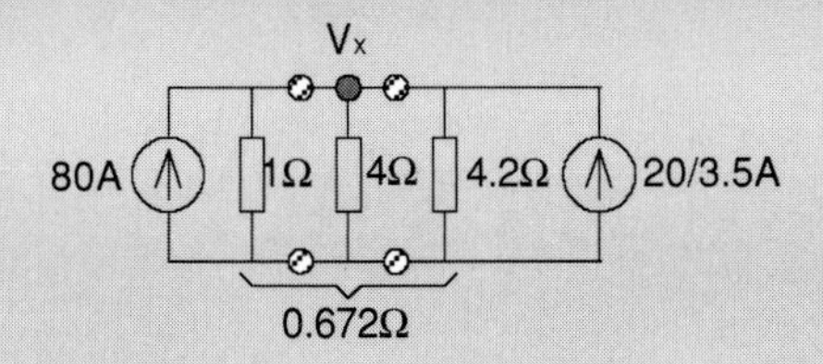

Figure 1.81

The redrawn circuit allows us to calculate the requested voltage

$V_x = IR = 85.7143 \times 0.672 = 57.6\text{V}$

PROBLEM 1.44 NORTON

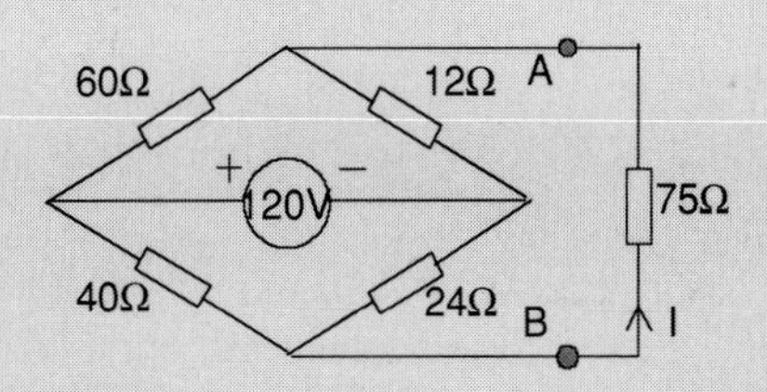

Figure 1.82

Find the value of I using Norton on the left of AB.

Answer: $I = 0.25\text{A}$

Norton resistance

$R_N = 60\|12 + 40\|24 = 10 + 15 = 25\Omega$

Norton current
This value is difficult to obtain in this case. We need the help of the next figure, to solve this problem. I_N has to flow from A to B. It is not the case on the left. It is the case on the right and for this we need Kirchhoff's law.

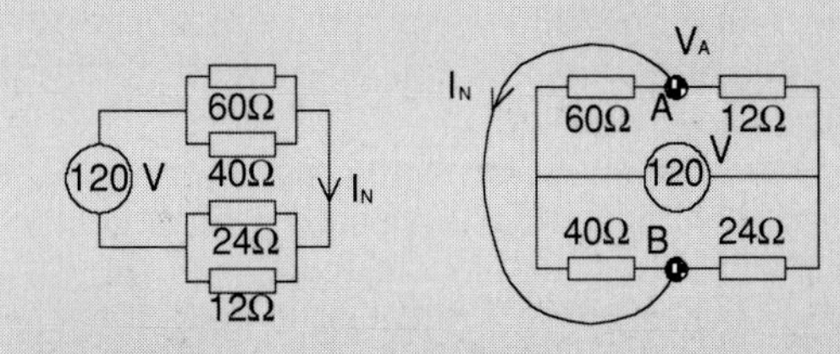

Figure 1.83

$$\frac{120 - V_A}{60} - I - \frac{V_A}{12} = 0 \qquad | \times 120$$

$$\frac{120 - V_A}{40} + I - \frac{V_A}{24} = 0 \qquad | \times 120$$

$$240 - 2V_A - 120I - 10V_A = 0$$

$$360 - 3V_A + 120I - 5V_A = 0$$

$$600 - 20V_A = 0 \qquad V_A = 30\,\text{V}$$

$$240 - 60 - 120I - 300 = 0$$

$$I = -1\,\text{A}$$

The current goes from B to A!

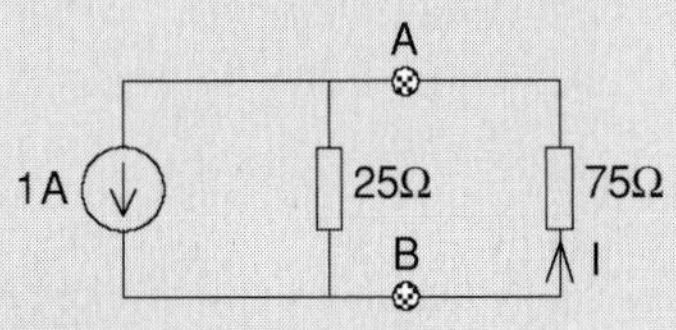

Figure 1.84

$$I = 1\frac{25}{25 + 75} = 0.25\,\text{A}$$

Thevenin–Norton conversions

In certain cases the conversion from Thevenin to Norton and vice versa can help us to obtain the solution to a problem.

A typical case is illustrated in Figure 1.85. This type of problem can be solved by mesh analysis, nodal analysis or by calculating the total impedance, total current, etc. Any of these methods will be quite complicated. The last method mentioned would be straightforward if it were not for the capacitor on the right-hand side. With the capacitor we will have imaginary and complex numbers to contend with.

We can use, with advantage, Thevenin–Norton conversions which are simple and progress to the right as shown in Figure 1.86. In the sequence shown, the values of voltage and current sources have not been indicated and, of course, they change.

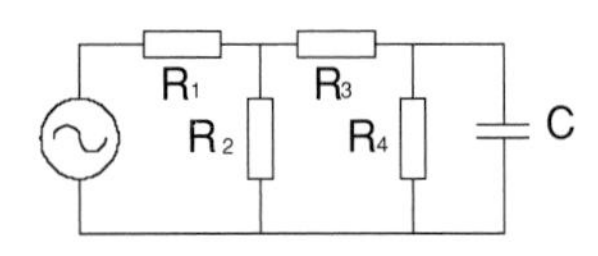

Figure 1.85

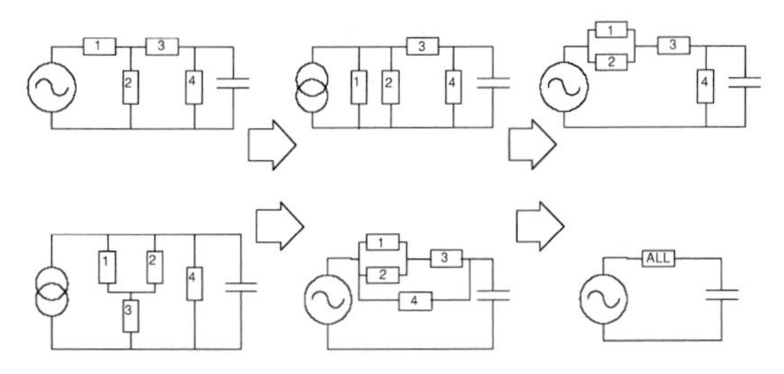

Figure 1.86

The resistors, when in parallel, can be combined and the series resistors added, but in Figure 1.86, have been left without combining to show the method of working the problem.

A useful though unexpected fact is that the resistance for a Norton to Thevenin or Thevenin to Norton, is the same, as shown in Figure 1.87, the resistor is 10Ω in one case and also in the other. Remember that to calculate the impedance in Norton or Thevenin we disable the supplies. That is to say, we short the voltage supply and open-circuit the current source. So, in both cases we end up with 10Ω.

Calculating the equivalent resistances will be no problem, in the case of Figure 1.86. We will only have to use parallel and series conversions.

We will now see how we calculate the value of the current source if we go from a Thevenin to a Norton equivalent (see Figure 1.88). According

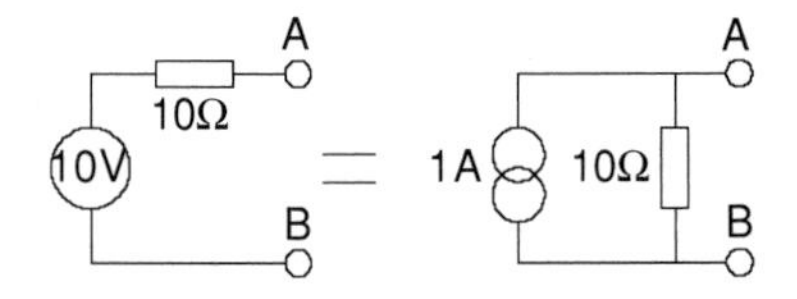

Figure 1.87

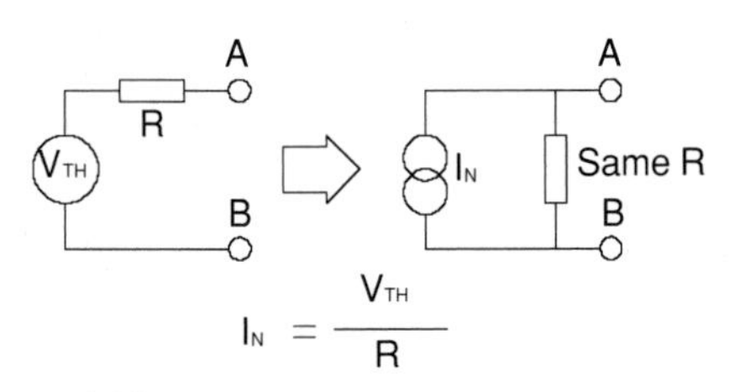

Figure 1.88

to the Norton theorem we have to short A and B and measure or calculate the short circuit current through AB. It is

$$I_N = \frac{V_{TH}}{R}$$

The only part now left, is how to calculate the value of the voltage source when we go from a Norton to a Thevenin equivalent. We now use Figure 1.89.

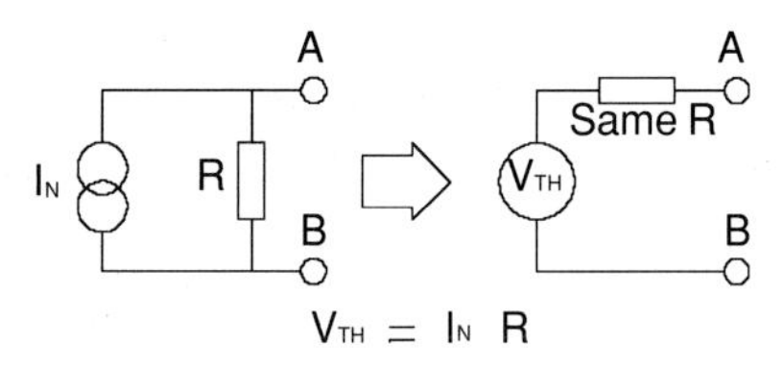

Figure 1.89

According to Thevenin's theorem, the voltage is obtained by measuring or calculating the voltage at points A and B, when they have been separated from the rest of the circuit. It is

$$V_{TH} = I_N R$$

There are several problems that will be solved by this technique in future chapters. You need to master it well in the examples that follow.

PROBLEM 1.45 THEVENIN–NORTON

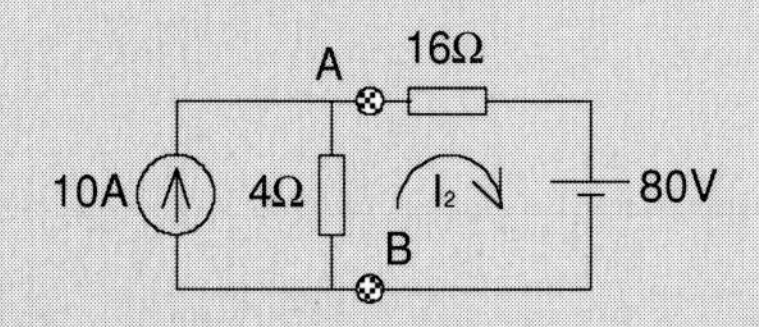

Figure 1.90

$I_2 = -2A$. Verify this value by converting the left of AB from Norton to Thevenin.

Answer: Yes, $I_2 = -2A$

Thevenin voltage

$$V_{TH} = 4 \times 10 = 40V$$

Thevenin resistance

$$R_{TH} = I_N = 4\Omega$$

We now redraw the circuit

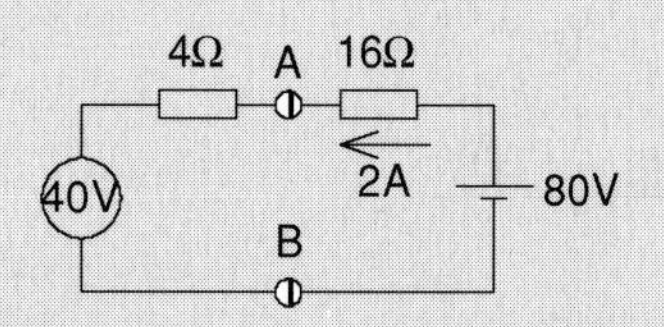

Figure 1.91

The voltage AB, calculated from both directions is

$$V_{AB} = 80 - 16 \times 2 = 48V$$

$$V_{AB} = 40 + 4 \times 2 = 48V$$

The current value is alright.

PROBLEM 1.46 THEVENIN–NORTON

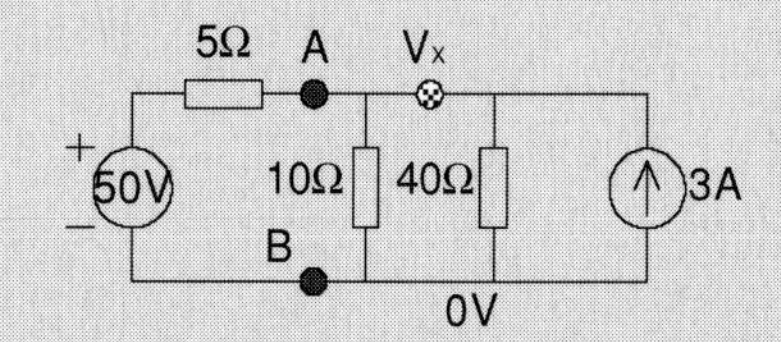

Figure 1.92

Find the value of V_X using Thevenin to Norton conversion at AB.

Answer: $V_X = 40V$

Norton resistance

$$R_N = R_{TH} = 5\Omega$$

Norton current

$$I_N = \frac{50}{5} = 10\,A$$

The new circuit is

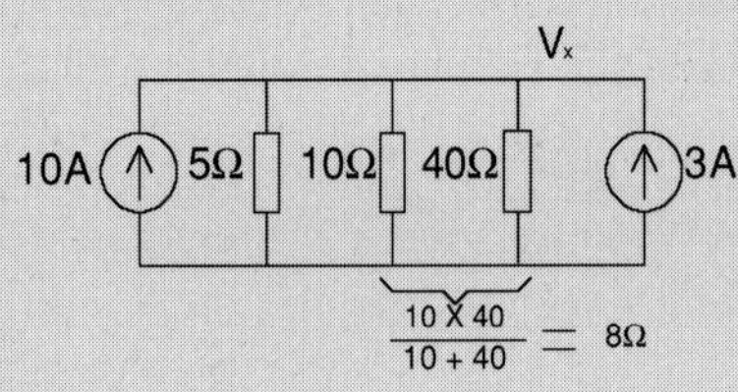

Figure 1.93

Resistance

$$8||5 = \frac{8 \times 5}{8+5} = \frac{40}{13}$$

Total current

$$10 + 3 = 13\,A$$

Voltage

$$V_X = \frac{13 \times 40}{13} = 40\,V$$

PROBLEM 1.47 THEVENIN–NORTON

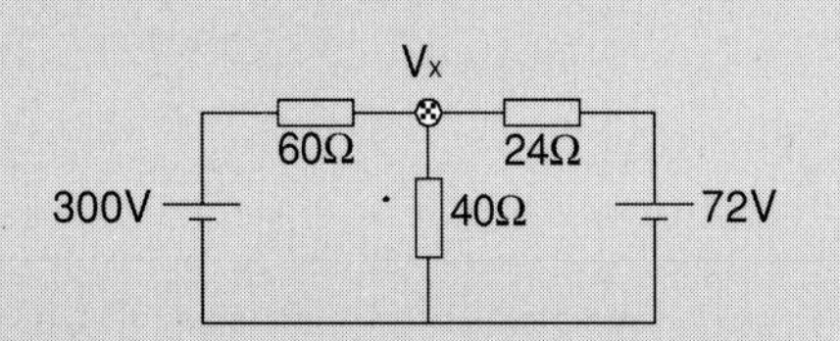

Figure 1.94

Find the value of V_X by changing Thevenin to Norton on both sides.

Answer: $V_X = 96V$

Left - hand side

Norton current

$$I_N = \frac{300}{60} = 5\,A$$

Norton resistance

$$R_N = 60\Omega$$

Right - hand side

Norton current

$$I_N = \frac{72}{24} = 3\,A$$

Norton resistance

$$R_N = 24\Omega$$

We now redraw the circuit

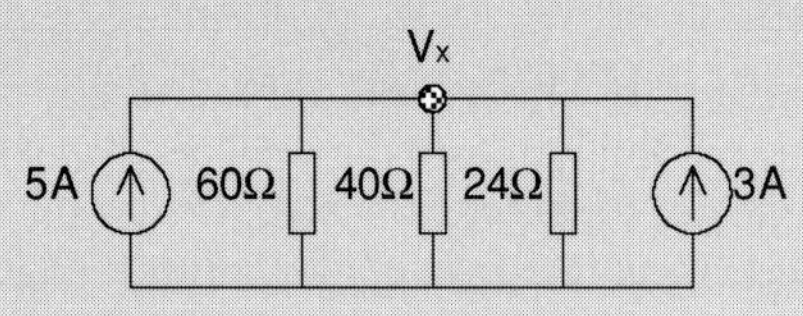

Figure 1.95

Total current

$$5 + 3 = 8\,A$$

Total resistance

$$60||40||24 = 12\Omega$$

and finally, the voltage

$$V_X = IR = (5+3)12 = 8 \times 12 = 96\,V$$

PROBLEM 1.48 THEVENIN–NORTON

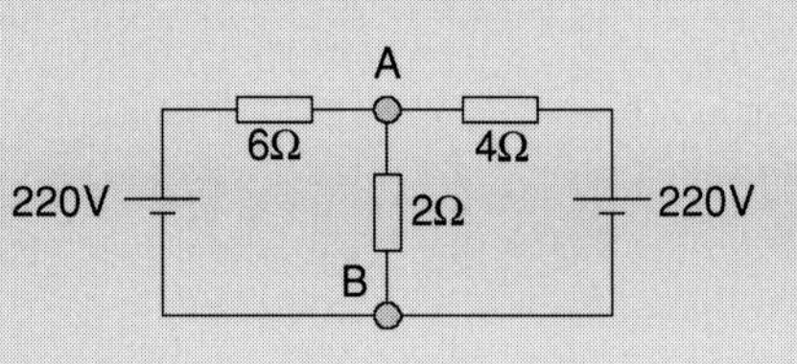

Figure 1.96

Find the voltage AB by converting Thevenin to Norton on both sides.

Answer: 100V

Left - hand side

$$I_N = \frac{220}{6} = 36\frac{2}{3}\,A$$

$$R_N = 6\Omega$$

Right - hand side

$$I_N = \frac{220}{4} = 55\,A$$

$$R_N = 4\Omega$$

We now redraw the circuit

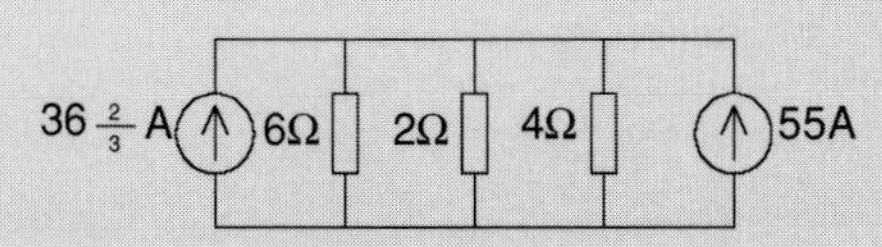

Figure 1.97

Total resistance

$$R_T = 6\|2\|4 = 1.0909\Omega$$

Voltage

$$V_{AB} = 91.666 \times 1.0909 = 100\,V$$

PROBLEM 1.49 THEVENIN–NORTON

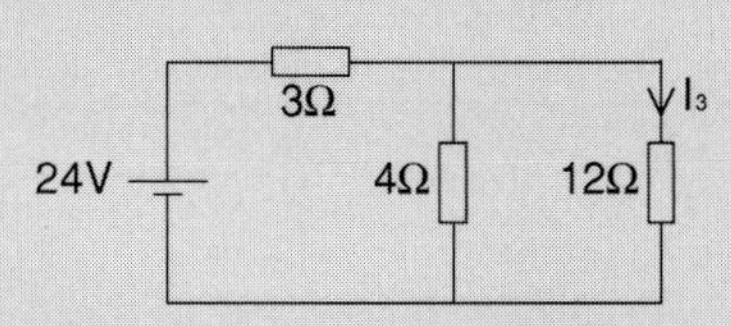

Figure 1.98

Calculate I_3 by Thevenin–Norton conversions left to right.

Answer: $I_3 = 1A$

Norton current

$$I_N = \frac{24}{3} = 8\,A$$

Norton resistance

$$R_N = 3\Omega$$

We now redraw the circuit

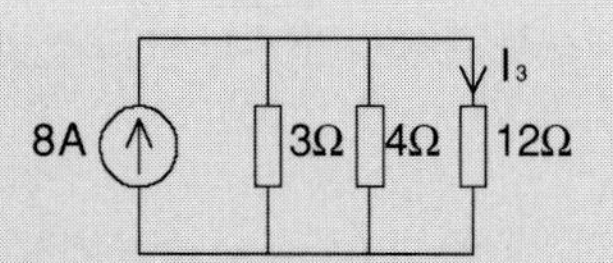

Figure 1.99

The 3 and 4Ω resistors can be joined together. Then we can go from Norton to Thevenin and get nearer to the solution

$$3\|4 = \frac{3 \times 4}{3+4} = \frac{12}{7}\Omega$$

Thevenin voltage

$$V_{TH} = IR = \frac{8 \times 12}{7} = \frac{96}{7}\,V$$

Thevenin resistance

$$R_{TH} = \frac{12}{7}\Omega$$

The circuit is redrawn once more.

Figure 1.100

The current can now be calculated

$$I_3 = \frac{V_{TH}}{\frac{12}{7} + 12} = \frac{96}{7} \times \frac{1}{\frac{12}{7} + \frac{84}{7}}$$

$$= \frac{96}{7}\frac{7}{96} = 1\,A$$

PROBLEM 1.50 THEVENIN–NORTON

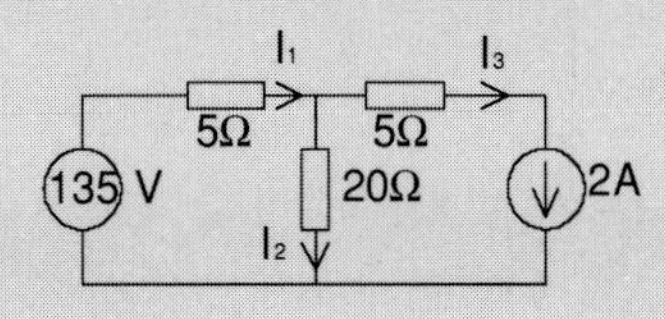

Figure 1.101

Find the value of I_2 by converting Thevenin to Norton.

Answer: $I_2 = 5A$

There is only one Thevenin circuit, the 135V source with the 5Ω resistor, we convert this one.

$$I_N = \frac{135}{5} = 27\,\text{A}$$

$$R_N = 5\Omega$$

The new circuit is

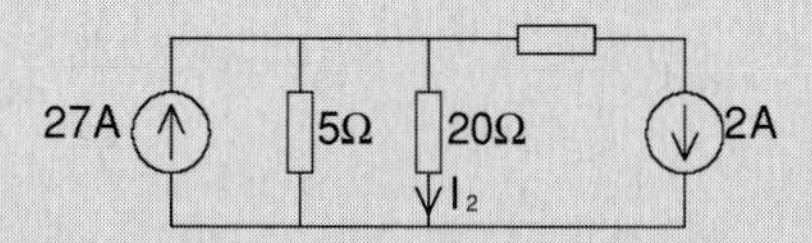

Figure 1.102

The currents are opposing. The equivalent current source is therefore 27 – 2 = 25A.
I_2 is obtained by current division.

$$I_2 = 25\frac{5}{20+5}$$

$$I_2 = 5\,\text{A}$$

PROBLEM 1.51 THEVENIN–NORTON

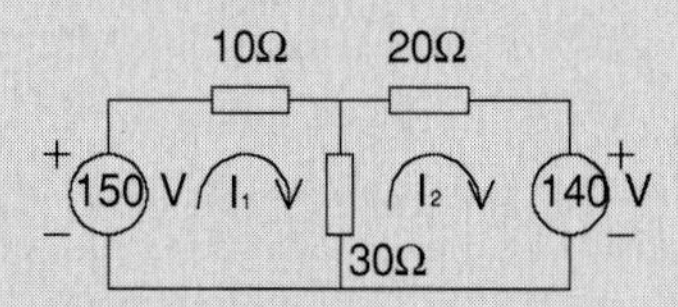

Figure 1.103

Find the value of I_2 by making Thevenin–Norton conversions from left to right until there are two voltage sources linked directly by resistors.

Answer: $I_2 = -1\text{A}$

Thevenin to Norton

$$I_N = \frac{150}{10} = 15\,\text{A}$$

$$R_N = 10\Omega$$

The result is shown

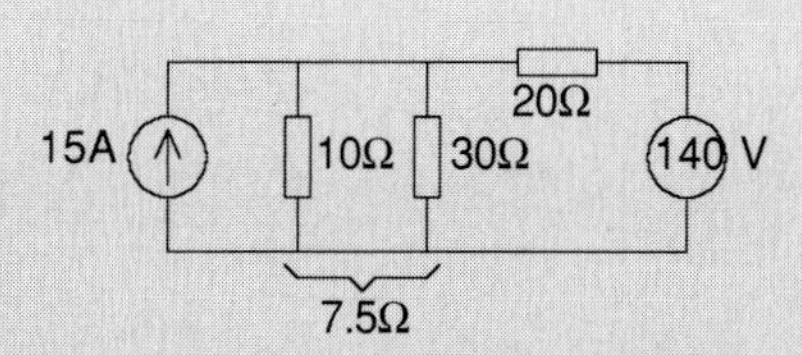

Figure 1.104

Norton to Thevenin

$$V_{TH} = 15 \times 7.5 = 112.5\,\text{V}$$

$$R_{TH} = 7.5\Omega$$

The new circuit is shown

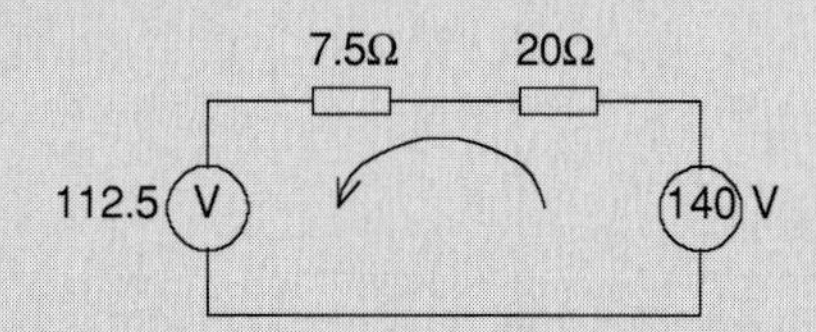

Figure 1.105

The value of I_2 can be calculated.

$$I_2 = \frac{140 - 112.5}{20 + 7.5} = \frac{27.5}{27.5} = 1\,\text{A}$$

Because the direction is opposite to the one above, we must call it $-1\,\text{A}$.

PROBLEM 1.52 THEVENIN–NORTON

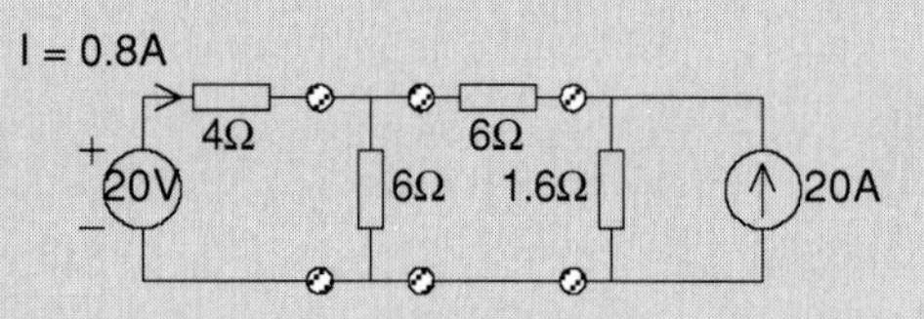

Figure 1.106

Use Norton–Thevenin conversions on the right-hand side to show that the current on the 20V battery is 0.8A

Answer: 0.8A

Thevenin voltage

$V_{TH} = 1.6 \times 20 = 32V$

Thevenin resistance

$R_{TH} = 1.6\Omega$

The new circuit is shown

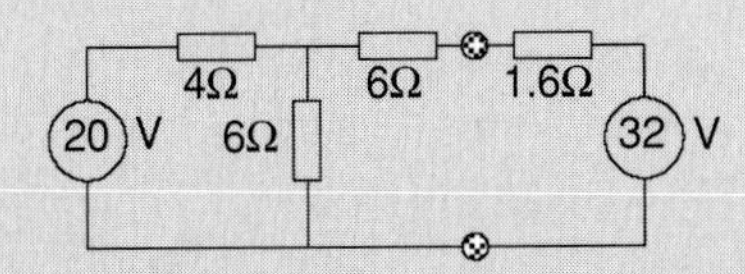

Figure 1.107

The next conversion is

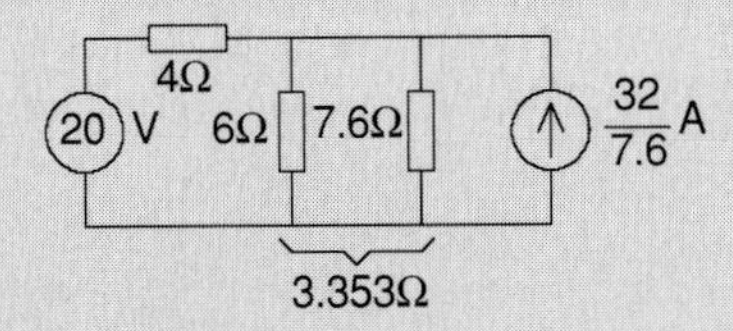

Figure 1.108

and the last transformation is

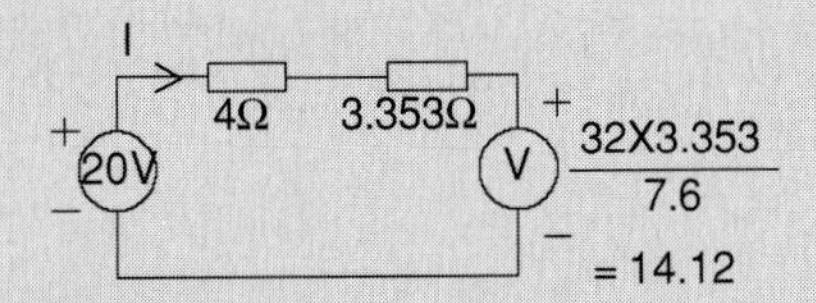

Figure 1.109

The current can now be evaluated

$$I = \frac{20 - 14.12}{4 + 3.353} = \frac{5.88}{7.353} = 0.8\,\text{A}$$

PROBLEM 1.53 THEVENIN–NORTON

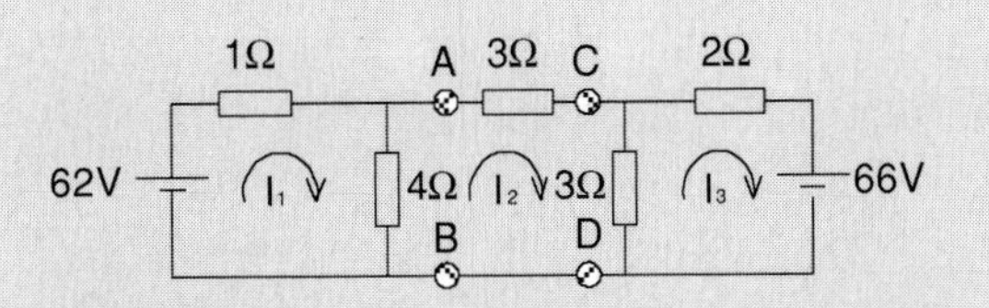

Figure 1.110

We know that in the circuit shown $I_1 = 14A$, $I_2 = 2A$ and $I_3 = -12A$.
Verify the value of I_2 by converting the left of AB to Thevenin and the right of CD to Norton. Then check if the value of I_2 is correct.

Answer: $I_2 = 2A$

Thevenin

$$V_{TH} = 62\frac{4}{1+4} = 49.6\,\text{V}$$

$$R_{TH} = 1 \| 4 = \frac{4}{5} = 0.8\Omega$$

Norton

$$I_N = \frac{66}{2} = 33\,\text{A} \text{ (the } 3\Omega \text{ is short circuited)}$$

$$R_N = 3 \| 2 = \frac{3 \times 2}{3+2} = \frac{6}{5} = 1.2\Omega$$

The new circuit is shown

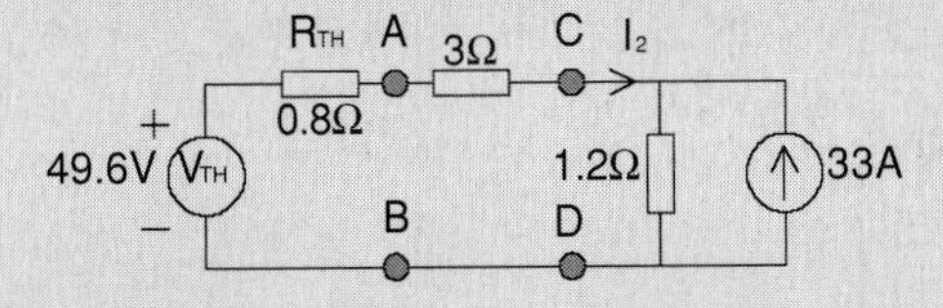

Figure 1.111

$V_{CD} = IR = (33 + 2)1.2 = 42V$

$V_{CD} = 49.6 - 2 \times 0.8 - 3 \times 2 = 42V$

Therefore $I_2 = 2A$ is alright.

PROBLEM 1.54 THEVENIN–NORTON

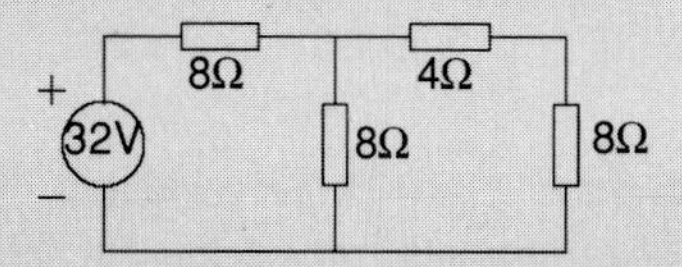

Figure 1.112

Use Thevenin–Norton conversions to demonstrate that the circuit shown is equivalent to a current source of 2A with a load of 4Ω.

Answer: Yes

Thevenin to Norton conversion

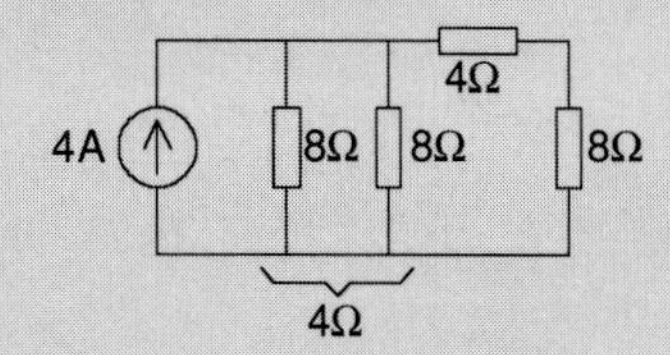

Figure 1.113

Norton to Thevenin conversion

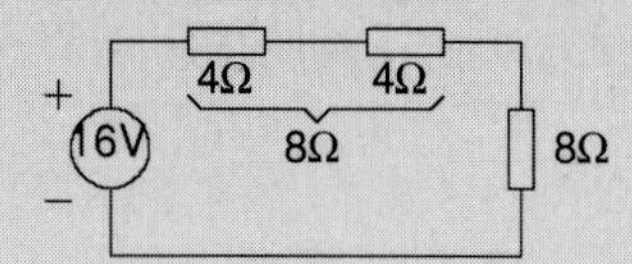

Figure 1.114

Thevenin to Norton conversion

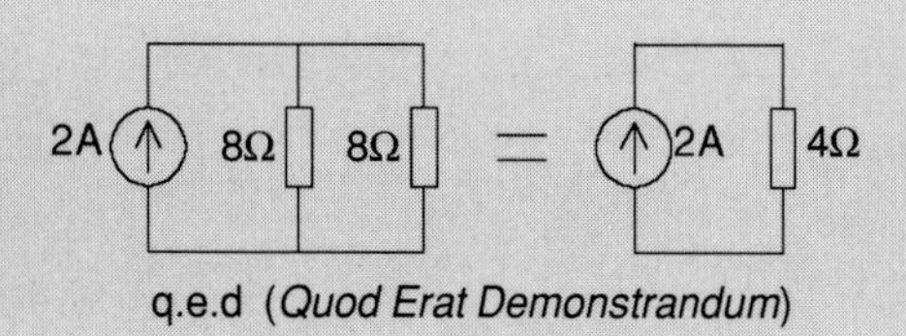

Figure 1.115

PROBLEM 1.55 THEVENIN–NORTON

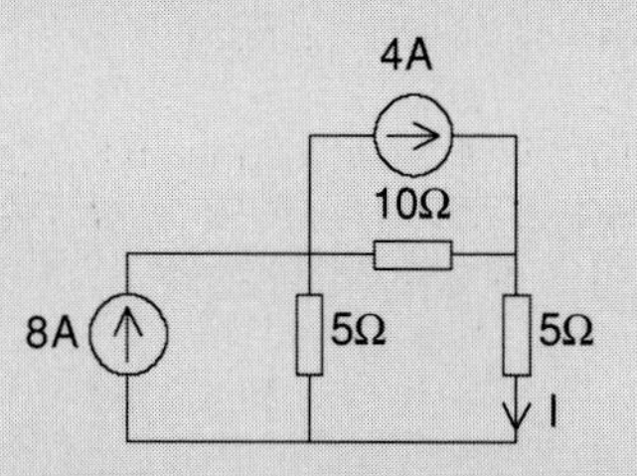

Figure 1.116

Convert Norton to Thevenin for both sources, to find the value of I.

Answer: $I = 4\text{A}$

Thevenin voltage

$V_{TH} = 8 \times 5 = 40\text{V}$

Thevenin resistance

$R_{TH} = 5\Omega$

Thevenin voltage

$V_{TH} = 4 \times 10 = 40\text{V}$

Thevenin resistance

$R_{TH} = 10\Omega$

The new circuit is shown

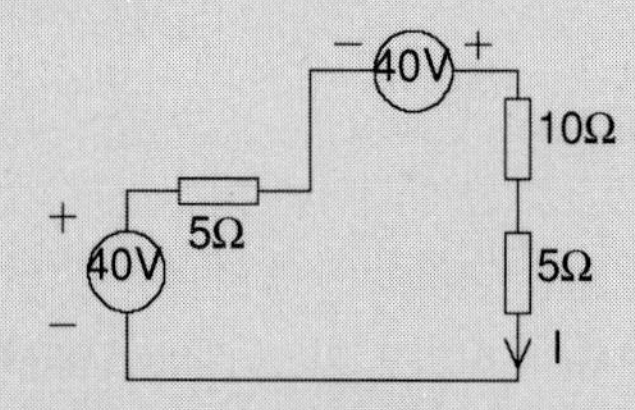

Figure 1.117

$$I = \frac{40+40}{5+10+5} = \frac{80}{20} = 4\,\text{A}$$

Superposition

Superposition is a very useful technique to solve complicated problems. We will certainly use it in the solution of operational-amplifier (Op–Amp) problems, where there is more than one input.

If the system has two inputs, you can consider the output of the system when one input is 'reduced to zero', leaving only the other input to produce an output.

Then you 'reduce to zero' the other input and consider the output with only the input that was 'reduced to zero' in the first place.

The total output is given by the sum of the two outputs:

$$V_0 = V_{01} + V_{02}$$

This is really an application of a more general principle, that of the cause and effect: two causes and two effects.

Care must be taken to make sure that we are not in saturation where cause and effect no longer applies, i.e., it must be a linear system that we are considering. Additionally, you must be careful with the 'reduction to zero' as this has two meanings. In the case of a voltage source, the voltage is 'reduced to zero' by replacing the (unwanted) voltage source with a *short circuit*. You could say short-circuiting the (unwanted) battery. In the case of a current source, the opposite is true. In order to 'reduce to zero' a current source, you replace the current source with an *open circuit*.

Figure 1.118 shows the superposition when the circuit has active sources of voltage sources. Here

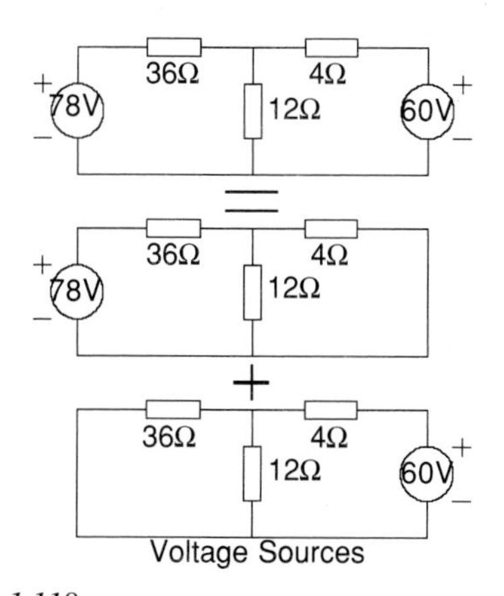

Figure 1.118

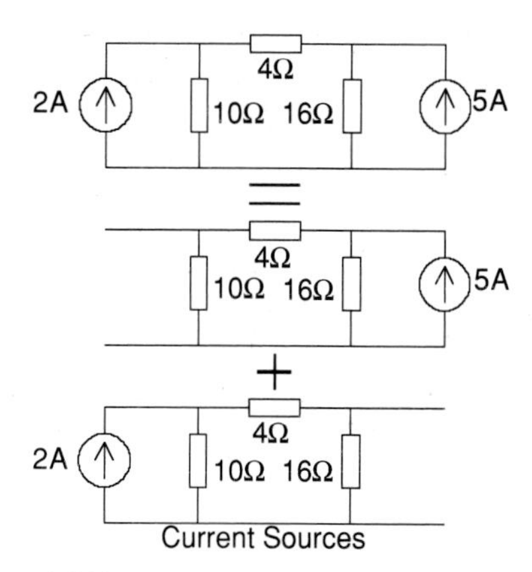

Figure 1.119

we see a simple example of how to deal with the voltage sources under superposition.

In Figure 1.119, we see how to deal with a problem under superposition when the source is a current source.

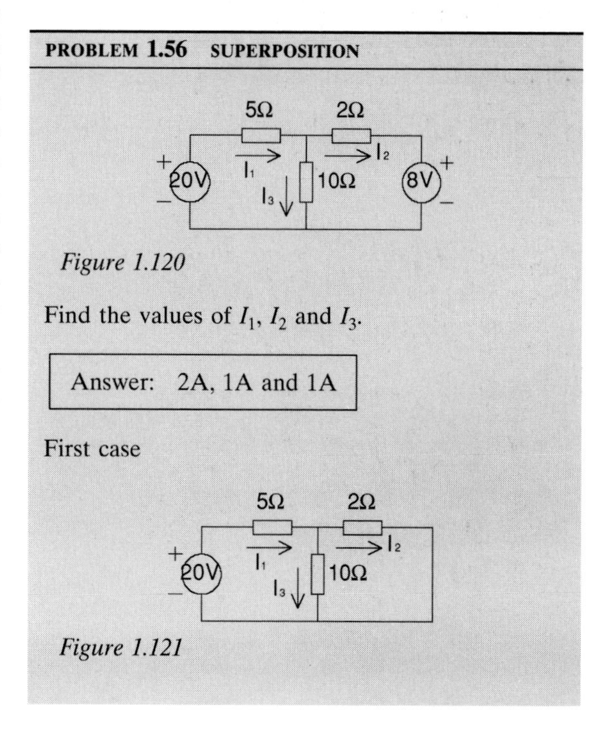

PROBLEM 1.56 SUPERPOSITION

Figure 1.120

Find the values of I_1, I_2 and I_3.

Answer: 2A, 1A and 1A

First case

Figure 1.121

Total resistance

$$R_T = 5 + 10\|2 = 5 + \frac{10 \times 2}{10+2} = \frac{20}{3}\Omega$$

Currents

$$I_1 = \frac{20}{\frac{20}{3}} = 3\,\text{A}$$

I_2 and I_3 obtained by current division

$$I_2 = 3\frac{10}{10+2} = \frac{10}{4} = \frac{5}{2}\,\text{A}$$

$$I_3 = 3\frac{2}{10+2} = \frac{2}{4} = \frac{1}{2}\,\text{A}$$

Second case

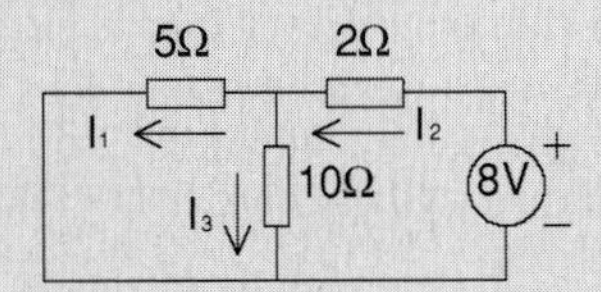

Figure 1.122

Total resistance

$$R_T = 2 + 5\|10 = 2 + \frac{5 \times 10}{5+10} = \frac{16}{3}\Omega$$

Currents

$$I_2 = \frac{8}{\frac{16}{3}} = \frac{3}{2}\,\text{A}$$

$$I_3 = \frac{3}{2}\frac{5}{10+5} = \frac{1}{2}\,\text{A}$$

$$I_1 = \frac{3}{2}\frac{10}{10+5} = 1\,\text{A}$$

Now we join the two results together

$$I_1 = 3 - 1 = 2\,\text{A}$$

$$I_2 = \frac{5}{2} - \frac{3}{2} = 1\,\text{A}$$

$$I_3 = \frac{1}{2} + \frac{1}{2} = 1\,\text{A}$$

PROBLEM 1.57 SUPERPOSITION

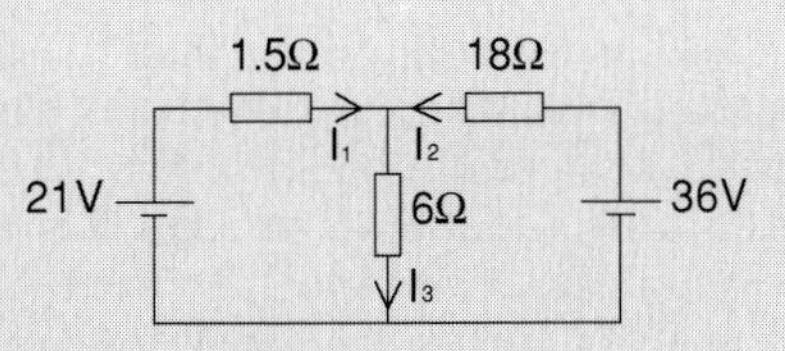

Figure 1.123

Use superposition to find the values of I_1, I_2 and I_3.

Answer: $I_1 = 2\text{A}$, $I_2 = 1\text{A}$ and $I_3 = 3\text{A}$

Superposition. Left-side supply on

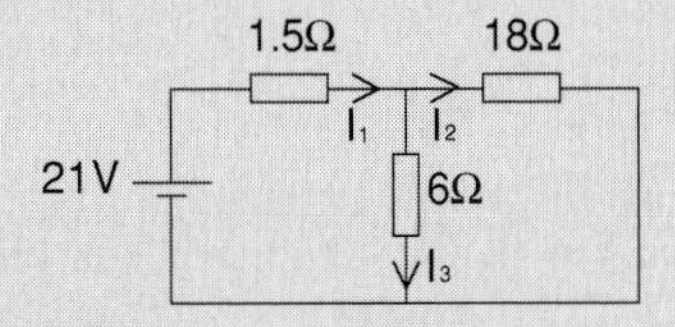

Figure 1.124

Total resistance

$$R_T = 6\|18 + 1.5 = \frac{6 \times 18}{6+18} + 1.5 = 6\Omega$$

Currents

$$I_1 = \frac{21}{6} = 3.5\,\text{A}$$

$$I_2 = -I_1\frac{6}{6+18} = -3.5\frac{6}{24} = -0.875$$

$$I_3 = I_1\frac{18}{6+18} = 3.5\frac{18}{6+18} = 2.625$$

Second case. From the other side

Total resistance

$$R_T = 6\|1.5 + 18 = 1.2 + 18 = 19.2\Omega$$

$$I_2 = \frac{36}{19.2} = 1.875\,\text{A}$$

$$I_3 = 1.875\frac{1.5}{1.5+6} = 0.375\,\text{A}$$

$$I_1 = -1.875\frac{6}{1.5+6} = -1.5\,\text{A}$$

Total values

$$I_1 = 3.5 - 1.5 = 2\,\text{A}$$

$$I_2 = -0.875 + 1.875 = 1\,\text{A}$$

$$I_3 = 2.625 + 0.375 = 3\,\text{A}$$

PROBLEM 1.58 SUPERPOSITION

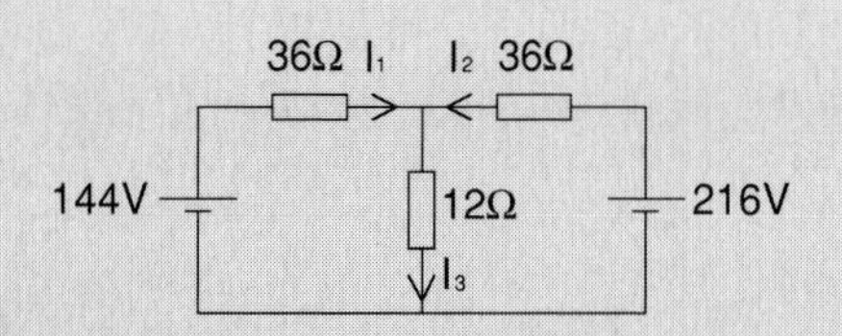

Figure 1.125

I_1 = 2A, I_2 = 4A and I_3 = 6A

Demonstrate this using superposition.

Answer: Yes

Case a)
Total resistance

$$R_T = 36 \| 12 + 36 = 9 + 36 = 45\Omega$$

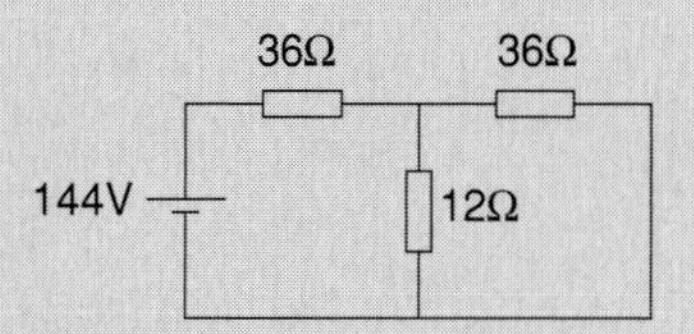

Figure 1.126

$$I_1 = \frac{144}{45} = 3.2\,\text{A}$$

$$I_3 = I_1 \frac{36}{36+12} = 3.2\frac{36}{48} = 2.4\,\text{A}$$

$$I_2 = I_1 \frac{12}{12+36} = -3.2\frac{12}{48} = -0.8\,\text{A}$$

Case b)
Total resistance, the same as before

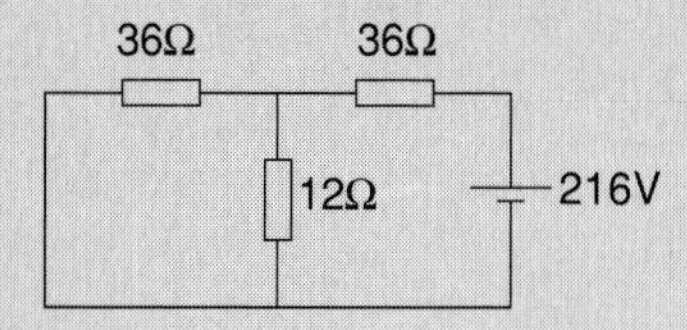

Figure 1.127

$$I_2 = \frac{216}{45} = 4.8\,\text{A}$$

$$I_3 = I_2 \frac{36}{12+36} = 4.8\frac{36}{48} = 3.6\,\text{A}$$

$$I_1 = -I_2 \frac{12}{12+36} = -4.8\frac{12}{48} = -1.2\,\text{A}$$

Result

$$I_1 = -1.2 + 3.2 = 2\,\text{A}$$

$$I_2 = -0.8 + 4.8 = 4\,\text{A}$$

$$I_3 = 2.4 + 3.6 = 6\,\text{A}$$

PROBLEM 1.59 SUPERPOSITION

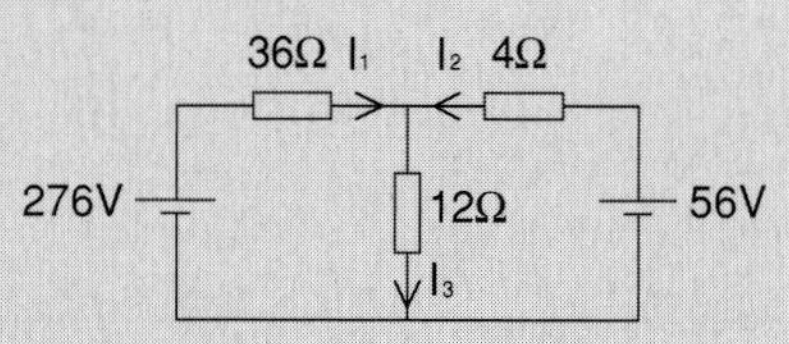

Figure 1.128

The values of current are: I_1 = 6A, I_2 = –1A and I_3 = 5A. Verify this using superposition.

Answer: Yes

Case a)

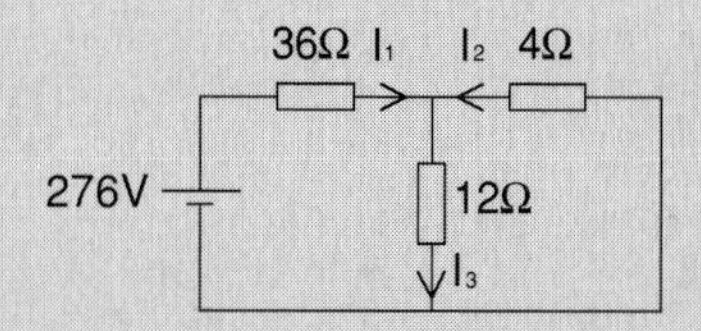

Figure 1.129

Total resistance

$$R_T = 12 \| 4 + 36 = 39\Omega$$

$$I_1 = \frac{276}{39} = 7.077\,\text{A}$$

$$I_3 = I_1 \frac{4}{12+4} = 7.077\frac{4}{16} = 1.77\,\text{A}$$

$$I_2 = -I_1 \frac{12}{12+4} = -7.077\frac{12}{16} = -5.31\,\text{A}$$

Case b)

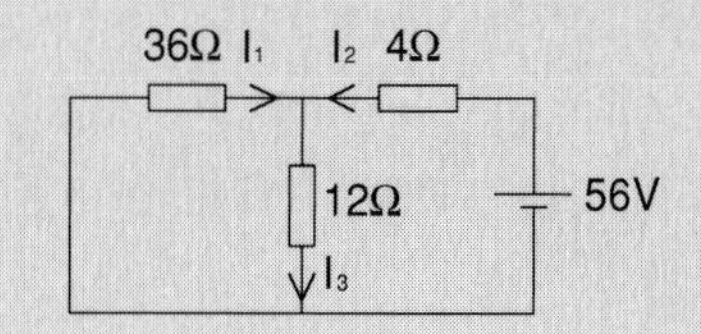

Figure 1.130

Total resistance

$$R_T = 36\|12 + 4 = 9 + 4 = 13\Omega$$

$$I_2 = \frac{56}{13} = 4.31\,\text{A}$$

$$I_3 = I_2 \frac{36}{36+12} = 4.31\frac{36}{48} = 3.23\,\text{A}$$

$$I_1 = -I_2 \frac{12}{12+36} = -4.31\frac{12}{48} = -1.077\,\text{A}$$

Result:

$$I_1 = 7.077 - 1.077 = 6\,\text{A}$$

$$I_2 = -5.31 + 4.31 = -1\,\text{A}$$

$$I_3 = 1.77 + 3.23 = 5\,\text{A}$$

PROBLEM 1.60 SUPERPOSITION

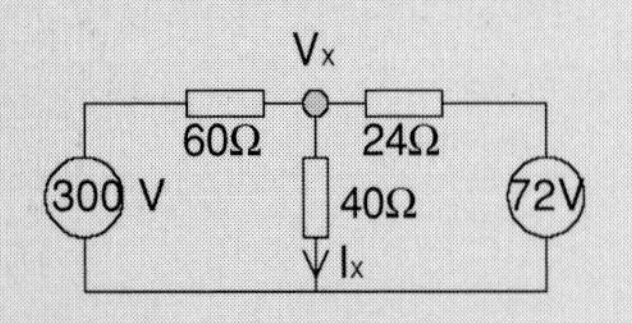

Figure 1.131

Using superposition find the value of the voltage at V_X. Calculate the current flowing through the 40Ω resistor first.

Answer: $V_X = 96\text{V}$

Case a)

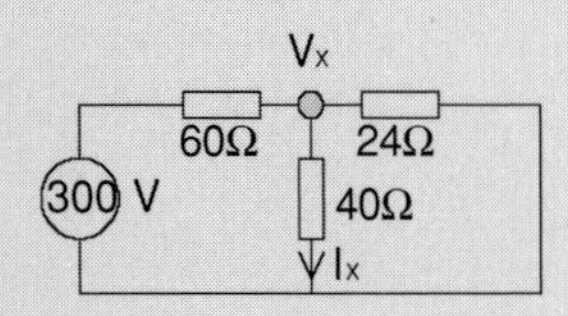

Figure 1.132

Total resistance

$$R_T = 60 + 40\|24 = 60 + 15 = 75\Omega$$

$$I_r = \frac{300}{75} = 4\,\text{A}$$

$$I_x = 4\frac{24}{40+24} = 4\frac{24}{64} = 1.5\,\text{A}$$

Case b)

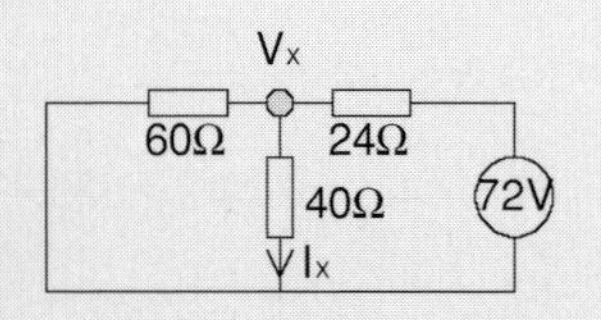

Figure 1.133

Total resistance

$$R_T = 24 + 60\|40 = 24 + 24 = 48\Omega$$

$$I_T = \frac{72}{48} = 1.5\,\text{A}$$

$$I_X = 1.5\frac{60}{60+40} = 0.9\,\text{A}$$

Result

$$I_X = 0.9 + 1.5 = 2.4\,\text{A}$$

$$V_X = IR = 2.4 \times 40 = 96\,\text{V}$$

PROBLEM 1.61 SUPERPOSITION

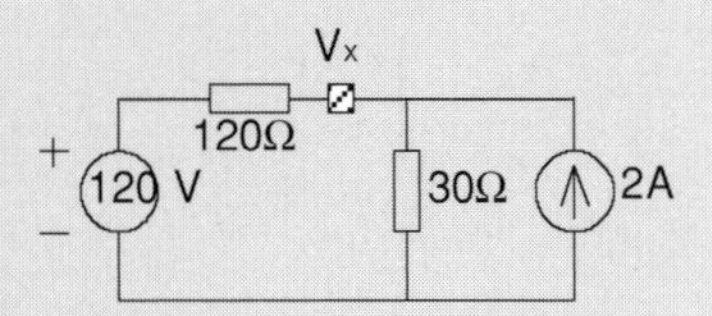

Figure 1.134

Using superposition find the value of V_X. Calculate the current through the 30Ω resistor to calculate V_X.

Answer: $V_X = 72\text{V}$

Case a)

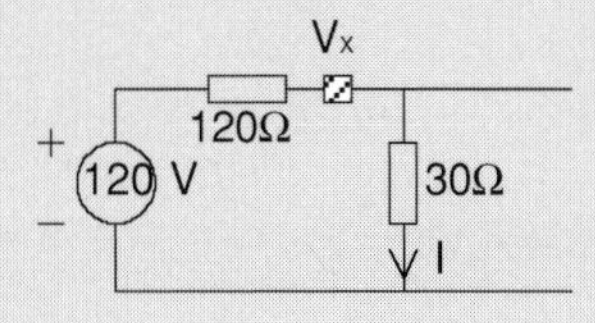

Figure 1.135

Current

$$I = \frac{120}{150} = 0.8\,\text{A}$$

Case b)

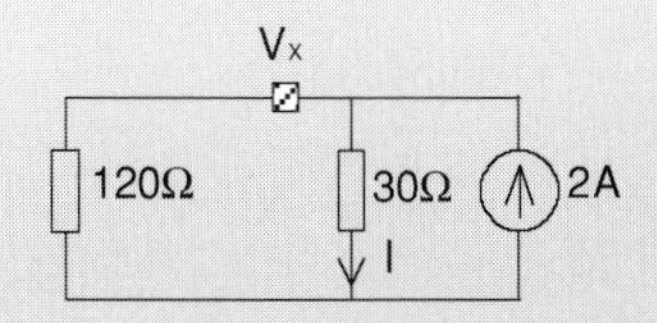

Figure 1.136

Current

$$I = 2\frac{120}{120+30} = 2\frac{120}{150} = 1.6\,\text{A}$$

Result

$$I = 0.8 + 1.6 = 2.4\,\text{A}$$

$$V_X = 30 \times 2.4 = 72\,\text{V}$$

PROBLEM 1.62 SUPERPOSITION

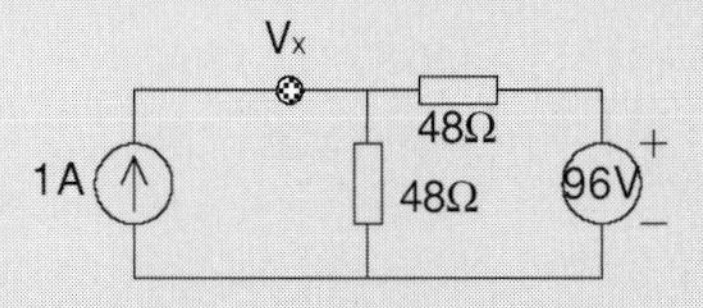

Using superposition to find V_X. Find the current through the middle branch first.

Answer: $V_X = 72\text{V}$

Case a)

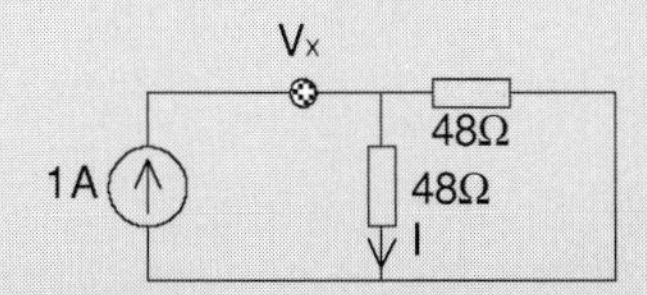

Figure 1.138

Current (both resistors being of the same value, the current is half the value of the generator).

$$I = 0.5\text{A}$$

Case b)

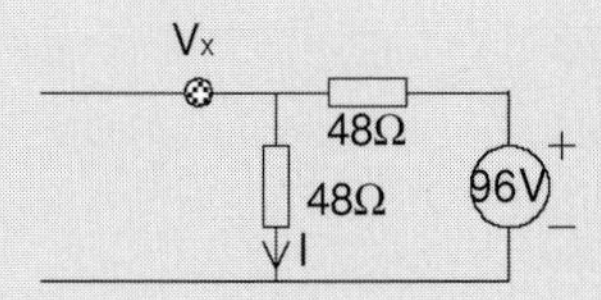

Figure 1.139

Current

$$I = \frac{96}{48+48} = 1\,\text{A}$$

Total

$$I = 1 + 0.5 = 1.5\text{A}$$

$$V_X = 1.5 \times 48 = 72\text{V}$$

PROBLEM 1.63 SUPERPOSITION

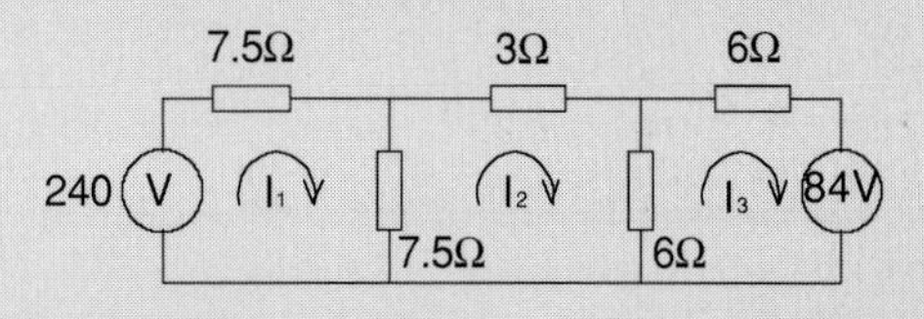

Figure 1.140

Find the values of I_1, I_2 and I_3 using superposition.

Answer: $I_1 = 20\text{A}$, $I_2 = 8\text{A}$ and $I_3 = -3\text{A}$

Case a)

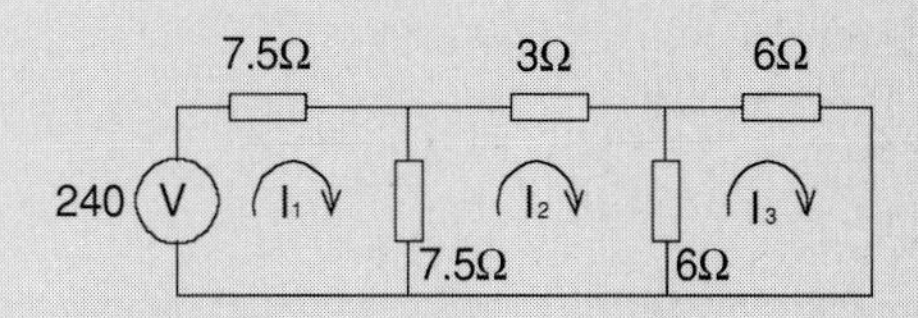

Figure 1.141

Total resistance

$$R_T = 7.5 + 7.5 \| (3 + 6 \| 6)$$

$$= 7.5 + 7.5 \| 6 = 7.5 + 3\frac{1}{3}$$

$$= 10\frac{5}{6}$$

Currents

$$I_1 = \frac{240}{10\frac{5}{6}} = \frac{240 \times 6}{65} = 22.153\,8\,\text{A}$$

$$I_2 = 22.1538\frac{7.5}{7.5+6} = 12.307\,7\,\text{A}$$

$$I_3 = \frac{I_2}{2} = 6.1538\,\text{A}$$

Case b)

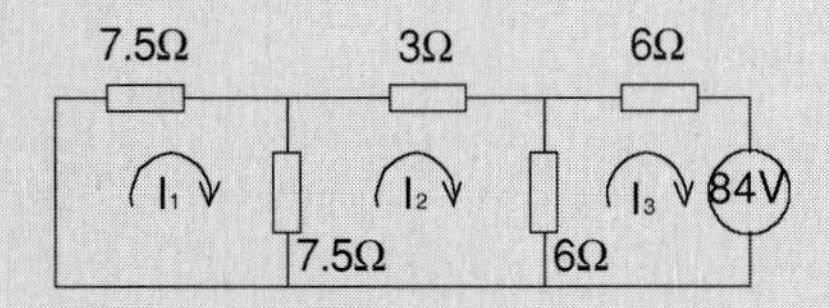

Figure 1.142

Resistance

$$R_T = 6 + 6 \| (3 + 7.5 \| 7.5)$$
$$= 6 + 6 \| 6.75 = 9.1765\Omega$$

Currents

$$I_3 = \frac{84}{9.1765} = 9.1538\,\text{A}$$

$$I_2 = 9.1538 \frac{6}{6 + 6.75} = 4.3077\,\text{A}$$

$$I_1 = \frac{I_2}{2} = 2.1538\,\text{A}$$

Result

$$I_1 = 22.1538 - 2.1538 = 20\,\text{A}$$
$$I_2 = 12.3077 - 4.3077 = 8\,\text{A}$$
$$I_3 = 6.1538 - 9.1538 = -3\,\text{A}$$

PROBLEM 1.64 SUPERPOSITION

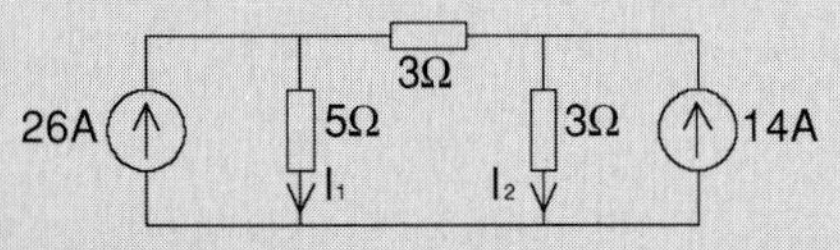

Figure 1.143

Use superposition to find the value of I_1 and I_2.

Answer: I_1 = 18A and I_2 = 22A

Case a)

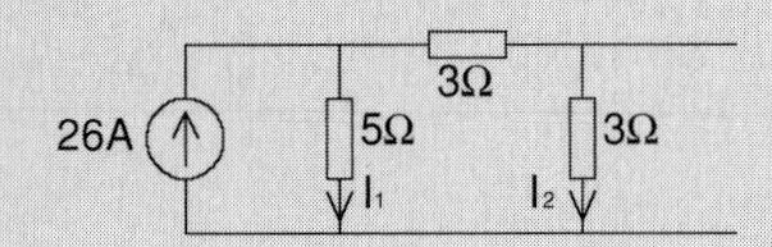

Figure 1.144

Currents (by current division)

$$I_1 = 26 \frac{6}{5 + 6} = 14.181818\,\text{A}$$

$$I_2 = 26 \frac{5}{5 + 6} = 11.818181\,\text{A}$$

Case b)

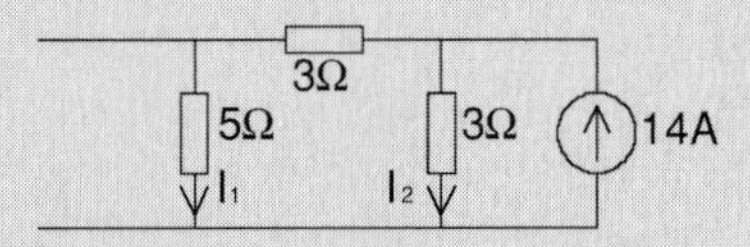

Figure 1.145

Currents

$$I_1 = 14 \frac{3}{3 + 8} = 3.818181\,\text{A}$$

$$I_2 = 14 \frac{8}{3 + 8} = 10.181818\,\text{A}$$

Result

$$I_1 = 14.181818 + 3.818181 = 18\,\text{A}$$
$$I_2 = 11.818181 + 10.181818 = 22\,\text{A}$$

PROBLEM 1.65 SUPERPOSITION

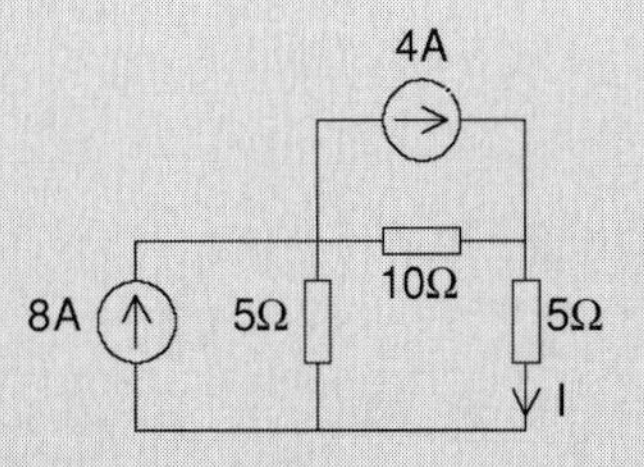

Figure 1.146

Use superposition to find the value of I.

Answer: I = 4A

Case a)

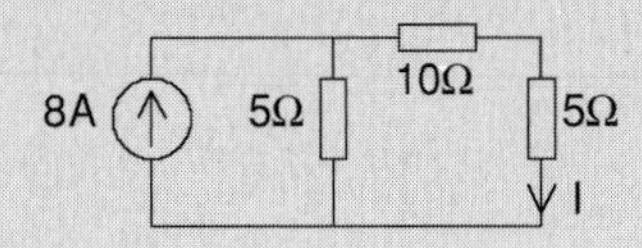

Figure 1.147

Current (by current division)

$$I = 8\frac{5}{15+5} = \frac{40}{20} = 2\,\text{A}$$

Case b)

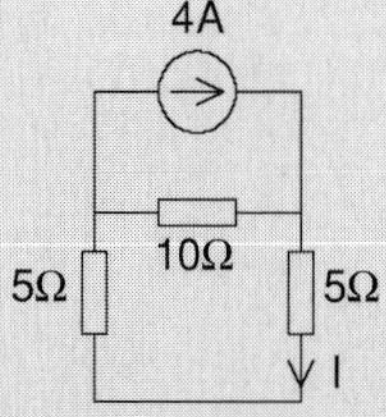

Figure 1.148

Current

$$I = 4\frac{10}{10+10} = 2\,\text{A}$$

Result

$$I = 2+2 = 4\,\text{A}$$

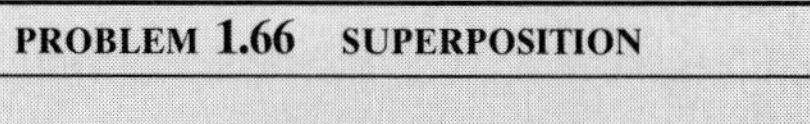

PROBLEM 1.66 SUPERPOSITION

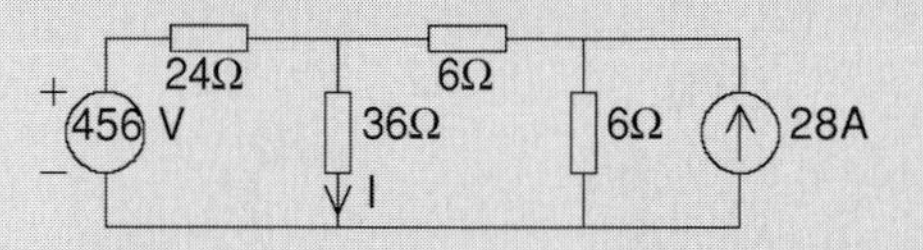

Figure 1.149

Use superposition to find the value of I.

Answer: $I = 6\text{A}$

Case a)

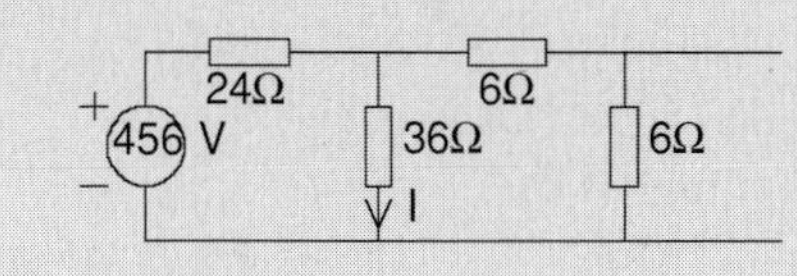

Figure 1.150

Resistance

$$R_T = 24 + 36 \| 12 = 24 + 9 = 33\,\Omega$$

Current

$$I_T = \frac{456}{33}\,\text{A} \qquad I = \frac{456}{33}\frac{12}{36+12} = 3.4545\,\text{A}$$

Case b)

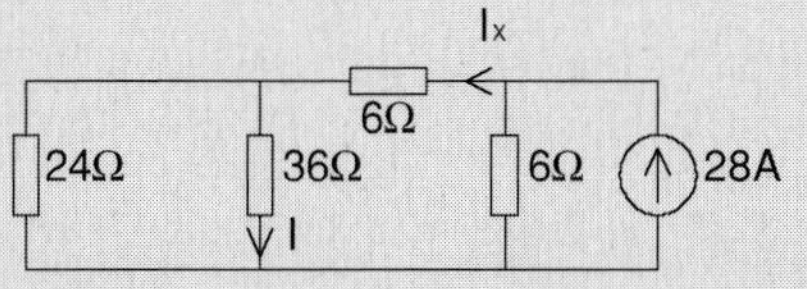

Figure 1.151

Resistance (only the three resistances on the left)

$$R = 6 + 24 \| 36 = 6 + 14.4 = 20.4\,\Omega$$

Current

$$I_X = 28\frac{6}{20.4+6} = 6.3636\,\text{A}$$

$$I = I_X\frac{24}{24+36} = 2.5454\,\text{A}$$

Total:

$$I = 2.5454 + 3.4545 = 6\,\text{A}$$

Nodal analysis

Nodal analysis is a direct application of Kirchhoff's current law (KCL). It is based on the fact that at a node or a junction point, the sum of the currents is equal to zero.

According to what we said before, currents going into a node are considered positive and currents going out are considered negative. A simple example is illustrated in Figure 1.152. In order to solve this problem we first indicate the nodes in the drawing. We have three nodes in the upper part, with one reference node at the bottom. Secondly we assume some voltages at the nodes, say, V_1, V_2 and V_3 for nodes A, B and C, respectively. Thirdly, we assign the currents, that is to say we give them names and directions. In the circuit shown in Figure 1.153, the currents

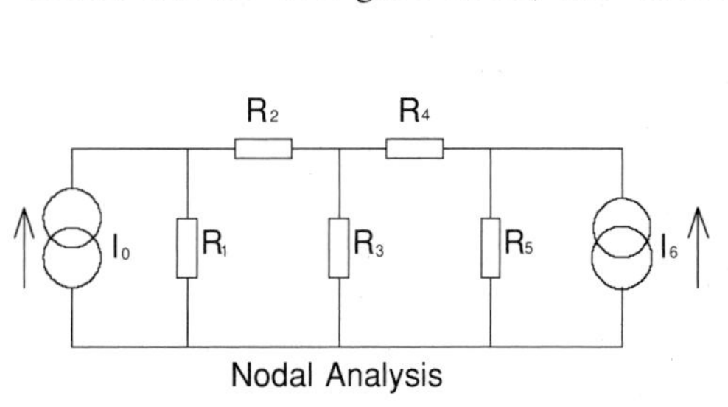

Figure 1.152

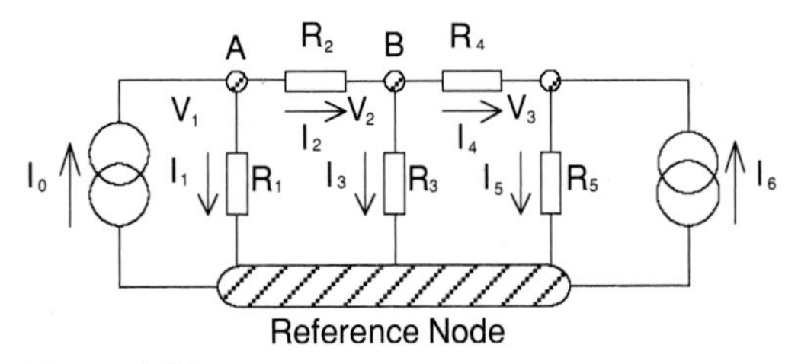

Figure 1.153

have the same number as the resistor where they circulate, but this is purely accidental, you could use any numbers that you want.

The sense of the current is not important at this stage, you can make an intelligent guess, but the important consideration, is to give a direction and then stick to it. If the sense of the current is not in the same sense as in reality, then you will get a minus sign in the result for this particular current.

Node A

$$I_0 - I_1 - I_2 = 0$$

Node B

$$I_2 - I_3 - I_4 = 0$$

Node C

$$I_6 + I_4 - I_5 = 0$$

Please note that this equation is true because it corresponds to the direction in Figure 1.153. If we change the direction of any current, then the equation will change the sign for that particular current.

The next stage is to replace the currents by their equivalent in voltages and resistances according to Ohm's law.

Node A

$$I_0 - \frac{V_0}{R_1} - \frac{V_1 - V_2}{R_2} = 0$$

Node B

$$\frac{V_1 - V_2}{R_2} - \frac{V_2 - V_3}{R_4} - \frac{V_2}{R_5} = 0$$

Node C

$$I_6 + \frac{V_2 - V_3}{R_4} - \frac{V_3}{R_5} = 0$$

In this case we know the current sources I_0 and I_6. We also know the resistors. The unknowns are V_1, V_2 and V_3. As we have three unknowns and three equations, we can solve the problem.

PROBLEM 1.67 NODAL ANALYSIS

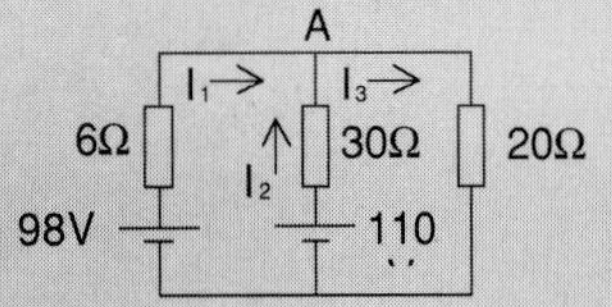

A, I_1, I_3, 6Ω, 30Ω, 20Ω, I_2, 98V, 110

Figure 1.154

Find the values of I_1, I_2 and I_3.

Answer: $I_1 = 3\text{A}$, $I_2 = 1\text{A}$ and $I_3 = 4\text{A}$

Currents

$$I_1 + I_2 = I_3$$

$$\frac{98 - V_A}{6} + \frac{110 - V_A}{30} = \frac{V_A}{20} \qquad | \times 60$$

$$980 - 10V_A + 220 - 2V_A = 3V_A$$

$$1200 = 15V_A$$

$$V_A = 80\,\text{V}$$

$$I_1 = \frac{18}{6} = 3\,\text{A}$$

$$I_2 = \frac{30}{30} = 1\,\text{A}$$

$$I_3 = \frac{80}{20} = 4\,\text{A}$$

PROBLEM 1.68 NODAL ANALYSIS

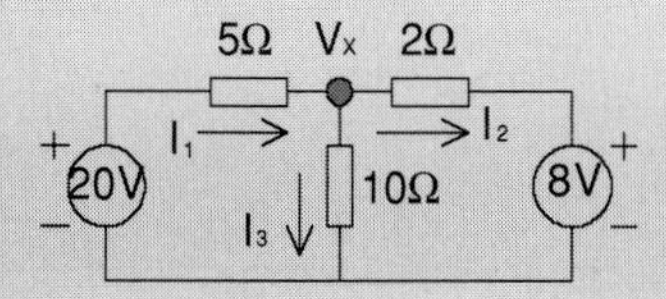

Figure 1.155

Find the value of the voltage V_X with respect to 0V and subsequently find the values of I_1, I_2 and I_3.

> Answer: $V_X = 10V$, $I_1 = 2A$, $I_2 = 1A$ and $I_3 = 1A$

Currents

$$I_1 = I_2 + I_3$$

$$\frac{20 - V_X}{5} = \frac{V_X - 8}{2} + \frac{V_X}{10} \quad | \times 10$$

$$40 - 2V_X = 5V_X - 40 + V_X$$

$$80 = 8V_X$$

$$V_X = 10\text{V}$$

Now the values of currents

$$I_1 = \frac{20 - V_X}{5} = \frac{20 - 10}{5} = 2\text{A}$$

$$I_2 = \frac{V_X - 8}{2} = \frac{10 - 8}{2} = 1\text{A}$$

$$I_3 = \frac{V_X}{10} = \frac{10}{10} = 1\text{A}$$

PROBLEM 1.69 NODAL ANALYSIS

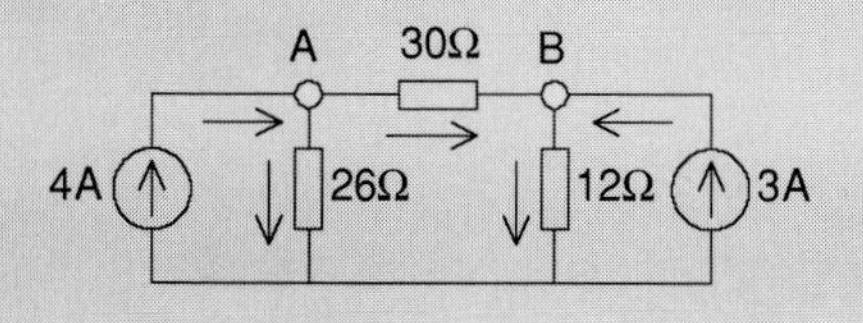

Figure 1.156

Using nodal analysis calculate V_A and V_B.

> Answer: $V_A = 78V$ and $V_B = 48V$

Node A (+ into node; – out of node)

$$4 - \frac{V_A}{26} - \frac{V_A - V_B}{30} = 0 \quad | \times 390$$

$$1560 - 15V_A - 13V_A + 13V_B = 0$$

$$1560 - 28V_A + 13V_B = 0 \quad (1)$$

Node B

$$3 - \frac{V_B}{12} + \frac{V_A - V_B}{30} = 0 \quad | \times 60$$

$$180 - 5V_B + 2V_A - 2V_B = 0$$

$$180 - 7V_B + 2V_A = 0 \quad (2)$$

Add equation (1) and (2) (multiplied by 14)

$$1560 - 28V_A + 13V_B = 0$$

$$2520 - 98V_B + 28V_A = 0$$

$$4080 - 85V_B = 0 \qquad V_B = 48\text{V}$$

Replace in equation (2)

$$-180 + 336 - 2V_A = 0 \qquad V_A = 78\text{V}$$

PROBLEM 1.70 NODAL ANALYSIS

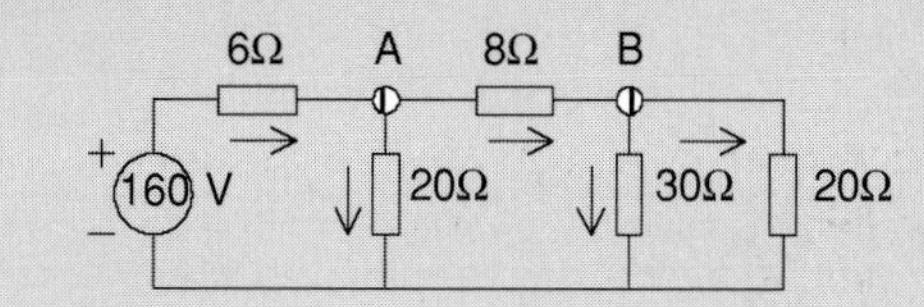

Figure 1.157

Find V_A and V_B.

> Answer: 100V and 60V

The sum of the currents into the node are equal to the sum of the currents out of the node.

Node A (1)

$$\frac{160 - V_A}{6} = \frac{V_A}{20} + \frac{V_A - V_B}{8} \qquad | \times 120$$

Node B (2)

$$\frac{V_A - V_B}{8} = \frac{V_B}{30} + \frac{V_B}{20} \qquad | \times 120$$

There are two equations and two unknowns. We can solve it. Multiplied by 120.

$$3200 - 20V_A = 6V_A + 15V_A - 15V_B$$

$$3200 = 41V_A - 15V_B \qquad (3)$$

$$15V_A - 15V_B = 4V_B + 6V_B$$

$$15V_A = 25V_B$$

$$3V_A = 5V_B$$

$$9V_A = 15V_B \qquad (4)$$

Replace equation (4) in equation (3)

$$3200 = 41V_A - 9V_A$$

$$3200 = 32V_A \qquad V_A = 100\,V$$

Replace V_A in equation (4)

$$900 = 15V_B \qquad V_B = 60\,V$$

PROBLEM 1.71 NODAL ANALYSIS

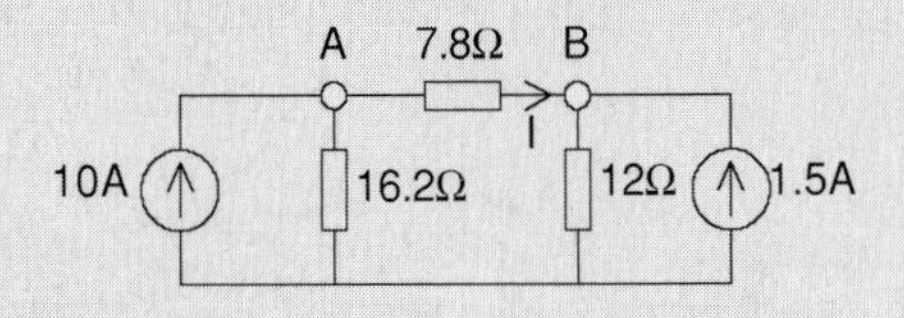

Figure 1.158

Find the voltages V_A, V_B and then the value of I.

Answer: $V_A = 97.2V$, $V_B = 66V$ and $I = 4A$

Node A

$$10 - \frac{V_A}{16.2} - \frac{V_A - V_B}{7.8} = 0 \qquad | \times 7.8 \times 16.2$$

$$1263.6 - 7.8V_A - 16.2V_A + 16.2V_B = 0$$

$$1263.6 - 24V_A + 16.2V_B = 0 \qquad (1)$$

Node B

$$1.5 - \frac{V_B}{12} + \frac{V_A - V_B}{7.8} = 0 \qquad | \times 7.8 \times 12$$

$$140.4 - 7.8V_B + 12V_A - 12V_B = 0$$

$$140.4 + 12V_A - 19.8V_B = 0 \qquad (2)$$

Equation (1), plus equation (2) multiplied by 2.

$$1263.6 - 24V_A + 16.2V_B = 0$$

$$280.8 + 24V_A - 39.6V_B = 0$$

$$1544.4 - 23.4V_B = 0 \qquad V_B = 66\,V$$

Replace V_B in equation (2).

$$140.4 + 12V_A - 1306.8 = 0$$

$$12V_A - 1166.4 = 0 \qquad V_A = 97.2\,V$$

and finally

$$I = \frac{97.2 - 66}{7.8} = \frac{31.2}{7.8} = 4\,A$$

PROBLEM 1.72 NODAL ANALYSIS

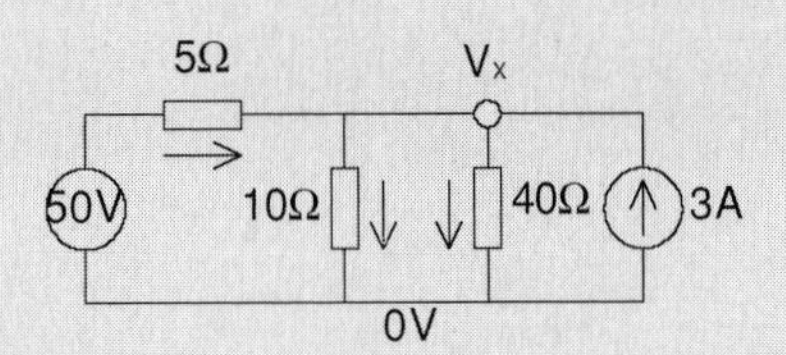

Figure 1.159

Find the value of V_X with respect to 0V using nodal analysis.

Answer: $V_X = 40V$

Assume currents as shown.
The sum of currents going in is equal to the sum of currents going out.

$$\frac{50 - V_X}{5} + 3 = \frac{V_X}{10} + \frac{V_X}{40} \qquad | \times 40$$

$$8(50 - V_X) + 120 = 4V_X + V_X$$

$$520 = 13V_X$$

$$V_X = 40\,V$$

PROBLEM 1.73 NODAL ANALYSIS

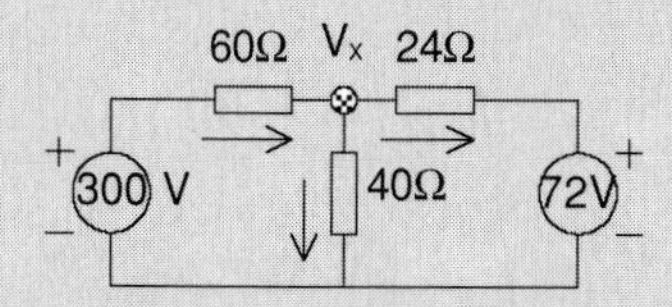

Figure 1.160

Find V_X by nodal analysis.

Answer: $V_X = 96\text{V}$

(Currents in positive, currents out negative.)

$$\frac{300-V_X}{60}+\frac{72-V_X}{24}-\frac{V_X}{40}=0 \quad | \times 120$$

$$600-2V_X+360-5V_X-3V_X=0$$

$$960-10V_X=0$$

$$V_X=96\text{V}$$

PROBLEM 1.74 NODAL ANALYSIS

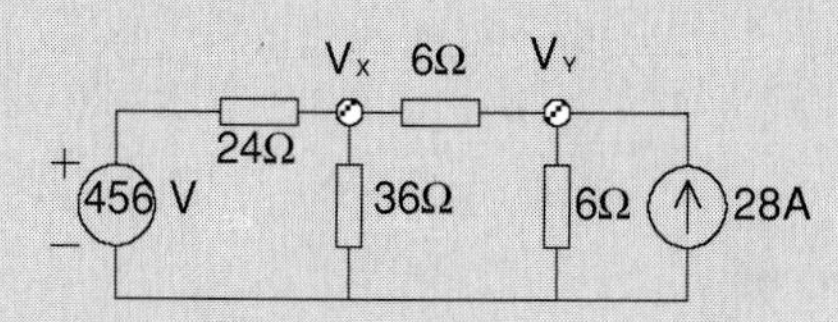

Figure 1.161

Use nodal analysis to find the value of V_X and V_Y.

Answer: $V_X = 216\text{V}$ and $V_Y = 192\text{V}$

$$-\frac{V_X}{36}+\frac{V_Y-V_X}{6}+\frac{456-V_X}{24}=0 \quad | \times 72$$

$$-2V_X+12V_Y-12V_X+1368-3V_X=0$$

$$-17V_X+12V_Y+1368=0 \quad (1)$$

$$-\frac{V_Y}{6}+28+\frac{V_X-V_Y}{6}=0 \quad | \times 6$$

$$-V_Y+168+V_X-V_Y=0$$

$$-\frac{V_X}{36}+\frac{V_Y-V_X}{6}+\frac{456-V_X}{24}=0 \quad | \times 72$$

$$-2V_X+12V_Y-12V_X+1368-3V_X=0$$

$$-17V_X+12V_Y+1368=0 \quad (1)$$

$$-\frac{V_Y}{6}+28+\frac{V_X-V_Y}{6}=0 \quad | \times 6$$

$$-V_Y+168+V_X-V_Y=0$$

$$V_X-2V_Y+168=0 \quad (2)$$

Equation (1), plus equation (2) times 6.

PROBLEM 1.75 NODAL ANALYSIS

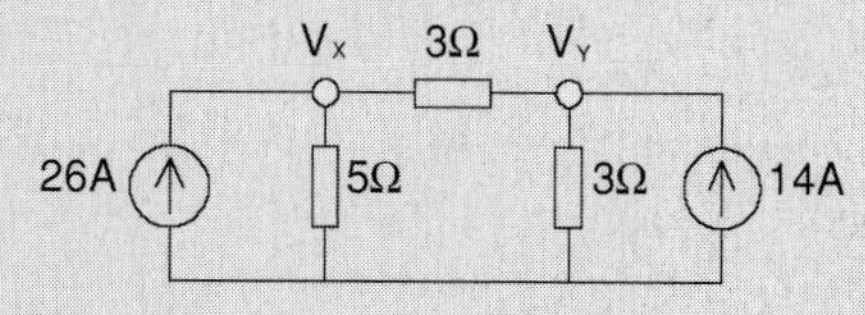

Figure 1.162

Use nodal analysis to find the value of V_X and V_Y.

Answer: $V_X = 90\text{V}$ and $V_Y = 66\text{V}$

$$26-\frac{V_X}{5}-\frac{V_X-V_Y}{3}=0 \quad | \times 15$$

$$390-3V_X-5V_X+5V_Y=0$$

$$390-8V_X+5V_Y=0 \quad (1)$$

$$14-\frac{V_Y}{3}-\frac{V_Y-V_X}{3}=0 \quad | \times 3$$

$$42-V_Y-V_Y+V_X=0$$

$$42-2V_Y+V_X=0 \quad (2)$$

Equation (1), plus equation (2) multiplied by 8

$$390-8V_X+5V_Y=0$$

$$336-16V_Y+8V_X=0$$

$$726-11V_Y=0 \qquad V_Y=66\text{V}$$

Replace V_Y in equation (2)

$$42-2V_Y+V_X=0$$

$$42-132+V_X=0$$

$$-90+V_X=0 \qquad V_X=90\text{V}$$

PROBLEM 1.76 NODAL ANALYSIS

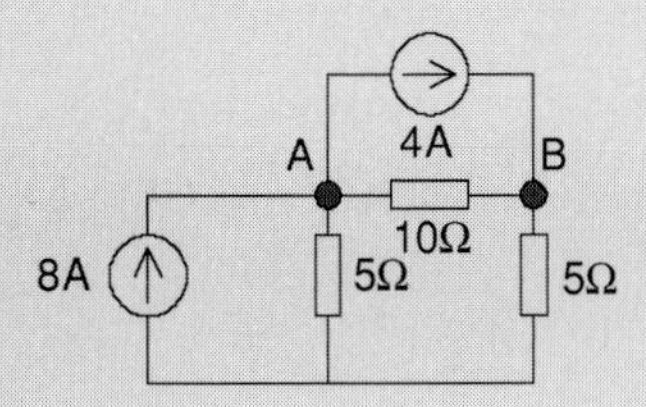

Figure 1.163

Using nodal analysis find the voltages V_A and V_B.

Answer: 20V and 20V

Node A (sum of currents in equals zero)

$$8-4-\frac{V_A}{5}+\frac{V_B-V_A}{10}=0$$

$$4-\frac{V_A}{5}+\frac{V_B}{10}-\frac{V_A}{10}=0$$

$$40-3V_A+V_B=0 \qquad (1)$$

Node B

$$-4+\frac{V_B}{5}+\frac{V_B-V_A}{10}=0 \qquad | \times 10$$

$$-40+2V_B+V_B-V_A=0$$

$$3V_B-V_A=40 \qquad (2)$$

Equation (1) times (–3), plus equation (2)

$$-3V_B+9V_A=120$$

$$3V_B-V_A=40$$

$$8V_A=160 \qquad V_A=20\text{V}$$

Replace V_A in equation (2)

$$3V_B-20=40 \qquad V_B=20\text{V}$$

PROBLEM 1.77 NODAL ANALYSIS

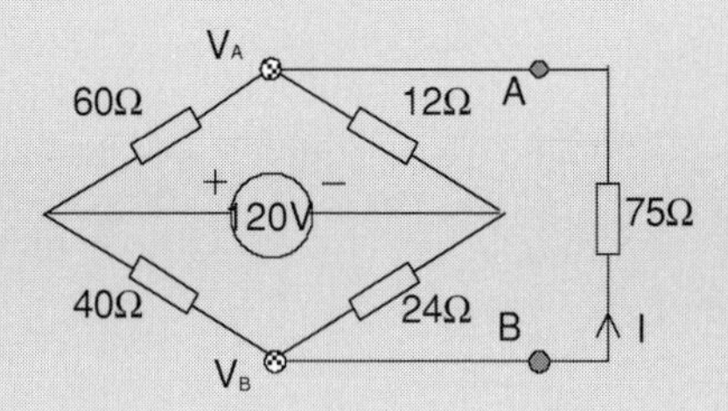

Figure 1.164

Find V_A, V_B and the value of I.

Answer: 22.5V, 41.25V and 0.25A

$$\frac{V_A}{12}+\frac{V_A-120}{60}-\frac{V_B-V_A}{75}=0 \qquad | \times 600$$

$$50V_A+10V_A-1200+8V_A-8V_B=0$$

$$68V_A-8V_B-1200=0 \qquad (1)$$

$$\frac{V_B}{24}+\frac{V_B-120}{40}-\frac{V_A-V_B}{75}=0 \qquad | \times 600$$

$$25V_B+15V_B-1800+8V_B-8V_A=0$$

$$-8V_A+48V_B-1800=0 \qquad (2)$$

Equation (2) plus equation (1) times 6

$$-8V_A+48V_B-1800=0$$

$$408V_A-48V_B-7200=0$$

$$400V_A-1980=0 \qquad V_A=22.5\text{V}$$

Replace in equation (2)

$$-180+48V_B-1800=0$$

$$48V_B-1980=0 \qquad V_B=41.25\text{V}$$

Current

$$I=\frac{V_A-V_B}{75}=\frac{41.25-22.5}{75}=0.25\text{A}$$

Mesh analysis

Mesh analysis is a direct application of Kirchhoff's voltage law (KVL). Mesh analysis comes in handy in the derivation of a transistor model (more of that in Chapter 4). It is based on the fact that the sum of the voltages around a

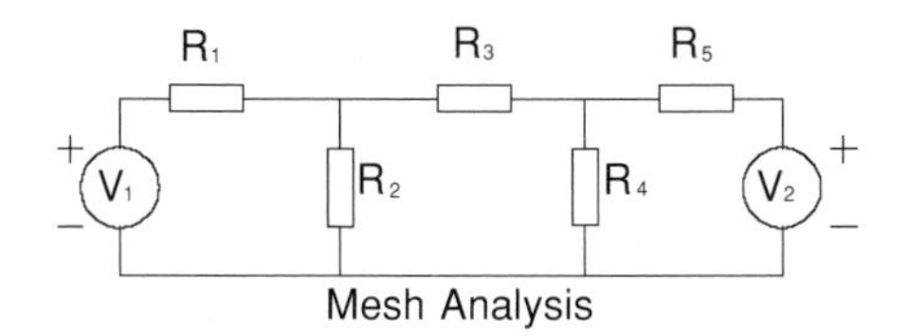

Figure 1.165

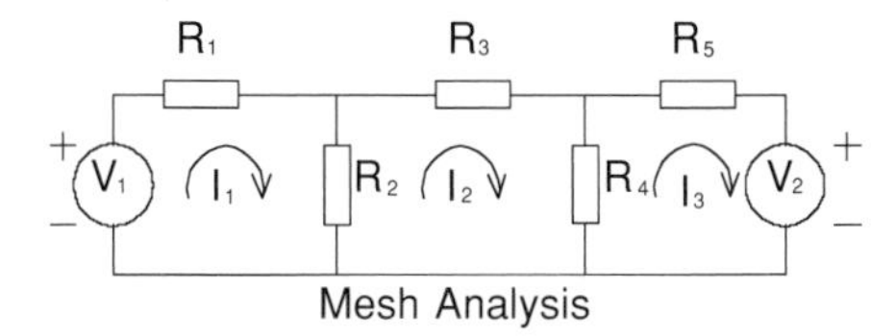

Figure 1.166

closed loop is equal to zero. A typical circuit is shown in Figure 1.165.

First of all we identify, arbitrarily, the sense of the currents in the different meshes. This is not a triviality, but as we saw at the beginning, the sense of the current dictates the sense of the voltage that appears in the resistors. The chosen direction for currents can be seen in Figure 1.166. From the author's experience, students prefer to use a modified KVL, but still a KVL, that is to say:

> The voltage sources around a closed loop are equal to the voltage drops in the loop.

NOTE: If the voltage source in the loop is against the sense of the current in that loop, as it is the case in the third loop, then the value of the voltage source is written as negative.

We can now write the equations:

Mesh 1

$$V_1 = I_1R_1 + I_1R_2 - I_2R_2$$

Mesh 2

$$0 = I_2R_2 + I_2R_3 + I_2R_4 - I_1R_2 - I_3R_4$$

Mesh 3

$$-V_2 = I_3R_4 + I_3R_5 - I_2R_4$$

This can be simplified and tidied up, giving:

Mesh 1

$$V_1 = I_1(R_1 + R_2) - I_2R_2$$

Mesh 2

$$0 = -I_1R_2 + I_2(R_2 + R_3 + R_4) - I_3R_4$$

Mesh 3

$$-V_2 = -I_2R_4 + I_3(R_4 + R_5)$$

This is more presentable and looks better. It can easily be transferred into matrix notation. Matrix notation is only an abbreviated form of writing the equation, but we are not interested in matrices at this moment.

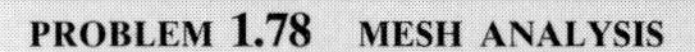

PROBLEM 1.78 MESH ANALYSIS

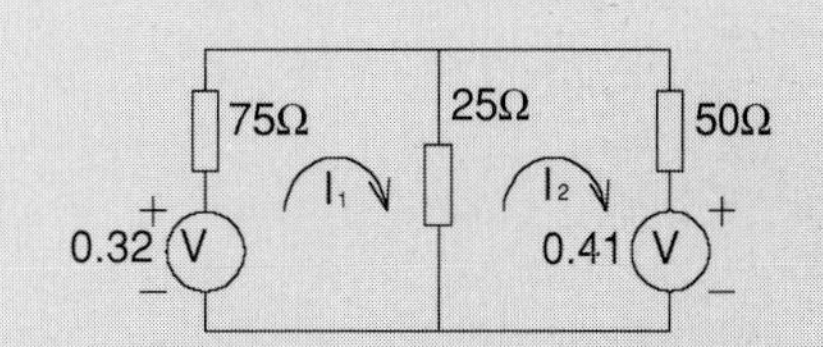

Figure 1.167

Find the value of I_1 and I_2.

Answer: I_1 = 2mA and I_2 = –4.8mA

Mesh 1

$$0.32 = (75 + 25)I_1 - 25I_2 \qquad (1)$$

$$-0.41 = -25I_1 + 75I_2 \qquad (2)$$

$$0.32 = 100I_1 - 25I_2 \qquad | \times 3$$

$$0.96 = 300I_1 - 75I_2$$

$$-0.41 = -25I_1 + 75I_2$$

$$0.55 = 275I_1 \qquad I_1 = 2\text{mA}$$

$$-0.41 = -0.05 + 75I_2$$

$$-0.36 = 75I_2 \qquad I_2 = -4.8\text{mA}$$

PROBLEM 1.79 MESH ANALYSIS

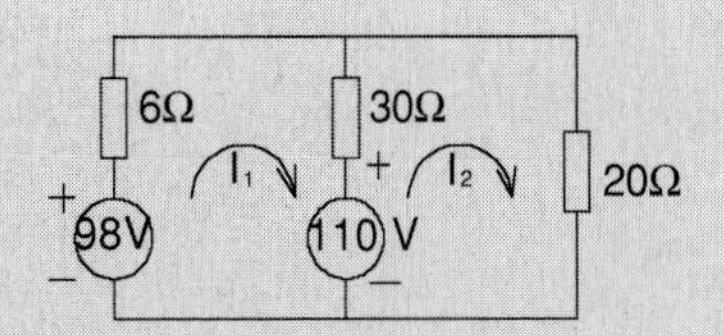

Figure 1.168

Find the values of I_1 and I_2.

Answer: $I_1 = 3A$ and $I_2 = 4A$

Mesh 1

$98 - 100 = I_1(6 + 30) - I_2 30$

$-12 = 36I_1 - 30I_2 \quad | \div 6$

$-2 = 6I_1 - 5I_2 \quad (1)$

Mesh 2

$110 = -I_1 30 + I_2(30 + 20)$

$110 = -30I_1 + 50I_2$

$11 = -3I_1 + 5I_2 \quad (2)$

Add equations (1) and (2)

$9 = 3I_1 \quad I_1 = 3A$

Replace in equation (2)

$11 = -9 + 5I_2$

$20 = 5I_2 \quad I_2 = 4A$

PROBLEM 1.80 MESH ANALYSIS

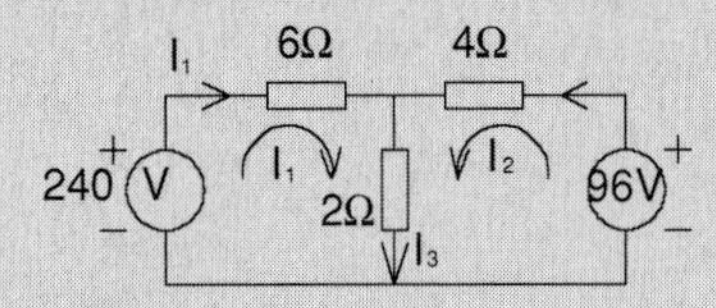

Figure 1.169

Find the values of I_1, I_2 and I_3.

Answer: 28.36A, 6.55A and 34.91A

Mesh 1

$240 = 8I_1 + 2I_2 \quad | \times 3 \quad (1)$

$96 = 2I_1 + 6I_2 \quad (2)$

$720 = 24I_1 + 6I_2$

$96 = 2I_1 + 6I_2$

$624 = 22I_1 \quad I_1 = 28.36A$

Replace in equation (1)

$2I_2 = -226.9 + 240$

$= 13.09 \quad I_2 = 6.55A$

$I_3 = I_1 + I_2 = 28.36 + 6.55 = 34.91A$

PROBLEM 1.81 MESH ANALYSIS

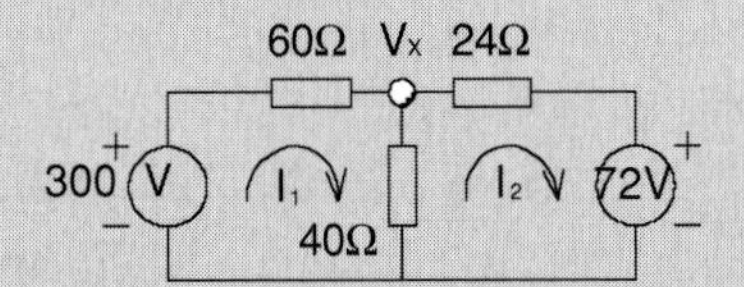

Figure 1.170

Find the value of V_X, by first calculating I_1 and I_2.

Answer: 96V

$300 = I_1(60 + 40) - I_2 40$

$300 = 100I_1 - 40I_2 \quad (1)$

$-72 = I_2(40 + 24) - I_1 40$

$-72 = -40I_1 + 64I_2 \quad (2)$

Equation (1) divided by 10 and equation (2) divided by 8

$30 = 10I_1 - 4I_2$

$-9 = -5I_1 + 8I_2$

$60 = 20I_1 - 8I_2$

$-9 = -5I_1 + 8I_2$

$$51 = 15I_1 \qquad I_1 = \frac{51}{15}\text{A}$$

$$-9 = -\frac{51}{3} + 8I_2 \qquad | \times 3$$

$$-27 = -51 + 24I_2$$

$$24 = 24I_2 \qquad I_2 = 1\text{A}$$

Voltage

$$V_X = (I_1 - I_2)R = \left(\frac{51}{15} - \frac{15}{15}\right)40$$

$$= \frac{36}{15}40 = 96\text{V}$$

PROBLEM 1.82 MESH ANALYSIS

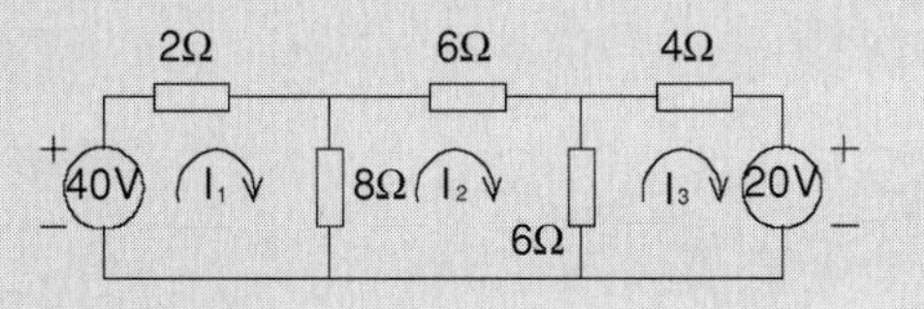

Figure 1.171

Find the mesh currents I_1, I_2 and I_3.

Answer: $I_1 = 5.6\text{A}$, $I_2 = 2\text{A}$ and $I_3 = -0.8\text{A}$

Mesh equations

$$40 = I_1(2 + 8) - I_2 8$$

$$0 = -8I_1 + I_2(8 + 6 + 6) - 6I_3$$

$$-20 = -I_2 6 + I_3(6 + 4)$$

$$40 = 10I_1 - 8I_2 \qquad (1)$$

$$0 = -8I_1 + 20I_2 - 6I_3 \qquad (2)$$

$$-20 = -6I_2 + 10I_3 \qquad (3)$$

Eliminating I_1 from equations (1) and (2)

$$160 = 40I_1 - 32I_2$$

$$0 = -40I_1 + 100I_2 - 30I_3$$

$$160 = 68I_2 - 30I_3$$

$$-60 = -18I_2 + 30I_3$$

$$100 = 50I_2 \qquad I_2 = 2\text{A}$$

Replace I_2 in equation (1)

$$40 = 10I_1 - 16 \qquad I_1 = 5.6\text{A}$$

From equation (3)

$$-20 = -12 + 10I_3 \qquad I_3 = -0.8\text{A}$$

Checking mesh currents

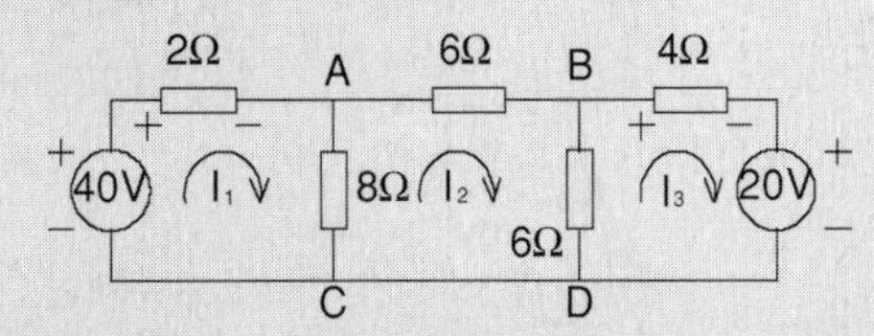

Figure 1.172

Having established from mesh analysis the values of I_1, I_2 and I_3, it is easy to verify these values. At node A for instance V_A is given by

$$V_B + 6I_2 = V_A \qquad (1)$$

$$(I_1 - I_2)8 = V_A \qquad (2)$$

$$40 - 2I_1 = V_A \qquad (3)$$

From equation (2)

$$(5.6 - 2)8 = 28.8\text{V}$$

From equation (3)

$$4 - 2 \times 5.6 = 28.8\text{V}$$

From equation (1)

$$V_B = V_A - 6I_2$$

$$= 28.8 - 12 = 16.8\text{V}$$

V_B is also equal to

$$V_B = 20 + 4I_1$$

$$= 20 + 4(-0.8)$$

$$= 20 - 3.2 = 16.8\text{V}$$

V_B is also (at BD)

$$V_B = (I_2 - I_3)6$$

$$= (2 - (-0.8))6$$

$$= 2.8 \times 6 = 16.8\text{V}$$

Any discrepancies would be revealed with these tests.

PROBLEM 1.83 MESH ANALYSIS

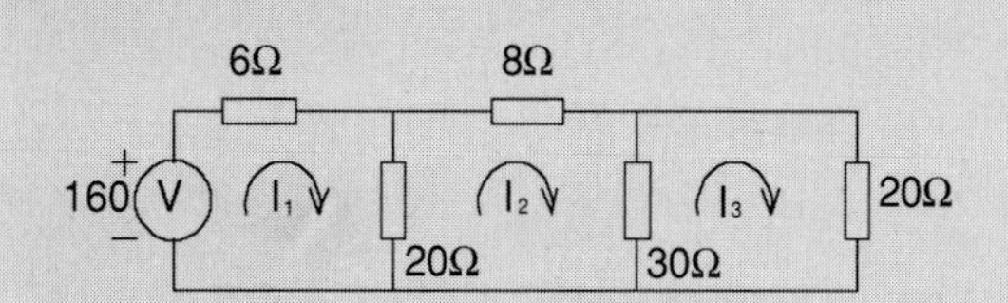

Figure 1.173

Find the values of I_1, I_2 and I_3.

Answer: 10A, 5A and 3A

Mesh equations

$$160 = I_1(6+20) - 20I_2$$
$$0 = -20I_1 + I_2(20+8+30) - 30I_3$$
$$0 = -30I_2 + I_3(30+20)$$

Rearranged

$$160 = 26I_1 - 20I_2 \qquad (1)$$
$$0 = -20I_1 + 58I_2 - 30I_3 \qquad (2)$$
$$0 = -30I_2 + 50I_3 \qquad (3)$$

From equation (3)

$$30I_2 = 50I_3$$
$$I_2 = \frac{5}{3}I_3 \qquad I_3 = \frac{3}{5}I_2$$

Replace I_3 in (2)

$$0 = -20I_1 + 58I_2 - 30\frac{3}{5}I_2$$
$$0 = -20I_1 + 40I_2$$
$$0 = -I_1 + 2I_2 \qquad | \times 26$$
$$0 = -26I_1 + 52I_2 \qquad | \text{ plus equation (1)}$$
$$160 = 26I_1 - 20I_2$$

$$160 = 32I_2 \qquad I_2 = 5\,\text{A}$$
$$I_3 = \frac{3}{5}I_2 \qquad I_3 = 3\,\text{A}$$
$$I_1 = 2I_2 \qquad I_1 = 10\,\text{A}$$

PROBLEM 1.84 MESH ANALYSIS

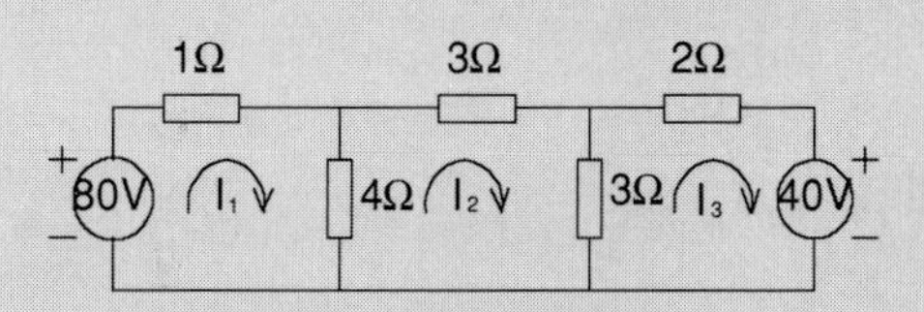

Figure 1.174

Find I_1, I_2 and I_3.

Answer: 22.4A, 8A and –3.2A

Mesh equations

$$80 = 5I_1 - 4I_2 \qquad (1)$$
$$0 = -4I_1 + 10I_2 - 3I_3 \qquad (2)$$
$$-40 = -3I_2 + 5I_3 \qquad (3)$$

From equation (1)

$$I_1 = \frac{80+4I_2}{5} = 16 + \frac{4}{5}I_2$$

Replace I_1 in equation (2)

$$0 = -4\left(16 + \frac{4}{5}I_2\right) + 10I_2 - 3I_3 \qquad | \times 5$$
$$0 = -320 - 16I_2 + 50I_2 - 15I_3$$
$$320 = 34I_2 - 15I_3$$
$$-40 = -3I_2 + 5I_3$$
$$320 = 34I_2 - 15I_3$$
$$-120 = -9I_2 + 15I_3$$

$$200 = 25I_2 \qquad I_2 = 8\,\text{A}$$

From equation (1)

$$80 = 5I_1 - 4I_2 = 5I_1 - 32$$
$$5I_1 = 80 + 32 = 112 \qquad I_1 = 22.4\,\text{A}$$

From equation (3)

$$-40 = 5I_3 - 3I_2 = 5I_3 - 24$$
$$5I_3 = -16 \qquad I_3 = -3.2\,\text{A}$$

PROBLEM 1.85 MESH ANALYSIS

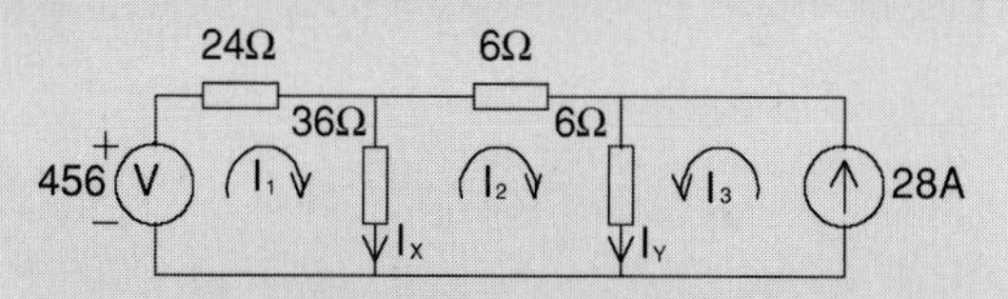

Figure 1.175

Use mesh analysis to find I_X and I_Y.

Answer: $I_X = 6A$, $I_Y = 32A$

I_3 is known to be 28A. We make equations for I_1 and I_2.

$$456 = I_1(24 + 36) - I_2 36$$

$$456 = 60I_1 - 36I_1$$

$$152 = 20I_1 - 12I_2 \qquad (1)$$

$$0 = I_2(36 + 6 + 6) - 36I_1 + 6 \times 28$$

$$0 = 48I_2 - 36I_1 + 168$$

$$-14 = -3I_1 + 4I_2 \qquad (2)$$

Add equation (1) plus equation (2) times 3

$$152 = 20I_1 - 12I_2$$

$$-42 = -9I_1 + 12I_2$$

$$110 = 11I_1 \qquad I_1 = 10A$$

$$-42 = -90 + 12I_2$$

$$48 = 12I_2 \qquad I_2 = 4A$$

The currents are

$$I_X = I_1 - I_2 = 10 + 4 = 6A$$

$$I_Y = I_2 + I_3 = 4 + 28 = 32A$$

PROBLEM 1.86 MESH ANALYSIS

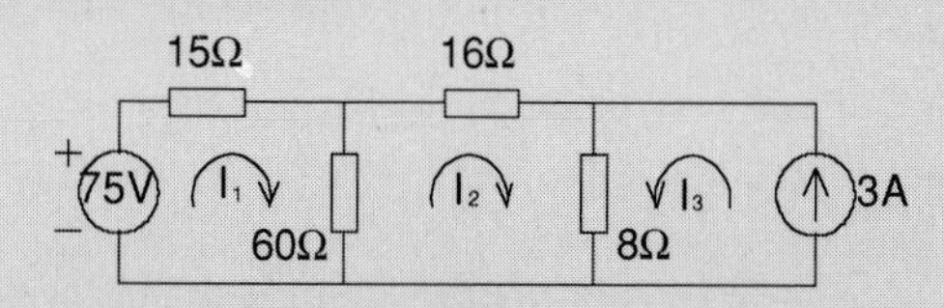

Figure 1.176

Find I_1 and I_2.

Answer: $I_1 = 1.8A$ and $I_2 = 1A$

$I_3 = 3\,A$

Mesh 1

$$75 = 75I_1 - 60I_2 \qquad | \div 15$$

$$5 = 5I_1 - 4I_2 \qquad (1)$$

Mesh 2

$$0 = -60I_1 + 84I_2 + 24 \qquad | \div 12$$

$$0 = -5I_1 + 7I_2 + 2$$

$$-2 = -5I_1 + 7I_2 \qquad (2)$$

Adding equations (1) and (2)

$$3 = 3I_2 \qquad I_2 = 1\,A$$

Replace I_2 in equation (2)

$$-2 = -5I_1 + 7$$

$$5I_1 = 9 \qquad I_1 = 1.8\,A$$

PROBLEM 1.87 MESH ANALYSIS

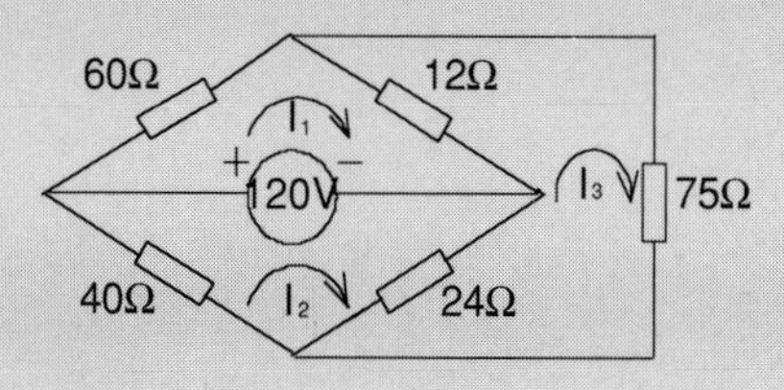

Figure 1.177

Find the values of I_1, I_2 and I_3 using mesh analysis.

Answer: $I_1 = 1.625A$, $I_2 = -1.96875A$ and $I_3 = -0.25A$

Mesh 1

$$120 = I_1(60 + 12) - I_3 12$$
$$120 = 72I_1 - 12I_1$$
$$10 = 6I_1 - I_3 \quad (1)$$

Mesh 2

$$-120 = I_2(40 + 24) - I_3 24$$
$$-120 = 64I_2 - 24I_1$$
$$-15 = 8I_2 - 3I_3 \quad (2)$$

Mesh 3

$$0 = I_3(75 + 12 + 24) - I_2 24 - I_1 12$$
$$0 = -12I_1 - 24I_2 + 111I_3$$
$$0 = -4I_1 - 8I_2 + 37I_3 \quad (3)$$

From equation (1), I_3 is

$$I_1 = 6I_1 - 10$$

Replace I_3 in equation (2)

$$-15 = 8I_2 - 3I_1$$
$$-15 = 8I_2 - 3(6I_1 - 10)$$
$$-15 = 8I_2 - 18I_1 + 30$$
$$-45 = -18I_1 + 8I_2 \quad (4)$$

Replace I_3 in equation (3)

$$0 = -4I_1 - 8I_2 + 37(6I_1 - 10)$$
$$0 = -4I_1 - 8I_1 + 222I_1 - 370$$
$$370 = 218I_1 - 8I_2 \quad (5)$$

Add equations (4) and (5)

$$325 = 200I_1 \qquad I_1 = 1.625\text{A}$$

Replace in equation (5)

$$370 = 354.25 - 8I_2$$
$$15.75 = -8I_2 \qquad I_2 = -1.96875\text{A}$$
$$I_3 = 9.75 - 10 \qquad I_3 = -0.25\text{A}$$

PROBLEM 1.88 MESH ANALYSIS

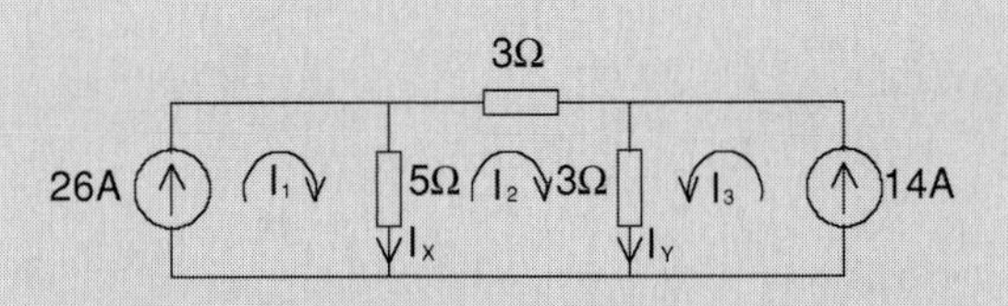

Figure 1.178

Use mesh analysis to find I_X and I_Y.

Answer: $I_X = 18\text{A}$ and $I_Y = 22\text{A}$

I_1 and I_3 are known, 26A and 14A sources. We can make an equation for I_2.

Mesh 2

$$0 = I_2(5 + 3 + 3) - 5I_1 + 3I_3$$

Note that the sign of I_3, is positive as I_3 and I_2 have the same direction in the 3Ω resistor.

$$0 = 11I_2 - 130 + 42$$
$$11I_2 = 88 \qquad I_2 = 8\text{A}$$

Currents

$$I_X = I_1 - I_2 = 26 - 8 = 18\text{A}$$
$$I_Y = I_2 + I_3 = 8 + 14 = 22\text{A}$$

Miller's theorem

Miller's theorem states that a given impedance can be replaced by two equivalent impedances. In order to illustrate this we refer to Figure 1.179, where if the behaviour of the circuits is the same, then the circuits are equivalent.

NOTE: Because of the negative sign that will appear in the formula, the Miller theorem only works on an inverting amplifier and it becomes meaningless otherwise.

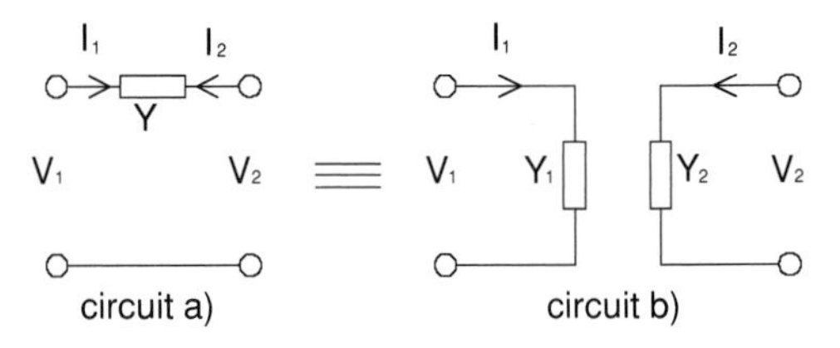

Figure 1.179

In circuit (a) we have

$$I_1 = (V_1 - V_2)Y$$

We factorise V_1:

$$I_1 = YV_1\left(1 - \frac{V_2}{V_1}\right)$$

The same I_1, in circuit (b) is given by

$$I_1 = Y_1V_1$$

Equating both of them

$$YV_1\left(1 - \frac{V_2}{V_1}\right) = Y_1V_1$$

Since V_1 is common to both circuits, they can be eliminated

$$Y_1 = Y\left(1 - \frac{V_2}{V_1}\right)$$

If we know the voltage amplification ratio

$$A_V = \frac{V_2}{V_1}$$

then

$$\boxed{Y_1 = Y(1 - A_V)}$$

We now look at the V_2 side

$$I_2 = YV_2\left(1 - \frac{V_1}{V_2}\right)$$

In circuit b)

$$I_2 = V_2Y_2$$

Equating both of them

$$V_2Y_2 = YV_2\left(1 - \frac{V_1}{V_2}\right)$$

V_2 can be simplified

$$Y_2 = Y\left(1 - \frac{V_1}{V_2}\right)$$

Using the ratio $A_V = V_2 / V_1$, we have

$$\boxed{Y_2 = Y\left(1 - \frac{1}{A_V}\right)}$$

In a transistor configuration (such as a common emitter), the amplification is negative and this will get rid of the negative sign. Care should be taken to ensure that the amplification A_V is taken at the points where Miller's theorem is applied. This will be different from the total amplification of the transistor in most cases.

The usefulness of this theorem consists in separating input and output circuits of the transistor, obtaining two circuits, instead of one. These two circuits can be tackled independently in a much simpler way, as we will see in Chapter 10. Without Miller's theorem, the analysis of the transistor under high frequency would be much more complex.

Special case: circuit with resistors

In the case of a circuit with resistors we can say that

$$Y_1 = \frac{1}{R_1} \qquad Y = \frac{1}{R}$$

Therefore

$$\frac{1}{R_1} = \frac{1}{R}(1 - A_V)$$

rearranging

$$R_1 = \frac{R}{1 - A_V}$$

For the other side

$$Y_2 = Y\left(1 - \frac{1}{A_V}\right)$$

replacing

$$Y_2 = \frac{1}{R_2} \quad \text{and} \quad Y = \frac{1}{R}$$

$$\frac{1}{R_2} = \frac{1}{R}\left(1 - \frac{1}{A_V}\right)$$

rearranging

$$R_2 = \frac{R}{1 - \dfrac{1}{A_V}}$$

In certain situations it is possible to neglect the value $1/A_V$ because we expect a high amplification. If this is the case we can say that:

$$Y_2 \approx Y \text{ and } R_2 \approx R$$

PROBLEM 1.89 MILLER

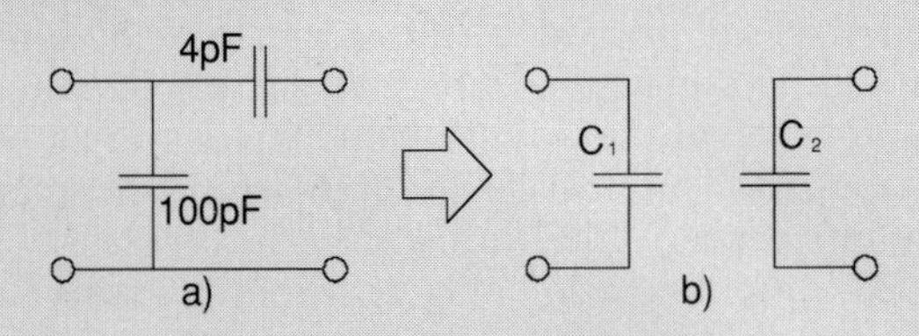

Figure 1.180

Convert the following circuit using Miller's theorem from a) to b), when $A_V = -100$.

Answer: C_1 = 504pF and $C_2 \approx$ 4pF

Input side

$$C_{M1} = 4(1 - A_v)$$
$$= 4(101) = 404\text{pF}$$

Output side

$$C_{M2} = 4\left(1 - \frac{1}{A_v}\right)$$
$$= 4\left(1 + \frac{1}{100}\right) \approx 4\text{pF}$$

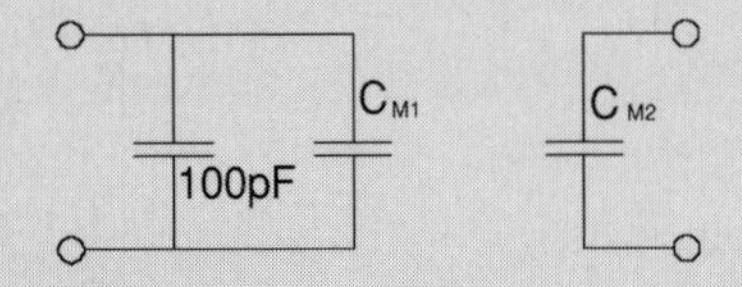

Figure 1.181

$$C_1 = 100 + C_{M1} = 100 + 404$$
$$= 504\text{pF}$$
$$C_2 = C_{M2} \approx 4\text{pF}$$

PROBLEM 1.90 MILLER

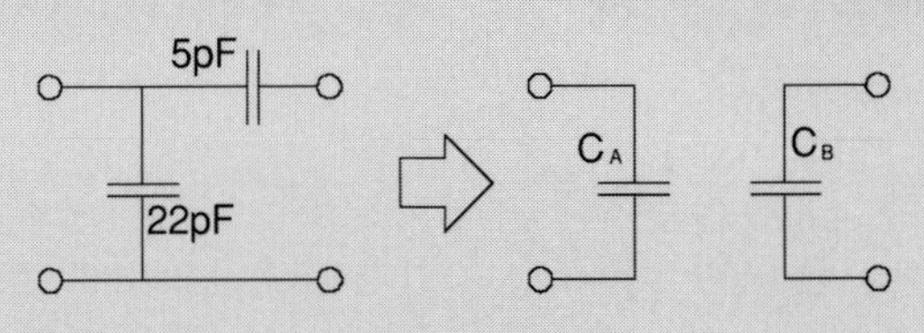

Figure 1.182

Find the values of C_A and C_B if the amplification is $A_V = -180$.

Answer: C_A = 927pF and C_B = 5pF

$$C_{M1} = 5(1 - A_v)$$
$$= 5(1 + 180) = 905\text{pF}$$
$$C_{M2} = 5\left(1 - \frac{1}{A_v}\right)$$
$$= 5\left(1 + \frac{1}{180}\right) \approx 5\text{pF}$$
$$C_A = 22\text{pF} + C_{M1}$$
$$= 22\text{pF} + 905\text{pF} = 927\text{pF}$$
$$C_B = C_{M2} \approx 5\text{pF}$$

PROBLEM 1.91 MILLER

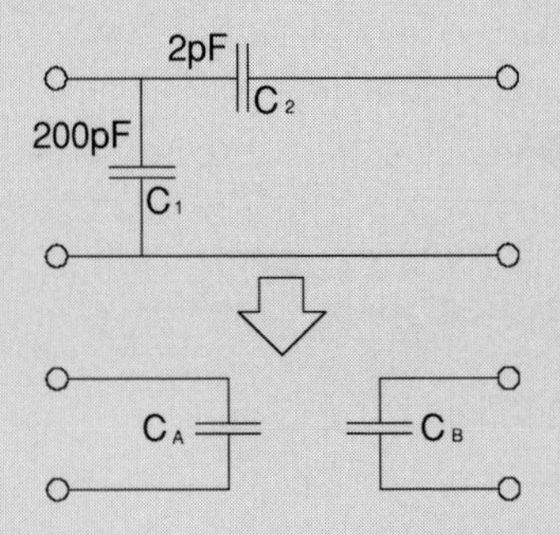

Figure 1.183

Find the value of C_A. The amplification is $A_V = -g_m R_L$. ($R_C >> R_L$ and is not taken into account). $g_m = 0.5\text{S}$, $R_L = 400\Omega$.

Answer: C_A = 602pF

$$C_{M1} = 2(1 - A_v)$$
$$= 2(1 - (-0.5 \times 400))$$
$$= 2(1 + 200)$$
$$= 402\text{pF}$$
$$C_A = C_1 + C_{M1} = 200 + 402$$
$$= 602\text{pF}$$

PROBLEM 1.92 MILLER

$C_{B'C} = 10\text{pF}$ $\quad C_{M1} = 1200\text{pF}$

Find the amplification

Answer: $A_V = -119$

$$C_{M1} = C_{B'C}(1 - A_V)$$
$$1200 = 10(1 - A_V)$$
$$1200 = 10 - 10A_V$$
$$1190 = 10A_V$$
$$A_V = -119$$

PROBLEM 1.93 MILLER

$g_m = 0.385\text{S}$, R_C and $R_L = 1200\Omega$, $C_{B'C} = 4.5\text{pF}$

Find the Miller equivalent capacitors.

Answer: $C_{M1} = 1044\text{pF}$ and $C_{M2} \approx 4.5\text{pF}$

The amplification is

$$A_V = -(R_C \| R_L)/r_e$$
$$= -g_m(R_C \| R_L)$$
$$= -0.385(1200 \| 1200)$$
$$= -0.385 \times 600 = -231$$
$$C_{M1} = 4.5(1 - A_V)$$
$$= 4.5(1 + 231) = 1044\text{pF}$$
$$C_{M2} \approx 4.5\text{pF}$$

PROBLEM 1.94 MILLER

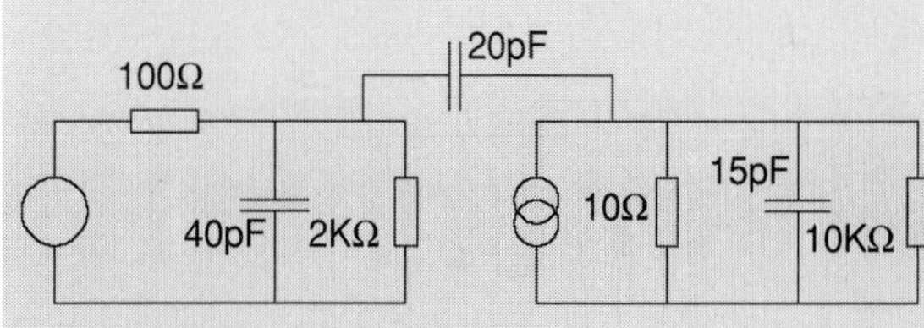

Figure 1.184

The amplification of this circuit is –200.

Find the Miller equivalent.

Answer: $C_1 = 4060\text{pF}$, $C_2 \approx 35\text{pF}$

The equivalent circuit is

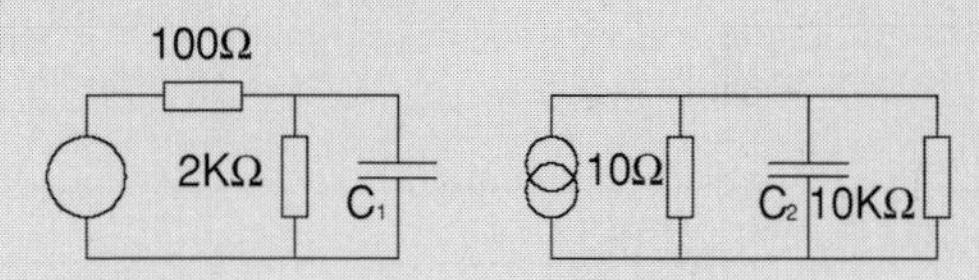

Figure 1.185

$$C_1 = 40\text{pF} + C_{M1}$$
$$C_2 = 15\text{pF} + C_{M2}$$
$$C_{M1} = 20(1 + 200) = 4020\text{pF}$$
$$C_{M2} = 20\left(1 + \frac{1}{200}\right) \approx 20\text{pF}$$
$$C_1 = 4020 + 40 = 4060\text{pF}$$
$$C_2 = 15 + 20 \approx 35\text{pF}$$

PROBLEM 1.95 MILLER

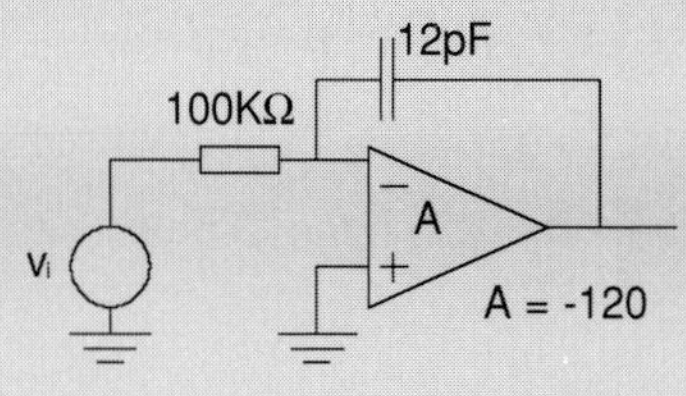

Figure 1.186

Find the values of the equivalent circuit.

Answer: $C_{M1} = 1452\text{pF}$, $C_{M2} \approx 12\text{pF}$

The circuit can be redrawn

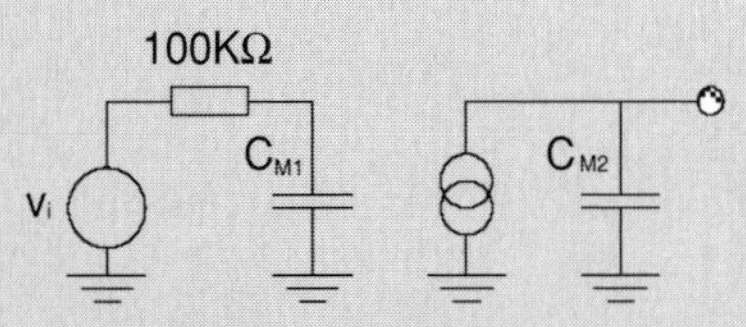

Figure 1.187

$$C_{M1} = 12(1 - A)$$
$$= 12(1 + 120) = 1452\text{pF}$$
$$C_{M2} = 12\left(1 - \frac{1}{A}\right)$$
$$= 12\left(1 + \frac{1}{120}\right) \approx 12\text{pF}$$

PROBLEM 1.96 MILLER

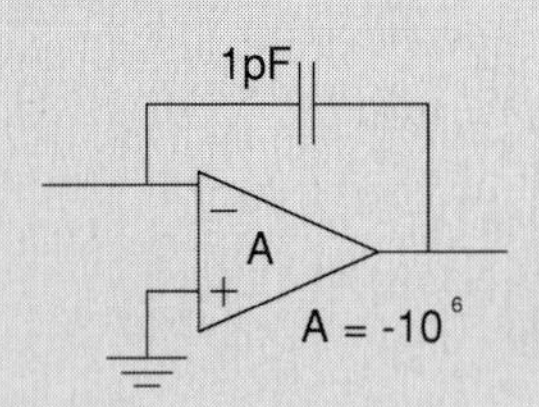

Figure 1.188

Find the value of the input capacitance of the amplifier and the value of amplification that would give an input capacitance of 0.

Answer: C_{M1} = 1μF, A = 1

Input capacitance

$$C_{M1} = 1\text{pF}(1 - A)$$
$$= 1(1 + 1\,000\,000) \approx 1\,000\,000\text{pF} = 1\mu\text{F}$$

The value of capacitance 0

$$0 = 1\text{pF}(1 - A)$$
$$1 - A = 0$$
$$A = 1$$

PROBLEM 1.97 MILLER

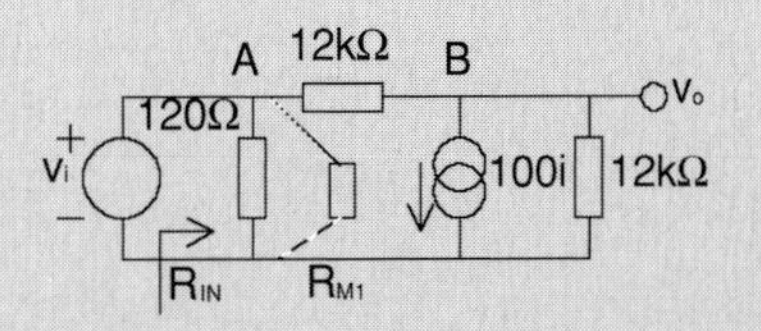

Figure 1.189

Find the value of R_{IN} using Miller's theorem.

Answer: R_{IN} = 2.35Ω

The main problem here is to find the voltage amplification of the transistor, before we can apply Miller's theorem. We can try nodal analysis.

Node B

$$\frac{V_o}{12000} + 100i + \frac{V_o - V_i}{12000} = 0 \qquad | \times 12\text{k}$$
$$V_o + 1\,200\,000i + V_o - V_i = 0$$
$$2V_o - V_i = -1\,200\,000i$$

At node A

$$V_i = 120i$$
$$2V_o - 120i = -1\,200\,000i$$
$$2V_o = -1\,199\,880i$$
$$V_o = -599\,940i$$

Amplification

$$\frac{V_o}{V_i} = \frac{-599\,940i}{120i} = -4999.5 \qquad \text{Say } 5000$$

Miller

$$R_{M1} = 12\,000\left(\frac{1}{1 + 5000}\right) = 2.4\Omega$$

Input resistance

$$R_{IN} = 2.4 || 120 = 2.35\Omega$$

PROBLEM 1.98 MILLER

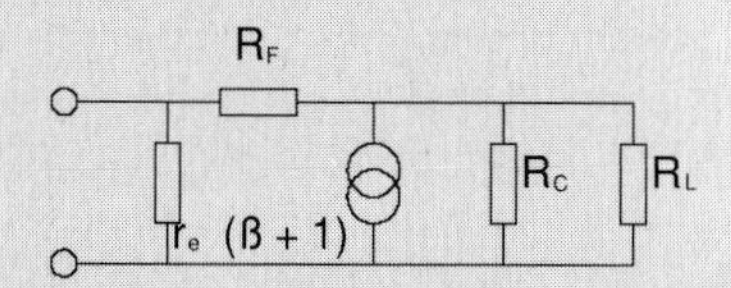

Figure 1.190

The circuit shown is the equivalent circuit of a transistor with collector feedback biasing. The amplification is A_V and it is a negative number. Use Miller's theorem to split the circuit in two at R_F.

Answer: $R_{M1} = \dfrac{R_F}{1 - A_V}$ and $R_{M2} \dfrac{R_F}{1 - \dfrac{1}{A_V}}$

The new circuit is shown in Figure 1.191. Note that we are using impedances now. The formula was developed for admittances.

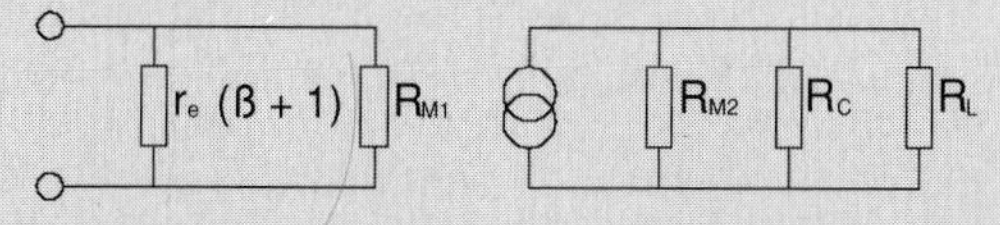

Figure 1.191

$$\frac{1}{R_{M1}} = \frac{1}{R_F}(1 - A_V)$$

$$\frac{1}{R_{M2}} = \frac{1}{R_F}\left(1 - \frac{1}{A_V}\right)$$

$$R_{M1} = \frac{R_F}{1 - A_V}$$

$$R_{M2} = \frac{R_F}{1 - \dfrac{1}{A_V}}$$

Gain impedance formula

This is a very interesting formula which relates the current gain, the voltage gain and the power gain of amplifiers.

It doesn't matter how many stages there are in a system. We only need to know the details of the input side of the first stage and the details of the output of the last stage, as we see in Figure 1.192.

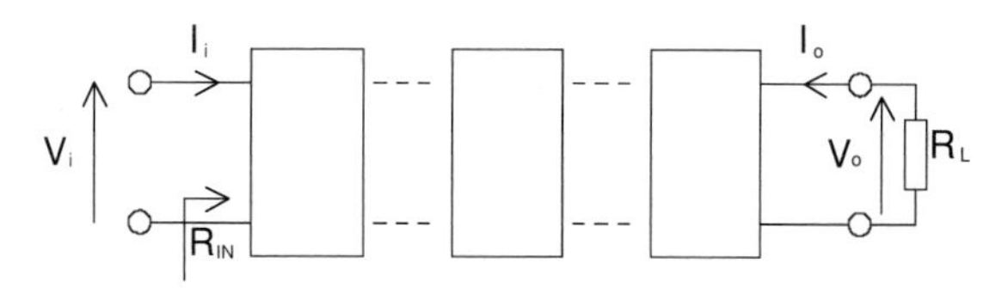

Figure 1.192

$$A_i = \frac{I_o}{I_i} \qquad A_v = \frac{V_o}{V_i}$$

$$I_o = \frac{V_o}{R_L} \qquad I_i = \frac{V_i}{R_{IN}}$$

Replacing

$$A_i = \frac{I_o}{I_i} = -\frac{V_o}{R_L} \times \frac{R_{IN}}{V_i} = -\frac{V_o}{V_i} \times \frac{R_{IN}}{R_L}$$

Therefore:

$$A_i = -A_v \frac{R_{IN}}{R_L} \qquad A_v = -A_i \frac{R_L}{R_{IN}}$$

NOTE: Use absolute value if you are not interested in the sign.

2

Biasing

We examine three types of biasing of the common emitter configuration. The three types are shown in Figure 2.1. As the transistor is a current controlled device with I_B setting the level of I_C, the object of the biasing is to set an appropriate level of I_C by controlling I_B. The three types are: supply resistor biasing, collector feedback biasing and potential divider biasing.

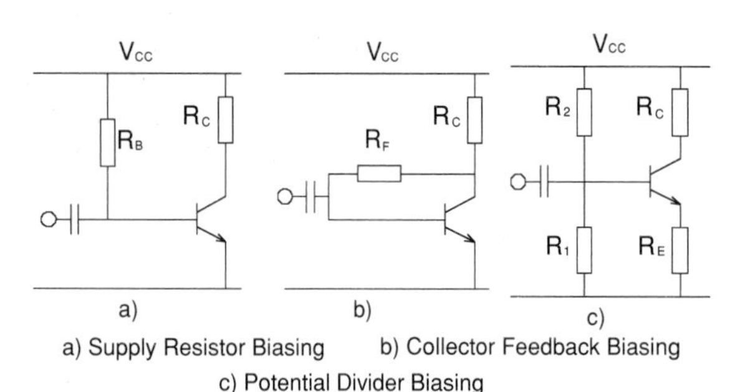

Figure 2.1

Supply resistor biasing

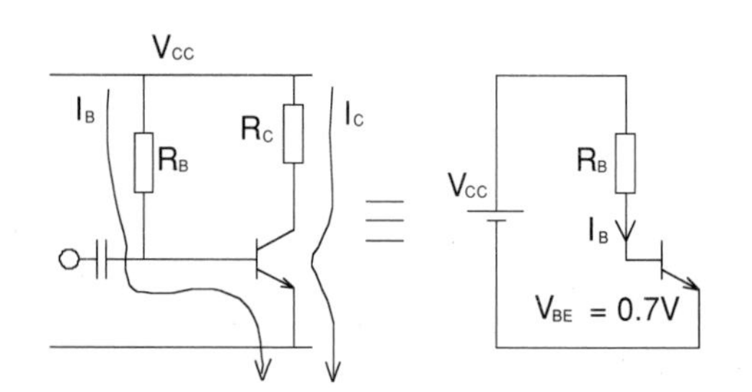

Figure 2.2

We now refer to Figure 2.2. We can easily see the voltage loop and we can apply KVL. In the modified form (sum of voltages equal to sum of voltage drops)

$$V_{CC} - 0.7 = I_B R_B$$

$$\boxed{I_B = \frac{V_{CC} - V_{BE}}{R_B}}$$

If the circuit is more complicated and it includes R_E, then the loop is extended to include this new voltage drop which would be $I_E R_E$.

The rest of the values such as I_C and the other voltages can be calculated after finding the value of I_B using the β factor.

Collector feedback biasing

This type of biasing, shown in Figure 2.3, is very interesting from the academic point of view, but performancewise it is not reliable because of its dependence on β. β is a difficult parameter to control in production. A small variation in β will vary I_B and I_C sufficiently to cause problems. The analysis, however, is interesting.

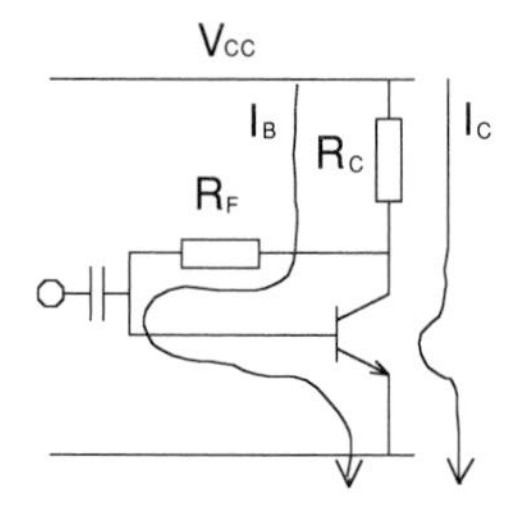

Figure 2.3

In this case the base current will flow through R_C. R_C therefore carries I_C and I_B so we can say that I_E flows through R_C in this case. Nevertheless it is only the collector current that goes through the collector of the transistor.

An alternative circuit is shown in Figure 2.4.

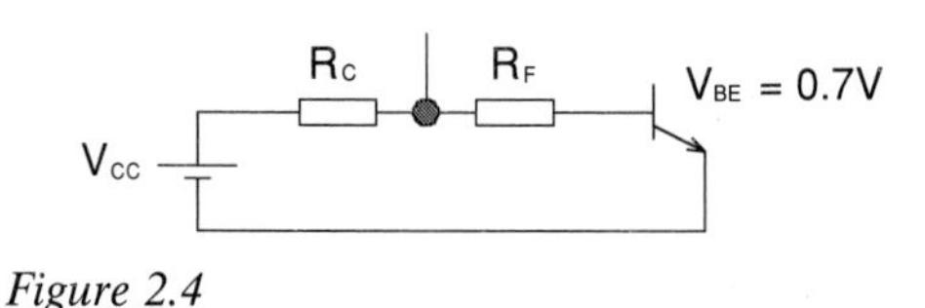

Figure 2.4

We apply KVL:

$$V_{CC} - V_{BE} = (I_C + I_B)R_C + I_B R_F$$

but

$$I_C = \beta I_B$$

$$V_{CC} - V_{BE} = I_B[(\beta + 1)R_C + R_F]$$

$$\boxed{I_B = \frac{V_{CC} - V_{BE}}{(\beta + 1)R_C + R_F}}$$

If the circuit is more complicated, for instance it includes R_E, then this has to be included in the equation.

GENERAL ADVICE: The best way to tackle the problems in general is not to memorise the formula, but rather to construct a simpler circuit from the original circuit and from the simpler circuit to set out your equations.

Potential divider biasing

There are two methods to calculate the DC conditions under this type of biasing. They are both relatively simple, but the simpler of the two is the approximate method.

Approximate method

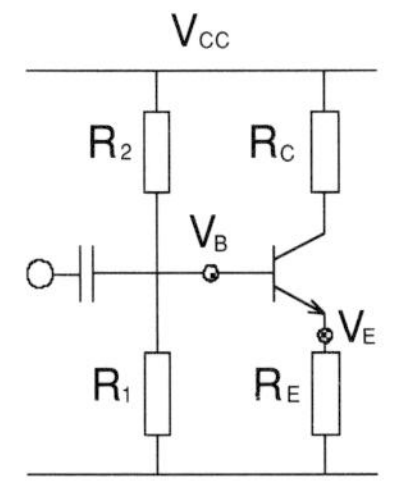

Figure 2.5

See Figure 2.5. This method assumes that the transistor has no effect on the potential divider. Therefore we can calculate

$$V_B = V_{CC} \frac{R_1}{R_1 + R_2} \quad (1)$$

Then we calculate V_E

$$V_E = V_B - V_{BE} \quad (2)$$

And finally I_E

$$I_E = \frac{V_E}{R_E} = \frac{V_B - V_{BE}}{R_E} \quad (3)$$

In reality, V_B will be affected by the impedance of the transistor. The following criteria can be used to decide on the approximate method:

$$\boxed{10R_1 < \beta R_E}$$

If this is true we use the approximate method, if not we use the Thevenin method.

Thevenin method for potential divider biasing

The Thevenin method helps us to reduce the transistor to a simpler form. Part of the transistor is converted to a circuit with a voltage source and a resistor. In order to better understand which part of the transistor is converted and how this is done, we have a sequence of four drawings to show this method (Figures 2.6–2.9).

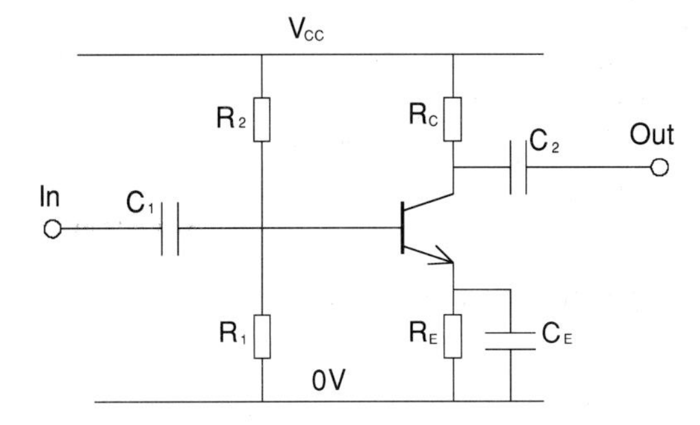

Figure 2.6

First of all we need to find the points A and B through which we are going to separate part of the transistor circuit to be converted into a Thevenin equivalent.

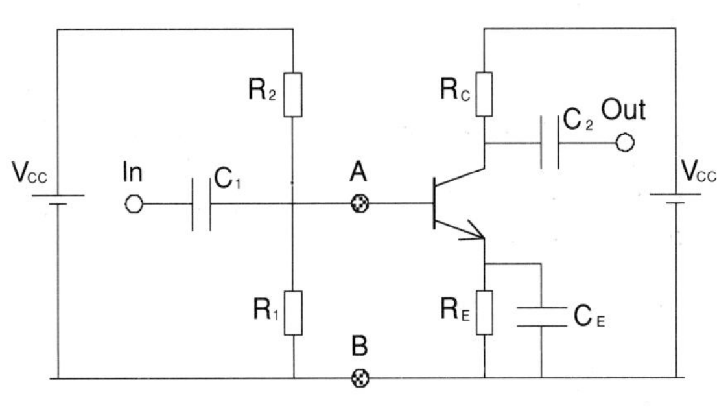

Figure 2.7

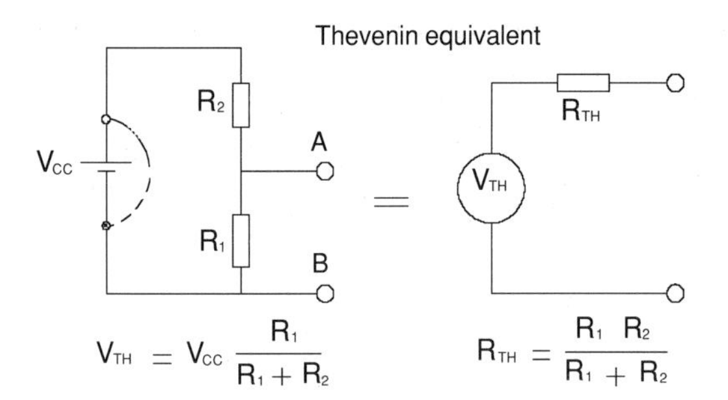

Figure 2.8

As V_{CC} would get in the way of this separation, we have Figure 2.7 where we have separated the supplies in a totally equivalent circuit, but a circuit where it is easier to find the points A and B. We can even separate it with a pair of scissors, without cutting V_{CC}.

In Figure 2.8 the left of AB is transformed to the Thevenin equivalent. We calculate V_{TH} as a potential divider, then we calculate R_{TH}. Remember that we have to 'reduce to zero' the sources in the circuit. That is why we have the dotted line shorting the power supply in Figure 2.8. The value of R_{TH} is then, R_1 and R_2 in parallel.

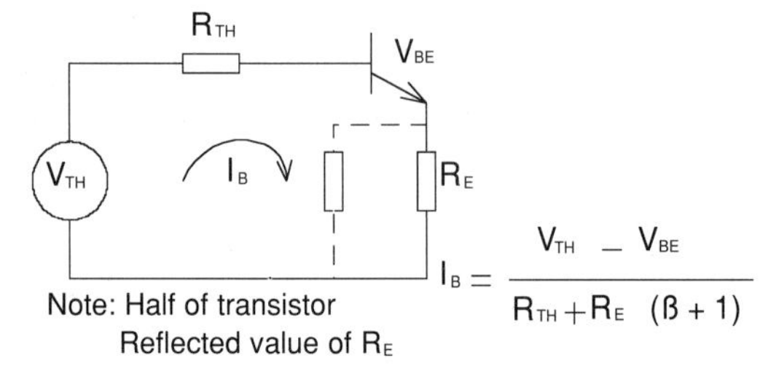

Figure 2.9

In Figure 2.9, we show the circuit with the Thevenin equivalent in place. We have only shown half of the transistor as I_B does not use the top part. In this way we concentrate our attention on what really matters.

We also bring to your attention the idea of the reflected value of R_E. We simply assume that R_E has a current of I_B, instead of the real I_E which is much larger. Then to compensate for this assumption, we just multiply R_E by (β + 1). This will get us to the solution more quickly. You might prefer a more mathematical approach, but this is commonly accepted in electronics.

PROBLEM 2.1 BIASING

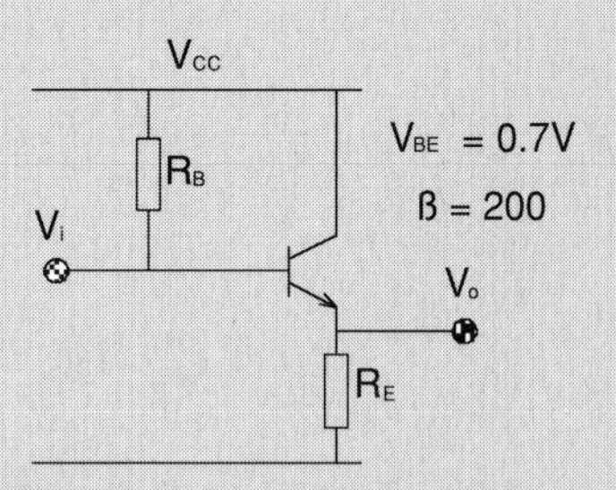

Figure 2.10

If I_C = 2 mA and V_O is 6 V at quiescent conditions, find the values of R_E and R_B.

Answer: R_E = 2985 Ω, R_B = 530 kΩ

Emitter current

$$I_E = 2\frac{201}{200} = 2.01\text{ mA}$$

$$R_E = \frac{6}{2.01} = 2985\ \Omega$$

$$I_B = \frac{I_C}{\beta} = 2\text{ mA} / 200 = 10\ \mu\text{A}$$

$$V_B = V_E + 0.7 = 6.7\text{ V}$$

$$R_B = \frac{V}{I} = \frac{12 - 6.7}{10\ \mu\text{A}} = \frac{5.3}{10\ \mu\text{A}} = 0.53\text{ M}\Omega$$

PROBLEM 2.2 BIASING

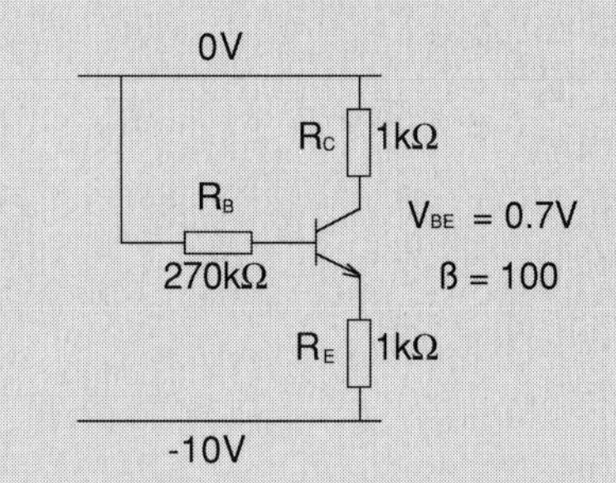

Figure 2.11

Find I_B, I_C and V_{CE}.

Answer: 25.07 μA, 2.507 mA, 4.96 V

KVL

$$0 - 0.7 - (-10) = I_B \times 27\,000 + I_B(\beta + 1)1\text{k}$$

$$9.3 = I_B(270\,000 + 101\,000)$$

$$I_B = \frac{9.3}{371\,000} = 25.07\ \mu\text{A}$$

$$I_C = \beta I_B = 2.507\ \text{mA}$$

$$I_E = (\beta + 1)I_B = 2.532\ \text{mA}$$

$$V_{CE} = 10 - 1\text{k}I_C - 1\text{k}I_E$$

$$= 10 - 2.507 - 2.532$$

$$= 4.961\ \text{V}$$

PROBLEM 2.3 BIASING

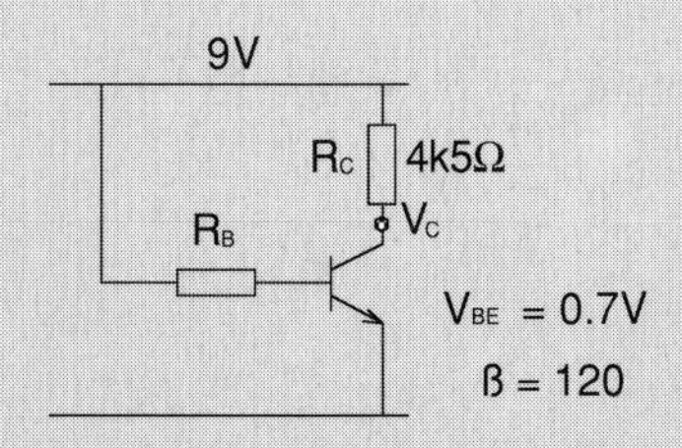

Figure 2.12

If the base current is 10 μA, find the values of R_B and V_C.

Answer: 830 kΩ, 3.6 V

$$R_B = \frac{9 - 0.7}{10 \times 10^{-6}} = \frac{8.3}{10 \times 10^{-6}} = 830\ \text{k}\Omega$$

$$I_C = \beta I_B = 120 \times 10\ \mu\text{A}$$

$$= 1.2\ \text{mA}$$

$$V_C = 9 - 1.2\ \text{mA} \times 4\text{k}5$$

$$= 9 - 5.44 = 3.6\ \text{V}$$

PROBLEM 2.4 BIASING

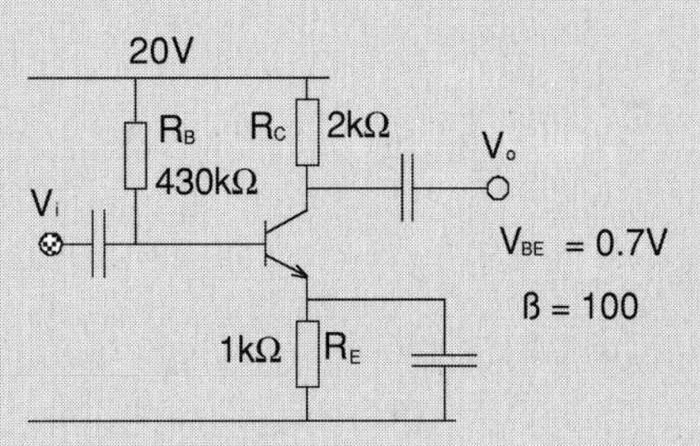

Figure 2.13

Find I_C and V_{CE}

Answer: 3.63 mA, 9.06 V

KVL

$$20 - 0.7 = I_B \times 430k + I_B(\beta + 1)1\text{k}$$

$$19.3 = I_B(430\,000 + 101\,000)$$

$$I_B = \frac{19.3}{531\,000}$$

$$I_C = \frac{19.3 \times 100}{531\,000} = 3.635\ \text{mA}$$

$$V_{CE} = 20 - I_C R_C - I_E R_E$$

$$= 20 - 3.635 \times 2 - 3.671 \times 1$$

$$I_E = \frac{19.3 \times 101}{531\,000} = 3.671\ \text{mA}$$

$$V_{CE} = 20 - 7.27 - 3.671$$

$$= 9.06\ \text{V}$$

PROBLEM 2.5 BIASING

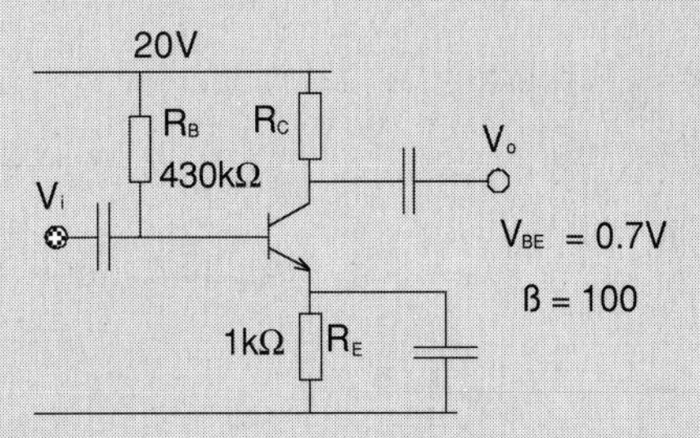

Figure 2.14

Calculate R_C to give $V_C = 10$ V

Answer: 2k75 Ω

KVL

$$20 - 0.7 = I_B \times 430k + I_B(\beta + 1)1k$$

$$19.3 = I_B(430\,000 + 101\,000)$$

$$I_B = \frac{19.3}{531\,000}$$

$$I_C = \frac{19.3 \times 100}{531\,000} = 3.635 \text{ mA}$$

For V_C = 10, the voltage drop in R_C is 10 V, therefore:

$$R_C = \frac{10}{3.635 \times 10^{-3}} = 2\text{k}75\ \Omega$$

PROBLEM 2.6 BIASING

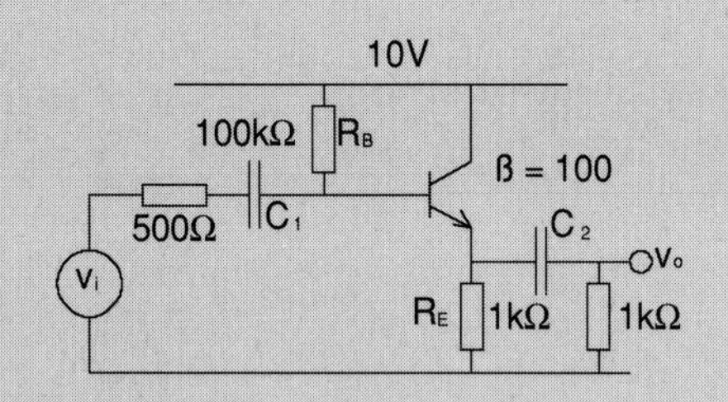

Figure 2.15

Find I_B and I_E.

Answer: 46.27 μA, 4.67 mA

KVL

$$10 - 0.7 = I_B R_B + I_E R_E$$

$$9.3 = I_B R_B + I_B(\beta + 1) R_E$$

$$9.3 = I_B 100\text{k} + I_B 101\,000$$

$$9.3 = I_B 201\,000$$

$$I_B = 46.27\ \mu\text{A}$$

$$I_E = (\beta + 1) I_B = 46.27 \times 10^{-6} \times 101$$

$$= 4.67 \text{ mA}$$

PROBLEM 2.7 BIASING

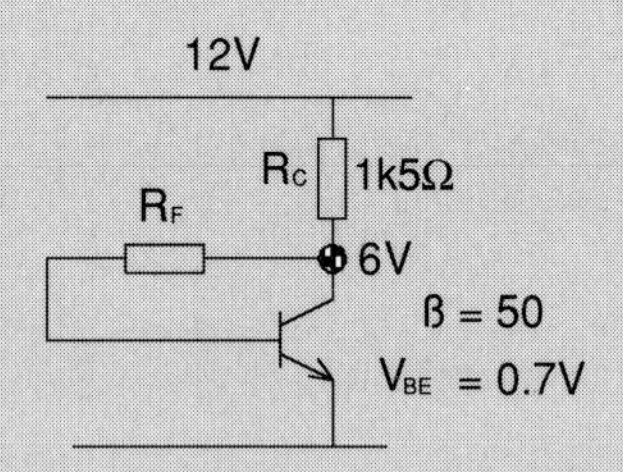

Figure 2.16

Find the value of R_F that will result in the voltage at the collector terminal being equal to half the supply voltage, i.e., 6 V.

Answer: 66k25 Ω

$$V_C = 6 \text{ V}$$

$$I_C = \frac{V_{CC} - V_C}{R_C} \quad \text{(from Ohm' s law)}$$

$$= \frac{12 - 6}{1.5 \times 10^3} = 4 \text{ mA}$$

$$I_B = \frac{I_C}{\beta} = \frac{4 \times 10^{-3}}{50} = 80\ \mu\text{A}$$

The voltage drop in R_F, which is $I_B R_F$ should be equal to $V_C - V_{BE}$:

$$R_F \times 80 \times 10^{-6} = 6 - 0.7$$

$$R_F = \frac{5.3 \times 10^6}{80} = 66\text{k}25\ \Omega$$

PROBLEM 2.8 BIASING

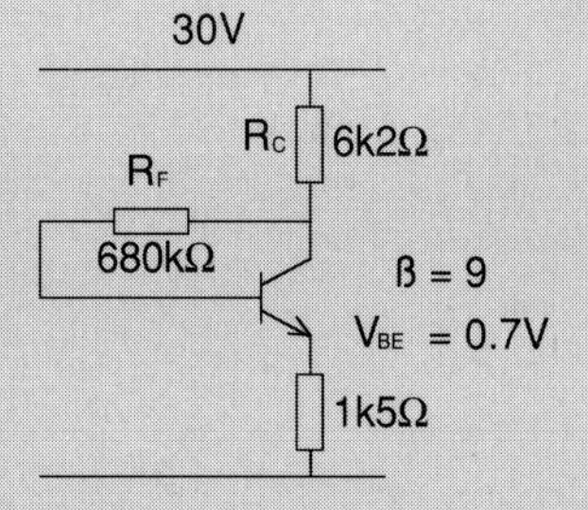

Figure 2.17

Calculate I_B and V_C.

Answer: 38.7 μA, 27.6 V

KVL

$$30 = 6k2(I_C + I_B) + 1k5(I_C + I_B) + 680kI_B + 0.7$$

but

$$(I_C + I_B) = \beta I_B + I_B = (\beta + 1)I_B = 10I_B$$

$$30 - 0.7 = 6k2 \times 10I_B + 1k5 \times 10I_B + 680k \times I_B$$

$$29.3 = I_B(62\,000 + 15\,000 + 680\,000)$$

$$I_B = \frac{29.3}{757\,000} = 38.7\ \mu A$$

$$V_C = 30 - (I_B + I_C)6k2$$

$$= 30 - 10I_B \times 6k2$$

$$= 30 - 2.4 = 27.6\ V$$

PROBLEM 2.9 BIASING

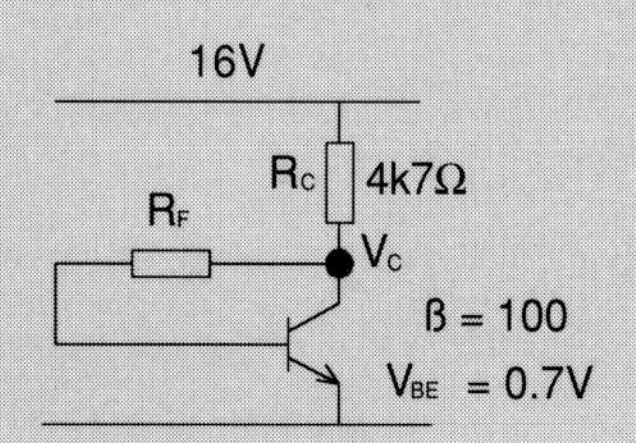

Figure 2.18

Find the value of R_F to give $V_C = 10$ V.

Answer: 736 kΩ

$$V_C = 10\ V$$

$$10 - 0.7 = I_B R_F$$

$$I_{RC} = \frac{16 - 10}{4k7} = \frac{6}{4700}$$

$$I_{RC} = I_B(\beta + 1) \quad (\text{as it is } I_C + I_B)$$

$$= \frac{6}{4700}$$

$$I_B = \frac{6}{4700 \times 101}$$

$$R_F = \frac{V_{RF}}{I_B} = \frac{(10 - 0.7) \times 4700 \times 101}{6}$$

$$= \frac{9.3 \times 4700 \times 101}{6} = 736\ k\Omega$$

PROBLEM 2.10 BIASING

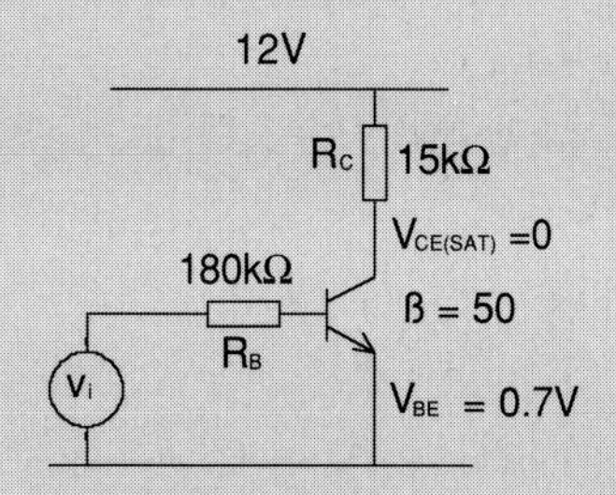

Figure 2.19

Determine I_C, I_B and V_i that will result in the saturation of the transistor.

NOTE: During saturation, the transistor voltage drops. The collector–emitter voltage is very low, typically 0.2 V or 0.3 V, or as indicated by the manufacturer. In this case it is reasonable to assume 0 V.

Answer: 2.88 V

$$I_C = \frac{12}{15k} = 0.8\ mA$$

$$I_B = \frac{0.8 \times 10^{-3}}{50} = 16\ \mu A$$

$$V_i = I_B R_B + 0.7$$

$$= 16 \times 10^{-6} \times 180 \times 10^3 + 0.7$$

$$= 2.88\ V$$

PROBLEM 2.11 BIASING

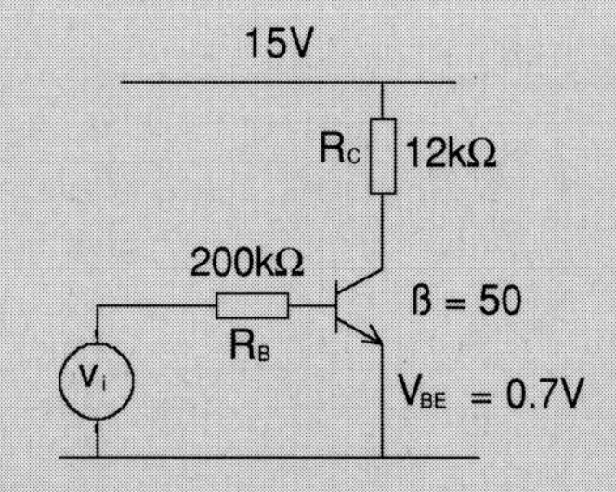

Figure 2.20

If V_i is 3 V, find the values of I_B and V_{CE}.

Answer: 0.575 mA, 8.1 V

KVL

$$3 - 0.7 = I_B R_B$$
$$2.3 = I_B \times 200\,000$$
$$I_B = 11.5\ \mu A$$
$$I_C = \beta I_B = 50 \times 11.5 \times 10^{-6}$$
$$= 0.575\ mA$$
$$V_{CE} = 15 - I_C R_C$$
$$= 15 - 12\,000 \times 0.575 \times 10^{-3}$$
$$= 15 - 6.9$$
$$= 8.1\ V$$

PROBLEM 2.12 BIASING

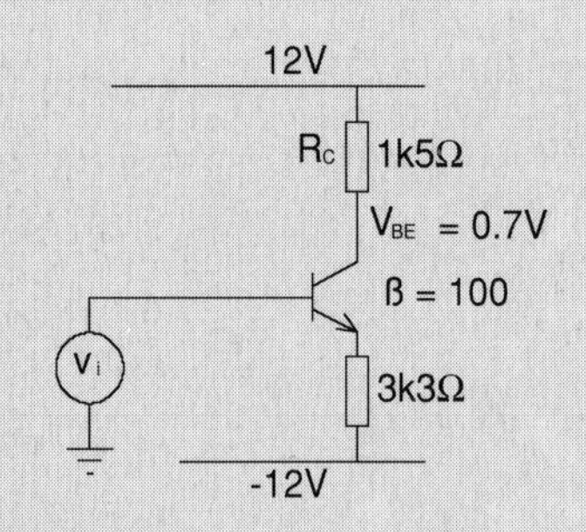

Figure 2.21

Find the values of I_C and V_{CE} for v_i equal to 0 V and 3 V.

Answer: 3.39 mA, 7.82 V, 3.99 mA, 4.72 V.

For 0 V input

$$V_E = 0 - 0.7 = -0.7\ V$$
$$I_E = \frac{-0.7 - (-12)}{3k3} = 3.42\ mA$$
$$\frac{I_C}{I_E} = \frac{\beta}{\beta + 1}$$
$$I_C = \frac{100}{101} \times 3.42 \times 10^{-3} = 3.39\ mA$$
$$V_{CE} = 24 - I_C R_C - I_E R_E$$
$$= 24 - 3.39 \times 1.5 - 3.42 \times 3.3 = 7.82\ V$$

For 3 V input

$$V_E = 2 - 0.7 = 1.3\ V$$

(with respect to − 12 V)

$$I_E = \frac{1.3 - (-12)}{3k3} = \frac{13.3}{3k3} = 4.03\ mA$$
$$I_C = 4.03\frac{100}{101} = 3.99\ mA$$
$$V_{CE} = 24 - I_C R_C - I_E R_E$$
$$= 24 - 3.99 \times 1.5 - 4.03 \times 3.3$$
$$= 4.72\ V$$

Note: for 0 V, the situation is

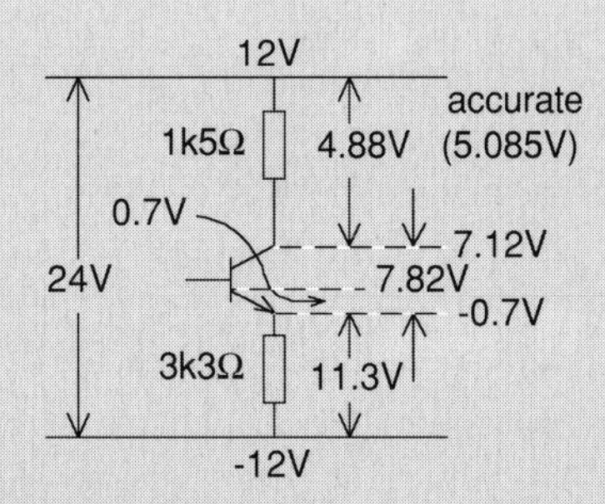

Figure 2.22

I_C in this case should be

$$I_C = \frac{4.88}{1k5} = 3.25\ mA$$

We had calculated 3.39 mA. There is a small difference here that needs explaining. The value of I_E is 3.424 242... . Using this value, instead of the one we used, we get V_{CE} = 7.615 V. 4.88 V is revised to 5.085 V and then the check is the correct 3.39 mA for I_C. Bear this in mind in the future.

PROBLEM 2.13 BIASING

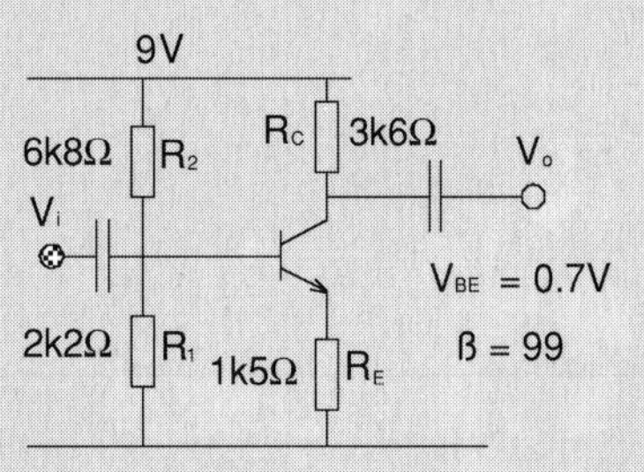

Figure 2.23

Find I_{CE} using the Thevenin method.

Answer: 4 V

$$V_{TH} = V_{CC}\frac{R_1}{R_1 + R_2} = 9\frac{2k2}{2k2 + 6k8}$$

$$= 9\frac{2.2}{9} = 2.2 \text{ V}$$

$$R_{TH} = \frac{2k2 \times 6k8}{2k2 + 6k8} = 1662\ \Omega$$

$$I_B = \frac{V_{TH} - V_{BE}}{R_{TH} + R_E(\beta + 1)} = \frac{2.2 - 0.7}{1662 + 1500 \times 100}$$

$$= \frac{1.5}{151\,662} = 9.89 \times 10^{-6} \text{ A}$$

Voltage drop across 3k6 resistor

$$0.979 \times 10^{-3} \times 3600 = 3.52 \text{ V}$$

Voltage drop across 1k5 resistor

$$I_E = (\beta + 1)I_B = 9.89 \times 10^{-6} \times 100$$

$$= 0.989 \text{ mA}$$

therefore

$$V_{CE} = 9 - 3.523 - 0.989 \text{ mA} \times 1k5$$

$$= 9 - 3.52 - 1.48$$

$$= 4 \text{ V}$$

PROBLEM 2.14 BIASING

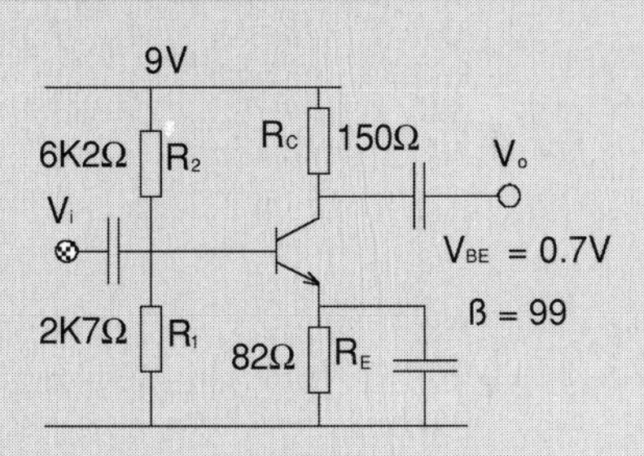

Figure 2.24

Find V_C with respect to 0 V.

> Answer: 6.01 V (Thevenin method)
> 5.32 V (approximate method)

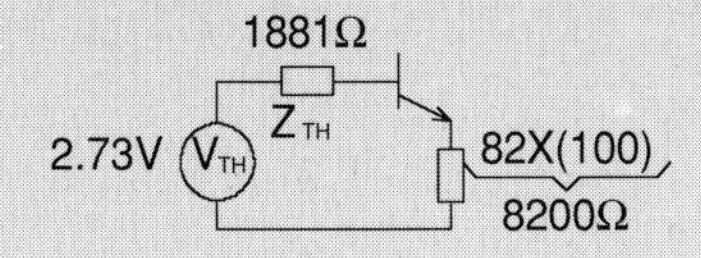

Figure 2.25

Thevenin

$$V_{TH} = 9\frac{2k7}{8k9} = 2.73 \text{ V}$$

$$R_{TH} = 6k2 \| 2k7 = 1881\ \Omega$$

$$I_B = \frac{2.73 - 0.7}{1881 + 8200} = \frac{2.03}{10\,081}$$

$$I_C = h_{FE} I_B = 99\frac{2.03}{10\,081} = 19.94 \text{ mA}$$

$$V_{drop} = I_C R_C = 19.94 \times 0.150$$

$$= 2.99 \text{ V}$$

$$V_C = 9 - 2.99 = 6.01 \text{ V}$$

PROBLEM 2.15 BIASING

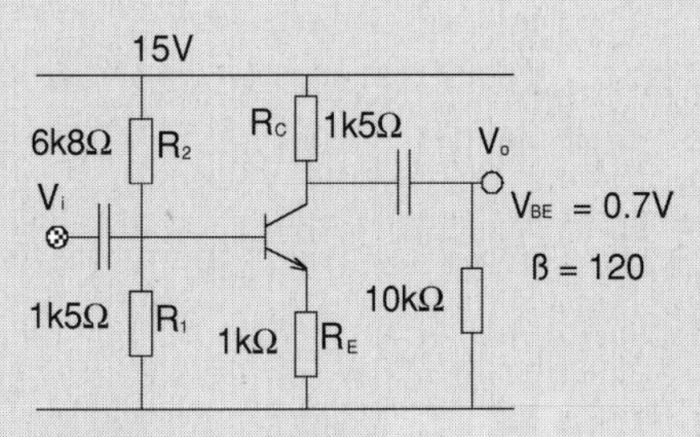

Figure 2.26

Find the percentage difference in I_C calculating it with the approximate method compared to the Thevenin method.

> Answer: 1%

Approximate method

$$V_B = 15\frac{1k5}{1k5 + 6k8} = 15\frac{1.5}{8.3} = 2.71 \text{ V}$$

$$V_E = V_B - V_{BE} = 2.71 - 0.7 = 2.01 \text{ V}$$

$$I_E = \frac{2.01}{1k} = 2.01 \text{ mA}$$

$$I_C = I_E \frac{\beta}{\beta + 1} = 1.99 \text{ mA}$$

Thevenin method

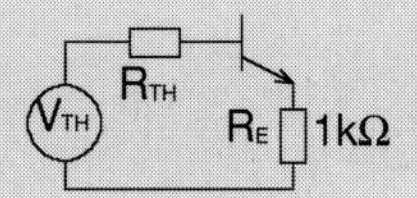

Figure 2.27

$$V_{TH} = 15\frac{1.5}{8.3} = 2.71\ \text{V}$$

$$R_{TH} = 1\text{k}5 \| 6\text{k}8 = 1229\ \Omega$$

$$I_B = \frac{V_{TH} - 0.7}{R_{TH} + R_E(\beta + 1)}$$

$$= \frac{2.71 - 0.7}{1229 + 121\,000}$$

$$= \frac{2.01}{122\,229}$$

$$= 16.44\ \mu\text{A}$$

$$I_C = \beta I_B = 16.44 \times 10^{-6} \times 120 = 1.97\ \text{mA}$$

$$\text{difference} = \frac{1.99 - 1.97}{1.97} = \frac{0.02}{1.97} = 1\%$$

PROBLEM 2.16 BIASING

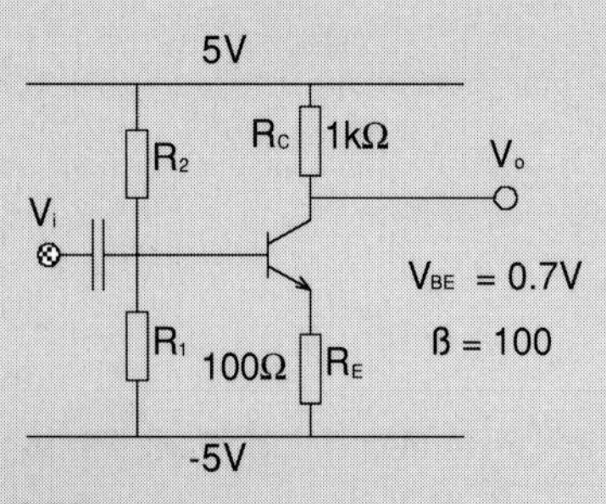

Figure 2.28

If R_2 is 12k, find R_1 so that $V_O = 0$ V, Use the approximate method, i.e. 100 Ω resistor has no loading effect.

Answer: 1714 Ω

$$I_C = \frac{5}{1\text{k}} = 5\ \text{mA}$$

$$I_E = 5\frac{101}{100} \quad \left(\frac{I_C}{I_E} = \frac{\beta}{\beta + 1}\right)$$

$$I_E = 5.05\ \text{mA}$$

$$V_E = I_E R_E - 5$$

$$= 0.505 - 5 = -4.45\ \text{V}$$

$$V_B = V_E + 0.7 = -3.75\ \text{V}$$

This is with respect to -5 V.

The voltage difference is $-3.75 - (-5)$.

$$V_B = 1.25 = 10\frac{R_1}{R_1 + R_2}$$

$$10R_1 = 1.25R_1 + 15\,000$$

$$8.75R_1 = 15\,000$$

$$R_1 = 1714\ \Omega$$

PROBLEM 2.17 BIASING

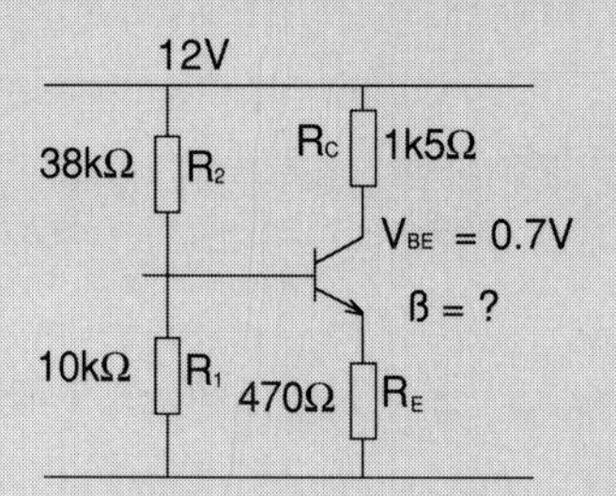

Figure 2.29

For which value of β is $I_C = 3$ mA in this circuit? Use Thevenin's method.

Answer: 64.54

$$V_{TH} = 12\frac{10\text{k}}{10\text{k} + 38\text{k}} = 2.5\ \text{V}$$

$$Z_{TH} = 10\text{k} \| 38\text{k} = 7.92\ \text{k}\Omega$$

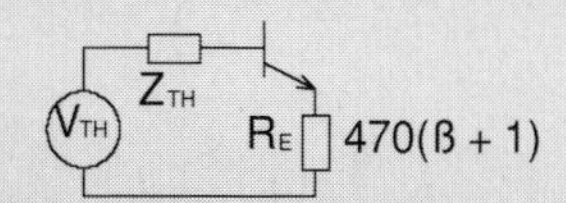

Figure 2.30

$$I_B = \frac{3\ \text{mA}}{\beta} = \frac{2.5 - 0.7}{7920 + 470(\beta + 1)}$$

$$\frac{3 \times 10^{-3}}{\beta} = \frac{1.8}{7920 + 470 + 470\beta}$$

$$1.8\beta = 25.17 + 1.41\beta$$

$$0.39\beta = 25.17$$

$$\beta = \frac{25.17}{0.39} = 64.54$$

PROBLEM 2.18 BIASING

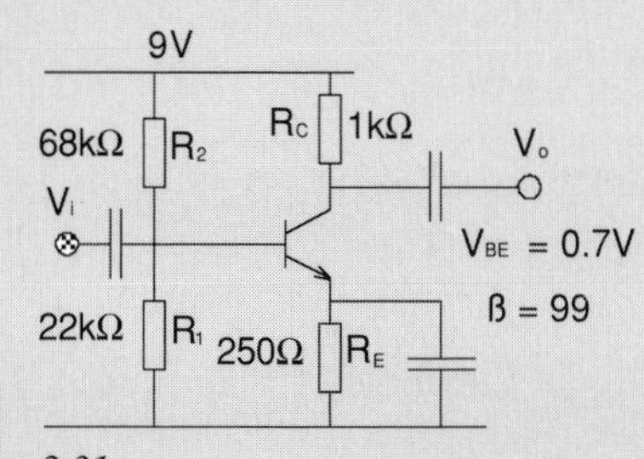

Figure 2.31

Determine I_{EQ} using the Thevenin method and the approximate method.

Answer: 3.6 mA, 6 mA

$$V_{TH} = V_{CC}\frac{R_1}{R_1 + R_2} = 9\frac{22}{22 + 68} = 2.2\ \text{V}$$

$$R_{TH} = \frac{R_1 R_2}{R_1 + R_2} = \frac{22 \times 68}{22 + 68} = 16\,622\ \Omega$$

$$I_B = \frac{2.2 - 0.7}{R_{TH} + R_E(\beta + 1)} = \frac{1.5}{16\,622 + 25\,000}$$

$$= 36\ \mu\text{A}$$

$$I_E = (\beta + 1) \times 36\ \mu\text{A} = 3.6\ \text{mA}$$

Approximate method

$$V_B = V_{CC}\frac{R_1}{R_1 + R_2} = 9\frac{22}{22 + 68} = 2.2\ \text{V}$$

$$V_E = V_B - V_{BE} = 2.2 - 0.7 = 1.5\ \text{V}$$

$$I_E = \frac{1.5}{250} = 6\ \text{mA}$$

NOTE: The approximation is not very good in this case. The criteria which allows us to use the approximate method, 22k × 10 < 250β or in this case, 220k < 25k is far from true. The Thevenin method should be used in this case.

PROBLEM 2.19 BIASING

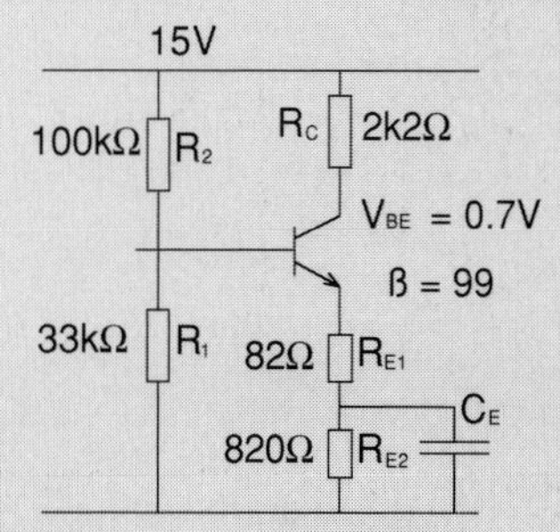

Figure 2.32

Calculate I_C and V_{CQ} using the Thevenin method.

Answer: 2.6 mA, 9.28 V

$$R_{TH} = \frac{R_1 R_2}{R_1 + R_2} = \frac{100\text{k} \times 33\text{k}}{133\text{k}} = 24\text{k}81\ \Omega$$

$$V_{TH} = 15\frac{R_1}{R_1 + R_2} = \frac{15 \times 33}{133} = 3.72\ \text{V}$$

We now draw the new circuit

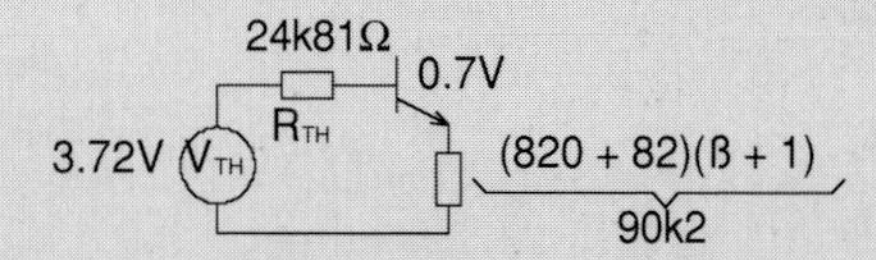

Figure 2.33

$$I_B = \frac{3.72 - 0.7}{24\text{k}81 + 90\text{k}2} = 26.26\ \mu\text{A}$$

$$I_C = 99 I_B = 2.6\ \text{mA}$$

$$V_{CQ} = 15 - 2.6 \times 2.2 = 9.28\ \text{V}$$

PROBLEM 2.20 BIASING

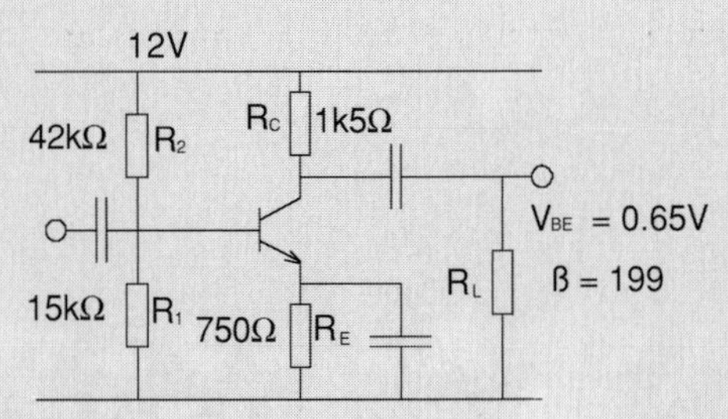

Figure 2.34

Use the Thevenin method to find the values of I_E and V_C under quiescent conditions.

Answer: $I_E = 3.12$ mA, $V_C = 7.35$ V

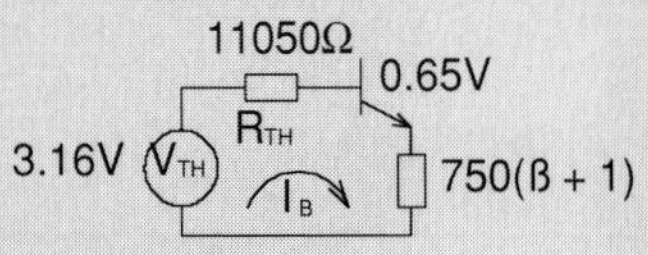

Figure 2.35

$$V_{TH} = 12\frac{15}{15+42} = 3.16 \text{ V}$$

$$R_{TH} = 15\text{k}\|42\text{k} = 11.05k\ \Omega$$

$$I_B = \frac{3.16-0.65}{11\,050+750(200)} = \frac{2.51}{161\,050}$$

$$= 15.585\ \mu\text{A}$$

$$I_E = (\beta+1)I_B = 15.585\times10^{-6}\times200 = 3.12 \text{ mA}$$

$$I_C = \beta I_B = 15.585\times10^{-6}\times199 = 3.10 \text{ mA}$$

$$V_C = 12 - I_C R_C$$

$$= 12 - 3.10\times10^{-3}\times1.5\times10^{3}$$

$$= 12 - 4.65$$

$$= 7.35 \text{ V}$$

3

Load lines

As we know, the transistor operation ranges from cut-off to the active region and saturation. This can be seen very clearly in a system of coordinates such as the one shown in Figure 3.1. Apart from the different states of the transistor we show the limitation due to power dissipation.

The load lines, similarly, present a visual form of what is happening with the transistor with no AC signal and, most important of all, we can see how much the AC signal can swing, before it hits the limits.

One word of warning. When we deal with load lines we forget a little about saturation and cut-off and we assume that everything is in the active region. In this way we transform the problem from one of electronics to one of geometry, that

Figure 3.1

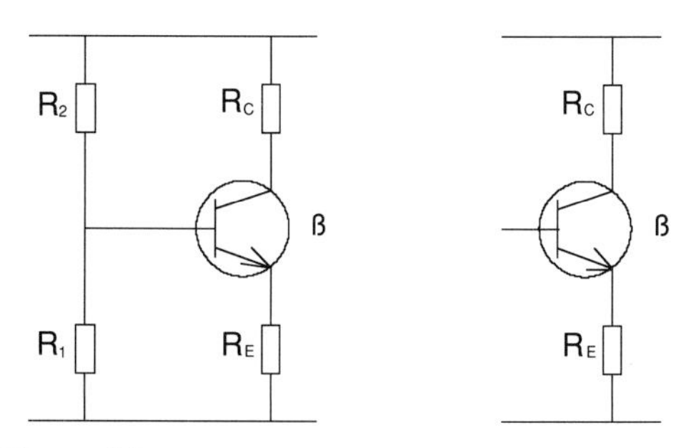

Figure 3.2

of a straight line. There will be a few other approximations and we will point them out as we go along. We start with a simple circuit as shown in Figure 3.2. You will find that most authors approximate $I_C = I_E$ when dealing with load lines, and we shall do the same.

We draw the system of coordinates as shown in Figure 3.3 on the left-hand side, then we find points A and B as follows:

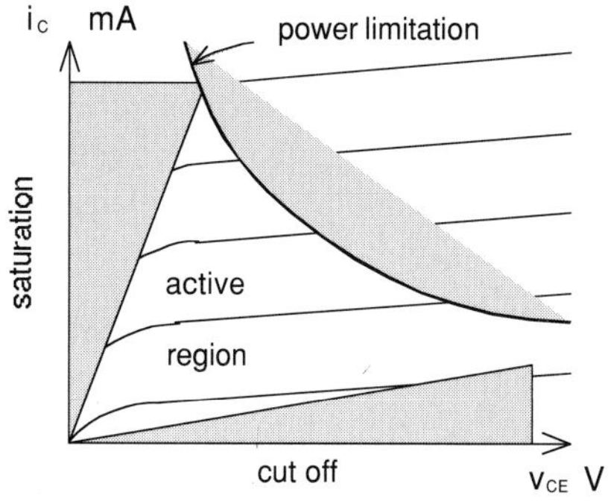

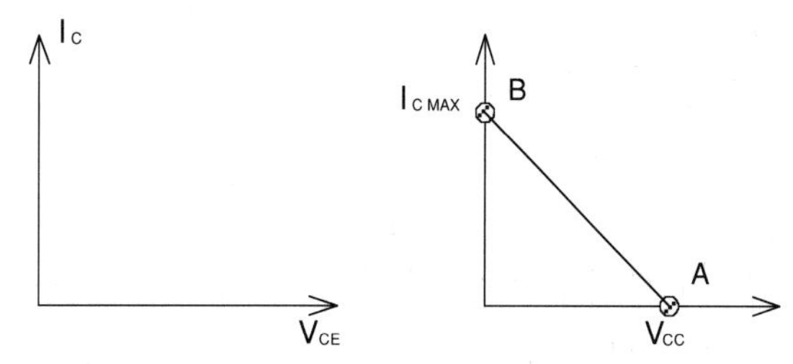

Figure 3.3

(A) This point is obtained when $I_C = 0$. The transistor is cut off. As there is no current, there is no voltage drop, and therefore the voltage applied to the transistor is the battery voltage V_{CC}.

(B) This point is obtained when $V_{CE} = 0$. Then we have the maximum current I_C. The transistor is saturated, i.e., the voltage V_{CE} of the transistor is 0 V.

This is another approximation. In reality, it is around 0.3 V, or as given by the manufacturers. Assuming $V_{CE} = 0$, then the voltage drops in R_C and R_E. So:

$$I_{cMAX} = \frac{V_{CC}}{R_C + R_E}$$

This last equation assumes that $I_C = I_E$. This simplifies your life and mine. These two points can be joined by a straight line which is the DC load line.

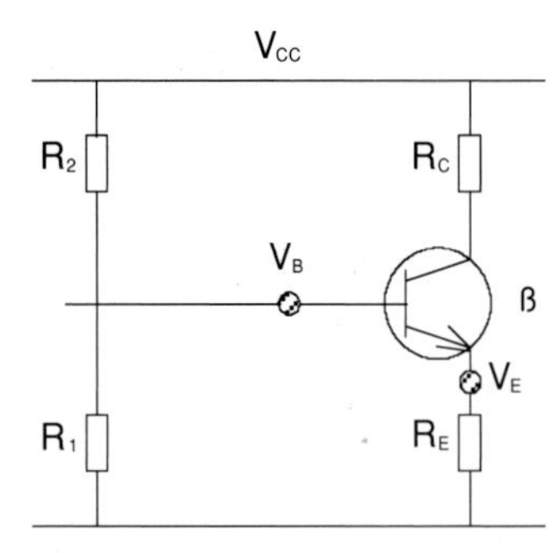

Figure 3.4

DC load line

The resistors R_1 and R_2, in Figure 3.4, will ensure that the transistor operating point is located at one point only along this line.

In Figure 3.4, the base voltage V_B is given by

$$V_B = V_{CC} \frac{R_1}{R_1 + R_2}$$

This is another approximation as the Thevenin method would have been more precise.

We know that the voltage V_{BE} is 0.7 V, another approximation.
Therefore:

$$V_E = V_B - 0.7$$

So:

$$I_E = \frac{V_E}{R_E}$$ (We use this as I_C)

In this way, we obtain the Q point in the straight line (see Figure 3.5). Q is short for quiescent, which is the situation of the transistor when it has DC power but no AC signal.

The Q point should also coincide with the value of V_{CE} calculated as

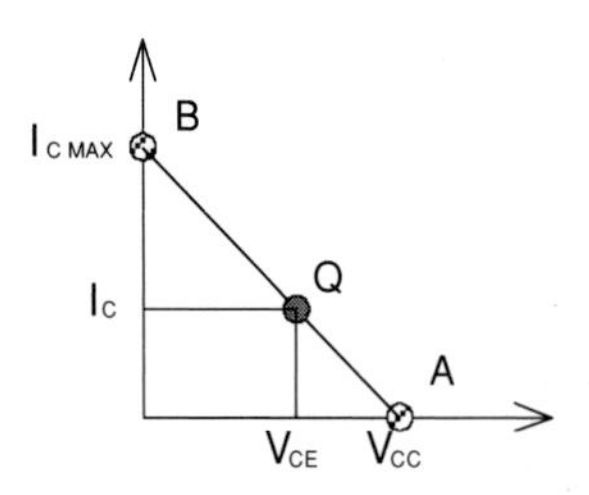

Figure 3.5

$$V_{CE} = V_{CC} - I_C(R_C + R_E)$$

Straight line review

A straight line is determined by either

- two points
- one point and the slope

The general equation is given by:

$$y = mx + c$$

x and y are the variables, m is the slope and c is the constant.

Two points

See the example shown in Figure 3.6

$1 = 0m + c$

$3 = 2m + c$ therefore $c = 1$ and $m = 1$

Another example is shown in Figure 3.7.

One point and slope

If we know the slope and a point (X_1, Y_1) then the equation of the line will be given by:

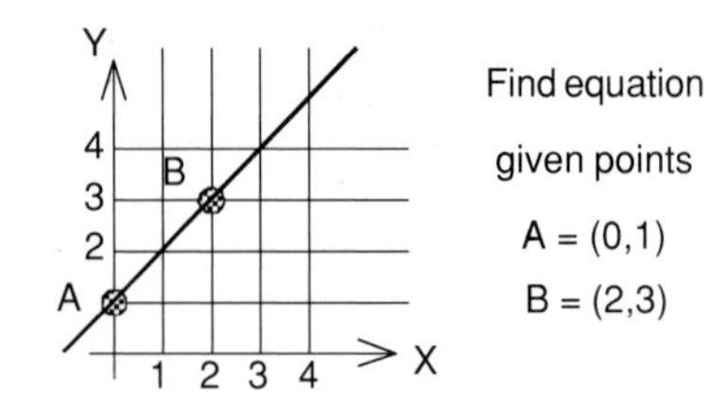

Figure 3.6

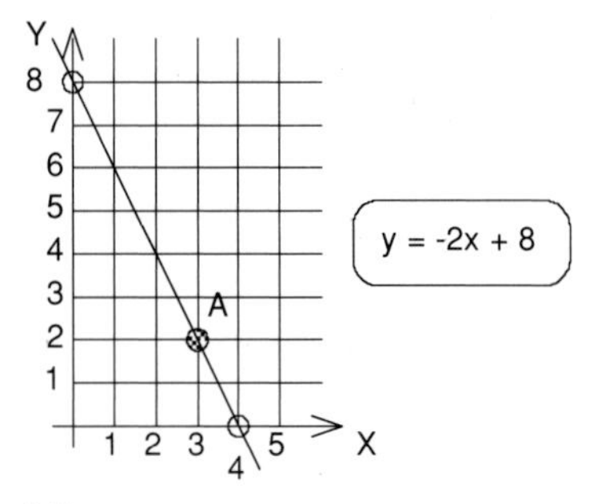

Figure 3.7

$$y - Y_1 = m(x - X_1)$$

So

$$y - 2 = -2(x - 3)$$
$$y = -2x + 6 + 2$$
$$y = -2x + 8$$

We will be using this second method in the calculation of the AC load line.

AC load line

We now move onto the AC load line. We consider Figure 3.8. Now that we are considering

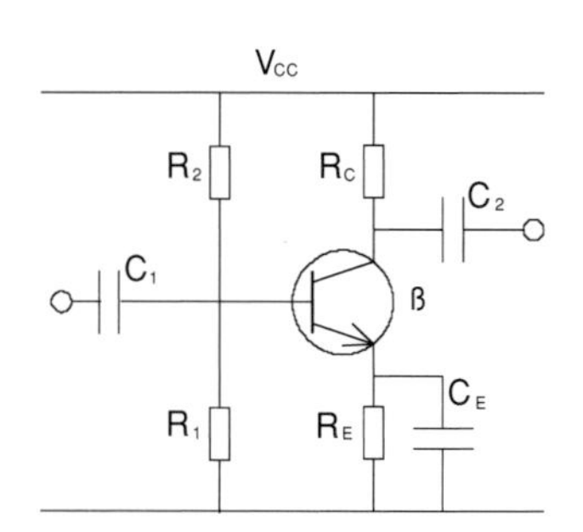

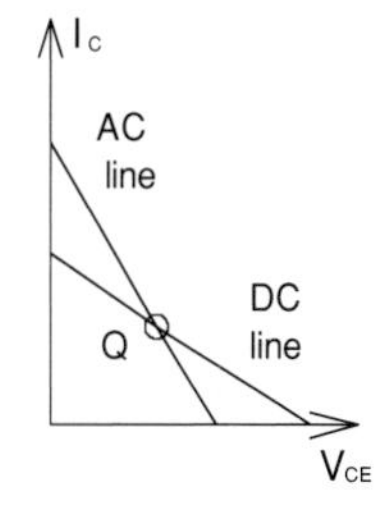

Figure 3.8

an AC signal, the capacitors play an important role. C_1 and C_2 let the AC signal through, but block the DC supplies. C_E bypasses the resistor R_E. This results in less resistance for the current. The current will be higher for a given voltage, when R_E is bypassed by C_E, and the slope will be higher.

The DC resistance was $R_C + R_E$ with a slope of $-1/(R_C + R_E)$. The AC resistance is only R_C and the slope is $-1/R_C$.

As we know the slope of the AC line and the Q point, we can now find the equation of this straight line as follows:

$$y - Y_1 = m(x - X_1)$$

Using some electronic terms, we obtain:

$$y - I_{CQ} = -\frac{1}{R_{AC}}(x - V_{CEQ})$$

We now replace x and y for the electronic term

$$i_C - I_{CQ} = -\frac{v_{CE} - V_{CEQ}}{R_{AC}}$$

The equation for i_c is

$$i_C - I_{CQ} = -\frac{v_{CE}}{R_{AC}} + \frac{V_{CEQ}}{R_{AC}}$$

Having this equation it is quite easy to find the the points where the straight line crosses the x and y axes:

For $V_{CE} = 0$ we have

$$i_{CMAX} = I_{CQ} = +\frac{V_{CEQ}}{R_{AC}}$$

For $i_c = 0$ we have

$$v_{CEmax} = V_{CEQ} + I_{CQ}R_{AC}$$

Load lines: equal swing

We now refer to the maximum equal swing in the load lines. This is illustrated in Figure 3.9.

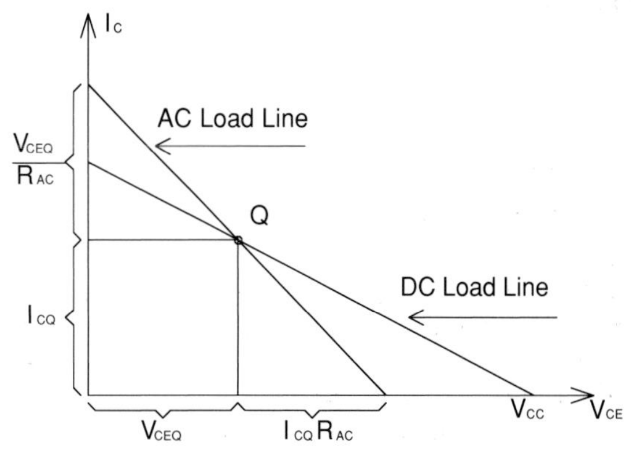

Figure 3.9

The values required to construct the load lines graph are:

For DC

$$V_{CEMAX} = V_{CC} \quad (1)$$

$$I_{CMAX} = \frac{V_{CC}}{R_{DC}} \quad (2)$$

For AC

$$v_{CEMAX} = V_{CEQ} + I_{CQ}R_{AC} \quad (3)$$

$$i_{CMAX} = I_{CQ} + \frac{V_{CEQ}}{R_{AC}} \quad (4)$$

For equal swing:

$$V_{CEQ} = I_{CQ}R_{AC} \quad (5)$$

$$I_{CQ} = \frac{V_{CEQ}}{R_{AC}} \quad (6)$$

Equation (4) can be rewritten as :

$$i_{CMAX} = 2I_{CQ}$$

The straight line equation for the DC load line is:

$$I_C = \frac{V_{CC} - V_{CE}}{R_{DC}}$$

At the Q point this becomes :

$$I_{CQ} = \frac{V_{CC} - V_{CEQ}}{R_{DC}} \quad (7)$$

From the drawing we can see that at equal swing:

$$I_{CQ} = \frac{V_{CEQ}}{R_{AC}} \quad (8)$$

This is also part of the AC load line. We can replace I_{CQ} in equation (7) for its value from equation (8):

$$\frac{V_{CEQ}}{R_{AC}} = \frac{V_{CC} - V_{CEQ}}{R_{DC}}$$

This can be rearranged as follows :

$$V_{CEQ}\frac{R_{DC}}{R_{AC}} = V_{CC} - V_{CEQ}$$

$$V_{CEQ} + V_{CEQ}\frac{R_{DC}}{R_{AC}} = V_{CC}$$

$$V_{CEQ}\left(1 + \frac{R_{DC}}{R_{AC}}\right) = V_{CC}$$

and finally:

$$V_{CEQ} = V_{CC}\frac{R_{AC}}{R_{DC} + R_{AC}} \quad (9)$$

From the drawing:

$$V_{CEQ} = I_{CQ}R_{AC}$$

Replacing this in the previous equation :

$$I_{CQ}R_{AC} = V_{CC}\frac{R_{AC}}{R_{DC} + R_{AC}}$$

Which finally gives :

$$I_{CQ} = \frac{V_{CC}}{R_{DC} + R_{AC}} \quad (10)$$

Equations (9) and (10) are very useful in dealing with problems of maximum equal swing as you will see in the following examples.

PROBLEM 3.1 LOAD LINES

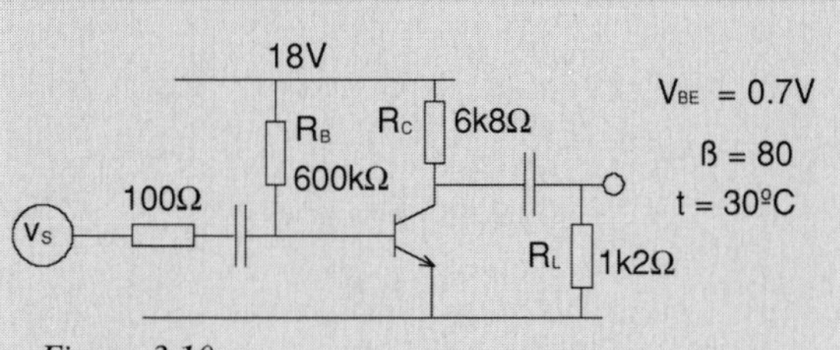

Figure 3.10

Find the maximum value of V_s to give the maximum swing on the 1k2 resistor.

Answer: 55.48 mV peak to peak.

We have to find the maximum swing, the voltage amplification, and then work back towards the source.

$$I_B = \frac{18 - 0.7}{600k} = \frac{17.3}{600k} = 28.83\ \mu A$$

$$I_C = \beta I_B = 28.82 \times 10^{-6} \times 80 = 2.306\ mA$$

$$I_E \approx I_C = 2.306\ mA$$

$$V_{CEQ} = 18 - 2.306 \times 6.8 = 18 - 15.68 = 2.32\ V$$

at 30°C

$$r_e = \frac{26}{2.306} = 11.27\ \Omega$$

Amplification

$$A_V = -\frac{R_C \| R_L}{r_e} = -\frac{1020}{11.27} = -90.5$$

Maximum swing

$$v_{CEMAX} = I_{CQ} + \frac{V_{CEQ}}{R_{AC}}$$

$$= 2.306 + \frac{2.32}{1.02} = 2.306 + 2.27$$

therefore, the maximum swing is

$$2 \times 2.27 = 4.54 \text{ V}$$

voltage at base is

$$\frac{4.54}{90.5} = 50 \text{ mV peak to peak (p} - \text{p)}$$

We still have to calculate the value of V_S to give V_B = 50 mV. We use the model

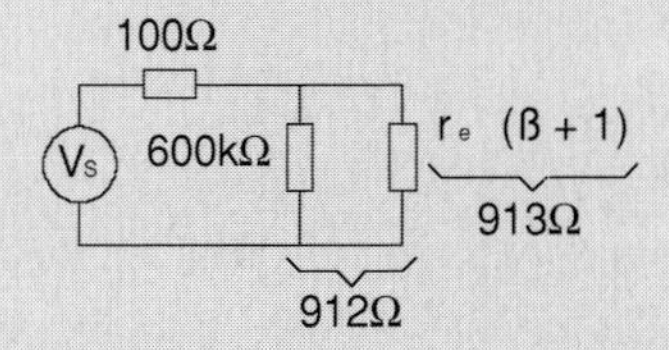

Figure 3.12

We have

$$v_B = 50 \times 10^{-3} = v_S \frac{912}{912 + 100}$$

$$v_S = 50 \times 10^{-3} \times \frac{1012}{912} = 55.48 \text{ mV p} - \text{p}$$

PROBLEM 3.2 LOAD LINES

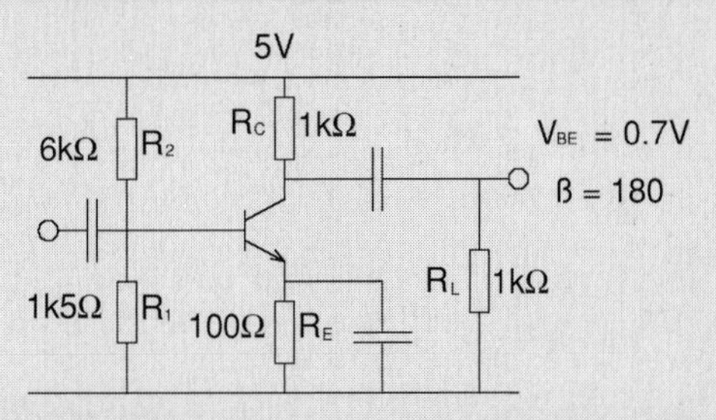

Figure 3.12

For the circuit shown, draw the load lines.

Answer: See Figure 3.14

DC conditions – approximate method

$$V_B = 5\frac{1.5}{7.5} = 1 \text{ V}$$

$$V_E = 1 - 0.7 = 0.3 \text{ V}$$

$$I_E = \frac{0.3}{100} = 3 \text{ mA}$$

$$I_{CQ} \approx 3 \text{ mA}$$

$$R_{DC} = R_C + R_E = 1100$$

$$R_{AC} = R_C \| R_L = 500$$

Note: on AC, R_E is bypassed by C_E.

$$V_{CEQ} = V_{CC} - I_{CQ} R_{DC}$$

$$= 5 - 3 \times 10^{-3} \times 1.1 \times 10^3$$

$$= 5 - 3.3 = 1.7 \text{ V}$$

DC conditions – Thevenin method

$$V_{TH} = 5\frac{1.5}{7.5} = 1 \text{ V}$$

$$R_{TH} = \frac{1\text{k}5 \times 6\text{k}}{1\text{k}5 + 6\text{k}} = 1\text{k}2$$

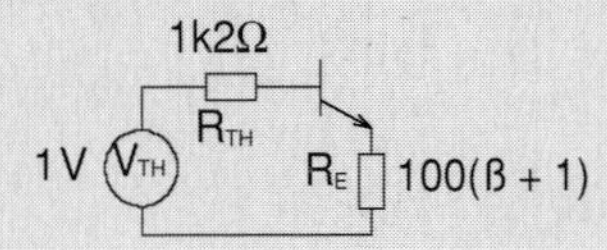

Figure 3.13

$$I_B = \frac{1-0.7}{1200+100\times 181} = \frac{0.3}{19300}$$

$$I_C = \frac{0.3\times 180}{19300} = 2.8\text{ mA}$$

$$I_{CQ} = 2.8\text{ mA}$$

$$V_{CEQ} = 5 - 2.8\text{ mA}\times 1\text{k}11 = 1.92\text{ V}$$

Values for plotting

$$V_{DCMAX} = V_{CC} = 5\text{ V}$$

$$I_{DCMAX} = \frac{V_{CC}}{R_{DC}} = \frac{5}{1\text{k}1} = 4.54\text{ mA}$$

$$v_{CEMAXAC} = V_{CEQ} + I_{CQ}R_{AC}$$

$$= 1.92 + 2.8\times 0.5 = 3.32\text{ V}$$

$$i_{CMAXAC} = I_{CQ} + \frac{V_{CEQ}}{R_{AC}}$$

$$= 2.8 + \frac{1.92}{0.5} = 6.64$$

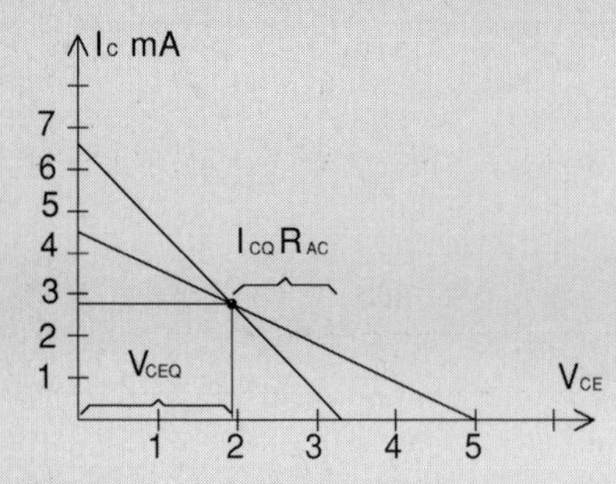

Figure 3.14

The maximum swing is 2 × 1.4 = 2.8 V

NOTE: The approximate method would have given 3 V as the maximum swing!

PROBLEM 3.3 LOAD LINES

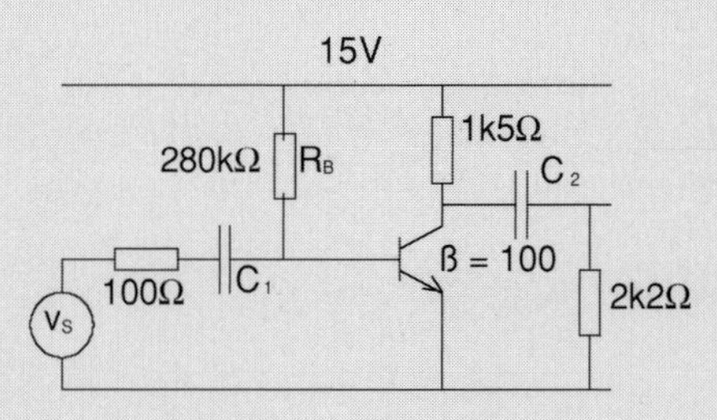

Figure 3.15

Find the maximum voltage swing on the 2k2 resistor.

Answer: 9.12 V

$$I_B = \frac{15-0.7}{280\text{k}} = 51.1\ \mu\text{A}$$

$$I_{CQ} = \beta I_B \times 100\times 10^{-6} = 5.11\text{ mA}$$

$$V_{CEQ} = 15 - I_C R_C = 7.34\text{ V}$$

The AC model is

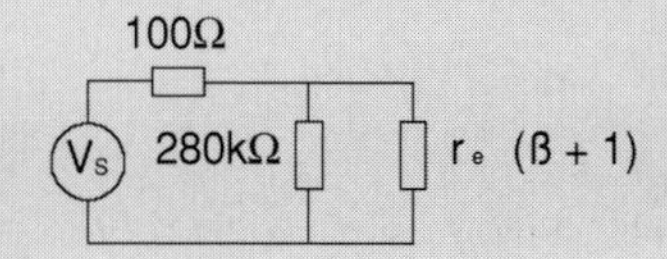

Figure 3.16

$$R_{AC} = 1\text{k}5 \| 2\text{k}2 = 892\ \Omega$$

Maximum values

$$V_{CEMAX} = 15\text{ V}$$

$$I_{CMAX} = \frac{15}{R_{DC}} = \frac{15}{1\text{k}5} = 10\text{ mA}$$

AC values

$$i_{CMAX} = I_{CQ} + \frac{V_{CEQ}}{R_{AC}} = 5.11 + \frac{7.34}{0.892}$$

$$= 5.11 + 8.42 = 13.53\text{ mA}$$

$$v_{CEMAX} = V_{CEQ} + I_{CQ}R_{AC}$$

$$= 7.34 + 4.56 = 11.9\text{ V}$$

Maximum swing

$$2\times 4.56 = 9.12\text{ V}$$

PROBLEM 3.4 LOAD LINES

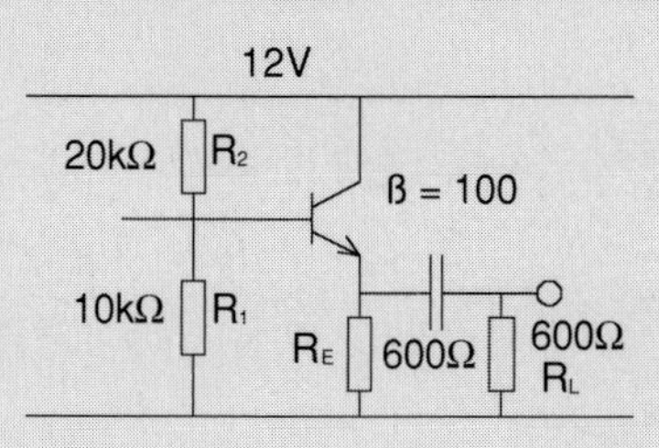

Figure 3.17

Find the Q point and the output voltage maximum swing.

Answer: I_{CQ} = 4.9 mA, maximum swing = 2.94 V

$$V_{TH} = 12\frac{R_1}{R_1 + R_2} = 12\frac{10k}{30k} = 4 \text{ V}$$

$$R_{TH} = R_1 \| R_2 = 6k66\ \Omega$$

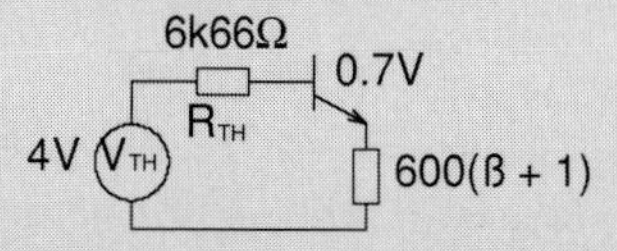

Figure 3.18

$$4 - 0.7 = I_B(6667 + R_E(\beta + 1))$$

$$= I_B(6667 + 60\,600)$$

$$I_B = \frac{3.3}{67\,267}$$

$$I_E = I_B(\beta + 1) = \frac{3.3 \times 101}{67\,267} = 4.95 \text{ mA}$$

$$I_{CQ} = I_E\frac{100}{101} = 4.9 \text{ mA}$$

$$R_{DC} = 600\ \Omega$$

$$R_{AC} = R_E \| R_L = 300\ \Omega$$

$$V_{CEQ} = V_{CC} - I_C R_C$$

$$= 12 - 4.90 \times 10^{-3} \times 0.6 \times 10^3$$

$$= 12 - 2.94 = 9.06 \text{ V}$$

Note: We have used $I_E = I_C$ here.

Values for plotting

$$V_{DCMAX} = V_{CC} = 12 \text{ V}$$

$$I_{DCMAX} = \frac{V_{CC}}{R_{DC}} = \frac{12}{600} = 20 \text{ mA}$$

$$v_{CEMAXAC} = V_{CEQ} + I_{CQ}R_{AC}$$

$$= 9.06 + 4.9 \times 0.3 = 10.53$$

$$i_{CMAXAC} = I_{CQ} + \frac{V_{CEQ}}{R_{AC}}$$

$$= 4.9 + \frac{9.06}{0.3} = 35.1 \text{ mA}$$

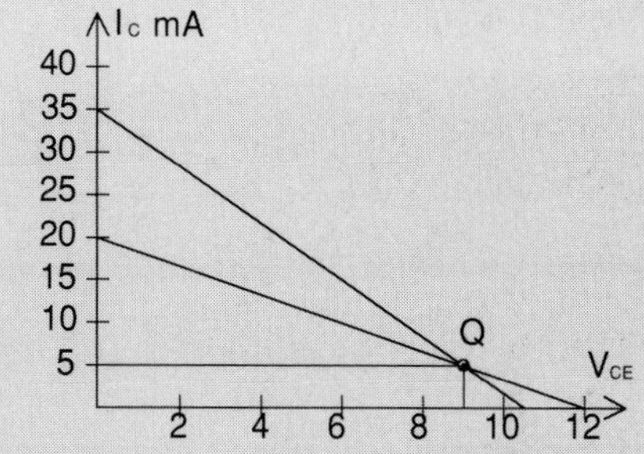

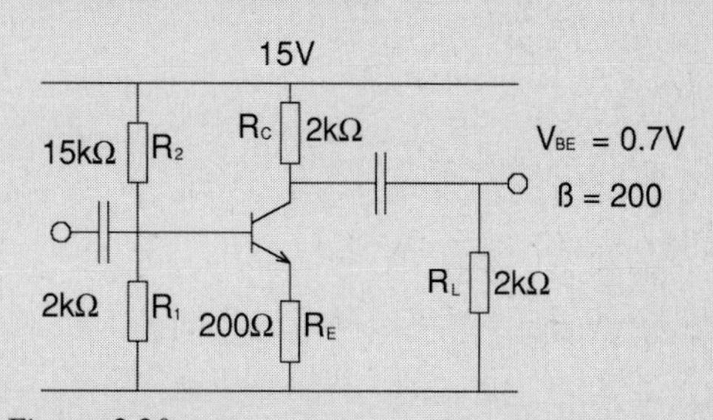

Figure 3.19

$$\text{Maximum swing} = 2I_{CQ}R_{AC}$$

$$= 2 \times 1.47$$

$$= 2.94 \text{ V}$$

PROBLEM 3.5 LOAD LINES

15V
15kΩ R2
Rc 2kΩ
VBE = 0.7V
ß = 200
RL 2kΩ
2kΩ R1
200Ω RE

Figure 3.20

Find the maximum swing, using the approximate method.

Answer: 6.592 V

$$V_B = 15\frac{2k}{17k} = 1.7647 \text{ V}$$

$$V_E = 1.0647 \text{ V}$$

$$I_E = \frac{1.0647}{200} = 5.32 \text{ mA}$$

$$I_{CQ} = 5.32 \text{ mA}$$

$R_{DC} = 2200\ \Omega$ $\qquad R_{AC} = 1200\ \Omega$

$$V_{CEQ} = 15 - 5.32 \times 2.2 = 3.296 \text{ V}$$

DC

$$I_{CMAX} = \frac{15}{2200} = 6.82 \text{ mA}$$

AC

$$i_{CMAX} = I_{CQ} + \frac{V_{CEQ}}{R_{AC}}$$

$$= 5.32 \times 10^{-3} + \frac{3.296}{1200}$$

$$= 5.32 + 2.747 = 8.067 \text{ mA}$$

Maximum swing $= V_{CEQ} \times 2 = 3.296 \times 2 = 6.592$ V

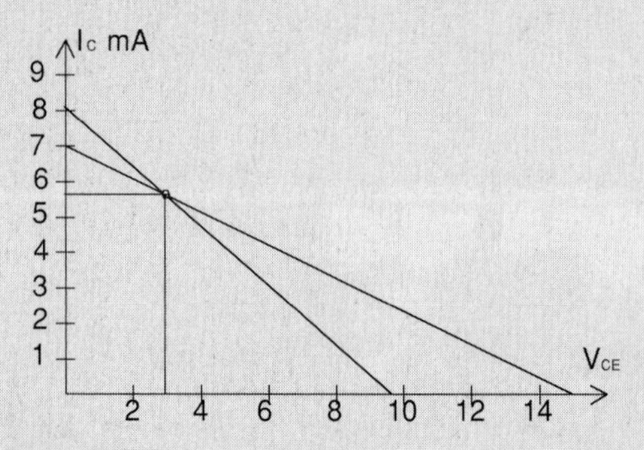

Figure 3.21

PROBLEM 3.6 LOAD LINES

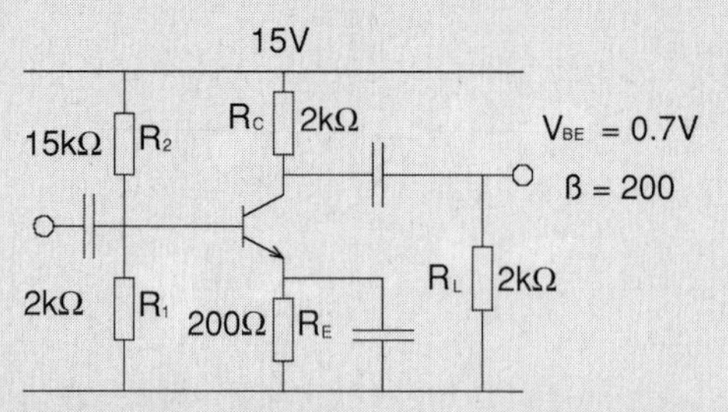

Figure 3.22

Find the peak to peak voltage output swing.

Answer: 7.68 V

$$V_{TH} = \frac{15 \times 2}{17} = 1.7647 \text{ V}$$

$$R_{TH} = \frac{2 \times 15k}{17} = 1.7647k\ \Omega$$

$$I_B = \frac{1.7647 - 0.7}{1764.7 + 40\,200} = \frac{1.0647}{41\,964.7}$$

$$I_{CQ} = \frac{1.0647 \times 200}{41\,964.7} = 5.074 \text{ mA}$$

$$V_{CEQ} = 15 - 5.074 \times 10^{-3} \times 2200 = 3.84 \text{ V}$$

Maximum swing either

$$2I_{CQ}R_{AC} = 2 \times 5.074 \times 10^{-3} \times 1k$$

$$= 10.148 \text{ V}$$

or

$$2V_{CEQ} = 2 \times 3.84 = 7.68 \text{ V}$$

Therefore, maximum swing = 7.68 V

NOTE: As seen in Figure 3.9 the right hand position of the Q point is $I_{CQ}R_{AC}$ which in this case is 5.074 V. The left hand side of the Q point is V_{CEQ} which in this case is 3.84 V.

The signal will swing to both sides of the Q point. That is why the maximum swing is either 10.148 V or 7.68 V.

We have to choose the smaller of the two because the signal, as it gets bigger and bigger, will reach the smaller value first, at one end. From then on, the signal will be clipped and that is not acceptable.

PROBLEM 3.7 LOAD LINES

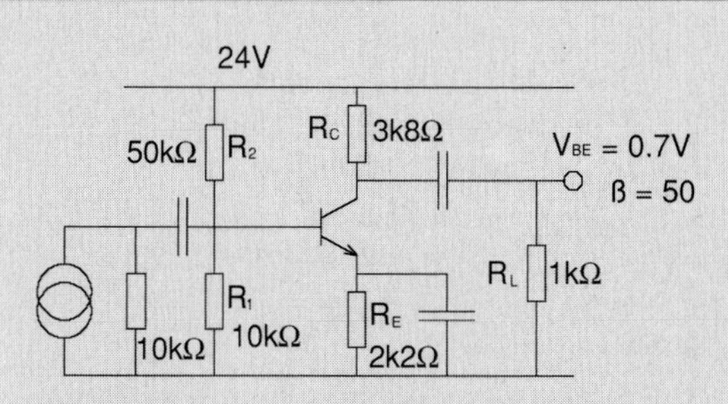

Figure 3.23

Draw the load lines, using the approximate method.

Answer: See Figure 3.24

$$V_B = 24\frac{10}{50+10} = 4\ \text{V}$$

$$V_E = 4 - 0.7 = 3.3\ \text{V}$$

$$I_E = \frac{3.3}{2k2} = 1.5\ \text{mA}$$

$$r_e = \frac{26}{1.5} = 17.33\ \Omega \qquad \text{(at } 30^\circ\text{C)}$$

$$R_{DC} = 6\ \text{k}\Omega$$

$$R_{AC} = 3k8 \| 1k = 792\ \Omega$$

$$V_{CEQ} = 24 - 1.5 \times 6 = 24 - 9 = 15\ \text{V}$$

Values for plotting

DC

$$V_{CEMAX} = 24\ \text{V}$$

$$I_{CMAX} = \frac{24}{6k} = 4\ \text{mA}$$

AC

$$i_{CEMAX} = V_{CEQ} + I_{CQ}R_{AC}$$

$$= 1.5 + \frac{15}{0.792} = 1.5 + 18.94$$

$$= 20.44\ \text{mA}$$

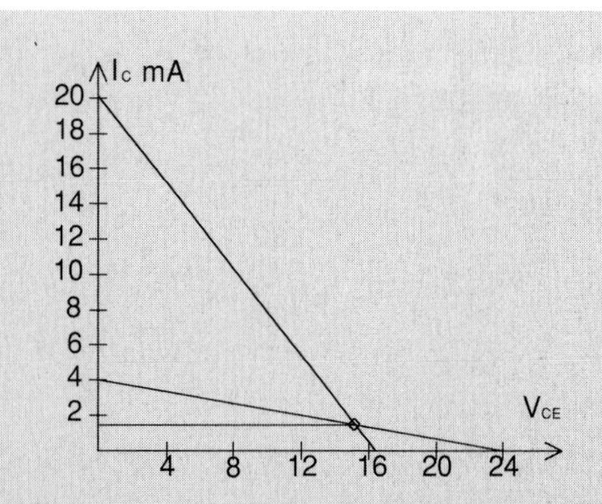

Figure 3.24

$$v_{CEMAX} = V_{CEQ} + I_{CQ}R_{AC}$$

$$= 15 + 1.5 \times 0.792 = 16.2\ \text{V}$$

Maximum swing $= 2 \times 1.2 = 2.4$ V

PROBLEM 3.8 LOAD LINES

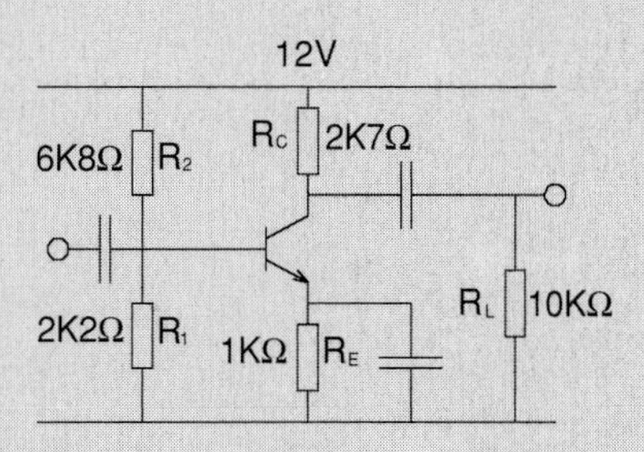

Figure 3.25

Draw the load lines and find the maximum swing on the load.

Answer: 7.48 V

Approximate method

$$V_B = 12\frac{2k2}{9k} = 2.933 \text{ V}$$

$$V_E = 2.933 - 0.7 = 2.233 \text{ V}$$

$$I_E = \frac{2.233}{1k} = 2.233 \text{ mA}$$

$$I_{CQ} = 2.233 \text{ mA}$$

$$V_{CEQ} = 12 - 2.233 \times 3.7 = 3.74 \text{ V}$$

$$R_{DC} = 2k7 + 1k = 3k7 \quad R_{AC} = 2k7 \| 10k = 2126\ \Omega$$

Values for plotting

DC

$$I_{CMAX} = \frac{12}{3k7} = 3.24 \text{ mA}$$

$$V_{CEMAX} = 12 \text{ V}$$

AC

$$v_{CEMAX} = V_{CEQ} + I_{CQ}R_{AC}$$
$$= 3.74 + 4.75 = 8.49 \text{ V}$$

$$i_{CMAX} = I_{CQ} + \frac{V_{CEQ}}{R_{AC}}$$
$$= 2.233 + 1.759 = 3.992 \text{ mA}$$

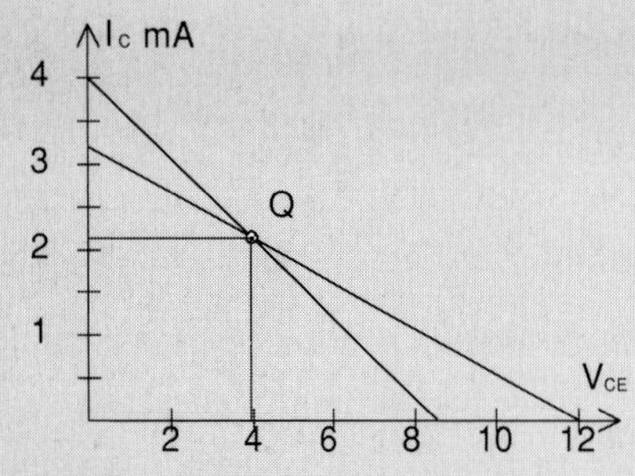

Figure 3.26

Maximum swing $= 3.74 \times 2$
$= 7.48$ V

PROBLEM 3.9 LOAD LINES

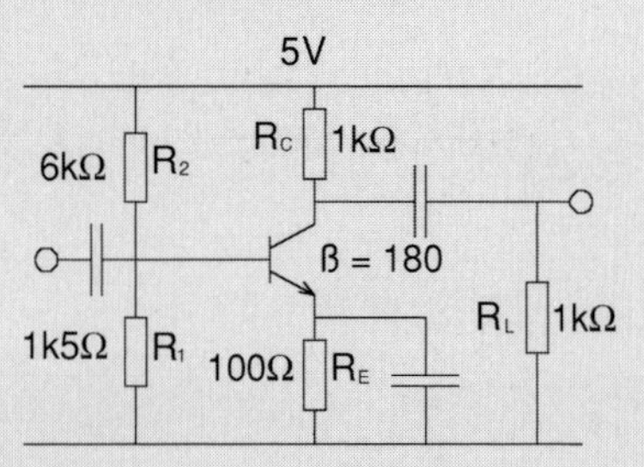

Figure 3.27

Find the Q-point.

Answer: 2.798 mA, 1.92 V

Using the Thevenin method

$$V_{TH} = 5\frac{R_1}{R_1 + R_2} = 5\frac{1.5}{7.5} = 1 \text{ V}$$

$$R_{TH} = \frac{R_1 R_2}{R_1 + R_2} = \frac{6 \times 1.5 \times 10^3}{7.5} = 1.2 \text{ k}\Omega$$

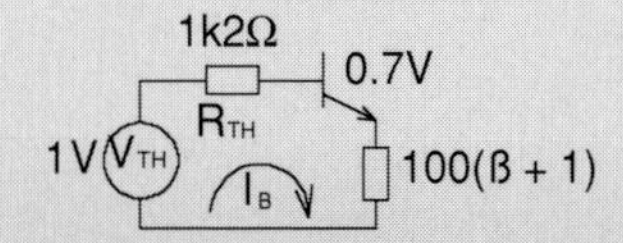

Figure 3.28

$$I_B = \frac{V_{TH} - V_{BE}}{1k2 + R_E(\beta + 1)} = \frac{0.3}{19300}$$

$$I_B(\beta + 1) = \frac{V_{TH} - V_{BE}}{\frac{1k2}{\beta + 1} + R_E}$$

$$I_C = \frac{0.3 \times 180}{19\,300} = 2.798 \text{ mA}$$

$$I_E = \frac{0.3 \times 181}{19\,300} = 2.813 \text{ mA}$$

$$V_{CE} = 5 - I_E R_E - I_C R_C = 5 - 0.2813 - 2.797$$
$$= 1.92 \text{ V}$$

PROBLEM 3.10 LOAD LINES

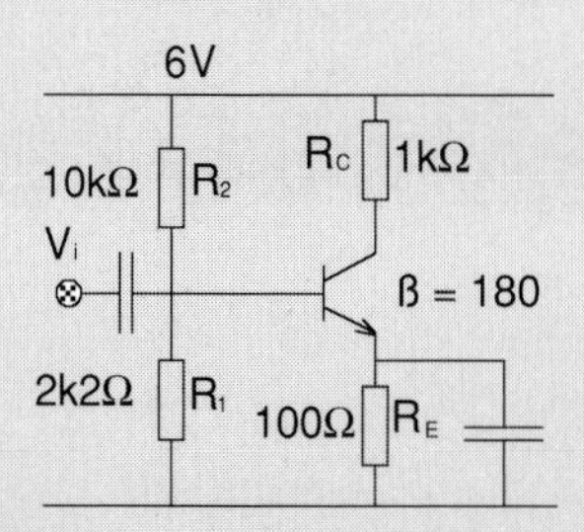

Figure 3.29

Sketch the load lines and find I_B, I_C, R_{AC} and R_{DC}.

> Answer: $I_B = 19.1\ \mu A$, $I_C = 3.44$ mA, $R_{AC} = 1k$, $R_{DC} = 1k1$

Use the Thevenin method, as 22k > 18k

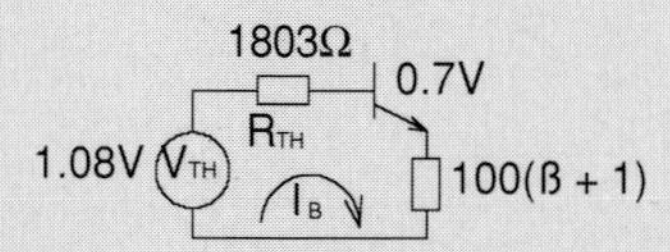

Figure 3.30

$$V_{TH} = 6\frac{2.2}{12.2} = 1.08\ \text{V}$$

$$R_{TH} = 2k2 \| 10k = 1803\ \Omega$$

$$I_B = \frac{1.08 - 0.7}{1803 + 100 \times 181}$$

$$= \frac{0.38}{19903} = 19.1\ \mu\text{A}$$

$$I_C = \beta I_B = 3.44\ \text{mA}$$

$$I_{CQ} = 3.44\ \text{mA}$$

$$V_{CEQ} = 6 - 3.44 \times 1.1 = 2.2\ \text{V}$$

Values for plotting

DC

$$V_{CEMAX} = V_{CC} = 6\ \text{V}$$

$$I_{CMAX} = \frac{V_{CC}}{R_{DC}} = \frac{6}{1.1} = 5.45\ \text{mA}$$

$$R_{DC} = 1k + 100 = 1100 \qquad R_{AC} = 1\ k\Omega$$

AC

$$v_{CEMAX} = V_{CEQ} + I_{CQ}R_{AC}$$

$$= 2.2 + 3.44 \times 1 = 5.64\ \text{V}$$

$$i_{CMAX} = I_{CQ} + \frac{V_{CEQ}}{R_{AC}}$$

$$= 3.44 + \frac{2.2}{1} = 5.64\ \text{mA}$$

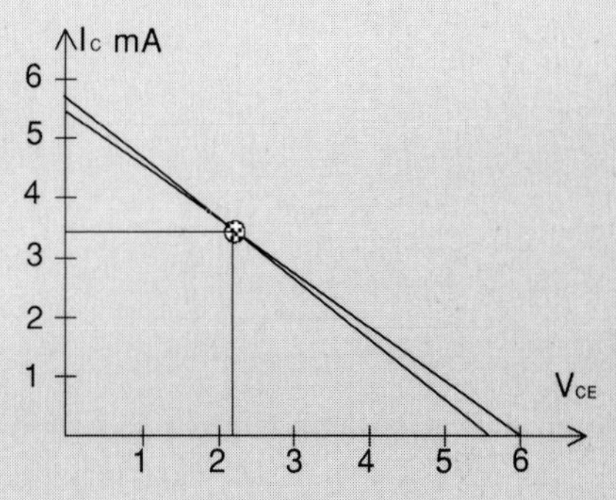

Figure 3.31

Maximum swing = $2V_{CEQ} = 2 \times 2.2 = 4.4$ V

PROBLEM 3.11 LOAD LINES

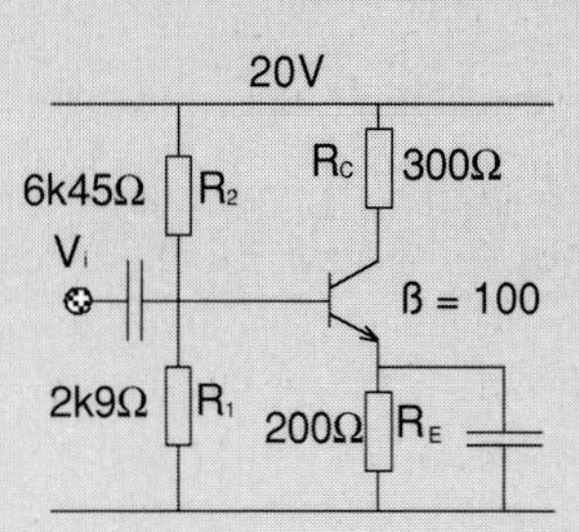

Figure 3.32

Using the approximate method, find the maximum swing.

> Answer: 12.5 V

$R_{DC} = 500\ \Omega,\quad R_{AC} = 300\ \Omega$

$$V_B = 20\frac{2.9}{9.35} = 6.2\ \text{V}$$

$$V_E = V_B - V_{BE} = 6.2 - 0.7 = 5.5\ \text{V}$$

$$I_E \approx I_C = \frac{5.5}{200} = 27.5\ \text{mA}$$

$$V_{CEQ} = 20 - 27.5\times10^{-3}\times500$$
$$= 20 - 13.75 = 6.25\ \text{V}$$

Values for plotting

DC

$$V_{CEMAX} = 20\ \text{V}$$

$$I_{CMAX} = \frac{20}{500} = 40\ \text{mA}$$

AC

$$v_{CEMAX} = V_{CEQ} + I_{CQ}R_{AC}$$
$$= 6.25 + 27.5\times10^{-3}\times300$$
$$= 6.25 + 8.25 = 14.5$$

$$i_{CMAX} = I_{CQ} + \frac{V_{CEQ}}{R_{AC}}$$
$$= 27.5\times10^{-3} + \frac{6.25}{300}$$
$$= (27.5 + 20.83)\times10^{-3} = 48.33\ \text{mA}$$

Maximum swing $= 2V_{CEQ} = 2\times6.25 = 12.5$ V

Note: Using the Thevenin method, you would get a maximum swing of around 14.86 V.

PROBLEM 3.12 LOAD LINES

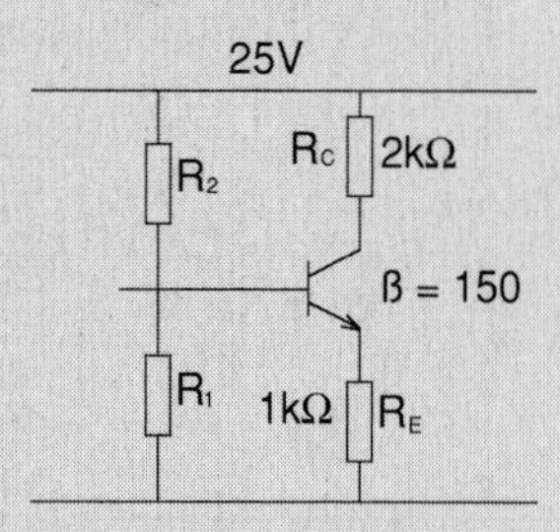

Figure 3.33

Find R_1 and R_2 to place the I_{CQ} in the middle of the DC load line.

> Answer: $R_1 = 10\ \text{k}\Omega$, $R_2 = 41\,377\ \Omega$

$$I_{CMAX} = \frac{25}{3\text{k}} = 8.333\ \text{mA}$$

Half $I_C = 4.166$ mA

$V_E = 4.1666$ V

$V_B = 4.8666$ V

$$\frac{25R_1}{R_1 + R_2} = 4.866$$

$$\frac{R_1}{R_1 + R_2} = \frac{4.866}{25}$$

$R_2 = 20.134 \qquad R_1 = 4.866$

If R_1 is 10 kΩ, then

$$\frac{10\text{k}}{10\text{k} + R_2} = \frac{4.866}{25}$$

$R_2 = 41377\ \Omega$

This is only one of the possible solutions.

PROBLEM 3.13 LOAD LINES

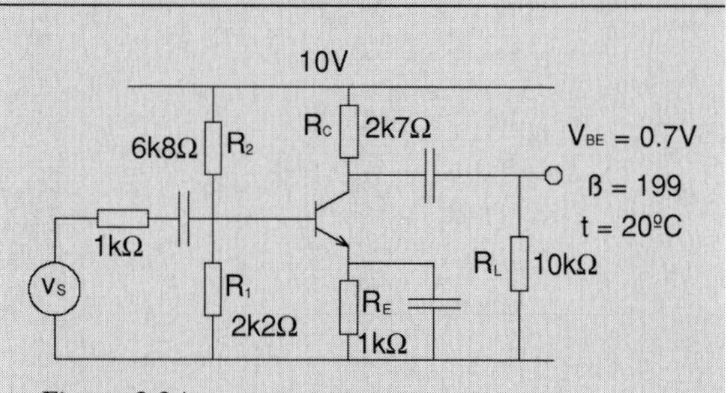

Figure 3.34

Draw the load lines and find the maximum swing on the load.

> Answer: 7.1 V

Approximate method

$$V_B = 10\frac{2\text{k}2}{9\text{k}} = 2.444\ \text{V}$$

$$V_E = 2.444 - 0.7 = 1.744\ \text{V}$$

$$I_E = \frac{1.744}{1\text{k}} = 1.744\ \text{mA}$$

$$r_e = \frac{25}{1.744} = 14.33\ \Omega$$

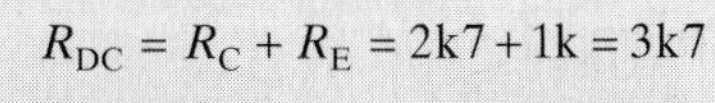

$$R_{DC} = R_C + R_E = 2k7 + 1k = 3k7$$

AC conditions

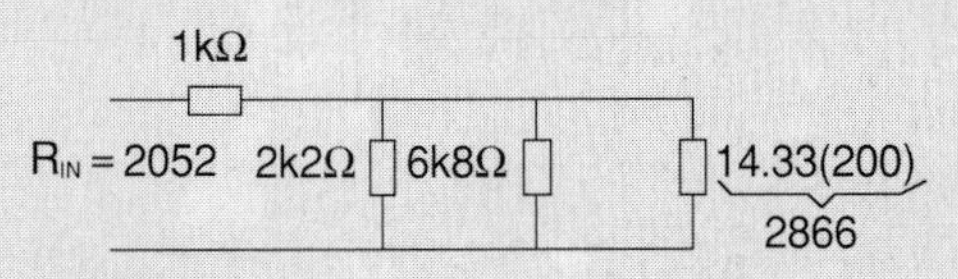

Figure 3.35

$$R_{AC} = 2k7 \| 10k = 2126\ \Omega$$

Values for plotting

$$V_{CEQ} = 10 - 1.744 \times 10^{-3} \times 3k7 = 3.55\ \text{V}$$

$$I_{CQ} \approx I_E = 1.744\ \text{mA}$$

DC

$$I_{CMAX} = \frac{10}{3k7} = 2.703\ \text{mA}$$

$$V_{CEMAX} = V_{CC} = 10\ \text{V}$$

AC

$$\begin{aligned} v_{CEMAX} &= V_{CEQ} + I_{EQ}R_{AC} \\ &= 3.55 + 2126 \times 1.744 \times 10^{-3} \\ &= 3.55 + 3.71 = 7.26\ \text{V} \end{aligned}$$

$$\begin{aligned} i_{CMAX} &= I_{EQ} + \frac{V_{CEQ}}{R_{AC}} \\ &= 1.744 + \frac{3.55}{2.126} = 1.744 + 1.67 \\ &= 3.414\ \text{mA} \end{aligned}$$

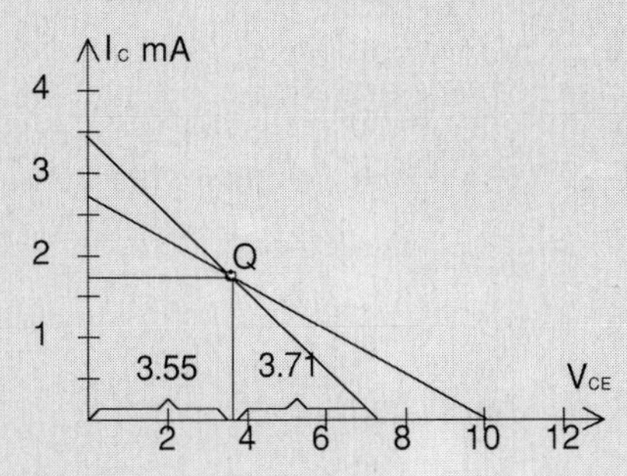

Figure 3.36

Maximum swing is 2 × V_{CEQ} = 2 × 3.55 = 7.1 V

PROBLEM 3.14 LOAD LINES

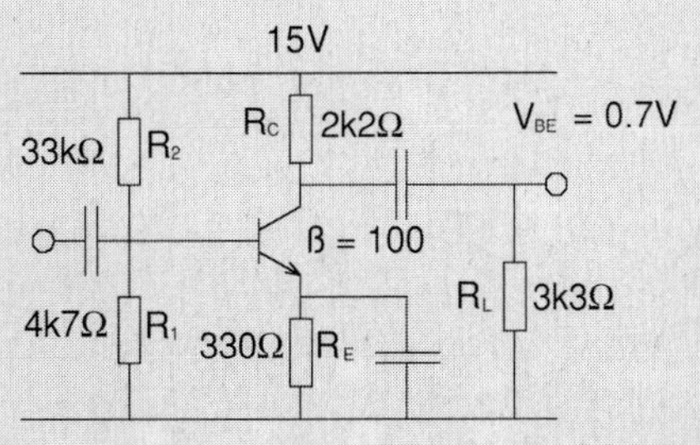

Figure 3.37

Sketch the load line diagram and find the maximum swing on the load.

Answer: swing = 8.34 V

$10R_1$ vs βR_E

47k vs 33k

Therefore, use Thevenin method

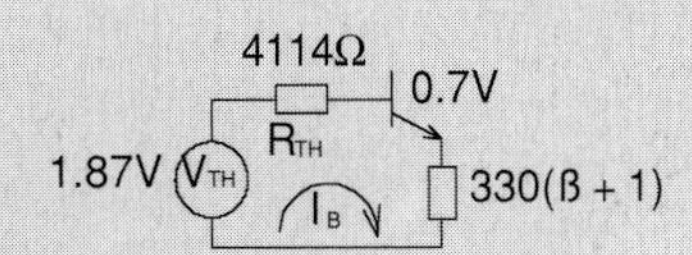

Figure 3.38

$$V_{TH} = 15\frac{4.7}{4.7 + 33} = 1.87\ \text{V}$$

$$R_{TH} = 4k7 \| 33k = 4144\ \Omega$$

$$\begin{aligned} I_B &= \frac{1.87 - 0.7}{4114 + 330(101)} \\ &= \frac{1.17}{37444} = 31.25\ \mu\text{A} \end{aligned}$$

$$I_{EQ} = \frac{1.17 \times 101}{37444} = 3.156\ \text{mA} \approx I_{CQ}$$

$$R_{DC} = 2k2 + 330 = 2530\ \Omega$$

$$R_{AC} = 2k2 \| 3k3 = 1320\ \Omega$$

$$V_{CEQ} = 15 - I_{EQ}R_{DC} = 15 - 3.156 \times 2.53 = 7.02\ \text{V}$$

Values for plotting

DC

$$V_{CEMAX} = V_{CC} = 15 \text{ V}$$

$$I_{CMAX} = \frac{V_{CC}}{R_{DC}} = \frac{15}{2350} = 5.93 \text{ mA}$$

AC

$$v_{CEMAX} = V_{CEQ} + I_{CQ}R_{AC}$$

$$= 7.02 + 3.156 \times 1.32 = 7.02 + 4.17$$

$$= 11.19 \text{ V}$$

$$i_{CMAX} = I_{CQ} + \frac{V_{CEQ}}{R_{AC}}$$

$$= 3.156 + \frac{7.02}{1.32} = 3.156 + 5.32$$

$$= 8.48 \text{ mA}$$

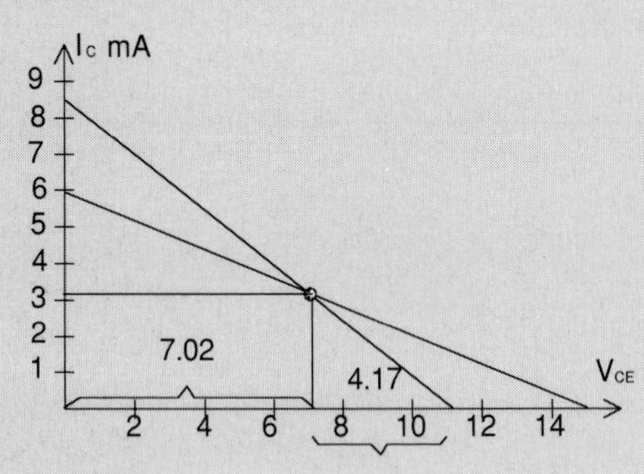

Figure 3.39

Maximum swing is therefore 2 × 4.17 = 8.34 V

PROBLEM 3.15 LOAD LINES

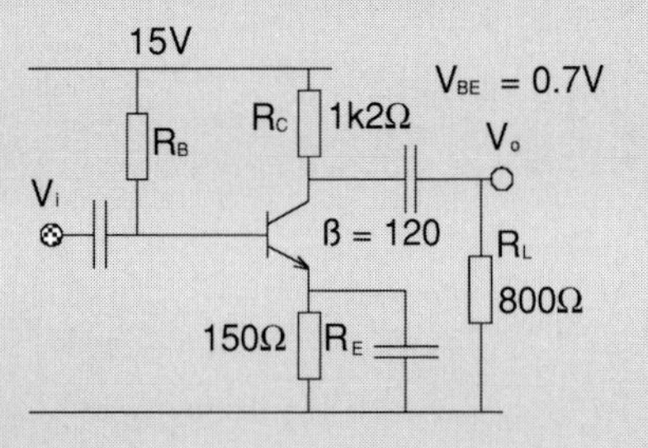

Figure 3.40

Find the value of V_{CEQ}, the maximum swing, and draw the load lines if I_B = 50 μA.

Answer: V_{CEQ} = 6.9 V, maximum swing = 5.76 V

$$R_{DC} = 1200 + 150 = 1350 \ \Omega$$

$$R_{AC} = 1\text{k}2 \| 800 = 480 \ \Omega$$

$$I_C = \beta I_B = 120 \times 50 \times 10^{-6} = 6 \text{ mA}$$

(Assuming $I_C \approx I_E$)

$$V_{CEQ} = V_{CC} - I_C R_{DC}$$

$$= 15 - 6 \times 10^{-3} \times 1350$$

$$= 15 - 8.1 = 6.9 \text{ V}$$

Plotting values

DC

$$V_{CEMAX} = V_{CC} = 15 \text{ V}$$

$$I_{CMAX} = \frac{V_{CC}}{R_{DC}} = \frac{15}{1350} = 11.11 \text{ mA}$$

AC

$$v_{CEQMAX} = V_{CEQ} + I_{CQ}R_{AC}$$

$$= 6.9 + +6.0 \times 0.48$$

$$= 6.9 + 2.88 = 9.78 \text{ V}$$

$$i_{CMAX} = I_{CQ} + \frac{V_{CEQ}}{R_{AC}}$$

$$= 6 + \frac{6.9}{0.480}$$

$$= 6 + 14.38 = 20.38 \text{ mA}$$

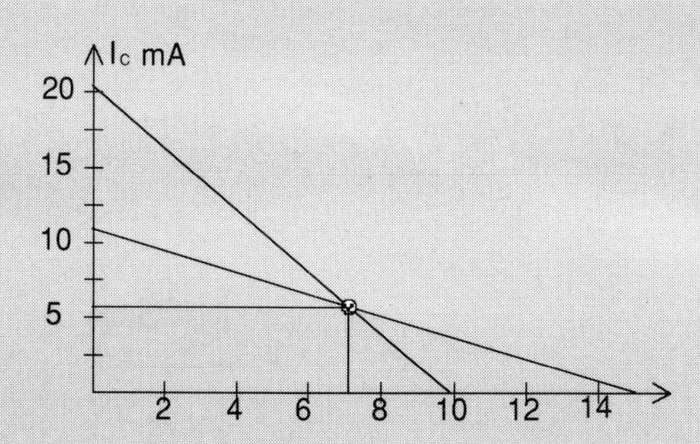

Figure 3.41

Maximum swing in this case is $2I_{CQ}R_{AC}$.

$= 2 \times 6 \times 0.48 = 2 \times 2.88$

$= 5.76$ V

PROBLEM 3.16 LOAD LINES

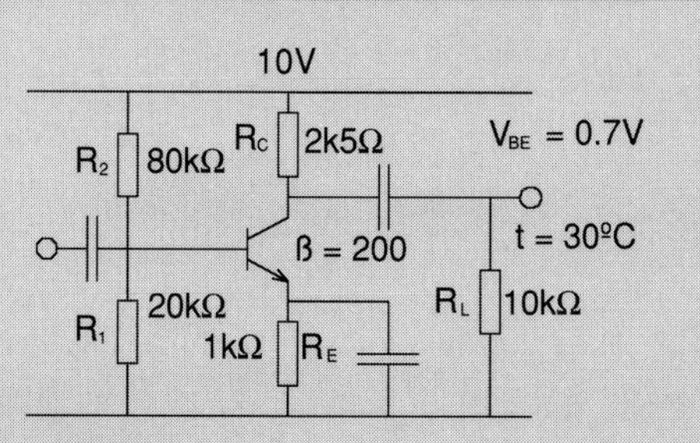

Figure 3.42

Draw the load lines, and find the maximum swing on the load.

Answer: 5.2 V

DC conditions – approximate method

$$V_B = 10\frac{20}{20+80} = 2.0 \text{ V}$$

$$V_E = 2.0 - 0.7 = 1.3 \text{ V}$$

$$I_E = \frac{1.3}{1\text{k}} = 1.3 \text{ mA} \approx I_C$$

$$R_{DC} = 2\text{k}5 + 1\text{k} = 3\text{k}5$$

$$R_{AC} = 2\text{k}5 \| 10\text{k} = 2 \text{ k}\Omega$$

$$V_{CEQ} = 10 - I_E \times 3\text{k}5$$

$$= 10 - 1.3 \times 3.5 = 10 - 4.55$$

$$= 5.45 \text{ V}$$

Values for plotting

DC

$$V_{CEMAX} = V_{CC} = 10 \text{ V}$$

$$I_{CMAX} = \frac{V_{CC}}{R_{DC}} = \frac{10}{3\text{k}5} = 2.86 \text{ mA}$$

AC

$$v_{CEMAX} = V_{CEQ} + I_{CQ}R_{AC}$$

$$= 5.45 + 1.3 \times 2 = 5.45 + 2.6$$

$$= 8.05 \text{ V}$$

$$i_{CMAX} = I_{CQ} + \frac{V_{CEQ}}{R_{AC}}$$

$$= 1.3 + \frac{5.45}{2} = 4.025 \text{ mA}$$

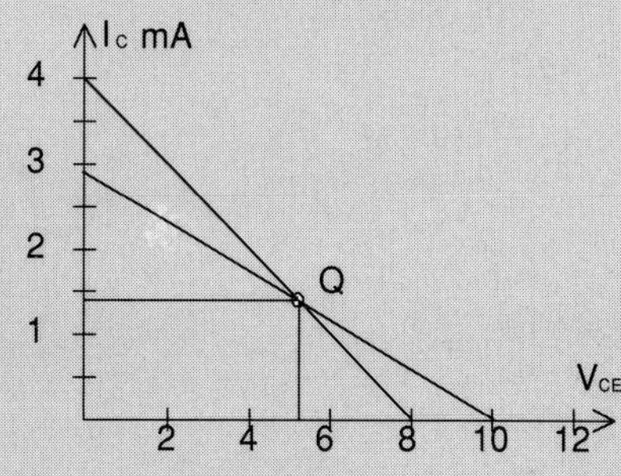

Figure 3.43

Maximum swing = $2I_{CQ}R_{AC}$

$= 2 \times 2.6 = 5.2$ V

PROBLEM 3.17 LOAD LINES

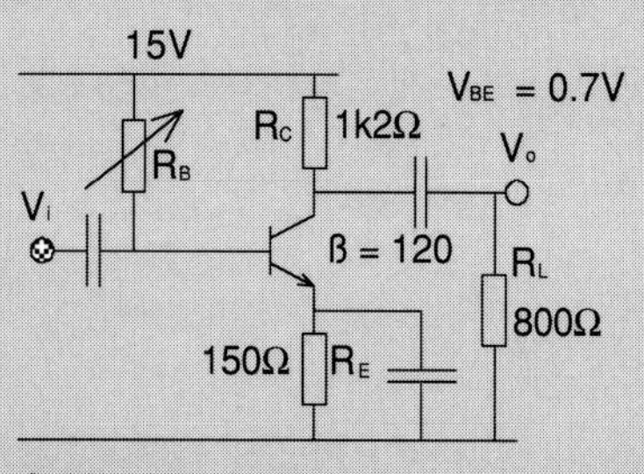

Figure 3.44

Draw the load lines and find the value of I_B that will cause the optimum maximum swing on the load.

Answer: $I_B = 68.33$ μA, swing = 7.87 V

For optimum maximum output swing

$$I_{CQ} = \frac{V_{CC}}{R_{DC} + R_{AC}}$$

$$R_{AC} = 1k2 || 800 = 480\ \Omega$$

$$R_{DC} = 1200 + 150 = 1350\ \Omega$$

$$I_{CQ} = \frac{15}{480 + 1350} = \frac{15}{1830} = 8.2\ \text{mA}$$

$$V_{CEQ} = V_{CC} - I_{CQ}R_{DC}$$

$$= 15 - 8.2 \times 1.35 = 3.93\ \text{V}$$

Plotting values

DC

$$V_{CEMAX} = V_{CC} = 15\ \text{V}$$

$$I_{CMAX} = \frac{V_{CC}}{R_{DC}} = \frac{15}{1350} = 11.11\ \text{mA}$$

AC

$$v_{CEMAX} = V_{CEQ} + I_{CQ}R_{AC}$$

$$= 3.93 + 8.2 \times 0.48$$

$$= 3.93 + 3.94$$

$$= 7.87\ \text{V}$$

$$i_{CMAX} = I_{CQ} + \frac{V_{CEQ}}{R_{AC}} = 8.2 + \frac{3.93}{0.48}$$

$$= 8.2 + 8.19$$

$$= 16.39\ \text{mA}$$

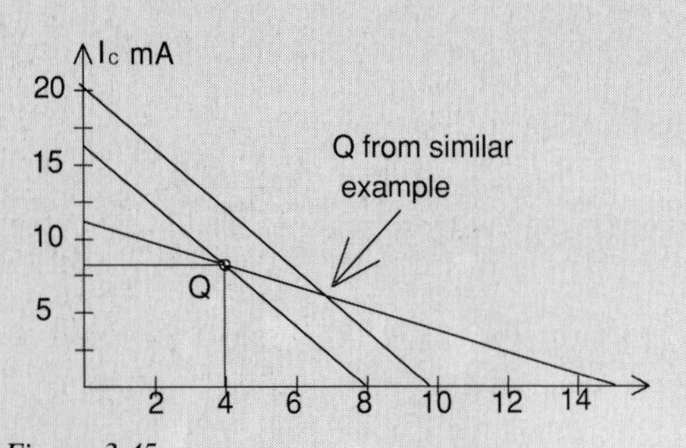

Figure 3.45

Maximum swing is 7.87 V.

Value of I_B.

$$I_B = \frac{I_{CQ}}{\beta} = \frac{8.2 \times 10^{-3}}{120} = 68.33\ \mu\text{A}$$

The location of Q in the DC load line can be moved by varying R_B. This in turn varies I_B which, in turn, varies I_C. We have worked the example to achieve maximum optimum equal swing. The values of 3.93 and 3.94 should be the same. The difference is due to rounding off in the calculation.

PROBLEM 3.18 LOAD LINES

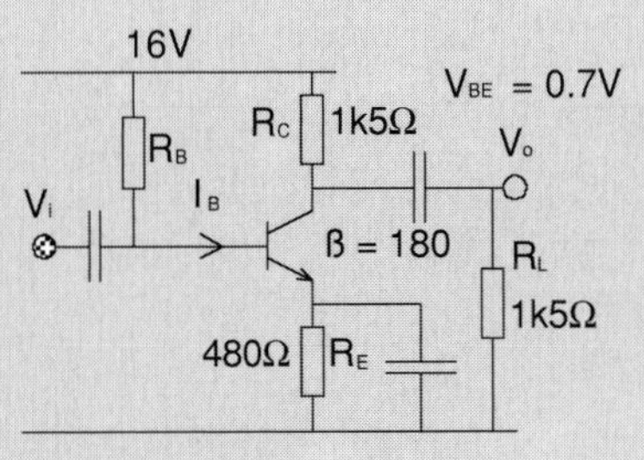

Figure 3.46

Find the values of I_B, R_B, and swing that will place the Q-point at the centre of the AC load line.

> Answer: $I_B = 32.55\ \mu\text{A}$, $R_B = 383\ 717\ \Omega$, swing = 8.8 V

$$R_{DC} = 480 + 1k5 = 1980\ \Omega$$

$$R_{AC} = 1k5 || 1k5 = 750\ \Omega$$

$$I_{CQ} = \frac{V_{CC}}{R_{DC} + R_{AC}} = \frac{16}{1980 + 750} = 5.86\ \text{mA}$$

$$V_{CEQ} = 16 - I_{CQ}R_{DC} = 16 - 5.86 \times 1.98$$

$$= 16 - 11.6 = 4.4\ \text{V}$$

Plotting values

DC

$$V_{CEMAX} = V_{CC} = 16\ \text{V}$$

$$I_{CMAX} = \frac{V_{CC}}{R_{DC}} = \frac{16}{1980} = 8.08\ \text{mA}$$

AC

$$v_{CEMAX} = V_{CEQ} + I_{CQ}R_{AC}$$

$$= 4.4 + 5.86 \times 0.75$$

$$= 4.4 + 4.4 = 8.8\ \text{V}$$

$$i_{CMAX} = I_{CQMAX} + \frac{V_{CEQ}}{R_{AC}}$$

$$= 5.86 + \frac{4.4}{0.75} = 5.86 + 5.86$$
$$= 11.72 \text{ mA}$$

Maximum swing $= 2 \times 4.4 = 8.8$ V

(approximation $I_C \approx I_E$)

$$V_E = I_E R_E$$
$$= 5.86 \times 0.480 = 2.81 \text{ V}$$
$$V_B = V_E + 0.7 \text{ V}$$
$$= 2.81 + 0.7 = 3.51 \text{ V}$$
$$I_B = \frac{I_C}{\beta} = \frac{5.86}{180}$$
$$= 32.55\ \mu\text{A}$$
$$V_{CC} - V_B = I_B R_B$$
$$16 - 3.51 = 32.55 \times 10^{-6} \times R_B$$
$$12.49 = 32.55 \times 10^{-6} \times R_B$$
$$R_B = \frac{12.49 \times 10^6}{32.55} = 383\,717\ \Omega$$

PROBLEM 3.19 LOAD LINES

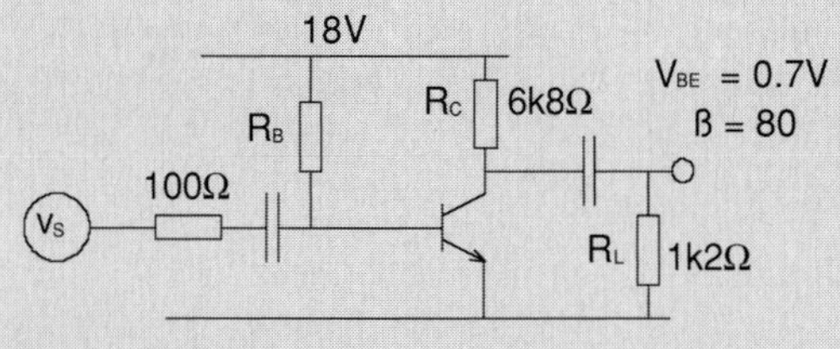

Figure 3.47

Find the value of R_B that results in the maximum output swing.

Answer: 609 kΩ

For maximum output, $I_{CQ} = \dfrac{V_{CC}}{R_{DC} + R_{AC}}$

$$R_{DC} = 6k8 \qquad R_{AC} = 6k8 \| 1k2 = 1020\ \Omega$$
$$I_{CQ} = \frac{18}{6k8 + 1020} = \frac{18}{7820} = 2.3 \text{ mA}$$
$$V_{CEQ} = 18 - I_{CQ} \times 6k8 = 18 - 15.64 = 2.36 \text{ V}$$
$$I_B = \frac{I_{CQ}}{\beta + 1} = \frac{2.3 \times 10^{-3}}{81} = 28.4\ \mu\text{A}$$
$$18 - 0.7 = I_B R_B$$
$$R_B = \frac{17.3}{28.4 \times 10^{-6}} = 609 \text{ k}\Omega$$

We draw the load lines to see what happened:

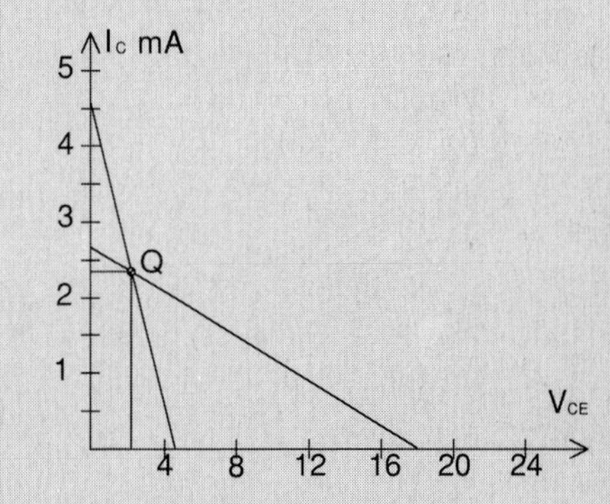

Figure 3.48

$$I_{CMAX} = \frac{18}{6k8} = 2.65 \text{ mA}$$
$$V_{CEMAX} = 18 \text{ V}$$
$$i_{CMAX} = I_{CQ} + \frac{V_{CEQ}}{R_{AC}}$$
$$= 2.36 + 2.35 = 4.71 \text{ mA}$$

Note: values should be equal. We have gained a small error due to rounding off in the calculation.

PROBLEM 3.20 LOAD LINES

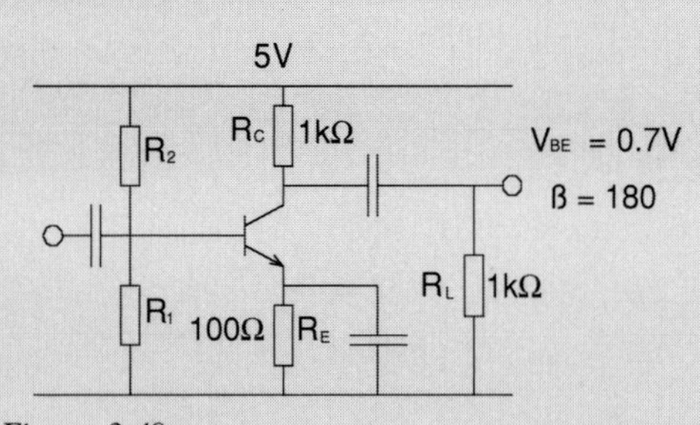

Figure 3.49

Find the values of R_1 and R_2 to achieve optimum maximum output voltage swing.

Answer: R_1 = 2k7 Ω, R_2 = 10 633 Ω

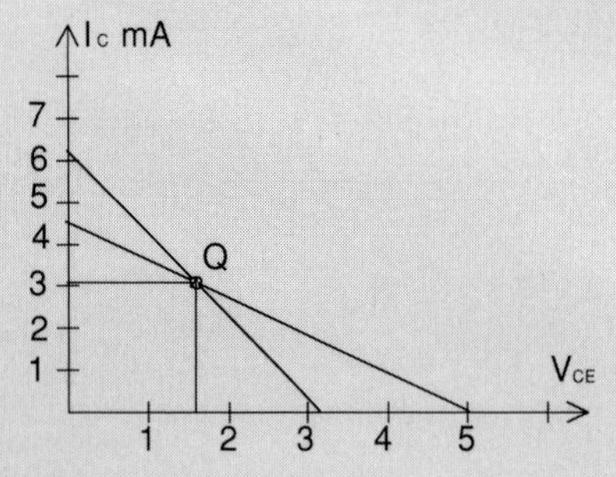

Figure 3.50

Maximum po int s

DC

$$V_{CEMAX} = V_{CC} = 5\text{ V}$$

$$I_{CMAX} = \frac{V_{CC}}{R_{DC}} = \frac{5}{1k1} = 4.5454\ \Omega$$

For optimum maximum output swing

$$I_{CQ} = \frac{V_{CC}}{R_{DC} + R_{AC}} = \frac{5}{1100 + 500} = 3.125\text{ mA}$$

$$V_{CEQ} = V_{CC}\frac{R_{AC}}{R_{DC} + R_{AC}} = 5\frac{500}{1600} = 1.562\,5\text{ V}$$

Maximum points

AC

$$i_{CMAX} = I_{CQ} + \frac{V_{CEQ}}{R_{AC}}$$

$$= 3.125 \times 10^{-3} + \frac{1.562\,5}{500}$$

$$3.125 \times 10^{-3} + 3.125 \times 10^{-3}$$

$$= 6.25\text{ mA}$$

Note that both terms are the same

AC

$$v_{CEMAX} = V_{CEQ} + I_{CQ}R_{AC}$$

$$= 1.562\,5 + 1.562\,5$$

$$= 3.125\text{ V}$$

Having found the Q-point successfully to achieve maximum equal swing, we now need to find the values of R_1 and R_2 that will produce that Q-point.

Solving by the approximate method:

$$V_E = 100 \times 3.125 \times 10^{-3}$$

$$= 0.312\,5\text{ V}$$

$$V_B = 1.012\,5\text{ V}$$

The balance criteria used before to decide between the Thevenin and the approximate method was

$$10R_1 = \beta R_E$$

Now to be on the safe side with the approximate method we add an extra 50%, that is to say:

$$10R_1 = 1.5\beta R_E$$

This new criteria will allow us to find the value of R_1, making the solution of the problem easier.

Solving the problem, we will arrive at the ratioR_1:R_2, but not at individual values. The above criteria reduces the problem.

$$10R_1 = 1.5\beta R_E$$

$$R_1 = 0.15 \times 180 \times 100$$

$$= 2k7$$

$$V_B = V_{CC}\frac{R_1}{R_1 + R_2}$$

$$1.0125 = 5\frac{2700}{2700 + R_2}$$

$$2733.75 + 1.012\,5R_2 = 13\,500$$

$$1.012\,5R_2 = 107\,66.25$$

$$R_2 = 10\,633\ \Omega$$

4

Transistor modelling

The transistor works with two types of signals, the DC signal and the AC signal.

The DC signal sets up the transistor at a convenient point within the working region, which is known as the quiescent point or simply the Q point. The AC signal works on top of the DC signal, but it is not superimposed. The AC signal follows a different path from the DC signal, as there are places where the AC signal can go, which are not accessible to the DC.

Because of this, it is better to completely separate the DC conditions which we call biasing, from the AC conditions which we call small signal conditions.

The DC analysis is straightforward and requires only a knowledge of circuit analysis, Ohm's law and Kirchhoff's laws. For the AC analysis we need two more steps:

- Transistor models
- Small signal models

Transistor model

In order to find out how we get a model of a transistor it is best to observe the following procedure from a general study of two-port networks in the order given:

- Open circuit parameters
- Short circuit parameters
- Hybrid parameters

Open circuit parameters

It is standard procedure in two-port network study to draw the currents going into the box, and the voltages pointing upwards, as in Figure 4.1. We shall do the same.

From mesh analysis we know that any passive network can be represented by the following equations

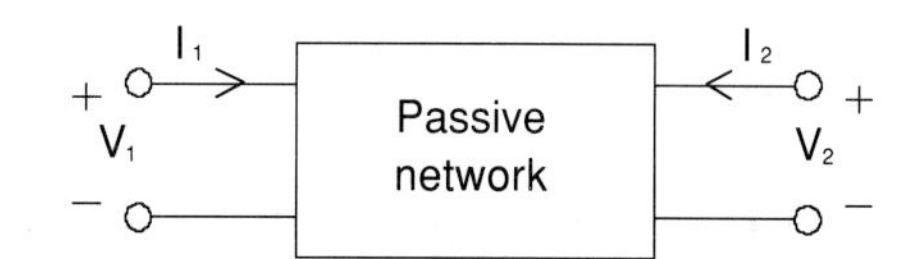

Figure 4.1

$$V_1 = I_1Z_{11} + I_2Z_{12}$$

$$V_2 = I_1Z_{21} + I_2Z_{22}$$

This lends itself to be represented in matrix form by

$$\begin{bmatrix} V_1 \\ V_2 \end{bmatrix} = \begin{bmatrix} Z_{11} & Z_{12} \\ Z_{21} & Z_{22} \end{bmatrix} \begin{bmatrix} I_1 \\ I_2 \end{bmatrix}$$

All we now need to do is to find the values of Z_{11}, Z_{12}, Z_{21} and Z_{22}.

Before we do that, we can relate the previous equation to an equivalent circuit, term by term. In Figure 4.2, Z_{11} is an impedance that produces a voltage I_1Z_{11}. $Z_{12}I_2$ is a current-dependent voltage generator. A voltage is produced in circuit 1 due to the current in circuit 2 and it is proportional to the current in circuit 2. We can now write:

$$V_1 = I_1Z_{11} + I_2Z_{12}$$

Similarly it can be found that the second part of the circuit corresponds to the second equation, i.e.

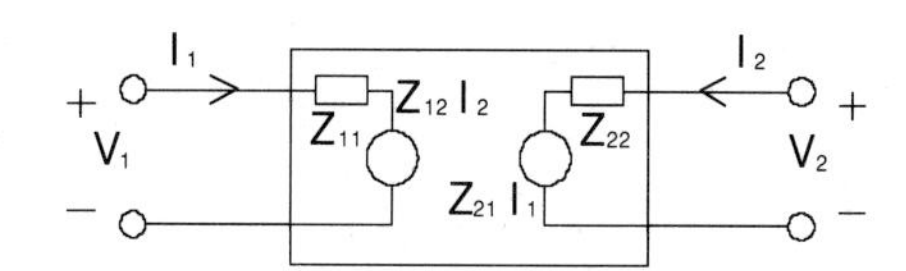

Figure 4.2

$$V_2 = I_1Z_{21} + I_2Z_{22}$$

Now that we have found this equivalent circuit, which is of the Thevenin type, we can go further to try to find the values of Z_{11}, Z_{12}, Z_{21} and Z_{22}.

In order to do this, we open circuit one of the terminals. Let us say that we make $I_2 = 0$. Because of this, the first equation will become:

$$V_1 = I_1Z_{11} \qquad (\text{as } I_2 = 0)$$

so

$$\boxed{Z_{11} = \frac{V_1}{I_1}}$$

The second equation becomes :

$$V_2 = I_1Z_{21}$$

so

$$\boxed{Z_{21} = \frac{V_2}{I_1}}$$

We can follow a similar procedure with the other circuit.

We make $I_1 = 0$. In this case, the first equation becomes:

$$V_1 = I_2Z_{12} \qquad (\text{as } I_1 = 0)$$

so

$$\boxed{Z_{12} = \frac{V_1}{I_2}}$$

The second equation becomes :

$$V_2 = I_2Z_{22}$$

so

$$\boxed{Z_{22} = \frac{V_2}{I_2}}$$

Because we have made the currents equal to zero by an open circuit to the terminal, the Z parameters are called open circuit parameters.

A similar definition of the two-port network can be carried out with the Y parameters. We are still far from the transistor model, but are getting nearer.

Short circuit parameters

We again draw the general circuit of a passive network in Figure 4.3.

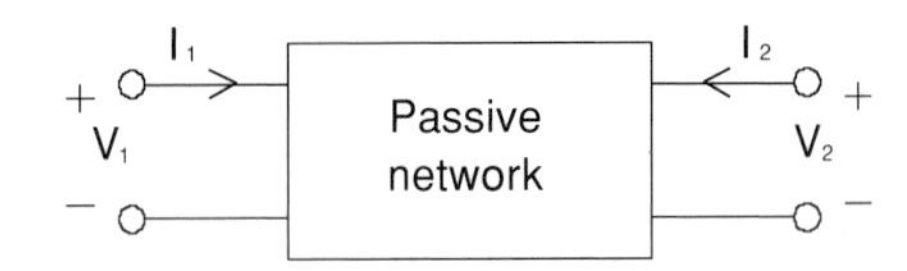

Figure 4.3

From nodal analysis we know that any passive network can be represented by the following equations:

$$I_1 = V_1Y_{11} + V_2Y_{12}$$

$$I_2 = V_1Y_{21} + V_2Y_{22}$$

This can also be represented in matrix form by:

$$\begin{bmatrix} I_1 \\ I_2 \end{bmatrix} = \begin{bmatrix} Y_{11} & Y_{12} \\ Y_{21} & Y_{22} \end{bmatrix} \begin{bmatrix} V_1 \\ V_2 \end{bmatrix}$$

We need to find the values of Y_{11}, Y_{12}, Y_{21} and Y_{22}. This can easily be done if we first identify a circuit that can represent the above equation.

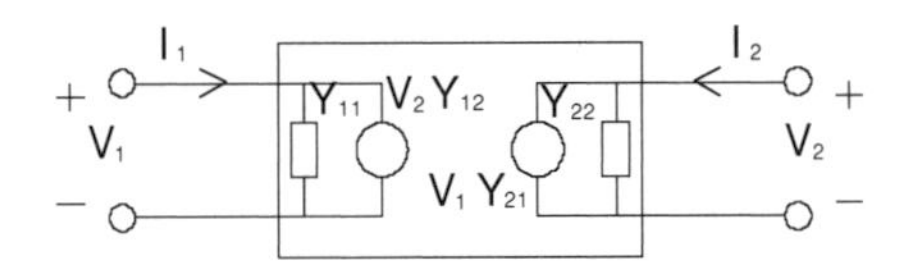

Figure 4.4

In Figure 4.4, Y_{11} is an admittance and the current going through it is V_1Y_{11}. V_2Y_{12} is a voltage-dependent current generator. The equation is of the type $I_a = I_b + I_c$

$$I_1 = V_1Y_{11} + V_2Y_{12}$$

Similarly we can identify every term in the second equation

$$I_2 = V_1Y_{21} + V_2Y_{22}$$

With the use of this equivalent circuit which is of the Norton type, we can go further to try to obtain the values of Y_{11}, Y_{12}, Y_{21} and Y_{22}. In order to do this we short-circuit one of the terminals. Let us say $V_2 = 0$. The first equation then becomes:

$$I_1 = V_1Y_{11}$$

$$I_1 = V_1 Y_{11}$$

so

$$\boxed{Y_{11} = \frac{I_1}{V_1}}$$

The second equation becomes :

$$I_2 = V_1 Y_{21}$$

so

$$\boxed{Y_{21} = \frac{I_2}{V_1}}$$

We can do the same at the other side of the two-port network.

$$V_1 = 0$$

The first equation becomes:

$$I_1 = V_2 Y_{12}$$

so

$$\boxed{Y_{12} = \frac{I_1}{V_2}}$$

and the other equation becomes :

$$I_2 = V_2 Y_{22}$$

so

$$\boxed{Y_{22} = \frac{I_2}{V_2}}$$

Because we have made the voltages equal to zero, by short-circuiting to the input and output, the Y parameters are called short circuit parameters.

There is another way to represent the two-port network, by another set of parameters and this is really of interest to the transistor model.

Hybrid parameters

We now repeat the general drawing Figure 4.5 of a passive network.

Using a mixture of current and voltages, hence the name hybrid, we can define a new set of equations:

$$V_1 = h_{11} I_1 + h_{12} V_2$$

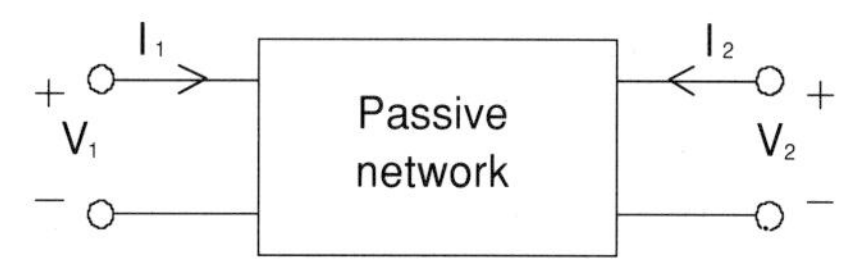

Figure 4.5

$$I_2 = h_{21} I_1 + h_{22} V_2$$

Analysing the first equation we see that on the left-hand side of the equals sign we have volts, then both terms on the right of the equal sign must also be of the dimension of volts. In order to achieve this, h_{11} has to be of ohm dimensions and h_{12} has to be dimensionless. In the second equation h_{21} is dimensionless whereas h_{22} has the dimension of Siemens (i.e. ohm^{-1}).

The above equation can also be written in matrix form:

$$\begin{bmatrix} V_1 \\ I_2 \end{bmatrix} = \begin{bmatrix} h_{11} & h_{12} \\ h_{21} & h_{22} \end{bmatrix} \begin{bmatrix} I_1 \\ V_2 \end{bmatrix}$$

And we can also find an equivalent circuit that corresponds term by term to the set of equations, as in Figure 4.6. It is interesting to note that what

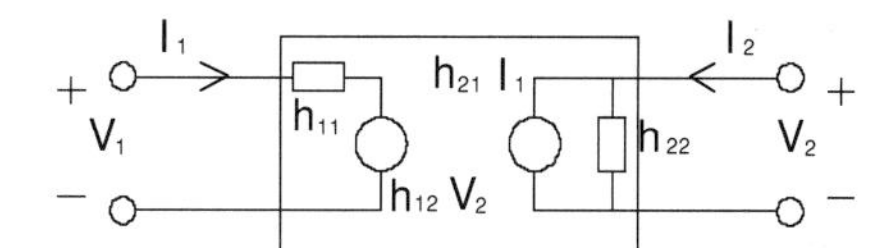

Figure 4.6

we have here is a Thevenin circuit followed by a Norton circuit and this is the form that we want for the transistor model.

All we now need is to use the symbols used in the transistor to have the transistor model. We redraw the circuit in Figure 4.7, with the nomenclature of a common emitter circuit. In Figure 4.7, the letters follow the following abreviations:

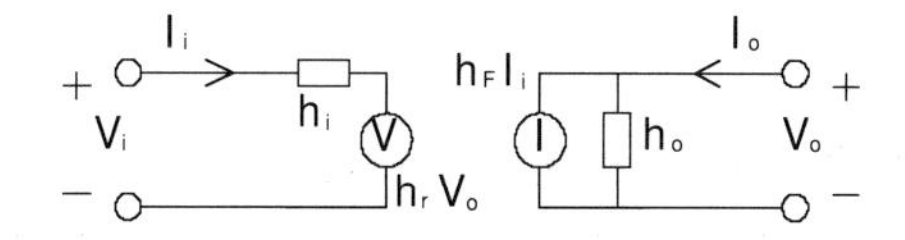

Figure 4.7

h_i input
h_r reverse transfer voltage ratio
h_f forward transfer current ratio
h_o output

The right-hand side of the subscript is reserved for a letter (e, b or c [transistor terminals])

Because the reverse transfer voltage ratio is small and because h_{oe} is high, both can be simplified, arriving at the 'approximate hybrid equivalent circuit'.

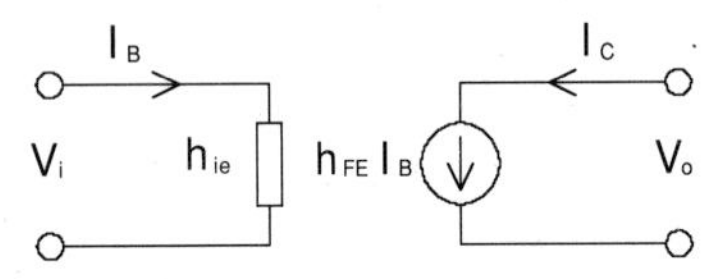

Figure 4.8

We see the approximate hybrid equivalent circuit in Figure 4.8. This approximate circuit can be used in a transistor amplifier as seen in Figure 4.9. This circuit is still not suitable for AC analysis. We still have to use the small signals equivalent circuit.

We haven't yet mentioned how to find the hybrid parameters, but as we are now in the realm of the transistor, we will do this with the help of the transistor characteristic curves.

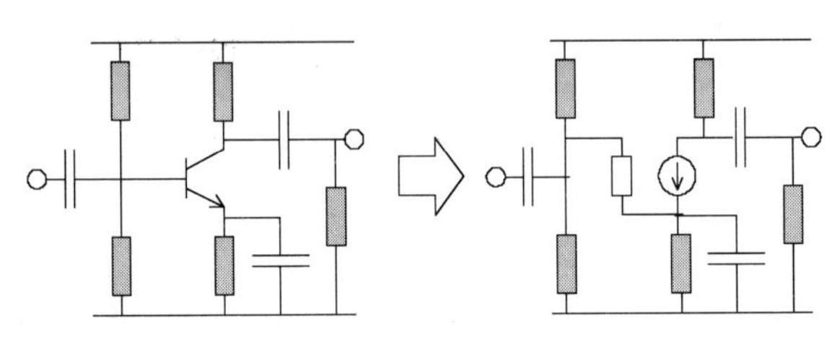

Figure 4.9

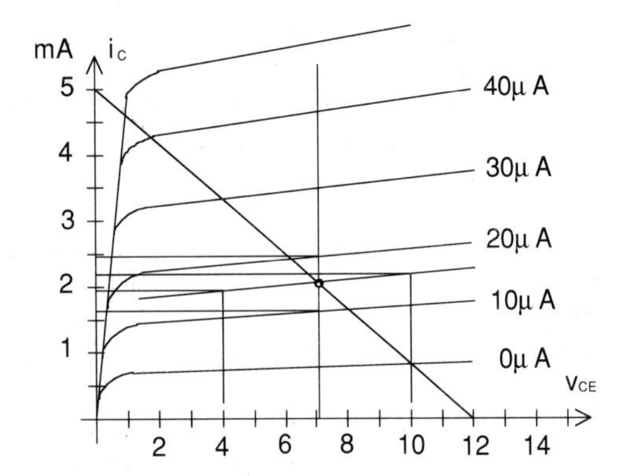

Figure 4.10

Figure 4.10, shows the output characteristics of a transistor. These curves are either given by the manufacturers or can be measured in the lab. There are special oscilloscopes that can give the curves from any transistor connected to its inputs.

$$h_{oe} = \frac{\Delta I_C}{\Delta V_{CE}} \quad \text{at } I_B \text{ constant}$$

$$= \frac{(2.3 - 1.9) \times 10^{-3}}{10 - 4} = 50\ \mu\text{S}$$

$$h_{fe} = \frac{\Delta I_C}{\Delta I_B} \quad \text{at } V_{CE} \text{ constant}$$

$$= \frac{(2.4 - 1.65) \times 10^{-3}}{(20 - 10) \times 10^{-6}} = 75$$

Figure 4.11, shows the input characteristics of a transistor. Looking at the input characteristics we can calculate h_{re} and h_{ie}.

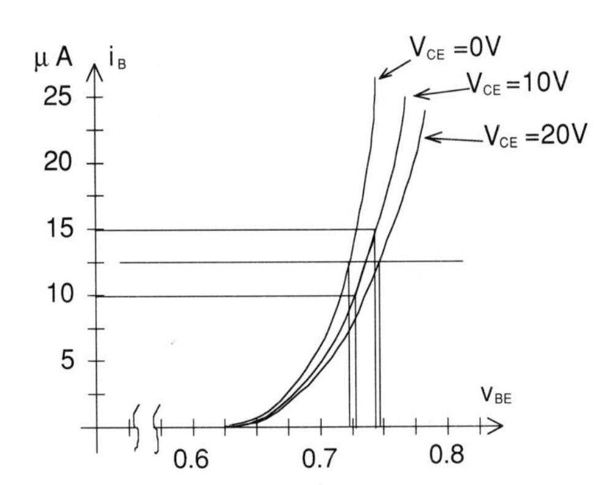

Figure 4.11

$$h_{re} = \frac{\Delta V_{BE}}{\Delta V_{CE}} \quad \text{at } I_B \text{ constant}$$

$$= \frac{0.747 - 0.72}{20 - 0}$$

$$= \frac{0.027}{20}$$

$$= 1.35 \times 10^{-3}$$

$$h_{ie} = \frac{\Delta V_{BE}}{\Delta I_B} \quad \text{at } V_{BE} \text{ constant}$$

$$= \frac{0.745 - 0.728}{(15 - 10) \times 10^{-6}}$$

$$= \frac{0.017}{5 \times 10^{-6}}$$

$$= 3400\ \Omega$$

Circuit transformations

Small signals

Due to the amplification of the transistor, a small signal at the input, will create a similar but larger signal at the output. If the input is sufficiently small, the output will neither go into saturation nor cut–off.

The transistor is said to be working in the active region. It is also said that when the transistor is operating in the active region, it is working under 'small signal conditions'.

Digital operation

The above section described the transistor as an amplifier. We can also use the transistor in a digital form, i.e., on and off, just like a switch. In this case the active region is of no use to us. In this mode of operation we specifically require the transistor to go from saturation to cut–off and vice versa. Digital operation is not within the scope of this book.

Circuit transformations

The transistor works with two types of signals, both are required at the same time. They perform different tasks and they follow different rules in their operation. We are referring to the DC signal or biasing on the one hand, and the AC signal on the other.

It is also important to realise that we are not talking about superposition, for DC and AC signals. Cause and effect can be added in superposition to obtain the resulting effect. In circuit transformations, the AC signal follows a different circuit to that of the DC, resulting in a complex system of two different operations in a single circuit.

The concept of two signals operating together to control the operation of the transistor is difficult to imagine and also difficult to explain even if you are experienced. Experiments performed in the laboratory where you can check your theoretical knowledge against practical results help considerably in understanding the operation. Fortunately, we can treat the signals separately and this appears to be a convenient way of solving the problem.

The DC conditions were tackled under biasing and this is straightforward. There are no complications, only some basic rules such as Ohm's law, Kirchhoff's law, and other theorems that you are familiar with and can be revised, if necessary, in the first chapter of this book.

The AC conditions, are more complicated to understand, because under AC conditions there are extra paths for the AC signal and the circuit changes quite a lot. When we move to AC conditions, we put DC conditions to one side, in the knowledge that the transistor will be properly biased.

In AC conditions, as the AC signal follows a different path to that of the DC circuit, we can redraw the circuit to make it easier for us to see what is happening and to apply the laws and rules of circuit analysis. The step of redrawing the circuit for AC conditions is very important in solving a problem. You must also remember that different circuits will have different models and therefore different equations. Don't use a formula if you don't know exactly which circuit it applies to.

There are two important factors in transforming a circuit:

(1) Capacitor effect
    Capacitors in an amplifier are designed to be fully conducting at the frequencies of interest. Intentionally, capacitors are placed strategically to block the DC signal and to let the AC signal through.
(2) DC supply transparency to AC
    An AC signal placed at one terminal of the DC power supply, will appear at the other side of the DC supply.

Following these two points, we can modify the circuit, in order to make it easier to analyse. We are trying to solve the problems presented by a transistor amplifier. With the modified circuit, we are a step nearer to solving these problems.

All we then need to do is to set the equation that corresponds to the modified circuit. The rest is just mathematics to get the final result.

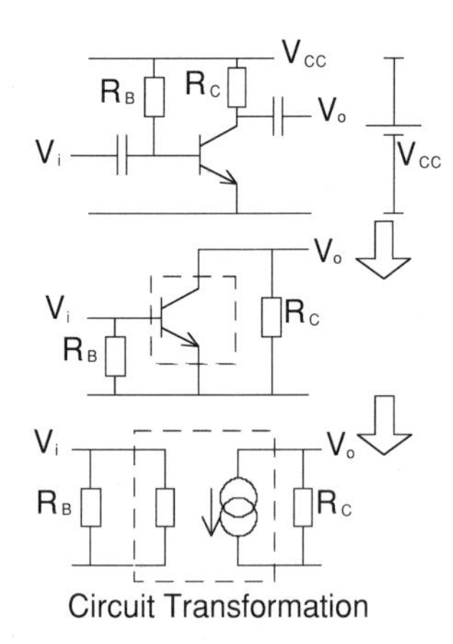

Figure 4.12

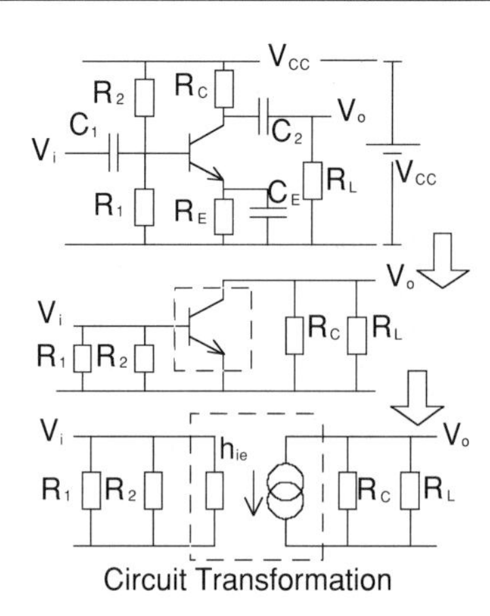

Figure 4.13

We are now ready to analyse the circuit transformations – under AC conditions only – of three typical examples, each with a sequence of three diagrams.

The first circuit to be transformed is a very simple common emitter circuit as shown in Figure 4.12. We have R_B in the input circuit and R_C in the output circuit. We can clearly see two circuits with a transistor in the middle.

This appears even clearer in the second of the three circuits. From the first to the second circuit, the capacitors have disappeared. They have been designed to be conducting at the frequency of interest. The frequency of interest in this case is the midband which is from around 100 Hz to around 20 000 Hz. Below this band, we have the low frequency and above this band, we have the high frequency.

The other important change from the first to the second diagram, which for AC conditions are totally equivalent, is that the resistors seem to have moved from the original position. The explanation for this is that the AC signal which is placed at v_i, will go up through R_B onto V_{CC}, which is the positive side of the DC power supply. This signal goes through the power supply indicated on the right hand side, to the zero volt side which is the lower line in each of the diagrams. The second of the three diagrams clearly shows the path followed by the AC signal which goes in at v_i and comes out at v_o.

The final circuit of Figure 4.12, shows the equivalent circuit of the transistor inserted in the circuit already modified by the AC conditions. It is easy to analyse the last circuit and calculate values such as the input impedance, output impedance, etc.

Note that the DC conditions, without which the transistor cannot work, do not appear in the diagrams. That was the agreement at the beginning. As it is too difficult to work with DC and AC signals together, we separate them completely. We deal with them separately, usually doing the biasing first.

The second circuit analysed is shown in Figure 4.13. This is also a common emitter circuit. It is the circuit with classical biasing, the potential divider biasing. It is also called β independent as we mentioned earlier.

In the first of the three diagrams, we have the standard circuit as it is normally drawn. Additionally we have shown the power supply on the right-hand side. This is important to visualise the extra path followed by the AC signal (through the power supply).

Together with the extra path provided by the power supply, the circuit is modified by the effect of the capacitors. Capacitors block the DC signal and allow the AC signal to go through.

On the second diagram we have the transistor at the centre. At the left-hand side of the transistor we have the input circuit. At the right-hand side of the transistor we have the output circuit.

The capacitors have disappeared as they are just a continuous line representing a short circuit in place of the capacitors. The capacitor C_E, shorts the resistance R_E, which will not appear any more in the transformation of the circuit. R_1 is now in parallel with R_2. R_C is in parallel with R_L. This is by a combination of C_2 being a short circuit and by the path provided by the power supply to ground for the AC signal.

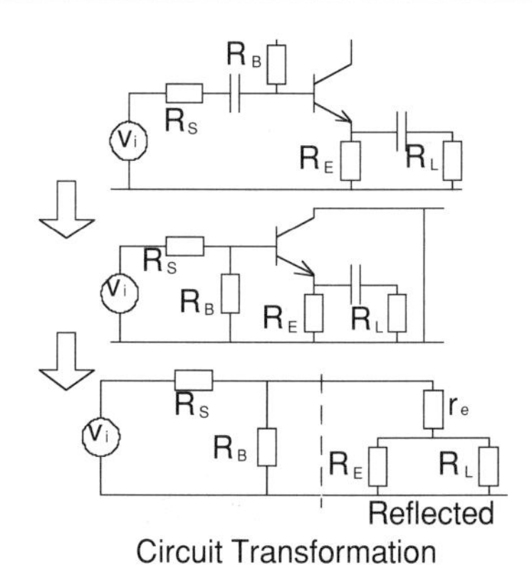

Figure 4.14

In the final of the three diagrams of Figure 4.13, we have also replaced the transistor with its equivalent model.

We end up with two circuits, which are very easy to analyse using the theorems that we know. Additionally, these circuits are linked by the fact that the current on the collector is ß times the current in the base.

The last circuit to be transformed is shown in Figure 4.14. It shows an emitter follower circuit. The input side shows the signal source with its own source resistance R_S. The second circuit of this sequence of three diagrams shows how the circuit is transformed by two conditions – the capacitor conducting and the power supply also conducting for the AC signal. In fact, all the components are part of the input circuit in this configuration.

There is one complication in this case. In this circuit there are two levels of current. R_S and R_B have the base current which is a small current, whereas r_e, R_E and R_L have a comparatively much larger current. Although different, these currents, are related by the factor (β + 1).

This complication can be overcome with a convenient trick. We use the reflected value for the emitter current circuit. We explained this as part of the biasing, but as it is important we will repeat it once more.

In order to simplify the equations, we use the reflected value of r_e, R_E and R_L. We assume that these three resistors are part of the input circuit where the base current circulates. To compensate for not using the emitter current which is larger by a factor of (β + 1), we use the base current, but we increase the value of the three resistors by a factor of (β + 1). The increased value of the resistors is the 'reflected' value of the resistors.

In this way, we set just one equation for the input circuit, which includes the emitter circuit components. The same result is achieved 'properly' by setting an equation with i_B and i_E, in the normal mathematical way. Due to the fact that they are two unknowns, mathematics requires another equation

$$i_E = i_B\ (\beta + 1)$$

There are now two equations and two unknowns, which means that we can solve the problem.

The reflected value method, however, represents a short cut to the set of equations, is user friendly and well established in electronic circles.

You might have noticed that we haven't used the transistor model in this final example. The equivalent circuit is not required in this case as there are no components in the collector side of the transistor which is only useful to provide a return path for the emitter current.

Little r_e model

Little r_e is the emitter resistance inside the emitter. This is an imaginary resistance which can be justified as there is a voltage on the base of the transistor and a current flowing through the emitter. This is seen in Figure 4.15.

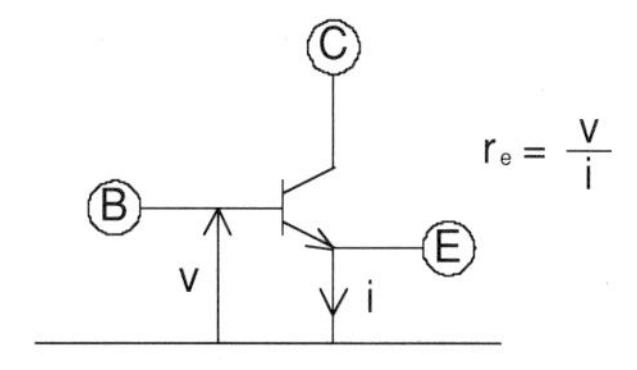

Figure 4.15

The definition of r_e would be

$$r_e = \frac{V_{BE}}{I_E} \quad \text{at a constant value of } v_{CE}$$

The transistor must be biased at a convenient Q point. The base emitter junction must be forward biased.

We can also imagine the transistor as a conducting diode and use the exponential relationship between voltage and current in its forward region. This results in the well known diode equation which can be used to find an approximate value for r_e:

$$r_e = \frac{V_T}{I_E} \qquad V_T = \frac{KT}{q} \quad \text{(thermal voltage)}$$

$$r_e = \frac{KT}{qI_E}$$

where

K Boltzman constant 1.38×10^{23} joule/K
(K is the absolute temperature in degrees Kelvin)
q electronic charge 1.602×10^{-19} coulombs
T absolute temperature (K = 273 + °C)
(The value of I_E in mA)

You can verify, with your calculator, that

$r_e = \frac{25}{I_E}$	at 20°C
$r_e = \frac{26}{I_E}$	at 30°C

As we watch Figure 4.16, showing the base current as a function of v_{BE}, we see that it rises very quickly and that at the Q point the voltage will be around 0.7 V. We use this value as an approximation whenever the transistor is conducting.

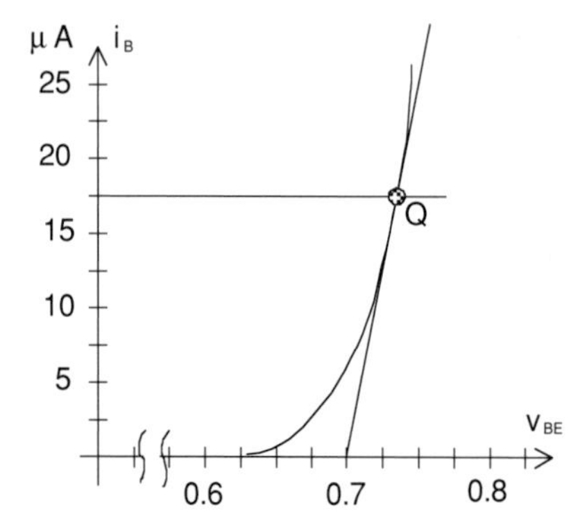

Figure 4.16

A satisfactory model of the transistor can be obtained by evaluating the DC conditions to establish the Q point and the value of I_E.

According to the ambient temperature, use the approximate formula, in the box above, to calculate little r_e. Remember that the value of I_E must be in mA. You can then use this value of r_e in an equivalent circuit, such as the one shown in Figure 4.17.

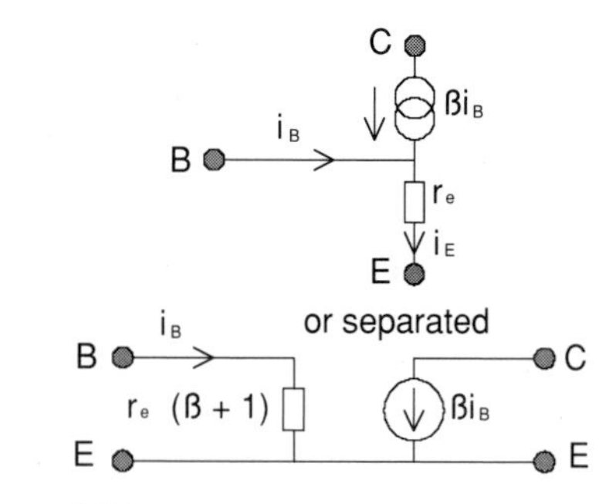

Figure 4.17

Transistor model for high frequency

For high frequencies a special version of the model is required for the following reasons:

(1) We first consider the 'base spreading resistance'. This is a resistance from the outside of the base to the inside of the base. So we need a new terminal B' that is not accessible outside, but is inside the transistor case.
(2) We need to consider the diffusion capacitance. This corresponds to the stored charge in the base region

$Q = CV$

In the active region electrons move from emitter to collector by diffusion, when the charge is injected from the emitter to the base region. It is a diffusion capacitance between the new terminal B' and the emitter terminal.
(3) The depletion region between the N and P regions gives rise to a capacitance in the ordinary sense of the word (i.e. parallel plate capacitance).
(4) We need to allow for the effect of reverse voltage feedback. For this, we add $R_{B'C}$ which is of the order of 200 MΩ.

Bearing these points in mind we obtain Figure 4.18, which is the full hybrid equivalent circuit

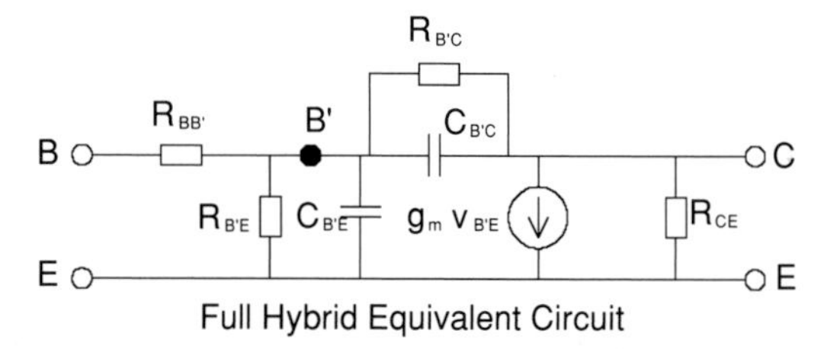

Figure 4.18

$C_{B'C}$ is the depletion capacitance
$C_{B'E}$ is the diffusion capacitance
$R_{B'C}$ is the reverse voltage feedback resistance
$R_{B'E}$ is our original h_{ie}
$R_{B'B}$ is the base spreading resistance
R_{CE} is the transistor output resistance

In order to simplify the calculations in high frequency problems, there are two simplifications which are acceptable (if you have to do calculations by hand, as opposed to having computer crunching power) to the full hybrid equivalent circuit

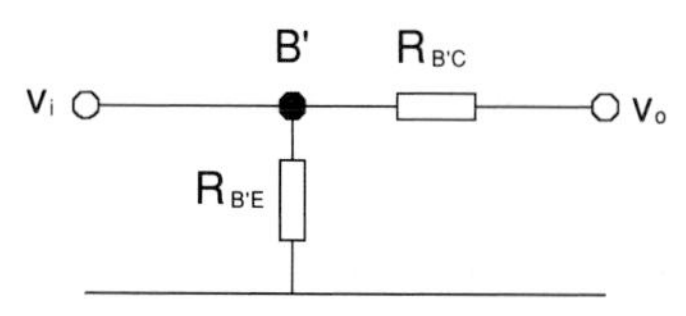

Figure 4.19

First of all $R_{B'C}$ is very high (see Figure 4.19). In this case, $R_{B'E}$ is 2 KΩ and h_{re} (the reverse voltage transfer ratio) is 10^{-5}. What is the value of $R_{B'C}$?

It is a case of voltage division

$$\frac{v_i}{v_o} = 10^{-5} = \frac{R_{B'E}}{R_{B'C} + R_{B'E}}$$

$$10^{-5} = \frac{2k}{2k + R_{B'C}}$$

$$R_{B'C} = 2 \times 10^8$$

Because of this high value we can simplify the circuit by eliminating $R_{B'C}$. We obtain the circuit shown in Figure 4.20. The second simplification comes from using Miller's theorem in the circuit of Figure 4.20.

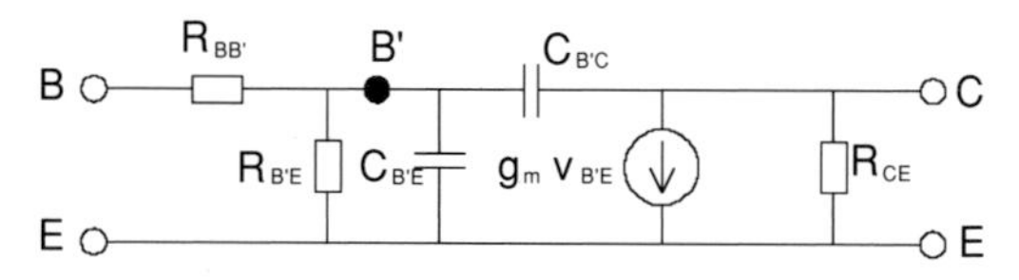

Figure 4.20

As a result we obtain the circuit shown in Figure 4.21.

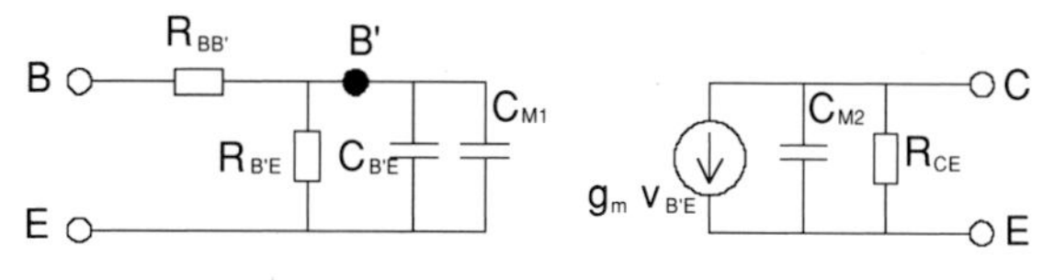

Figure 4.21

$$C_{M1} = C_{B'C}(1 - A_V)$$

$$C_{M2} = C_{B'C}\left(1 - \frac{1}{A_V}\right) \approx C_{B'C}$$

5

Current gain

Previous knowledge required:

Reflected value of r_e
Thevenin–Norton conversions
Transistor modelling
Current division rule
Small signals equivalent circuits

The circuit is shown in Figure 5.1. This is a common emitter configuration with the classical biasing (or beta independent configuration). We should remember that there are two circuits to the transistor, the primary or input circuit and the secondary or output circuit.

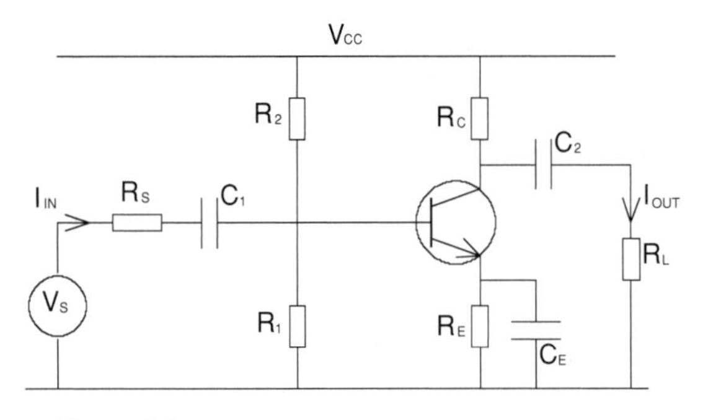

Figure 5.1

The definition of current amplification is A_i.

$$A_i = \frac{i_o}{i_i}$$

That is to say the current amplification is the ratio of the output current to the input current. The input current comes from the source, through the first coupling capacitor, as this is an AC signal. The output current goes through the second coupling capacitor and through resistor R_L, the load resistor.

In Figure 5.2, we see the transformation of the standard circuit into the AC model, which in this figure shows only the input side. At first it is difficult to recognise that these circuits are all equivalent and correspond to the input circuit of

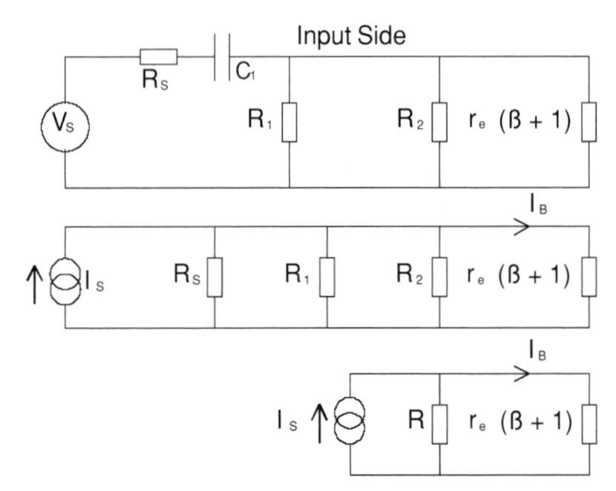

Figure 5.2

Figure 5.1. For this reason we will briefly mention the differences.

Let's concentrate on the input circuit first. The transistor has been replaced by its equivalent circuit, but we only use the base current circuit, as we are only considering the input circuit of the amplifier circuit. The other important transformation is due to the fact that the capacitors and DC supplies are transparent to AC signals. Therefore the capacitors disappear and the V_{CC} line is shorted to the earth line, for the AC model. We could have done the Thevenin equivalent circuit of part of the input, but we are not using that approach on this occasion.

On the source side we have used a Thevenin to Norton conversion, in Figure 5.2, from the first to the second circuit, although this is not essential to solving the problem.

The next transformation, to arrive at the final circuit of Figure 5.2, is easy to understand and is to simply combine three resistors together. We don't combine the four resistors together because we are interested in the base current which goes through the last resistor, r_e.

Finally we have used the reflected value of r_e. The real current going through r_e is I_E and not I_B as we have shown. But because I_E is $(\beta + 1)$ times

I_B, we can still use I_B provided that we use a larger value of r_e in the same proportion, that is $(\beta + 1)$. This is how we arrive at the value of $r_e(\beta + 1)$.

Using the final circuit of the input circuit of the amplifier it is easy to calculate the relationship between the source current and the current i_B, this is simply the current division rule:

$$i_B = i_S \frac{R}{r_e(\beta+1)+R}$$

Now, let us look at the output circuit, in Figure 5.3. The resistor R_E has disappeared due to the effect of C_E the bypass capacitor, which has an enormous effect on the feedback and stability of the circuit.

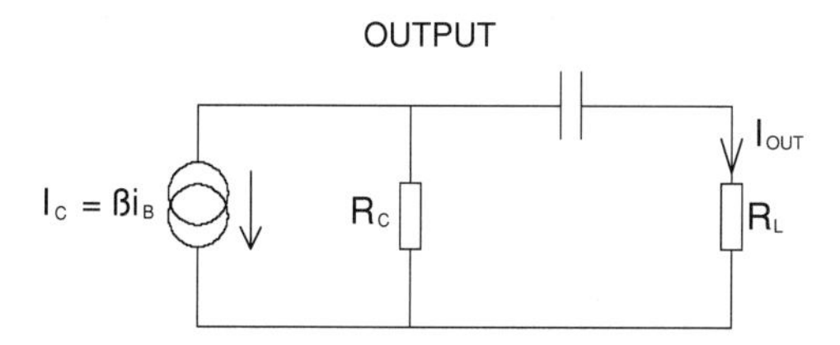

Figure 5.3

Due to the transparency of the DC supply to AC signals we find that R_C in parallel with R_L, this is helped by the coupling capacitor which becomes a short circuit to the AC signals. This, incidentally, is not accidental, the values of these capacitors have been chosen to be open to DC signals, but short-circuit to the frequencies of interest in the midband range of frequencies.

Once we have the final circuit it is easy to calculate the ratio of the output current to the collector current. This again is the current division rule.

The link between input and output circuits is the fact that the collector current is β times the base current.

$$i_B = i_S \frac{R}{R+r_e(\beta+1)}$$

$$i_C = \beta i_B$$

$$i_o = -i_C \frac{R_C}{R_C+R_L}$$

Joining these equations together , with $i_s = i_i$, we obtain

$$\frac{i_o}{i_i} = -\beta \frac{R}{R+r_e(\beta+1)} \frac{R_C}{R_C+R_L}$$

The minus sign above is due to the way in which the arrows are pointing in Figure 5.3.

PROBLEM 5.1 CURRENT GAIN

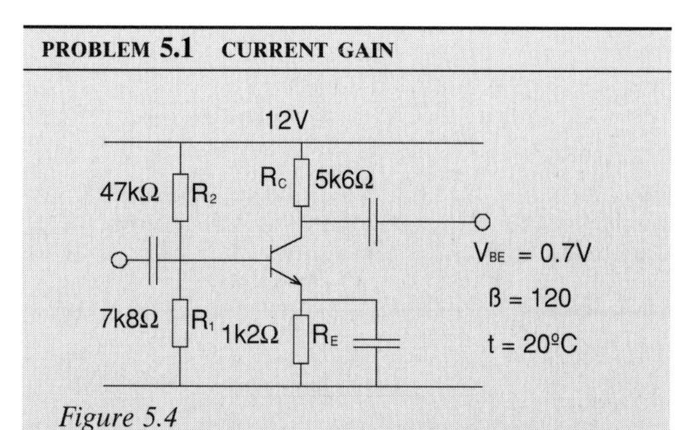

Figure 5.4

Using the approximate method for DC conditions, find the current gain A_i of the transistor.

Answer: $A_i = 78$

$$V_B = 12\frac{7.8}{7.8+47} = 12\frac{7.8}{54.8} = 1.708 \text{ V}$$

$$V_E = 1.708 - 0.7 = 1.008 \text{ V}$$

$$I_E = \frac{1.008}{1k2} = 0.84 \text{ mA}$$

$$r_e = \frac{25}{0.84} = 29.76 \,\Omega$$

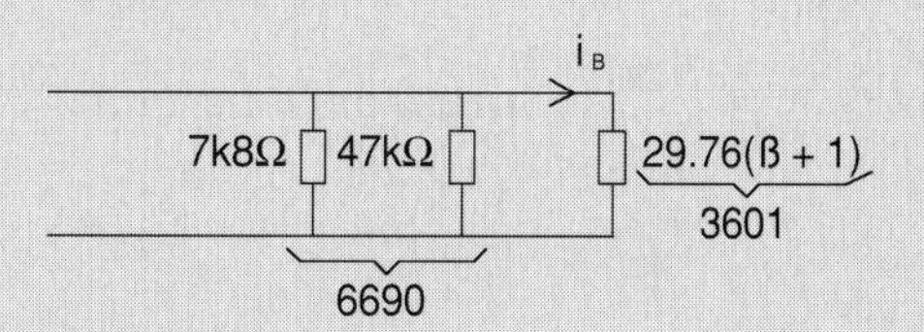

Figure 5.5

$$i_B = i_S \frac{6690}{6690+3601} = i_S \times 0.65$$

$$i_C = \beta i_B = 120 i_B$$

$$i_C = 120 \times 0.65 i_S = i_S 78$$

$$\frac{i_C}{i_S} = 78$$

PROBLEM 5.2 CURRENT GAIN

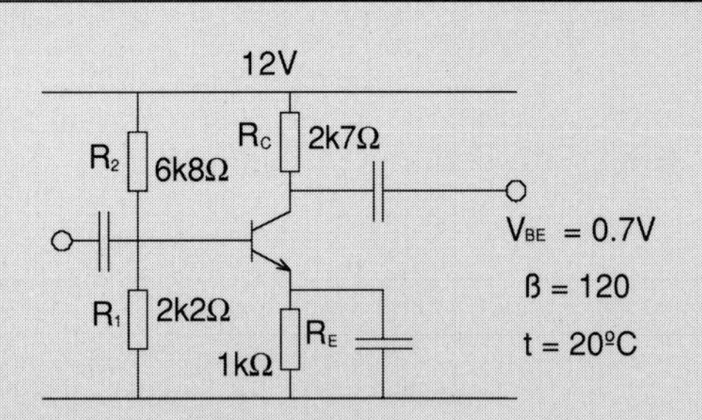

Figure 5.6

Find A_i for the circuit, using the approximate method.

Answer: $A_i = 66.12$

$$V_B = 12\frac{2.2}{2.2+6.8} = 12\frac{2.2}{9} = 2.93 \text{ V}$$

$$V_E = V_B - V_{BE} = 2.93 - 0.7$$

$$= 2.23 \text{ V}$$

$$I_E = \frac{V_E}{R_E} = \frac{2.23}{1k} = 2.23 \text{ mA}$$

$$r_e = \frac{25}{2.23} = 11.21\ \Omega$$

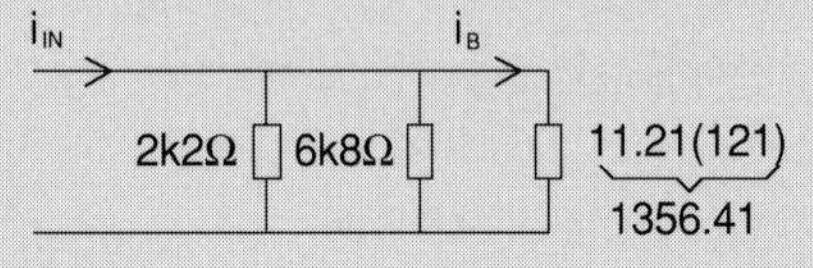

Figure 5.7

$$i_B = i_{IN}\frac{1662}{1662+1356.41} = i_{IN}0.551$$

$$A_i = \beta \times 0.551 = 66.12$$

PROBLEM 5.3 CURRENT GAIN

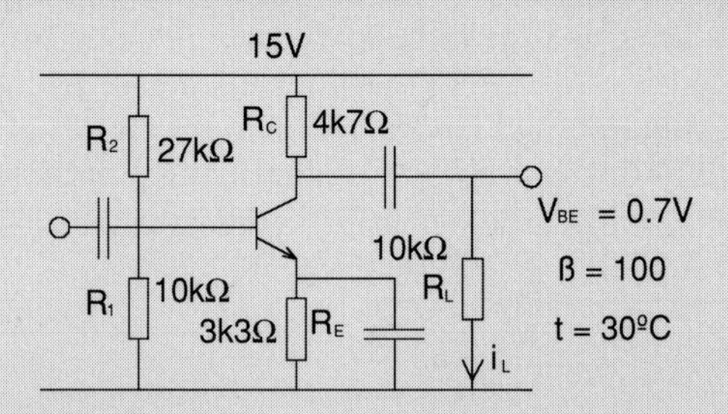

Figure 5.8

Find the current amplification A_i.

Answer: $A_i = -23.59$

Approximate method

$$V_B = 15\frac{10k}{10k+27k} = 4.05 \text{ V}$$

$$V_E = 4.05 - 0.7 = 3.35 \text{ V}$$

$$I_E = \frac{3.35}{3k3} = 1.015 \text{ mA}$$

$$r_e = \frac{26}{1.015} = 25.6\ \Omega$$

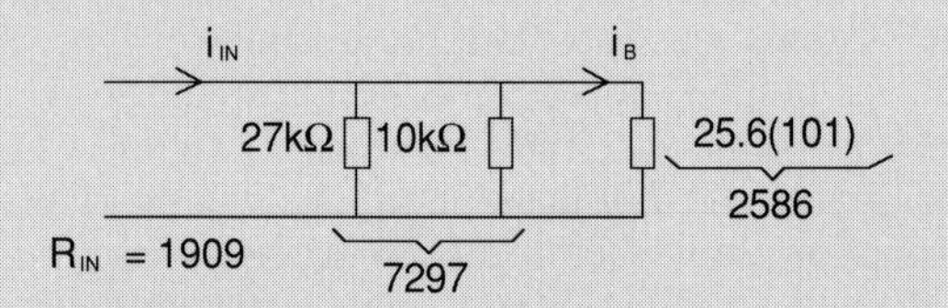

Figure 5.9

$$i_B = i_{IN}\frac{7297}{7297+2586} = i_{IN}0.738$$

$$i_C = i_{IN} \times 0.738 \times 100$$

$$i_L = -i_{IN} \times 0.738 \times 100 \times \frac{4k7}{4k7+10k}$$

$$i_L = -i_{IN} \times 23.59$$

$$A_i = \frac{i_L}{i_{IN}} = -23.59$$

PROBLEM 5.4 CURRENT GAIN

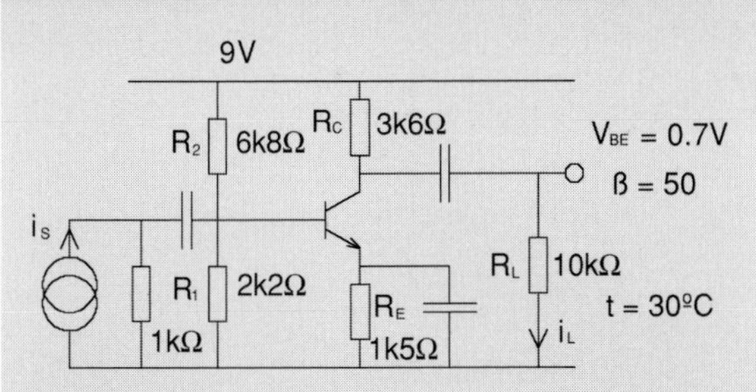

Figure 5.10

Find the current amplification, i_L/i_S.
Temperature = 30°C, β = 99, V_{BE} = 0.7 V.

Answer: $A_i = -5.06$

Approximate method

$$V_B = 9\frac{2.2}{6.8+2.2} = 2.2 \text{ V}$$

$$V_E = 2.2 - 0.7 = 1.5 \text{ V}$$

$$I_E = \frac{1.5}{1k5} = 1 \text{ mA}$$

$$r_e = \frac{26}{1} = 26\ \Omega$$

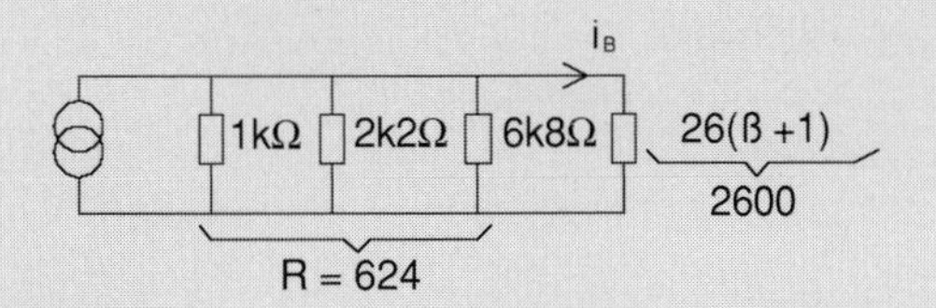

Figure 5.11

$$i_B = i_S\frac{624}{624+2600} = 0.193 i_S$$

$$i_C = \beta i_B = 99 \times 0.193 i_S = 19.11 i_S$$

$$i_L = -i_C\frac{R_C}{R_C+R_L} = -i_C\frac{3.6}{3.6+10}$$

$$= -i_C \times 0.2647 = -0.2647 \times 19.11 \times i_S$$

$$= -5.06 i_S$$

PROBLEM 5.5 CURRENT GAIN

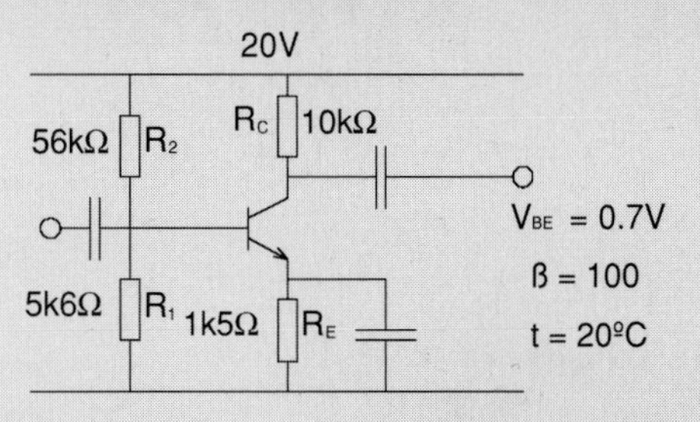

Figure 5.12

Use the Thevenin method to evaluate DC conditions and find the value of A_i.

Answer: $A_i = 59.3$

$$V_{TH} = 20\frac{5k6}{61k6} = 1.82 \text{ V}$$

$$R_{TH} = 5k6 \| 56k = 5091\ \Omega$$

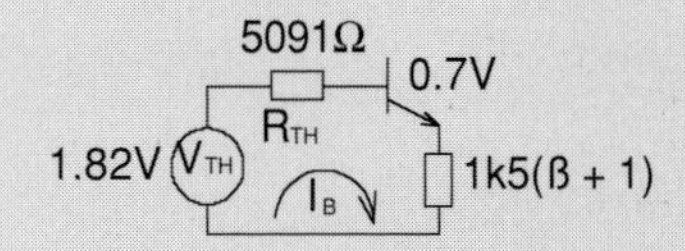

Figure 5.13

$$I_B = \frac{1.82-0.7}{5091+151500} = \frac{1.12}{156591}$$

$$I_E = \frac{1.12 \times 101}{156591} = 0.722 \text{ mA}$$

$$r_e = \frac{25}{0.722} = 34.63\ \Omega$$

AC conditions

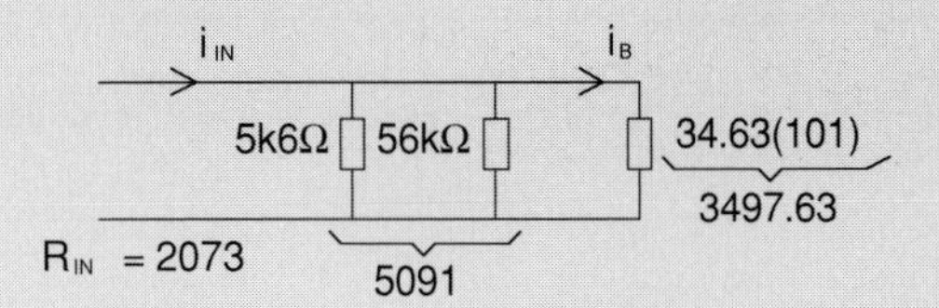

Figure 5.14

$$i_B = i_{IN}\frac{5091}{5091+3497.63} = i_{IN} 0.593$$

$$i_C = \beta i_B = i_{IN} \times 59.3$$

$$A_i = 59.3$$

PROBLEM 5.6 CURRENT GAIN

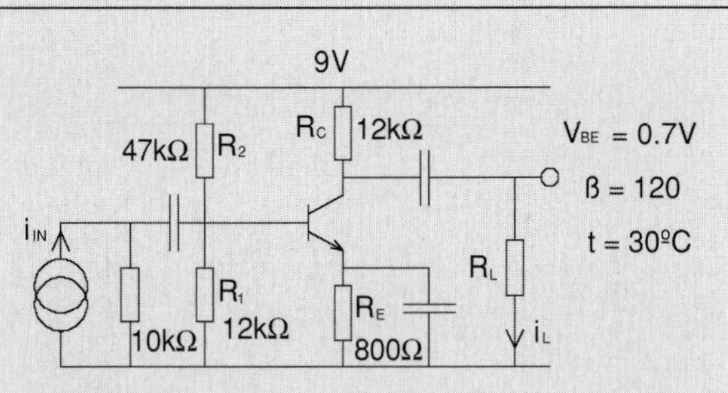

Figure 5.15

Find the value of R_L which will result in a current amplification of –50.

> Answer: $R_L = 7181\ \Omega$

$10R_1 = 120\text{k}$ vs $\beta R_E = 96\text{k}$, therefore use Thevenin

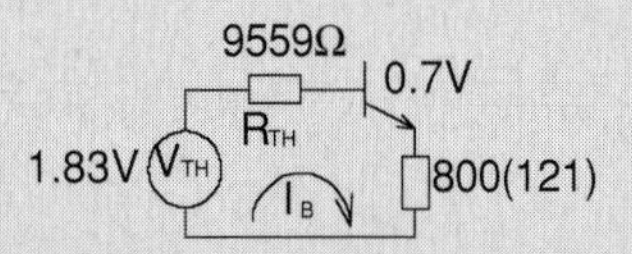

Figure 5.16

$$V_{TH} = 9\frac{12}{12+47} = 1.83\ \text{V}$$

$$R_{TH} = 12\text{k}\|47\text{k} = 9559\ \Omega$$

$$I_B = \frac{1.83-0.7}{9559+96\,800} = \frac{1.13}{106\,359} = 10.62\ \mu\text{A}$$

$$I_E = I_B(\beta+1) = 10.62\times10^{-6}\times121 = 1.285\ \text{mA}$$

$$r_e = \frac{26}{1.285} = 20.23\ \Omega$$

AC model

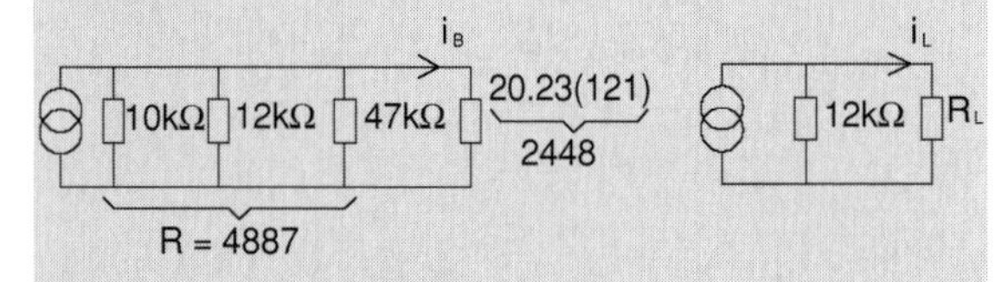

Figure 5.17

$$i_B = i_{IN}\frac{4887}{4887+2448}$$

$$= i_{IN}\frac{4887}{7335} = 0.666 i_{IN}$$

$$i_C = \beta i_B = 120\times0.666 i_{IN} = 79.92 i_{IN}$$

$$i_L = -i_C\times\frac{12\text{k}}{12\text{k}+R_L}$$

$$i_L = -79.92 i_{IN}\frac{12\text{k}}{12\text{k}+R_L}$$

$$\frac{i_L}{i_{IN}} = -79.92\frac{12\text{k}}{12\text{k}+R_L} = -50$$

$$959\,040 = 600\,000 + 50R_L$$

$$359\,040 = 50R_L$$

$$R_L = 7181\ \Omega$$

PROBLEM 5.7 CURRENT GAIN

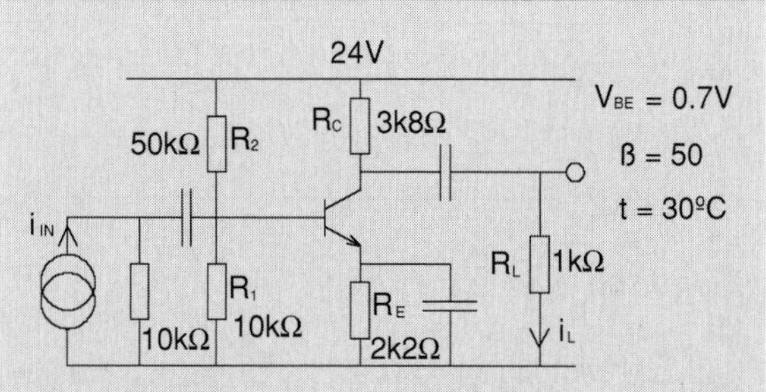

Figure 5.18

Find $A_i = i_L/i_{IN}$.

> Answer: $A_i = -32.79$

Thevenin

$$V_{TH} = 24\frac{10}{10+50} = 4\ \text{V}$$

$$R_{TH} = 10\text{k}\|50\text{k} = 8333\ \Omega$$

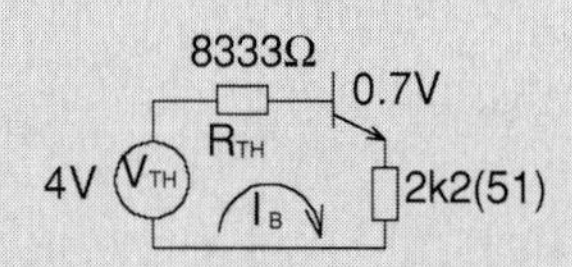

Figure 5.19

$$I_B - \frac{4-0.7}{8333+112\,200} = \frac{3.3}{120\,533} = 27.38\ \mu\text{A}$$

$$I_E = (\beta+1)I_B = 51\times27.38\times10^{-6} = 1.4\ \text{mA}$$

$$r_e = \frac{26}{1.4} = 18.57\ \Omega$$

AC model

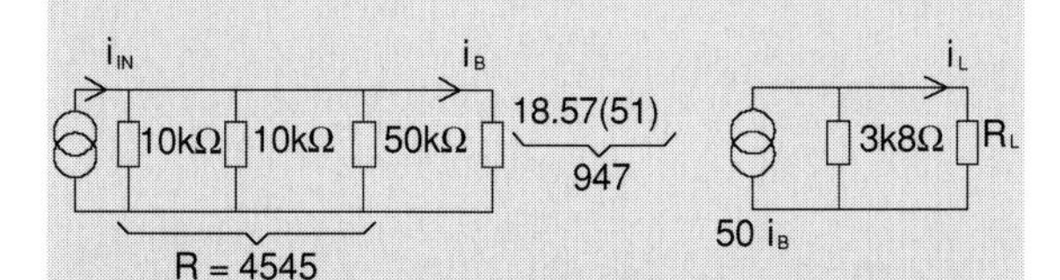

Figure 5.20

$$i_B = i_{IN}\frac{4545}{4545+947} = 0.83 i_{IN}$$

$$i_C = \beta i_B = 50 \times 0.83 i_{IN} = 41.5 i_{IN}$$

$$i_L = -i_C\frac{3.8}{3.8+1} = -i_C 0.79$$

$$i_L = -0.79 \times 41.5 \times i_{IN} = -32.79 i_{IN}$$

$$A_i = -32.79$$

PROBLEM 5.8 CURRENT GAIN

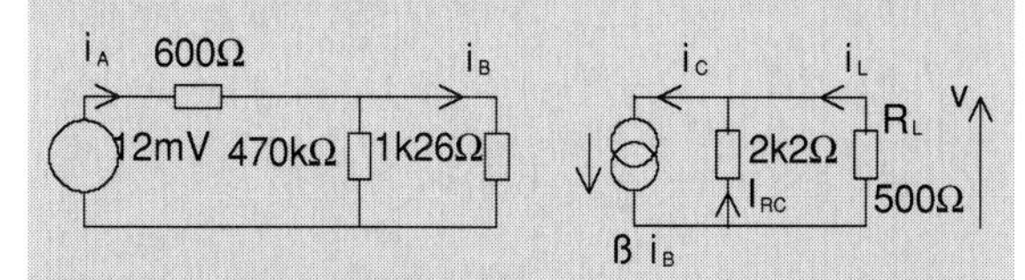

Figure 5.21

Find the values of I_A, I_B, I_C, I_{BC}, I_L, and V_O in the equivalent model. β = 100

Answer: 6.46 μA, 6.44 μA, 6.44 μA, 0.119 mA, 525 μA, and 0.26 V, respectively

$$I_A = \frac{12\times10^{-3}}{470\text{k}\|1\text{k}26 + 600}$$

$$= \frac{12\times10^{-3}}{1857} = 6.46\ \mu\text{A}$$

$$I_B = 6.46\times10^{-6}\frac{470\text{k}}{470\text{k}+1\text{k}26}$$

$$= 6.44\times10^{-6} = 6.44\ \mu\text{A}$$

$$I_C = I_B\beta = 6.44\times10^{-6}\times100 = 0.644\ \text{mA}$$

$$I_L = I_C\frac{2.2}{2.2+0.5} = I_C\frac{2.2}{2.7} = 0.815 I_C$$

$$I_L = 0.815\times0.644\times10^{-3} = 0.525\ \text{mA}$$

$$V_O = -I_L\times500 = -0.525\times10^{-3}\times500$$

$$= -0.26\ \text{V}$$

$$I_{RC} = I_C - I_L = 0.644\ \text{mA} - 0.525\ \text{mA}$$

$$= 0.119\ \text{mA}$$

NOTE: Current and voltage directions follow the rule defined in Figure 1.2.

PROBLEM 5.9 CURRENT GAIN

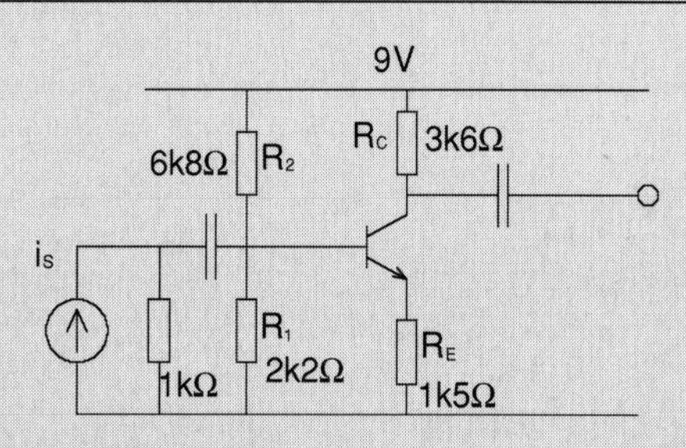

Figure 5.22

Find the current amplification i_C/i_S. t = 30°C, β = 99, V_{BE} = 0.7 V.

Answer: $A_i = 0.403$

Note: C_E not present.

Reflected value = $1k5 \times 100 = 150\ k\Omega$

Against $2k2 \times 10 = 22k$, therefore, use approximate method.

$$V_B = 9\frac{2.2}{6.8+2.2} = 2.2\ V$$

$$V_E = 2.2 - 0.7 = 1.5\ V$$

$$I_E = \frac{1.5}{1k5} = 1\ mA$$

$$r_e = \frac{26}{1} = 26\ \Omega$$

AC conditions

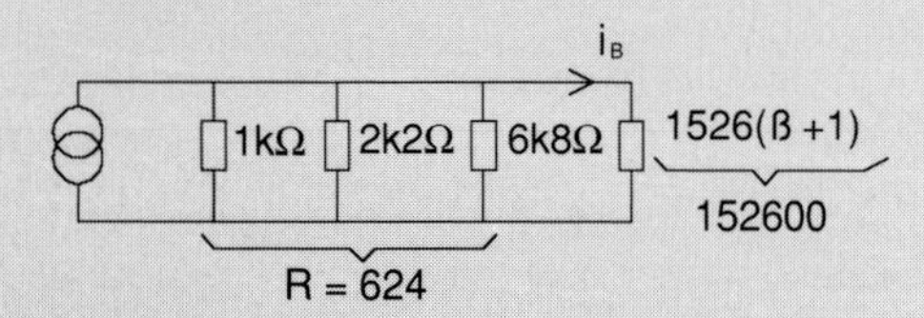

Figure 5.23

$$i_B = i_S\frac{624}{624+152\,600} = 4.072\times10^{-3}\times i_S$$

$$i_C = \beta_{iB} = 4.072\times10^{-3} i_S \times 99 = 0.403 i_S$$

$$A_i = \frac{i_C}{i_S} = 0.403$$

PROBLEM 5.10 CURRENT GAIN

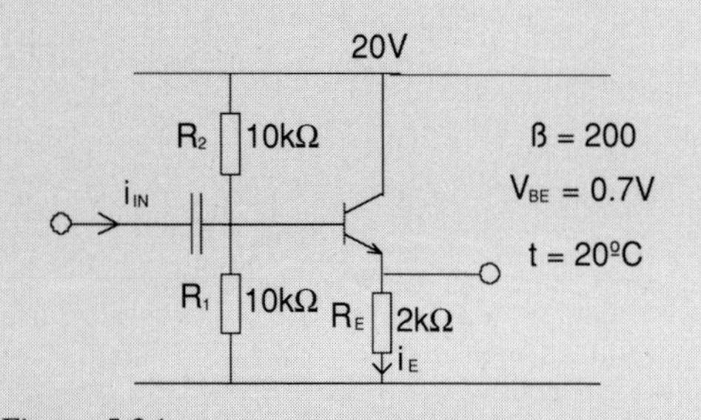

Figure 5.24

Find the current amplification i_E/i_{IN}.

Answer: $A_i = 2.462$

Approximate method

$$V_B = 10\ V, \qquad V_E = 9.3\ V$$

$$I_E = \frac{9.3}{2k} = 4.65\ mA$$

$$r_e = \frac{25}{4.65} = 5.376\ \Omega$$

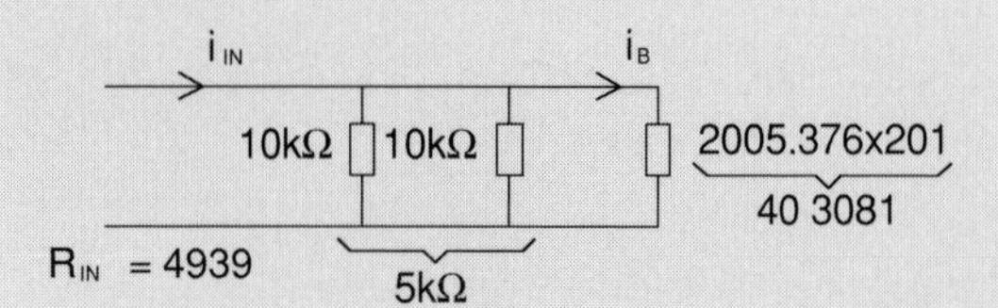

Figure 5.25

$$i_B = i_{IN}\frac{5k}{408\,081} = 12.25 i_{IN}\times10^{-3}$$

$$i_E = i_{IN}\times12.25\times201\times10^{-3} = 2.462 i_{IN}$$

$$A_i = \frac{i_E}{i_{IN}} = 2.462$$

PROBLEM 5.11 CURRENT GAIN

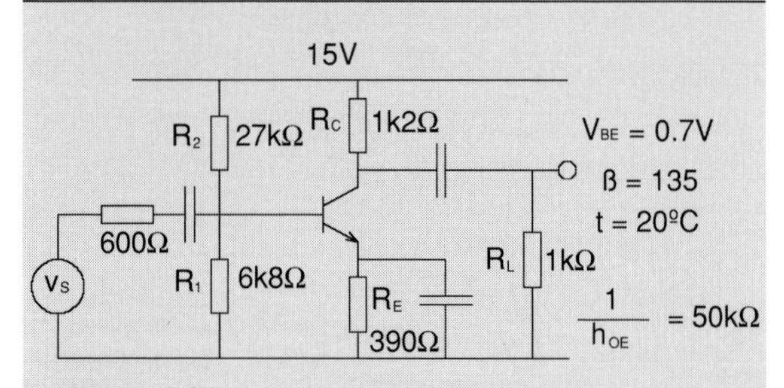

Figure 5.26

Find the current gain.

Answer: $A_i = -65.34$

$10R_1$ vs βR_E

68 000 52 650

Therefore, use Thevenin.

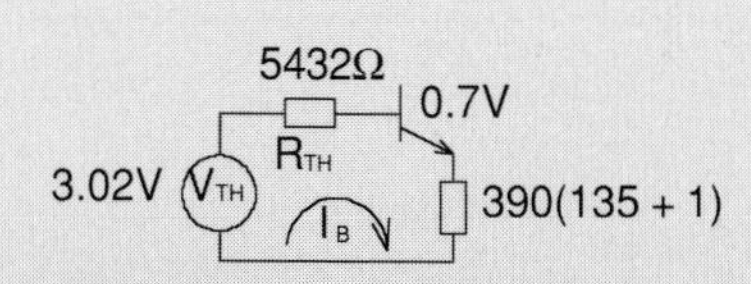

Figure 5.27

$$V_{TH} = 15\frac{6.8}{33.8} = 3.02 \text{ V}$$

$$R_{TH} = 6k8||27k = 5432\ \Omega$$

$$I_B = \frac{3.02 - 0.7}{5432 + 390(136)}$$

$$= \frac{2.32}{58\,472} = 39.68\ \mu\text{A}$$

$$I_E = 39.68 \times 10^{-6}(136) = 5.4 \text{ mA}$$

$$r_e = \frac{25}{5.4} = 4.63\ \Omega$$

AC model

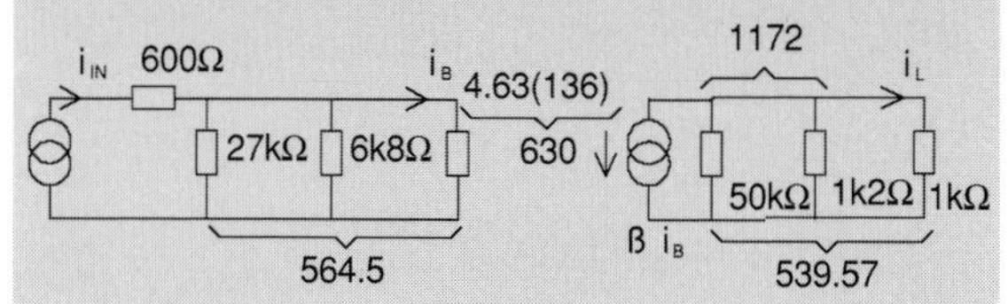

Figure 5.28

$$i_B = i_{IN}\frac{5432}{5432 + 630} = i_{IN} \times 0.896$$

$$i_C = \beta i_B$$

$$= 135 \times i_{IN} \times 0.896$$

$$= 121 \times i_{IN}$$

$$i_L = -i_C\frac{1172}{1172 + 1000} = -i_C\frac{1172}{2172}$$

$$= -121 \times i_{IN} \times 0.54$$

$$A_i = \frac{i_L}{i_{IN}} = -65.34$$

PROBLEM 5.12 CURRENT GAIN

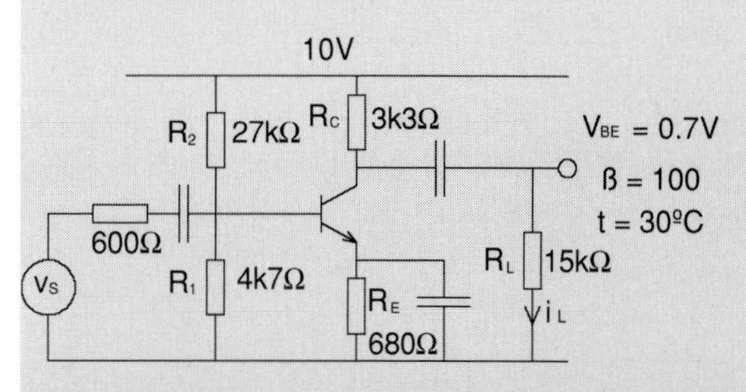

Figure 5.29

Find the current gain for the circuit.

Answer: $A_i = -0.954$

$$V_{TH} = 10\frac{4.7}{4.7 + 27} = 1.48 \text{ V}$$

$$R_{TH} = 4k7||27k = 4003\ \Omega$$

$$I_B = \frac{1.48 - 0.7}{4002 + 680(101)} = \frac{0.78}{72\,683}$$

$$I_E = \frac{0.78}{72683} \times 101 = 1.08 \text{ mA}$$

$$r_e = \frac{26}{1.08} = 24.07\ \Omega$$

AC model

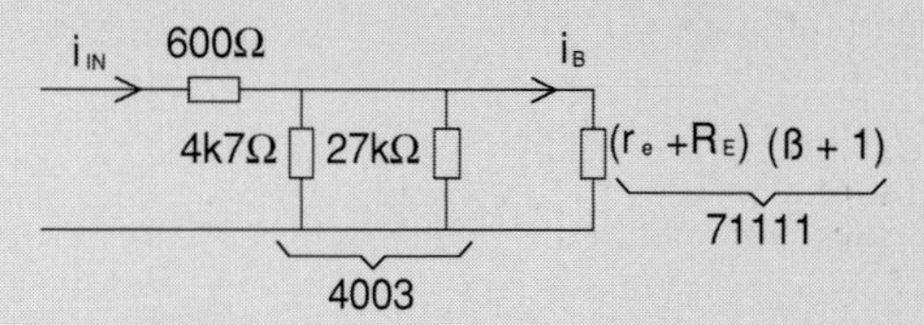

Figure 5.30

$$i_B = i_{IN}\frac{4003}{4003 + 71111} = i_{IN} \times 0.053$$

$$i_C = \beta i_B = 5.3 i_{IN}$$

$$i_L = -i_C\frac{3.3}{3.3 + 15} = -0.18 i_C$$

$$i_L = -0.18 \times 5.3 i_{IN} = -0.954 i_{IN}$$

$$A_i = \frac{i_L}{i_{IN}} = -0.954$$

PROBLEM 5.13 CURRENT GAIN

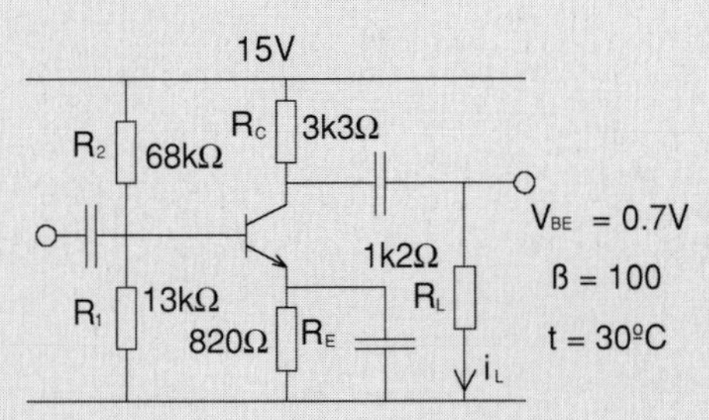

Figure 5.31

Use the Thevenin method to evaluate DC conditions. Find the value of current gain A_i.

Answer: $A_i = -64.8$

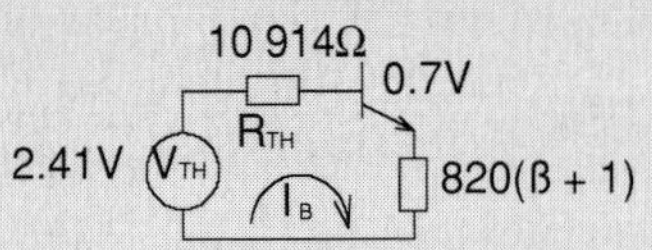

Figure 5.32

$$V_{TH} = 15\frac{13}{13+68} = 2.41\text{ V}$$

$$R_{TH} = 13\text{k}\|68\text{k} = 10\,914\ \Omega$$

$$I_B = \frac{2.41-0.7}{10\,914+82\,820} = \frac{1.71}{93\,734} = 18.24\ \mu\text{A}$$

$$I_E = I_B(\beta+1) = 18.24\times10^{-6}\times101 = 1.84\text{ mA}$$

$$r_e = \frac{26}{1.84} = 14.13\ \Omega$$

AC model

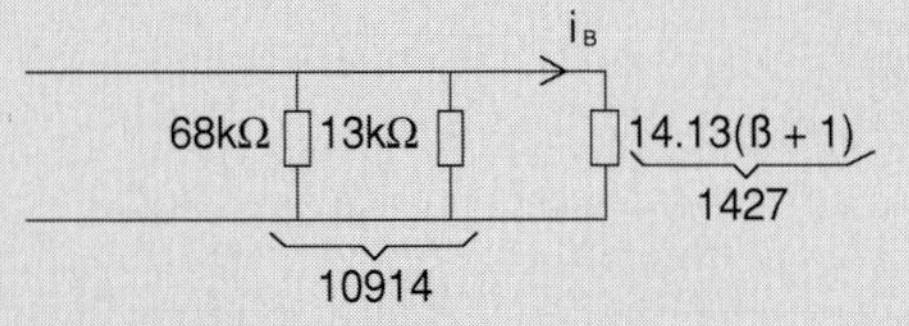

Figure 5.33

$$i_B = i_{IN}\frac{10914}{10914+1427} = i_{IN}\times0.884$$

$$i_C = \beta i_B = 100\times0.884 i_{IN} = 88.4 i_{IN}$$

$$i_L = -i_C\frac{3.3}{3.3+1.2} = -i_C\times0.733$$

$$i_L = -0.733\times88.4 i_{IN} = -64.8 i_{IN}$$

$$A_i = -64.8$$

PROBLEM 5.14 CURRENT GAIN

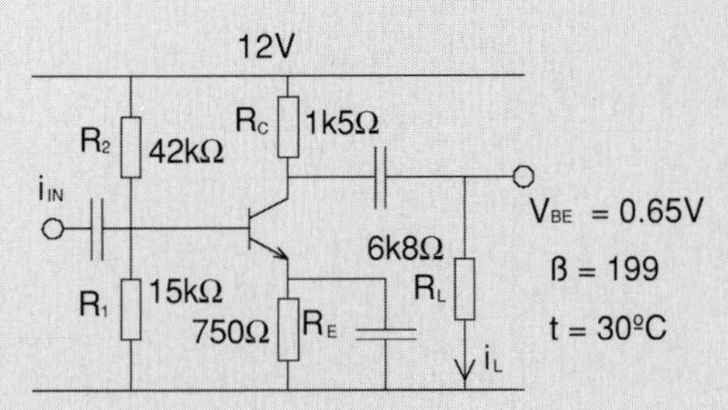

Figure 5.34

Find $A_i = i_L/i_{IN}$ by using the Thevenin method.

Answer: $A_i = -31.3$

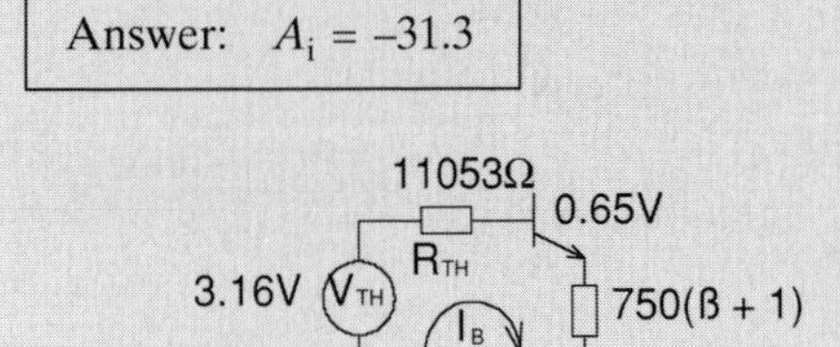

Figure 5.35

$$V_{TH} = 12\frac{15}{15+42} = 3.16\text{ V}$$

$$R_{TH} = 15\text{k}\|42\text{k} = 11053\ \Omega$$

$$I_B = \frac{3.16-0.65}{11\,053+150\,000} = \frac{2.51}{161\,053} = 15.58\ \mu\text{A}$$

$$I_E = (\beta+1)\times I_B = 15.58\times10^{-6}\times200 = 3.12\text{ mA}$$

$$r_e = \frac{26}{3.12} = 8.33\ \Omega$$

AC model

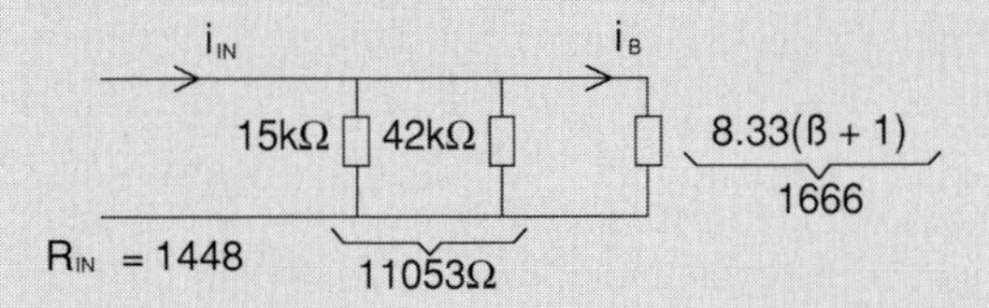

Figure 5.36

$$i_B = i_{IN}\frac{11\,053}{11\,053+1666} = 0.869 i_{IN}$$

$$i_L = -i_C\frac{1.5}{1.5+6.8} = -0.181 i_C$$

$$i_C = \beta i_B = 199\times0.869\times i_{IN}$$

$$i_L = -0.181\times199\times0.869\times i_{IN}$$

$$A_i = -31.3$$

PROBLEM 5.15 CURRENT GAIN

A transistor has the following parameters:

$h_{ie} = 1500\ \Omega$ $\qquad h_{re} = 5 \times 10^{-4}$

$h_{fe} = 99$ $\qquad h_{oe} = 25\ \mu S$

The collector resistance is 3kΩ. Sketch a small signal model and find the current gain, A_i.

Answer: $A_i = -92.1$

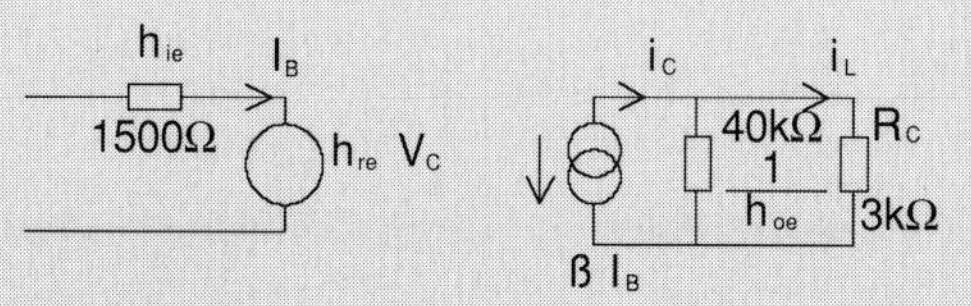

Figure 5.37

$$i_{IN} = i_B$$

$$i_C = \beta i_B$$

$$i_L = -i_C \frac{40k}{40k + 3k}$$

$$i_L = -\beta i_B \frac{40}{43}$$

$$= -i_{IN} \times 99 \times \frac{40}{43}$$

$$A_i = \frac{i_L}{i_{IN}} = -92.1$$

PROBLEM 5.16 CURRENT GAIN

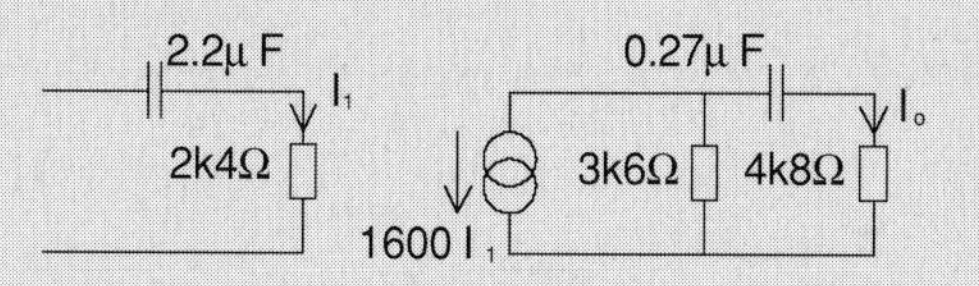

Figure 5.38

Find the current amplification in this equivalent circuit (remember, $A_i = I_o/I_i$).

Answer: $A_i = -688$

$$I_O = -I_C \frac{3.6}{3.6 + 4.8} = -I_C \times 0.43$$

$$I_C = 1600 I_i$$

$$I_O = -1600 \times I_i \times 0.43 = -688 I_i$$

$$\frac{I_O}{I_i} = -688$$

PROBLEM 5.17 CURRENT GAIN

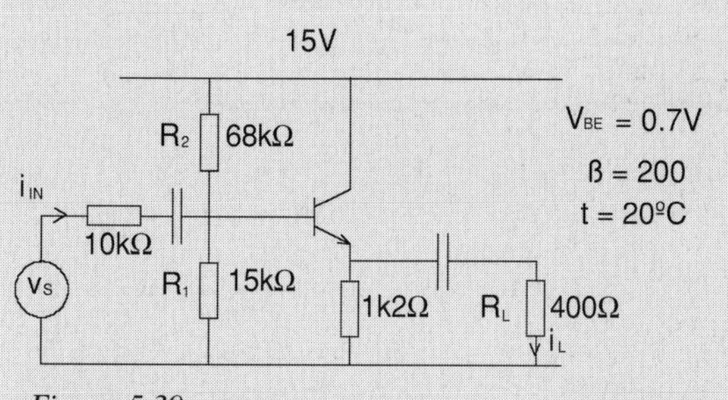

Figure 5.39

Find the value of the output i_L for an input of 5 mA.

Answer: 123.6 mA

Approximate method

$$V_B = 15\frac{15}{15+68} = 2.71 \text{ V}$$

$$V_E = 2.71 - 0.7 = 2.01 \text{ V}$$

$$I_E = \frac{2.01}{1k} = 2.01 \text{ mA}$$

$$r_e = \frac{25}{2.01} = 12.44\ \Omega$$

AC model

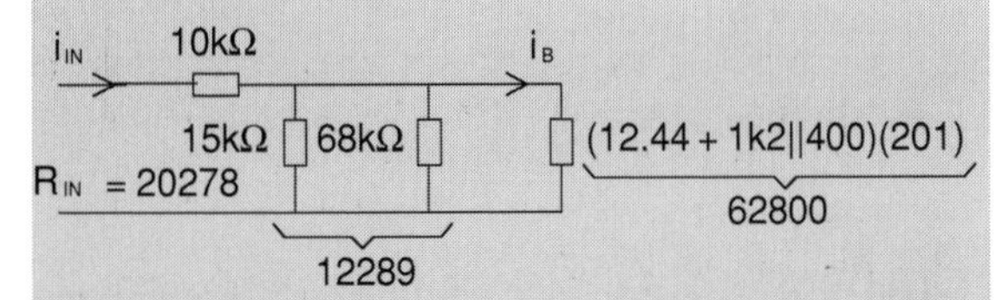

Figure 5.40

$$I_B = i_{IN}\frac{12\,289}{12\,289 + 62\,800} = i_{IN} \times 0.164$$

$$i_E = i_B(\beta + 1) = i_{IN} \times 0.164 \times 201$$

$$= i_{IN} \times 32.96$$

$$i_L = i_E\frac{1200}{1200+400} = 0.75 \times i_E$$

$$i_L = 0.75 \times 32.96 \times i_{IN} = 24.72 \times i_{IN}$$

$$i_L = 24.72 \times 5 \times 10^{-3} = 123.6 \text{ mA}$$

PROBLEM 5.18 CURRENT GAIN

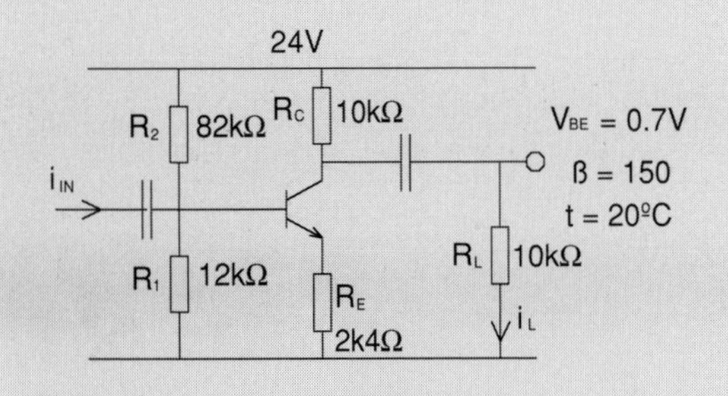

Figure 5.41

Find the output current if the input is 2 mA.

Answer: $i_L = -4.17$ mA

$10R_1$ vs βR_E

120k vs 360k

therefore, use approximate method

$$V_B = V_{CC}\frac{12}{12+82} = 24\frac{12}{94} = 3.06\text{V}$$

$$V_E = V_B - V_{BE} = 3.06 - 0.7 = 2.36 \text{ V}$$

$$I_E = \frac{V_E}{R_E} = \frac{2.36}{2400} = 0.98 \text{ mA}$$

$$r_e = \frac{25}{0.98} = 25.51\ \Omega$$

AC model

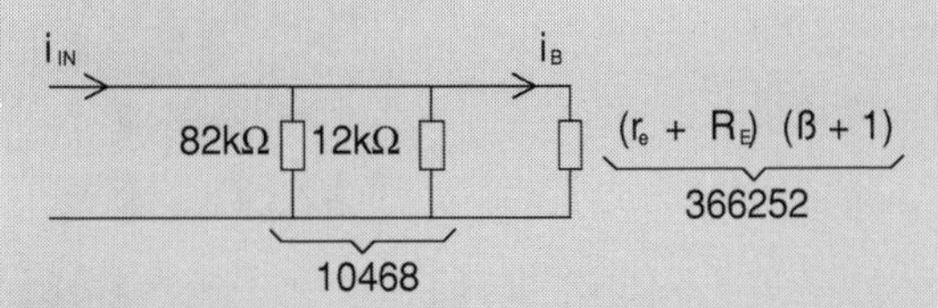

Figure 5.42

$$i_B = i_{IN}\frac{10\,468}{104\,68 + 366\,252} = i_{IN} \times 0.0278$$

$$i_C = \beta i_B = i_{IN} \times 4.17$$

$$i_L = -i_C\frac{10k}{10k + 10k} = 0.5 i_C$$

$$i_L = -2.085 i_{IN}$$

$$i_L = -2.085 \times 2 \times 10^{-3} = -4.17 \text{ mA}$$

PROBLEM 5.19 CURRENT GAIN

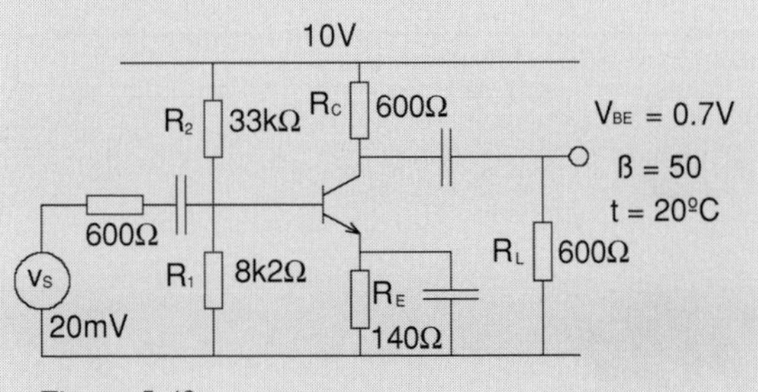

Figure 5.43

Find the value of current i_L.

Answer: $i_L = -0.561$ mA

$10R_1 = 82\text{k}$ vs $R_e\beta = 7\text{k}$,

Therefore, use Thevenin

Figure 5.44

$$V_{TH} = 10\frac{8.2}{41.2} = 1.99 \text{ V}$$

$$R_{TH} = 8k2 \| 33k = 6568\ \Omega$$

$$I_B = \frac{1.99 - 0.7}{6568 + 7140}$$

$$= \frac{1.29}{13708} = 94.1\ \mu\text{A}$$

$$I_E = (\beta + 1)I_B = 94.1 \times 10^{-6} \times 51 = 4.8 \text{ mA}$$

$$r_e = \frac{25}{4.8} = 5.21\ \Omega$$

AC model

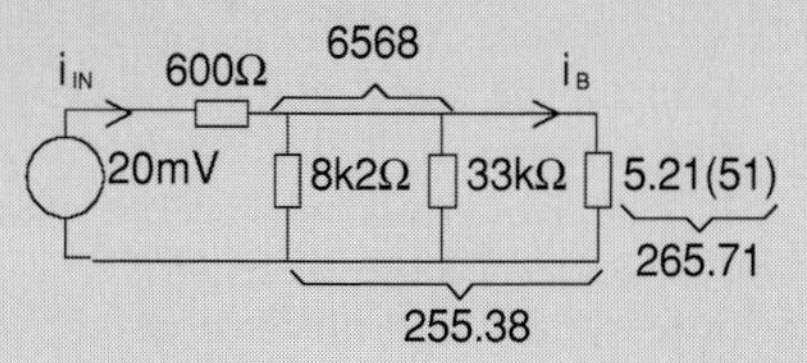

Figure 5.45

$$i_{IN} = \frac{20 \times 10^{-3}}{855.38} = 23.38\ \mu\text{A}$$

$$i_B = i_{IN}\frac{6568}{6568 + 265.71} = 0.96 \times i_{IN} = 22.44\ \mu\text{A}$$

$$i_C = \beta i_B = 50 \times 22.44 = 1.122 \text{ mA}$$

$$i_L = -\frac{i_C}{2} = -0.561 \text{ mA}$$

PROBLEM 5.20 CURRENT GAIN

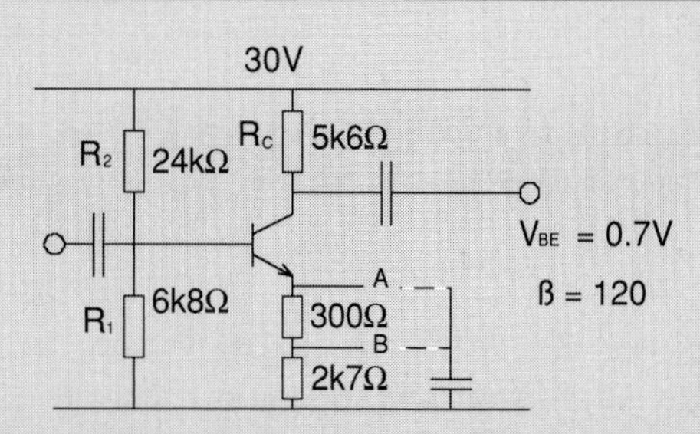

Figure 5.46

Find the amplification with full bypass and partial bypass (capacitor connected to A or B, respectively). Assume $I_C = 2$ mA, and $r_e = 13\ \Omega$.

Answer: 92.4, 14.76

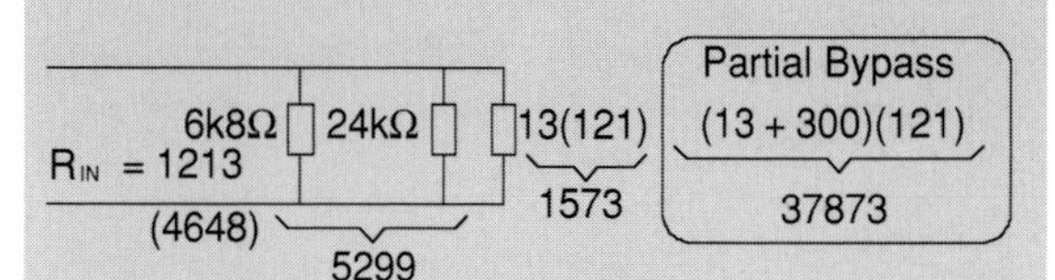

Figure 5.47

Full bypass

$$i_B = i_{IN}\frac{5299}{5299 + 1573} = 0.77 i_{IN}$$

$$i_C = \beta i_B = 120 \times 0.77 i_{IN} = 92.4 i_{IN}$$

$$\frac{i_C}{i_{IN}} = A_i = 92.4$$

Partial bypass

$$i_B = i_{IN}\frac{5299}{5299 + 1573} = 0.77 \times i_{IN}$$

$$i_C = \beta i_B = 120 \times 0.123 \times i_{IN} = 14.76 i_{IN}$$

$$\frac{i_C}{i_{IN}} = A_i = 14.76$$

6

Voltage gain

The typical circuit to calculate the voltage gain or voltage amplification is shown in Figure 6.1.

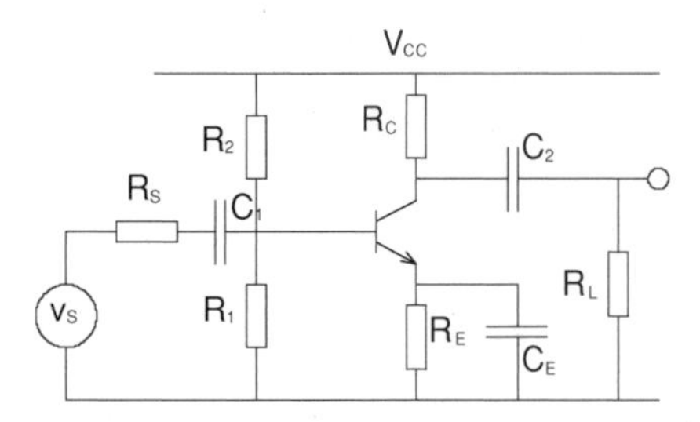

Figure 6.1

The method to calculate the voltage gain, in summary, is:

- Using DC conditions evaluate the currents. With I_E, using the little r_e model equation, calculate r_e.
- Using the modelling, transform the circuit to AC conditions and work out the voltages. The ratio v_o/v_i is the voltage amplification.

Figure 6.2, shows the primary of the transistor, or the input side. The capacitor C_E bypasses R_E

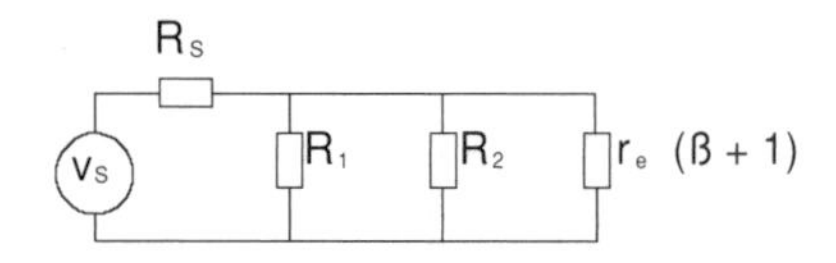

Figure 6.2

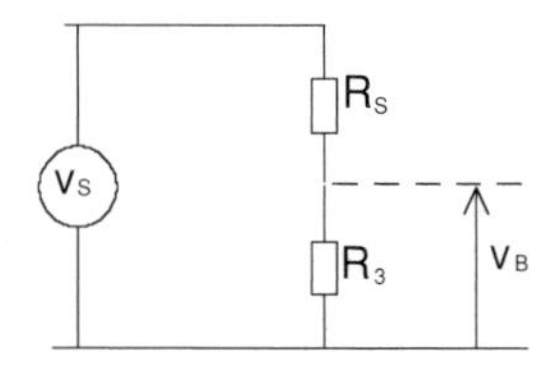

Figure 6.3

and this will not show anywhere. The three resistors R_1, R_2 and $r_e(\beta + 1)$ can be joined into one. We then obtain the simple circuit of Figure 6.3.

R_3 represent the three resistors together. We now see that the base voltage v_B is given by the potential divider rule:

$$v_B = v_S \frac{R_3}{R_S + R_3}$$

We notice here that although we are analysing the voltage gain or amplification, from the source to the base of the transistor we have a reduction in voltage.

The voltage at the base of the transistor will normally be around half the voltage of the source. The amplification will come from the base to the collector of the transistor.

The output side of the transistor can be seen in Figure 6.4.

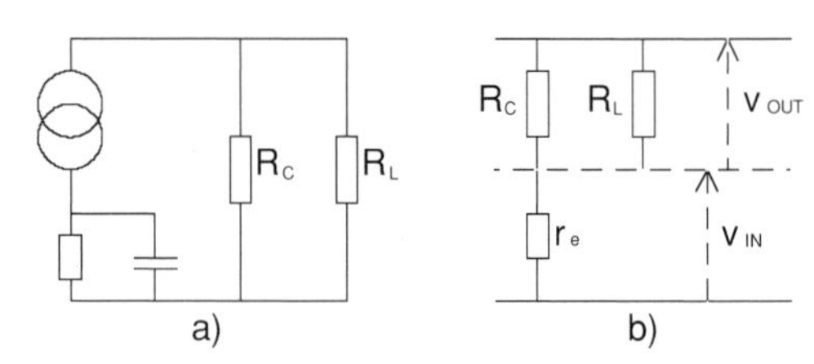

Figure 6.4

Circuit a) is the equivalent model of the output side only. R_E is bypassed, but shown under the current generator. For the equation of input and output, it is best to imagine the circuit in b). R_E has disappeared. The input voltage is applied to r_e. R_L and R_C are in parallel due to the transparency of the DC source (V_{CC}) to the AC signal and because capacitor C_2 is conducting. The capacitor is designed to be fully conducting at the frequencies of interest, whilst it should be open circuit for the DC.

$$v_{base} = i_E r_e$$

$$v_{out} = i_C(R_C \parallel R_L)$$

It can be seen immediately that a simple relation might result if we accept the approximation $i_E \approx i_C$. In this case:

$$\boxed{\frac{v_{out}}{v_{base}} = -\frac{R_C \| R_L}{r_e}}$$

so

$$\frac{v_o}{v_s} = -\frac{\frac{R_3}{R_S + R_3}(R_C \| R_L)}{r_e}$$

This equation is only valid for the circuit of Figure 6.1, which is under consideration.

If we are dealing with another type of circuit we must similarly work out the equation which corresponds to that circuit and which is a mathematical model of that particular circuit.

From the equation we can see that the gain can be varied greatly if R_E is introduced in the equation. As R_E is bypassed by C_E, the presence or absence of C_E is crucial.

During our problems we will encounter three different configurations for the R_E circuit, as seen in Figure 6.5. If we are considering DC conditions, all the resistors form part of the circuit.

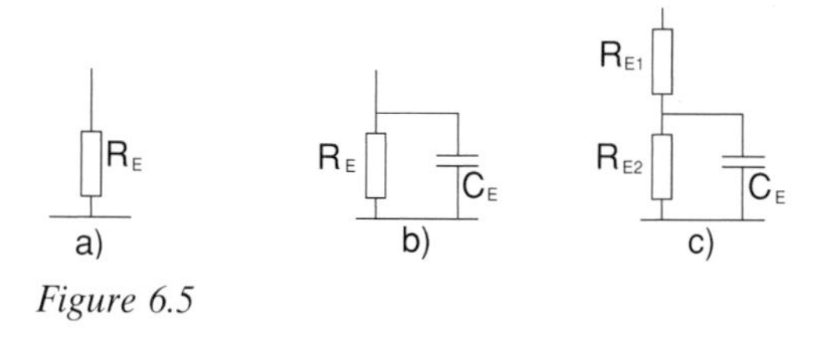

Figure 6.5

However, in AC conditions:

for circuit a) R_E will be present,
for circuit b) R_E will be absent,
for circuit c) R_{E1} will be present and R_{E2} will be absent

Emitter follower

The relevant part of an emitter follower circuit is shown in Figure 6.6. In this case the input voltage v_B is applied to the base of the transistor. The output voltage is taken from the emitter. Under

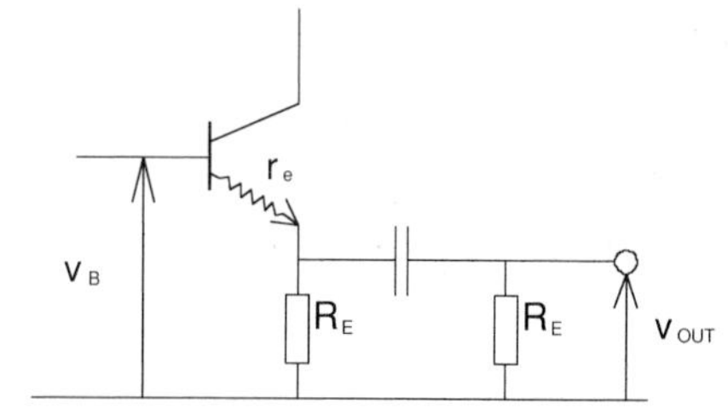

Figure 6.6

AC conditions, the capacitor is conducting so the output R_L is in parallel with R_E.

$$v_{out} = i_E(R_E \| R_L)$$

$$v_B = i_E(r_e + (R_E \| R_L))$$

$$\frac{v_{out}}{v_B} = \frac{R_E \| R_L}{r_e + (R_E \| R_L)}$$

The input and the output are in phase. The output, although near to 1 is a fraction less than 1.

In order to understand that the common emitter has an opposition phase between input and output and that the emitter follower output is in phase with the input, it is useful to examine Figure 6.7, where both outputs are allowed from a unique circuit called the phase splitter.

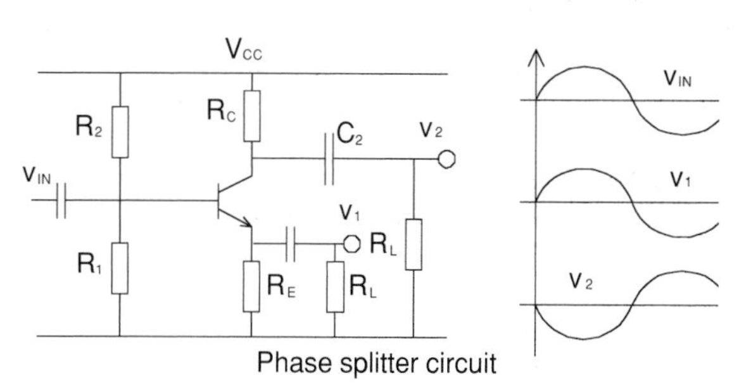

Figure 6.7

PROBLEM 6.1 VOLTAGE AMPLIFICATION

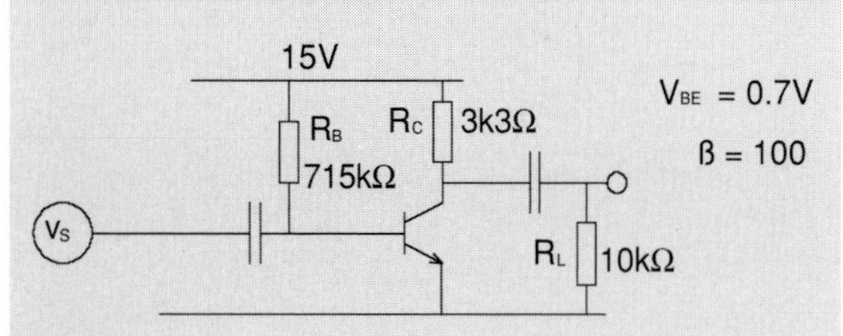

Figure 6.8

If v_i = 10 mV, find the voltage across the 10k resistor at 20°C.

Answer: –2 V

$$I_B = \frac{15-0.7}{715k} = 20\ \mu A$$

$$I_C = 20\ \mu A \times 100 = 2\ mA$$

$$I_E \approx I_C = 2\ mA$$

$$r_e = \frac{25}{2} = 12.4\ \Omega$$

$$A_V = -\frac{R_C || R_L}{r_e} = -\frac{3k3||10k}{12.5}$$

$$= -198.5$$

$$v_L = -10 \times 10^{-3} \times 198.5 = -1.98\ V$$

More accurately

$$I_E = I_C\frac{\beta+1}{\beta} = 2\frac{101}{100} = 2.02\ mA$$

$$r_e = \frac{25}{2.02} = 12.38$$

$$A_V = -\frac{3k3||10k}{12.38} = -\frac{2481.2}{12.38} = -200.42$$

$$v_L = -10 \times 10^{-3} \times 200.42 = -2\ V$$

PROBLEM 6.2 VOLTAGE AMPLIFICATION

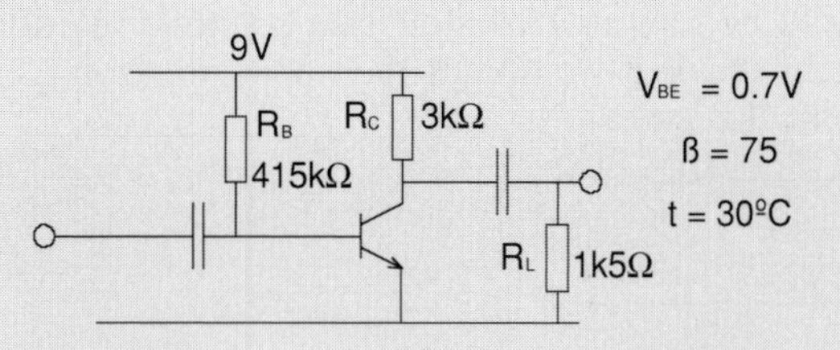

Figure 6.9

Find the voltage amplification for this transistor circuit.

Answer: –60.02

$$I_B = \frac{9-0.7}{415k} = 20\ \mu A$$

$$I_C = 20 \times 75 \times 10^{-6} = 1.5\ mA$$

$$I_C \approx I_E$$

$$r_e = \frac{25}{1.5} = 16.66\ \Omega$$

$$\text{Amplification} = -\frac{R_C || R_L}{r_e} = -\frac{3k||1k5}{16.66}$$

$$= -60.02$$

PROBLEM 6.3 VOLTAGE AMPLIFICATION

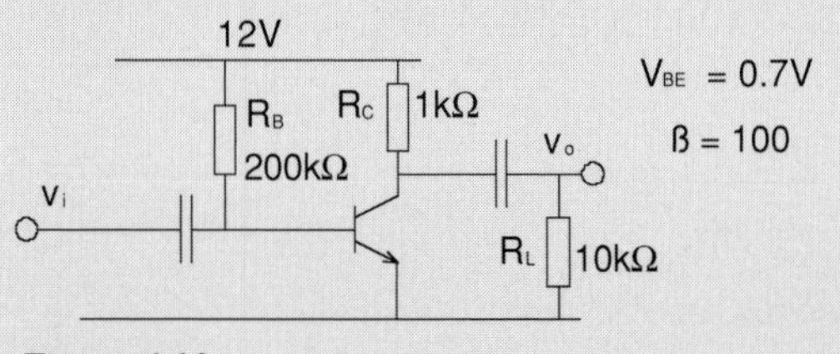

Figure 6.10

Find the value of r_e at 20°C, the voltage amplification, and the value of the output for an input of 15 mV RMS.

Answer: r_e = 4.38 Ω, A_V = –249.75, v_o = –3.75 V

$$I_B = \frac{V_{CC} - 0.7}{R_B} = \frac{12 - 0.7}{200k} = 56.5\ \mu A$$

$$I_E = I_B + I_C = I_B + \beta I_B = I_B(\beta + 1)$$

$$= 56.5 \times 10^{-6} \times (101) = 5.7\ mA$$

$$r_e = \frac{25}{I_E} \text{ at } 20°C$$

$$= \frac{25}{5.7} = 4.38\ \Omega$$

$$A_V = -\frac{R_C || R_L}{r_e} = -\frac{909.09}{3.64} = -249.75$$

$$v_o = v_i A_V = -3.75\ V$$

PROBLEM 6.4 VOLTAGE AMPLIFICATION

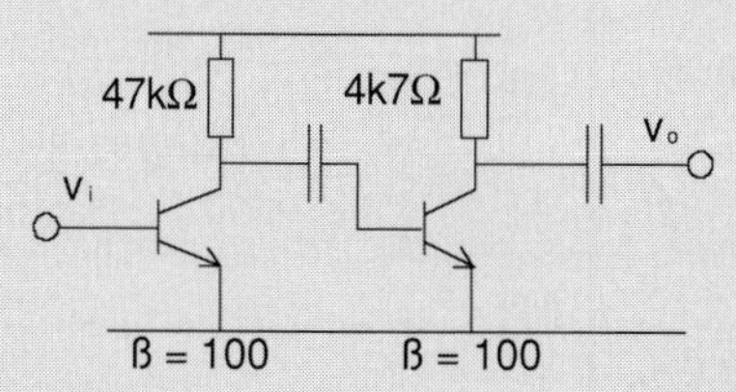

Figure 6.11

Assume I_{EQ1} = 100 µA, I_{EQ2} = 1 mA. Find R_{IN} and A_V.

Answer: 25k25 Ω, 1802

$$r_{e1} = \frac{25}{0.1} = 250\ \Omega$$

$$r_{e2} = \frac{25}{1} = 25\ \Omega$$

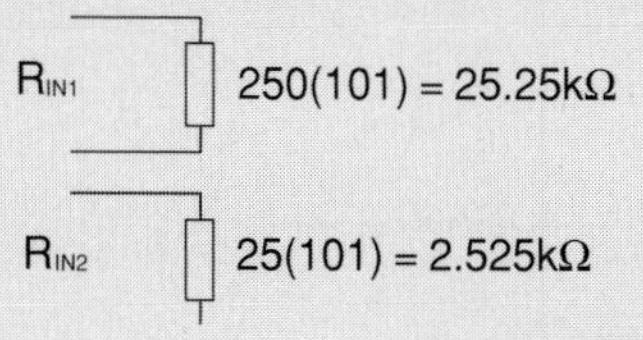

R_{IN1} 250(101) = 25.25kΩ

R_{IN2} 25(101) = 2.525kΩ

Figure 6.12

$$A_{V1} = \frac{47k || R_{IN2}}{250} = -9.585$$

$$A_{V2} = \frac{4k7}{25} = -188$$

$$A_V = 1802$$

PROBLEM 6.5 VOLTAGE AMPLIFICATION

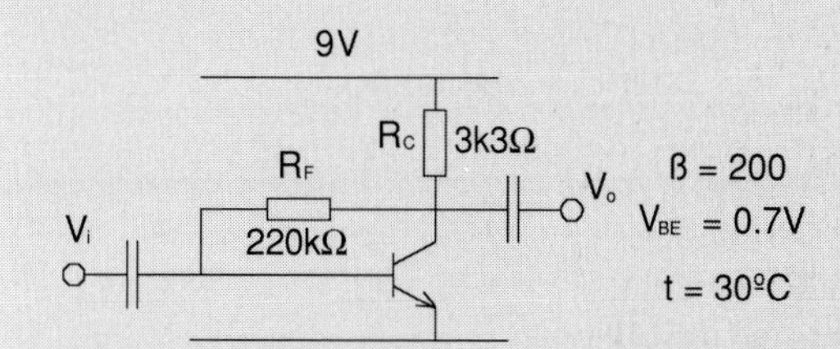

Figure 6.13

Find the voltage gain.

Answer: A_V = –240

$$I_B = \frac{9 - 0.7}{220k + (\beta + 1)3k3}$$

$$= \frac{8.3}{220\,000 + 663\,300}$$

$$= \frac{8.3}{883\,300} = 9.4\ \mu A$$

$$I_E = (\beta + 1)I_B = 201 \times 9.4 \times 10^{-6} = 1.89\ mA$$

$$r_e = \frac{26}{1.89} = 13.76\ \Omega$$

$$A_V = \frac{R_C}{r_e} = -\frac{3300}{13.76} = -240$$

PROBLEM 6.6 VOLTAGE AMPLIFICATION

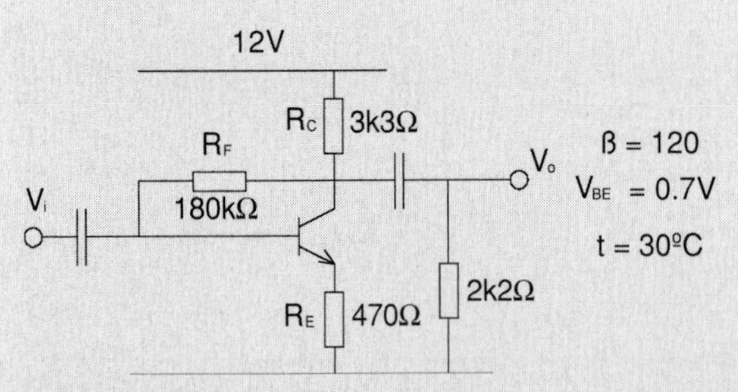

Figure 6.14

Find the voltage amplification, A_V.

> Answer: $A_V = -2.74$

We find I_B using KVL

$$12 - 0.7 = I_E(3k3 + 470) + I_B \times 180k$$

Note: 3k3 Ω resistor has I_E and I_B, therefore, it is I_E. (= I_C + I_B)

$$11.3 = I_B \times 121 \times 3770 + 180\,000 I_B$$

$$11.3 = I_B \times 456\,170 + 180\,000 I_B$$

$$I_B = \frac{11.3}{636\,170} = 17.76\ \mu A$$

$$I_E = I_B(\beta + 1) = 17.76 \times 10^{-6} \times 121$$

$$= 2.15\ mA$$

$$r_e = \frac{26}{2.15} = 12.09\ \Omega$$

$$A_V = -\frac{R_C \| R_L}{r_e + R_E} = -\frac{3k3 \| 2k2}{12.09 + 470}$$

$$= -\frac{1320}{482.09} = -2.74$$

PROBLEM 6.7 VOLTAGE AMPLIFICATION

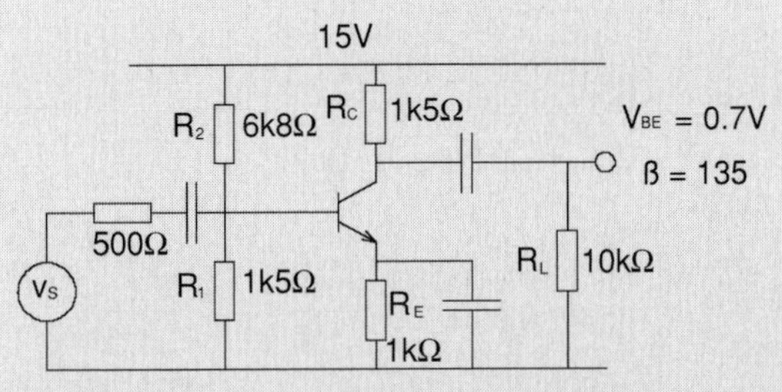

Figure 6.15

If the input is set to 20 mV, find the value of the voltage across the load:

(a) with C_E present
(b) with C_E removed (temperature = 30°C)

> Answer: −1.12 V and −18.28 mV, respectively.

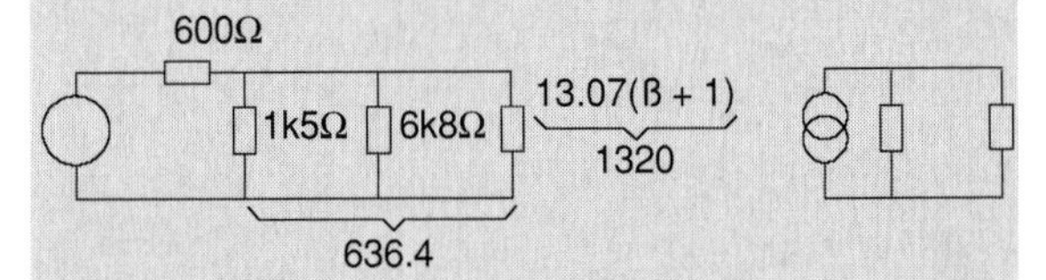

Figure 6.16

Use Thevenin for DC conditions

$$V_{TH} = 15\frac{1k5}{1k5 + 6k8} = 2.71\ V$$

$$R_{TH} = 1k5 \| 6k8 = 1229\ \Omega$$

$$I_B = \frac{2.71 - 0.7}{1229 + 10\,100} = \frac{2.01}{102\,229} = 19.66\ \mu A$$

$$I_E = I_B(\beta + 1) = 19.66 \times 10^{-6} \times 101 = 1.99\ mA$$

$$r_e = \frac{26}{1.99} = 13.07\ \Omega$$

Potential divider for source

$$v_B = v_S\frac{636.4}{500 + 636.4} = 20\frac{636.4}{1136.4} = 11.2\ mV$$

$$A_V = -\frac{R_C \| R_L}{r_e} = -\frac{1k5 \| 10k}{13.07}$$

$$= -\frac{1304}{13.07} = -99.77$$

$$V_L = 11.2 \times 10^{-3} \times (-99.77) = -1.12\ V$$

Now without a capacitor

The value of r_e will remain unchanged and it is based on the DC conditions. The potential divider will change.

The term $r_e(\beta + 1)$ becomes

$$(r_e + R_E)\times(\beta+1)$$

$$(13.07+1000)\times 101 = 102320$$

The three resistors in parallel become

$$1k5 \| 6k8 \| 102320 = 1214.33\ \Omega$$

$$v_B = v_S\frac{1214.33}{500+1214.33} = 14.17\ \text{mV}$$

The amplification changes a lot!

$$A_V = -\frac{R_C \| R_L}{r_e + R_E} = -\frac{1304}{1013} = -1.29$$

$$V_L = 14.17\times(-1.29) = -18.28\ \text{mV}$$

PROBLEM 6.8 VOLTAGE AMPLIFICATION

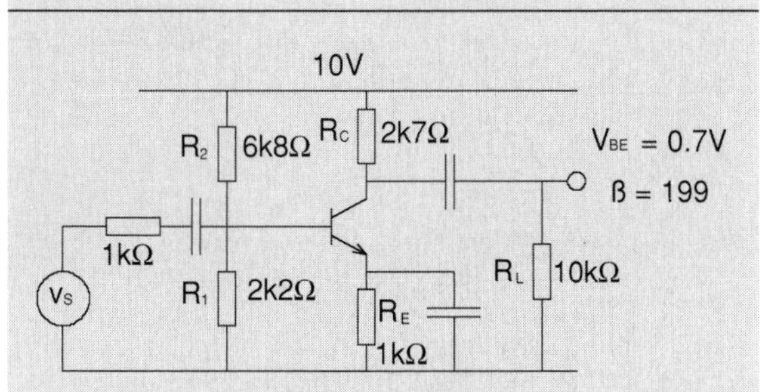

Figure 6.17

Calculate R_{IN} and $\frac{v_o}{v_s}$.

Answer: 1052 Ω, –76.11

DC conditions, approximate method

$$V_B = 10\frac{2k2}{9k} = 2.444\ \text{V}$$

$$V_E = 2.444 - 0.7 = 1.744\ \text{V}$$

$$I_E = \frac{1.744}{1k} = 1.744\ \text{mA}$$

$$r_e = \frac{25}{1.744} = 14.33\ \Omega$$

AC conditions

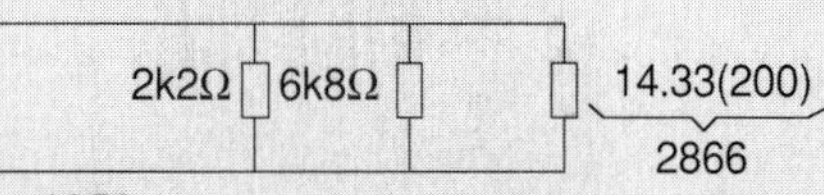

R_{IN} = 1052

Figure 6.18

$$\frac{v_B}{v_S} = \frac{R_{IN}}{R_{IN}+1k} = \frac{1052}{2052} = 0.513$$

$$\frac{v_O}{v_B} = -\frac{2k7 \| 10k}{14.33} = -148.36$$

$$A_V = -0.513\times 148.36 = -76.11$$

PROBLEM 6.9 VOLTAGE AMPLIFICATION

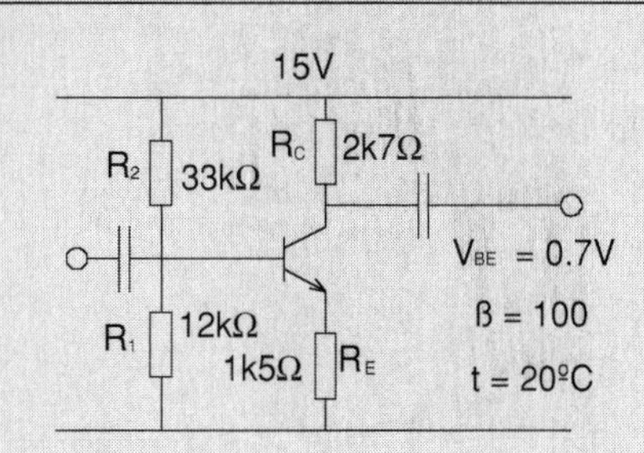

Figure 6.19

Find R_{IN} and A_V.

Answer: 8320 Ω, –1.786

Approximate method

$$V_B = 15\frac{12k}{45k} = 4 \text{ V}$$

$$V_E = 4 - 0.7 = 3.3 \text{ V}$$

$$I_E = \frac{3.3}{1k5} = 2.2 \text{ mA}$$

$$r_e = \frac{25}{2.2} = 11.36\ \Omega$$

AC conditions

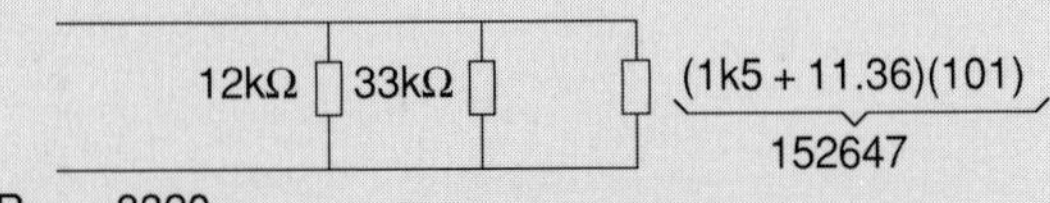

Figure 6.20

$$A_V = -\frac{2k7}{1511.36} = -1.786$$

NOTE: With bypass capacitor for R_E, the results are $R_{IN} = 1015\ \Omega$, $A_v = -237$.

PROBLEM 6.10 VOLTAGE AMPLIFICATION

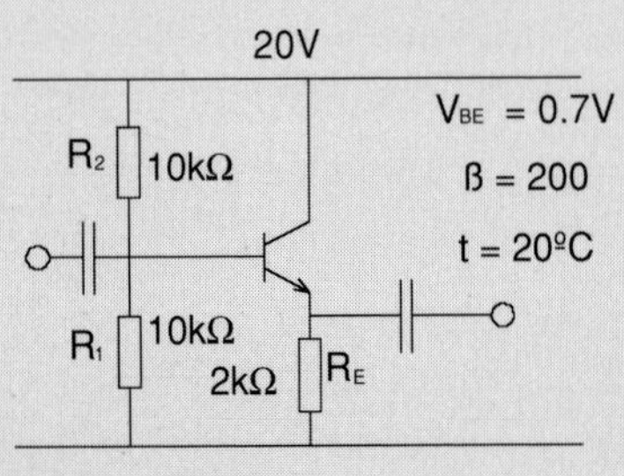

Figure 6.21

Calculate R_{IN} and A_V.

Answer: 4939 Ω, 0.977

Approximate method

$$V_B = 10 \text{ V} \quad V_E = 9.3 \text{ V}$$

$$I_E = \frac{9.3}{2k} = 4.65 \text{ mA}$$

$$r_e = \frac{25}{4.65} = 5.376\ \Omega$$

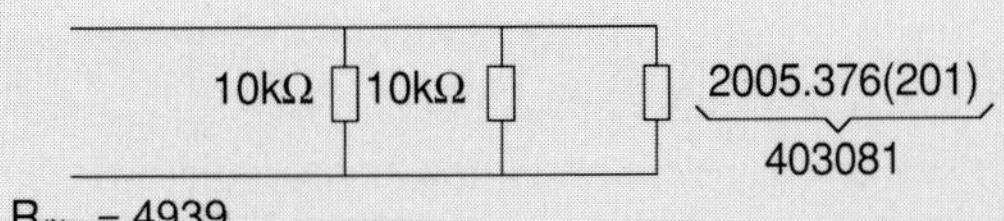

Figure 6.22

$$A_V = \frac{R_E}{R_E + r_e} = \frac{2k}{2k + 5.376} = 0.997$$

PROBLEM 6.11 VOLTAGE AMPLIFICATION

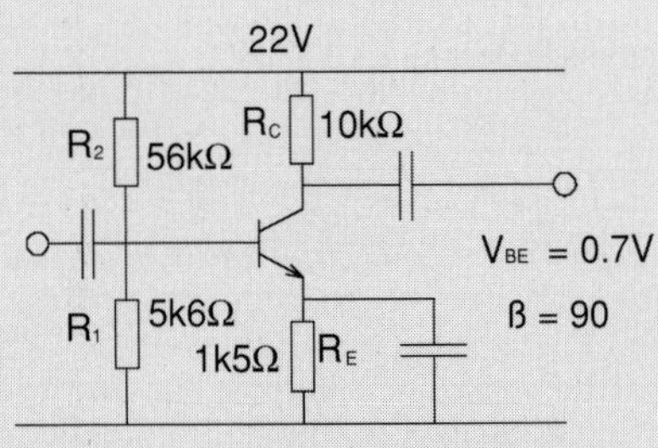

Figure 6.23

Use Thevenin to calculate the DC conditions and find the value of R_{IN} and A_V at 20°C and 30°C.

Answer: 1820 Ω, –321 (at 30°C)
1774 Ω, –334 (at 20°C)

DC conditions

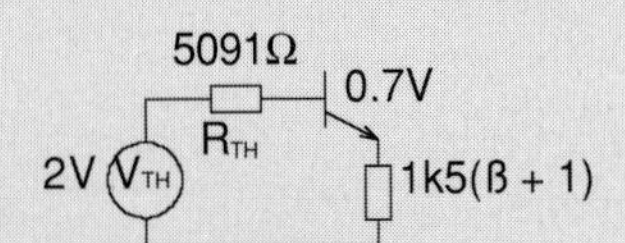

Figure 6.24

$$V_{TH} = 22\frac{5k6}{61k6} = 2 \text{ V}$$

$$R_{TH} = 5k6 \| 56k = 5091$$

$$I_B = \frac{2 - 0.7}{5091 + 136\,500} = \frac{1.3}{141\,591}$$

$$I_E = \frac{1.3 \times 91}{141\,591} = 0.835\,5 \text{ mA}$$

$$r_e = \frac{26}{0.835\,5} = 31.12\ \Omega$$

AC conditions

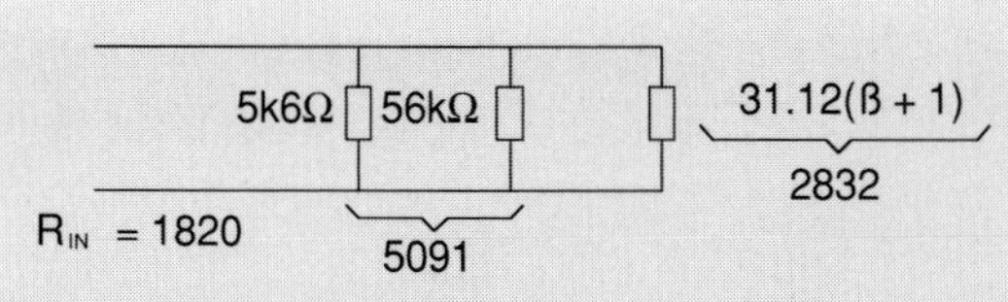

Figure 6.25

$$A_V = -\frac{10k}{31.12} = -321$$

NOTE: This is the solution at 30°C. You can verify the results for 20°C.

PROBLEM 6.12 VOLTAGE AMPLIFICATION

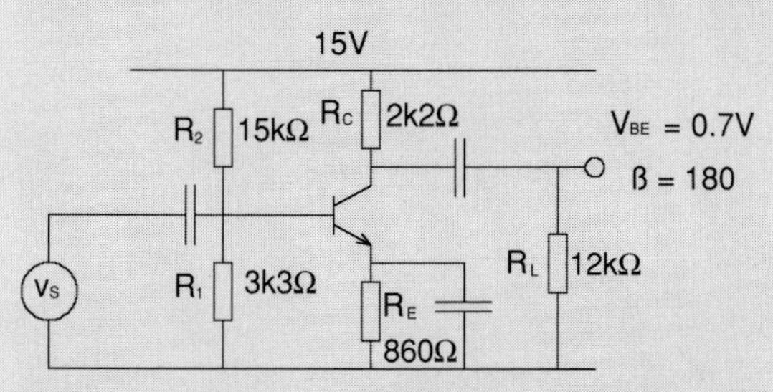

Figure 6.26

Find the voltage gain. Temperature = 20°C.

Answer: –173

Use the approximate method

$$V_B = 15\frac{3.3}{3.3+15} = 15\frac{3.3}{18.3} = 2.7 \text{ V}$$

$$V_E = 2.7 - 0.7 = 2.0 \text{ V}$$

$$I_E = \frac{2}{860} = 2.33 \text{ mA}$$

$$r_e = \frac{25}{2.33} = 10.73\ \Omega$$

$$A_V = -\frac{R_E \| R_L}{r_e} = -\frac{2k2 \| 12k}{r_e}$$

$$= -\frac{1859}{10.73} = -173$$

PROBLEM 6.13 VOLTAGE AMPLIFICATION

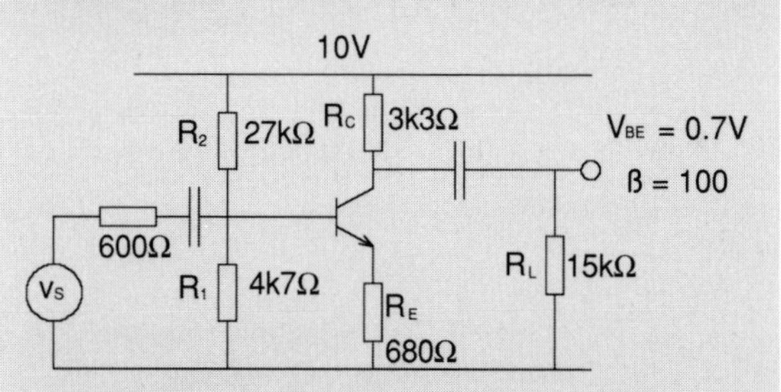

Figure 6.27

Find the voltage gain.
a) as shown
b) with bypass capacitor

Answer: a) –3.3, b) –80.81

$$V_{TH} = 10\frac{4.7}{4.7+27} = 1.48 \text{ V}$$

$$R_{TH} = 4k7 \| 27k = 4003\ \Omega$$

$$I_B = \frac{1.48 - 0.7}{4003 + 680(101)} = \frac{0.78}{72683}$$

$$I_E = \frac{0.78}{72683} \times 101 = 1.08 \text{ mA}$$

$$r_e = \frac{26}{1.08} = 24.07\ \Omega$$

AC circuit

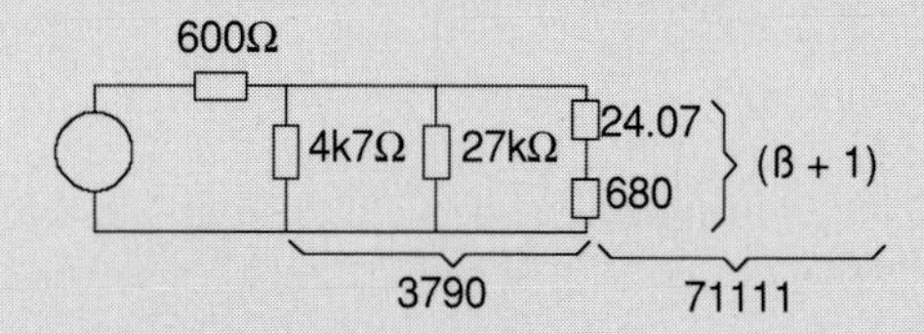

Figure 6.28

$$v_B = v_S\frac{3790}{3790+600}$$

$$= \frac{3790}{4390} = 0.86 v_S$$

Transistor gain

$$A_{VT} = -\frac{R_C \| R_L}{r_e + R_E}$$

$$= -\frac{2705}{704.07} = -3.84$$

Total gain $A_V = -0.86 \times 3.84 = -3.3$

With bypass, r_e is the same.

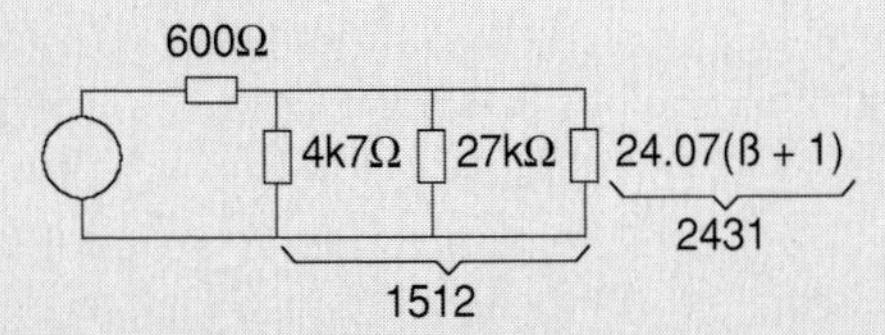

Figure 6.29

$$v_B = v_S \frac{1512}{1512 + 600} = \frac{1512}{2112} v_S = 0.72 v_S$$

$$A_{VT} = -\frac{R_C || R_L}{r_e}$$

$$= -\frac{2705}{24.07} = -112.38$$

Total gain $A_V = -0.72 \times 112.38 = -80.91$

PROBLEM 6.14 VOLTAGE AMPLIFICATION

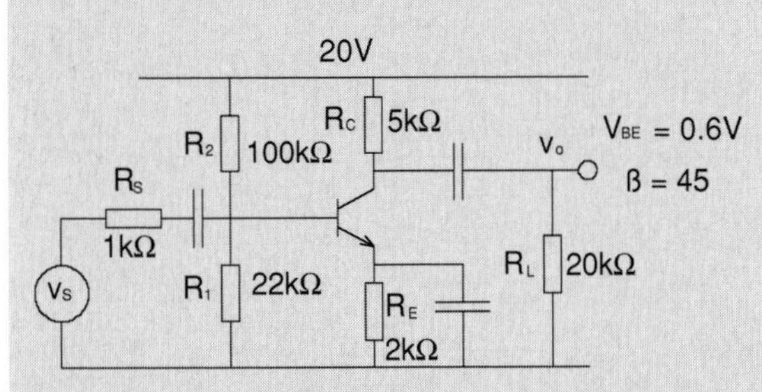

Figure 6.30

Find A_V, the voltage gain. Use the Thevenin method.

Answer: $A_V = -91.92$

$$V_{TH} = 20 \frac{22}{22 + 100} = 20 \frac{22}{122} = 3.61 \text{ V}$$

$$R_{TH} = 22\text{k} || 100\text{k} = 18033 \ \Omega$$

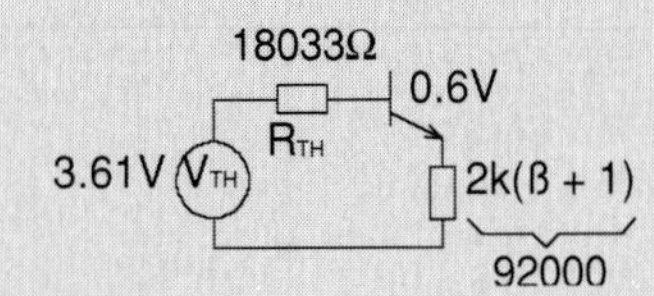

Figure 6.31

$$I_B = \frac{3.61 - 0.6}{18\,033 + 92\,000}$$

$$= \frac{3.01}{110\,033} = 27.355 \ \mu\text{A}$$

$$I_E = I_B \times 46 = 1.258 \text{ mA}$$

$$r_e = \frac{26}{1.258} = 20.67 \ \Omega$$

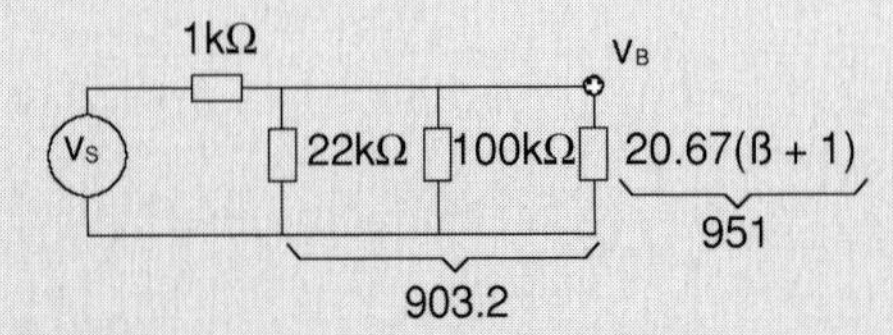

Figure 6.32

$$v_B = v_S \frac{903.2}{1\text{k} + 903.2} = v_S \times 0.475$$

$$\frac{v_O}{v_B} = -\frac{R_C || R_L}{r_e}$$

$$= -\frac{20\text{k}||5\text{k}}{20.67} = -193.52$$

$$A_V = \frac{v_O}{v_S} = \frac{v_B v_O}{v_S v_b}$$

$$= 0.475 \times (-193.52) = -91.92$$

PROBLEM 6.15 VOLTAGE AMPLIFICATION

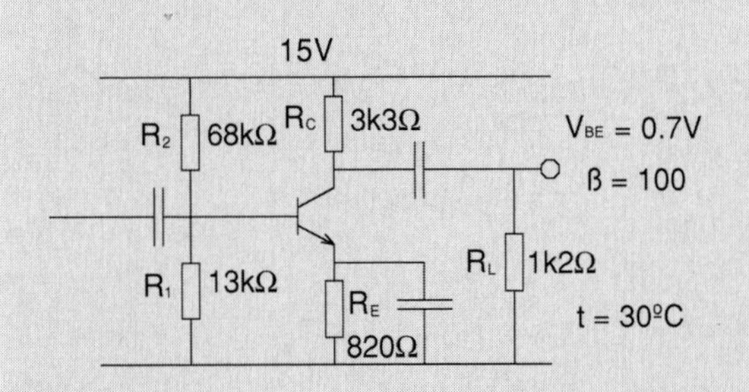

Figure 6.33

Find the voltage gain, using the Thevenin method.

Answer: −61.93

$$V_{TH} = 15\frac{13}{13+68} = 15\frac{13}{81} = 2.4 \text{ V}$$

$$R_{TH} = 13\text{k}||68\text{k} = 10.9 \text{ k}\Omega$$

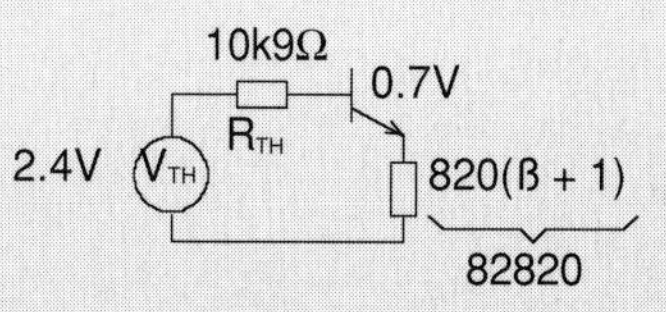

Figure 6.34

$$I_B = \frac{2.4-0.7}{10\,900+82\,820} = \frac{1.7}{93\,720} = 18.14\,\mu\text{A}$$

$$I_E = 18.14\times10^{-6}\times101 = 1.83 \text{ mA}$$

$$r_e = \frac{26}{1.83} = 14.21\,\Omega$$

$$A_V = -\frac{R_C||R_L}{r_e} = -\frac{3\text{k}3||1\text{k}2}{14.21}$$

$$= -\frac{880}{14.21} = -61.93$$

PROBLEM 6.16 VOLTAGE AMPLIFICATION

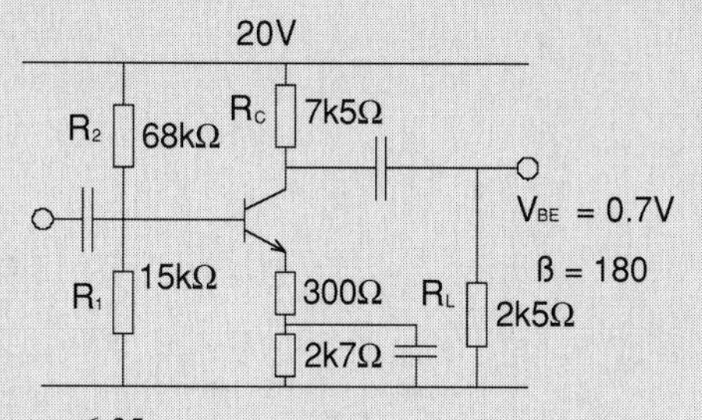

Figure 6.35

Assume I_C = 1.2 mA, and r_e = 21 Ω.
Find the voltage amplification, A_V.

Answer: $A_V = -5.84$

$$A_V = -\frac{R_C||R_L}{r_e+300} = -\frac{7\text{k}5||2\text{k}5}{321}$$

$$= -\frac{1875}{321} = -5.84$$

PROBLEM 6.17 VOLTAGE AMPLIFICATION

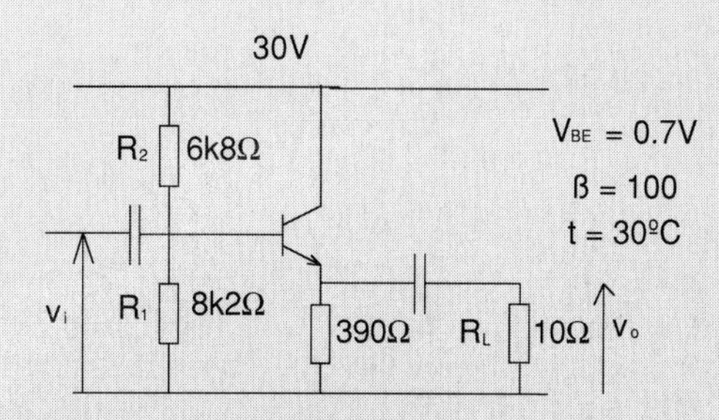

Figure 6.36

Find the voltage gain, using the approximate method.

Answer: $A_V = 0.938$

$$V_B = 30\frac{8.2}{8.2+6.8} = 16.4 \text{ V}$$

$$V_E = 16.4 - 0.7 = 15.7 \text{ V}$$

$$I_E = \frac{15.7}{390} = 40.26 \text{ mA}$$

$$r_e = \frac{26}{40.26} = 0.646\,\Omega$$

$$A_V = \frac{390||10}{(390||10)+r_e} = \frac{9.75}{9.75+0.646}$$

$$= \frac{9.75}{10.396}$$

$$A_V = 0.938$$

PROBLEM 6.18 VOLTAGE AMPLIFICATION

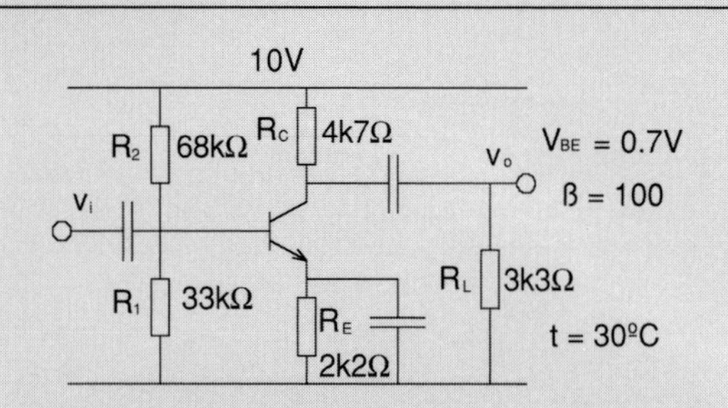

Figure 6.37

Find the voltage gain.

Answer: $A_V = -79$

$10R_1 = 330\text{k}$ vs $\beta R_E = 220\text{k}$

Therefore, use the Thevenin method

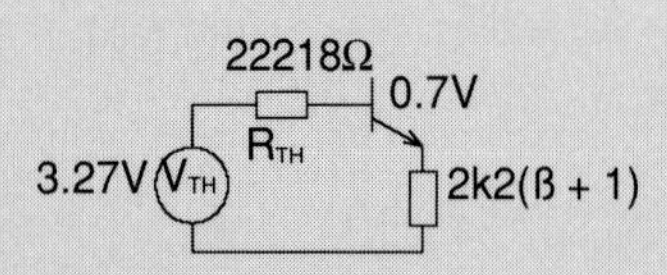

Figure 6.38

$$V_{TH} = 10\frac{33}{33+68} = 3.27\text{ V}$$

$$R_{TH} = 33\text{k}\|68\text{k} = 22218\ \Omega$$

$$I_B = \frac{3.27-0.7}{22\,218+222\,200}$$

$$= \frac{2.57}{244\,418} = 10.51\ \mu\text{A}$$

$$I_E = I_B(\beta+1) = 10.51\times10^{-6}\times101 = 1.06\text{ mA}$$

$$r_e = \frac{26}{1.06} = 24.53\ \Omega$$

$$A_V = -\frac{R_C\|R_L}{r_e} = -\frac{4\text{k}7\|3\text{k}3}{24.53}$$

$$= -\frac{1939}{24.53} = -79$$

PROBLEM 6.19 VOLTAGE AMPLIFICATION

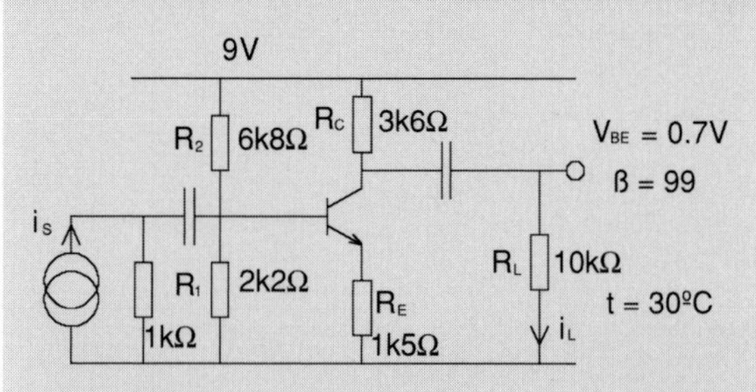

Figure 6.39

Find the voltage amplification.

Answer: $A_V = -1.73$

Approximate method

$$V_B = 9\frac{2.2}{2.2+6.8} = 2.2\text{ V}$$

$$V_E = 2.2-0.7 = 1.5\text{ V}$$

$$I_E = \frac{1.5}{1.5\text{k}} = 1\text{ mA}$$

$$r_e = \frac{26}{1} = 26\ \Omega$$

$$A_V = -\frac{R_C\|R_L}{R_E+r_e} = -\frac{3\text{k}6\|10\text{k}}{26+1500}$$

$$= -\frac{2647}{1526}$$

$$A_V = -1.73$$

PROBLEM 6.20 VOLTAGE AMPLIFICATION

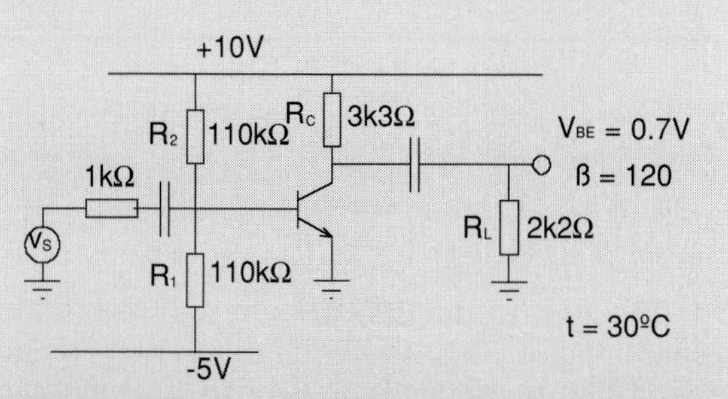

Figure 6.40

Find the voltage gain, A_V.

Answer: $A_V = -113.72$

We find I_B using Thevenin (different!)

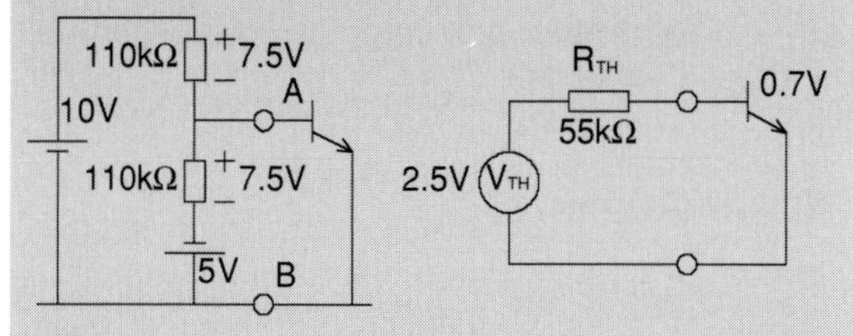

Figure 6.41

$$V_{TH} = 2.5 \text{ V}$$

$$R_{TH} = \frac{110\text{k}}{2} = 55 \text{ k}\Omega$$

$$I_B = \frac{2.5 - 0.7}{55\text{k}} = 32.73\ \mu\text{A}$$

$$I_E = (\beta + 1)I_B = 121 \times 32.73 \times 10^{-6} = 3.96 \text{ mA}$$

$$r_e = \frac{26}{3.96} = 6.57\ \Omega$$

AC model

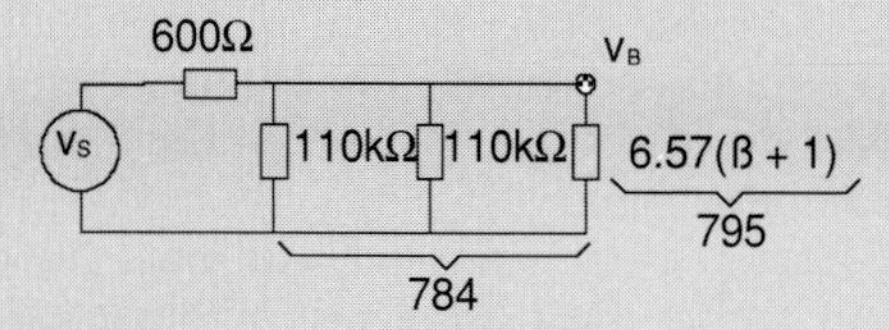

Figure 6.42

$$v_B = v_S \frac{784}{600 + 784} = v_S \frac{784}{1384} = 0.566 v_S$$

$$A_V = -0.566 \frac{3\text{k}3 \| 2\text{k}2}{6.57} = -113.72$$

7

Cascaded systems

One or more amplifying stages can be joined together to increase the overall amplification. The gain impedance formula, seen as part of the revision earlier in the book, is highly relevant. We are going to concentrate on two stages and we will see them in the amplification of the voltage and current.

Cascaded voltage gain

The typical circuit is seen in Figure 7.1. The method employed consists of breaking up the total voltage gain, into as many independent parts as possible. Then, we consider each separate part independently, one at a time.

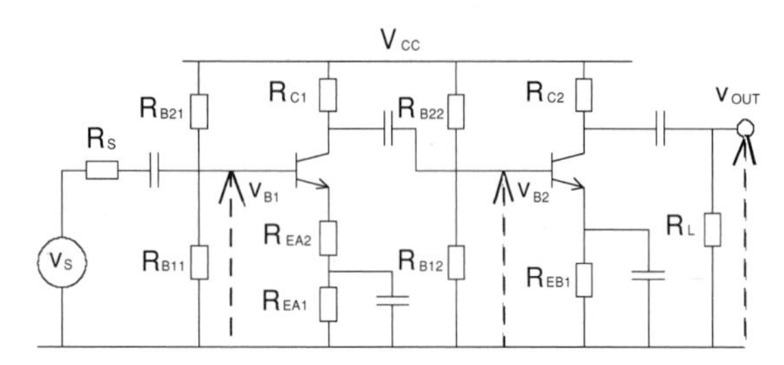

Figure 7.1

We can identify the following parts and therefore this will be the method to follow

$$\frac{v_o}{v_s} = \frac{v_{B1}}{v_s}\frac{v_{B2}}{v_{B1}}\frac{v_o}{v_{B2}}$$

In this case the total amplification is made up of three parts.

First from the source to the base of the first stage. As we know, from a single-stage amplifier, this value is usually a reduction to about 50% of the voltage source, but this depends on the source impedance and the input impedance of the first stage.

The second part of the above expression is the amplification of the first stage which is taken from the base of the first stage to the base of the second stage. This is normally a high value around 100, but depends on the parameters of the transistor amplifier.

The third part of the expression is the amplification of the output stage and this should be straightforward.

Before calculating the above values we will need to have the information resulting from evaluating the DC conditions. Because the capacitors block the DC supplies, the calculations of the DC conditions are completely independent for stage 1 and stage 2.

DC conditions

These are applicable to stage 1 and stage 2. There are two methods to evaluate the DC conditions. The approximate method and the Thevenin method. It should be noted that the Thevenin method to evaluate DC conditions on a potential divider biasing is the application of the Thevenin theorem to this particular circuit.

We also have a criteria to decide whether to use one or the other method. The criteria consists on comparing R_{B1} with R_E in the following way.

If $\boxed{10R_{B1} < \beta R_E}$ use the approximate method

Using either method we will eventually arrive at I_B, I_C and I_E. With the value of I_E we can obtain the value of r_e. It is also a good idea at this point to continue to obtain the input impedances of both stages, as they may be required later.

The circuit to be analysed will most likely be different and it must be clearly understood that the model that will follow is the model that corresponds to the circuit of Figure 7.1. So, when you do the equivalent circuit keep your eyes on your particular circuit. Our equivalent circuit is shown in Figure 7.2. This part will form a potential divider and the amplification will have a factor less than 1, from this part.

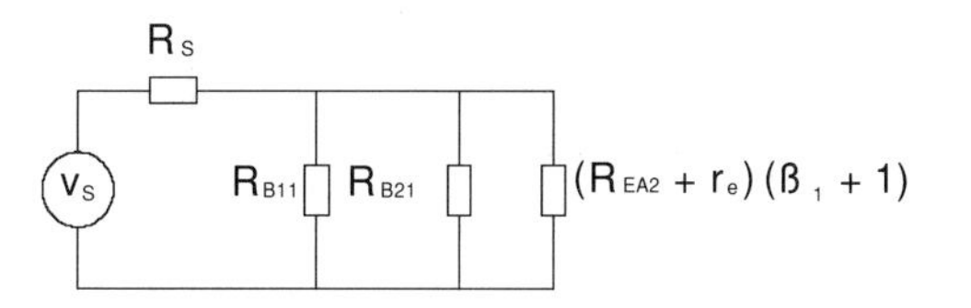

Figure 7.2

Care has to be taken to make sure that the emitter circuit is properly included, if it is not bypassed by a capacitor and also amplified by the factor (β + 1).

The most complicated part is the gain of the first stage, from base one to base two. The complication is that part of the loading of the first stage is the input of the second stage. So there are a lot of items to be considered as part of the equation.

The general equation is:

$$A_{V1} = -\frac{R_C || R_L}{r_e}$$

In connection with the circuit in question we have:

$$\frac{v_{B2}}{v_{B1}} = -\frac{R_{C1} || R_{B22} || R_{B12} || r_{e2}(\beta_2 + 1)}{R_{EA2} + r_{e1}}$$

We see this in Figure 7.3. In this circuit we have the collector resistor of stage 1, plus all the input resistances on stage 2. The circuit on the emitter of stage 2 is very important. In this case R_{EB1} is bypassed and it will not show.

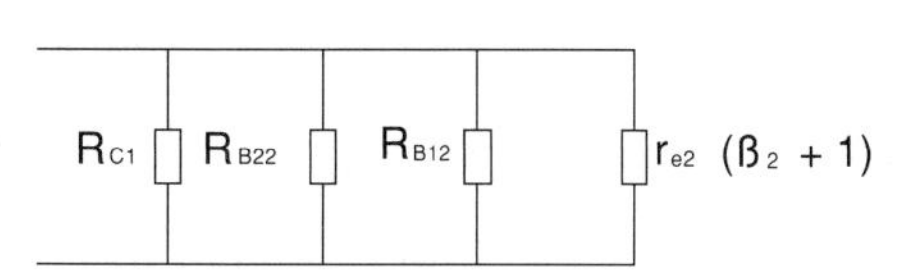

Figure 7.3

The denominator of the equation above correspond to the circuit on the emitter of stage 1. This is where the input voltage is applied. In this case R_{EA1} is bypassed and it does not appear in the equation. R_{EA2}, however, is not bypassed and it shows in the equation in series with r_{e1}.

The last part of the amplification corresponds to the gain of the last stage.

$$\frac{v_o}{v_{B2}} = -\frac{R_{C2} || R_L}{r_{e2}}$$

The total amplification will be

$$\frac{v_o}{v_s} = \frac{v_{B1}}{v_s}\frac{v_{B2}}{v_{B1}}\frac{v_o}{v_{B2}}$$

We mentioned earlier the gain impedance formula to convert from voltage to current amplification (and vice versa). This was:

$$A_i = A_v \frac{R_{IN}}{R_L}$$

(No sign as we are only interested in the magnitude.) R_L is clearly visible in the output of the second stage. R_{IN} is not so noticeable. So, we will evaluate it (see Figure 7.4).

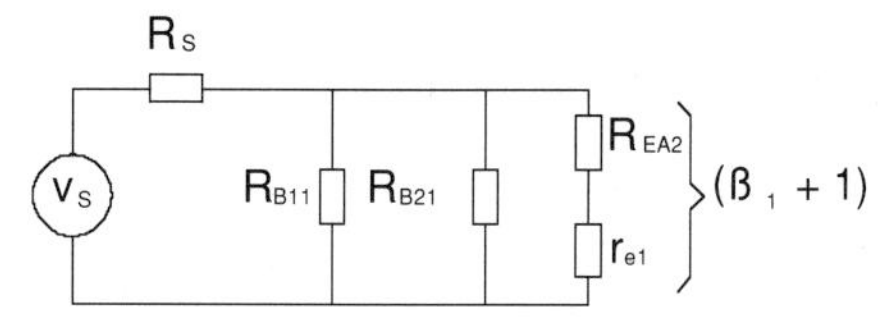

Figure 7.4

$$R_{IN} = R_S + R_{B11} || R_{B21} || (r_{e1} + R_{EA2})(\beta_1 + 1)$$

Cascaded current gain

The typical circuit is shown in Figure 7.5. First of all we evaluate the DC conditions for each of the

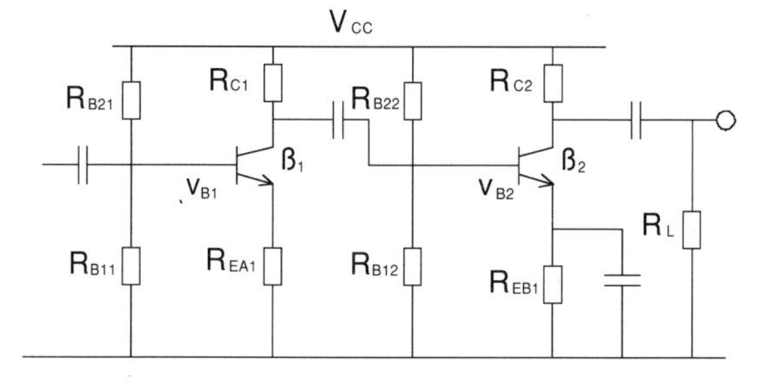

Figure 7.5

stages. As both stages are biased by potential divider types of biasing, we have two possibilities for evaluating the DC conditions, either the approximate method or the Thevenin method. The method we use depends on the loading provided by the transistor. The criterion that we use is:

If $10R_{B1} < \beta R_E$ we use the approximate method. Otherwise, we use the Thevenin method.

Using the method decided by the criterion we can find the value of I_E in each of the stages independently. With the value of I_E we can find the value of r_e according to the following alternatives:

$$r_e = \frac{25}{I_E} \quad \text{(at 20°C)} \qquad r_e = \frac{26}{I_E} \quad \text{(at 30°C)}$$

These formulae accept the value of I_E in mA.

For the calculation of the current amplification you need to use the current division rule. If you are not familiar with it at this moment, have a look at the first chapter. Otherwise continue with

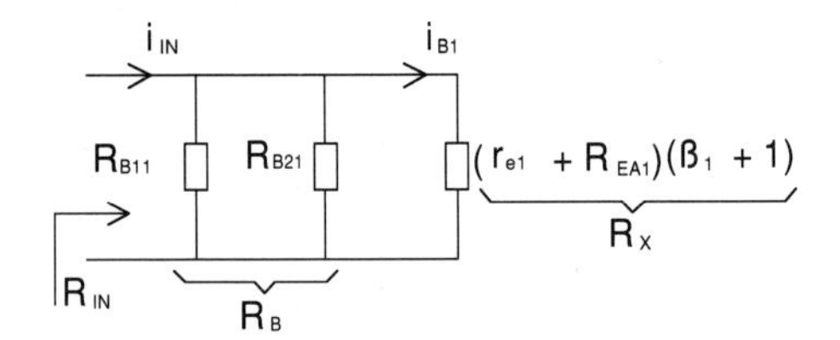

Figure 7.6

Figure 7.6. In this circuit we combine $R_{B11} \parallel R_{B21}$ which can be called for the moment R_B. We will call R_X the resistor associated with the emitter circuit of stage 1.

$$i_{B1} = \frac{i_{IN} R_B}{R_B + R_X}$$

The current out of the transistor 1 into the second stage is:

$$i_{IN2} = \beta_1 i_{B1}$$

The current coming out of the first stage will follow the equivalent circuit shown in Figure 7.7. Again we use the current division rule.

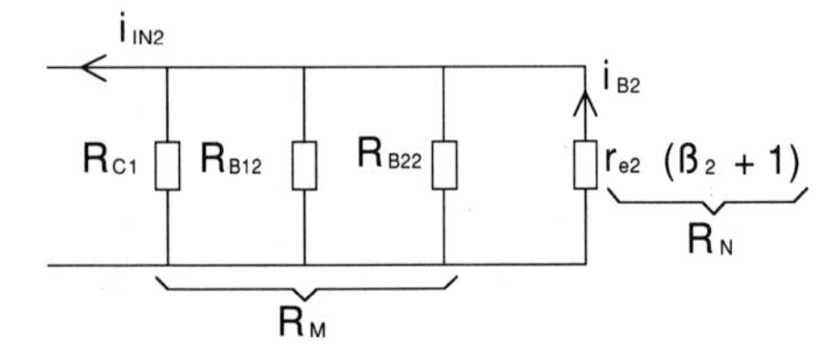

Figure 7.7

Before we do that, we join the three resistors into one, which we call R_M for the moment. We call all the resistors associated with the emitter circuit, R_N. This resistor could have been the result of a more complicated circuit, but this one corresponds to the circuit at the beginning and it is rather simple.

Remember that the model has to represent the circuit that you are dealing with. From the model you work out your equations.

$$i_{B2} = \frac{i_{IN2} R_M}{R_M + R_N}$$

From the base of the second transistor we can move to the output of the second transistor using the β factor.

$$i_{C2} = i_{B2}\beta_2$$

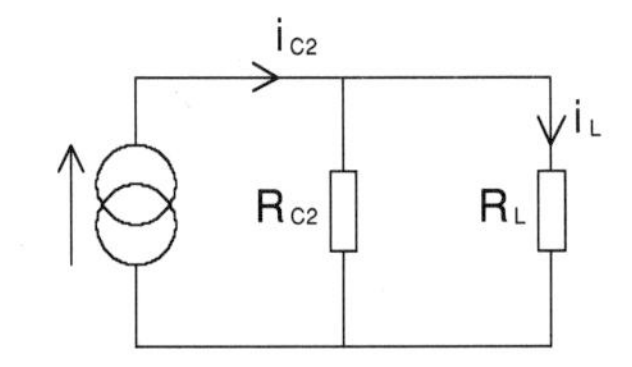

Figure 7.8

The circuit that we have now is shown in Figure 7.8. In this case we have the last application of the current division rule:

$$i_L = \frac{i_{C2} R_{C2}}{R_{C2} + R_L}$$

We need to join all the formulae into one, to obtain the total expression:

$$\frac{i_L}{i_{IN}} = \beta_1 \beta_2 \frac{R_B}{R_B + R_X} \frac{R_M}{R_M + R_N} \frac{R_{C2}}{R_{C2} + R_L}$$

It is best not to try to remember formulae of this type. It is best to work out the model from the circuit and work out the formulae from your model.

If we want to relate the current amplification to the voltage amplification, we can use the gain impedance formula:

$$A_V = \frac{A_I R_L}{R_{IN}}$$

For those of you that like anagrams note that the word RAIL should appear at one side of the

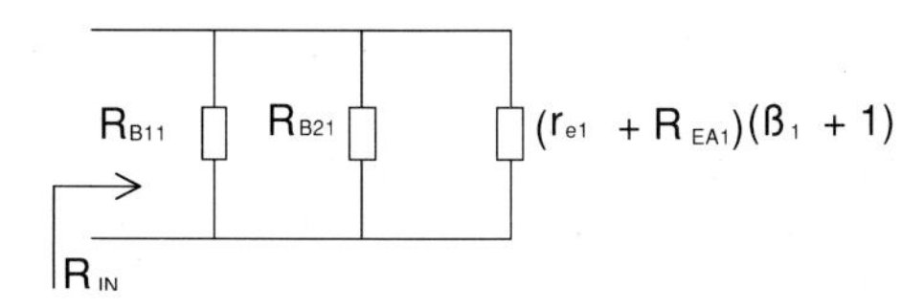

Figure 7.9

equation. You can resort to this mnemonic rule if you don't remember where the resistors go in the formula. Again R_L is easily identified. Note that if R_L is not present, then the ouput current would circulate only in R_{C2} and i_L would be identical with i_{C2}.

Let us work out R_{IN}. In this case we look at Figure 7.9. The value of R_{IN} to be used in the gain impedance formula will be:

$$R_{IN} = R_{B11} \parallel R_{B21} \parallel (r_{e1} + R_{EA1})(\beta_1 + 1)$$

PROBLEM 7.1 CASCADED SYSTEMS

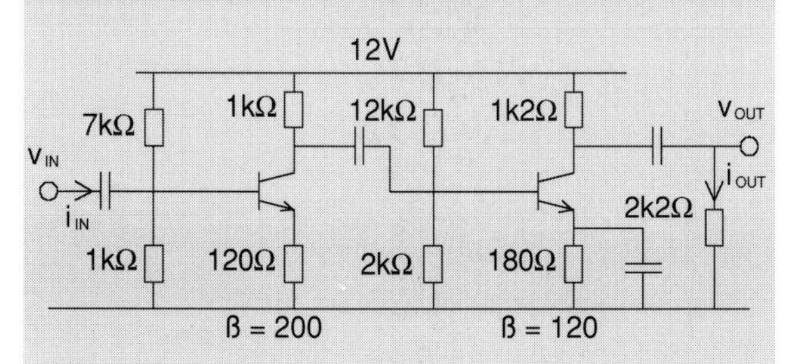

Figure 7.10

Find the overall voltage gain and overall current gain.

Answer: $A_V = 402$, $A_I = 152.5$

DC conditions

Criteria

$10R_1$	vs	βR_E	Use
10 000		24 000	Approx.
20 000		21 600	Approx.

Stage 1

$$V_{B1} = 12\frac{1\text{k}}{1\text{k} + 7\text{k}} = 12\frac{1}{8} = 1.5 \text{ V}$$

$$V_{E1} = 1.5 - 0.7 = 0.8 \text{ V}$$

$$I_{E1} = \frac{0.8}{120} = 6.67 \text{ mA}$$

$$r_{e1} = \frac{26}{6.67} = 3.9\ \Omega$$

Stage 2

$$V_{B2} = 12\frac{2\text{k}}{2\text{k} + 12\text{k}} = 1.71 \text{ V}$$

$$V_{E2} = 1.71 - 0.7 = 1.01 \text{ V}$$

$$I_{E2} = \frac{1.01}{180} = 5.61 \text{ mA}$$

$$r_{e2} = \frac{26}{5.61} = 4.63\ \Omega$$

AC conditions :

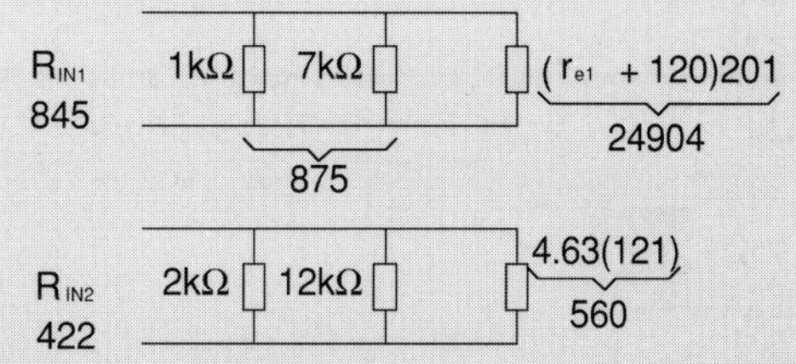

R IN1 845 — 1kΩ, 7kΩ (875), (r e1 + 120)201 (24904)

R IN2 422 — 2kΩ, 12kΩ, 4.63(121) (560)

Figure 7.11

$$\frac{v_{B2}}{v_{B1}} = -\frac{R_C \| R_{IN2}}{r_{e1} + R_{E1}}$$

$$= -\frac{1\text{k}\|422}{3.9 + 120} = -2.395$$

Second stage amplification

$$\frac{v_o}{v_{B2}} = -\frac{R_C \| R_L}{r_{e2}}$$

$$= -\frac{1\text{k}2\|2\text{k}2}{4.63} = -167.7$$

Total voltage gain

$$A_V = 2.395 \times 167.7 = 402$$

Gain impedance formula (used as an approximate check)

$$A_i = A_V\frac{R_{IN}}{R_L} = 402\frac{845}{2200} = 154.4$$

Current gain

Input side, current division

$$i_{B1} = i_{IN}\frac{875}{875 + 24\,904} = 0.034 i_{IN}$$

Output first stage

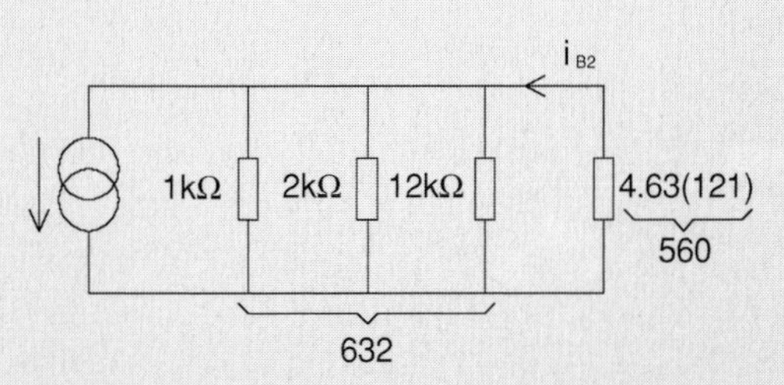

Figure 7.12

$$\beta i_{B1} = 200 \times 0.034 i_{IN} = 6.8 i_{IN}$$

$$i_{B2} = \beta i_{B1}\frac{632}{632+560} = 3.6 i_{IN}$$

Output second stage

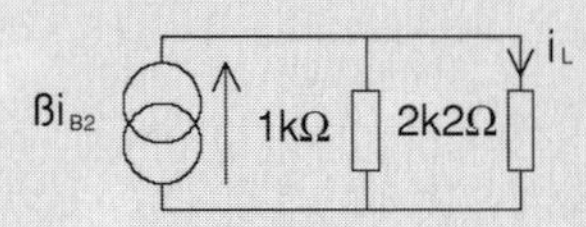

Figure 7.13

$$\beta i_{B2} = 120 \times 3.6 i_{IN} = 432 i_{IN}$$

$$i_L = \beta i_{B2}\frac{1k2}{1k2+2k2} = 152.5 i_{IN}$$

$$A_i = \frac{i_L}{i_{IN}} = 152.5$$

PROBLEM 7.2 CASCADED SYSTEMS

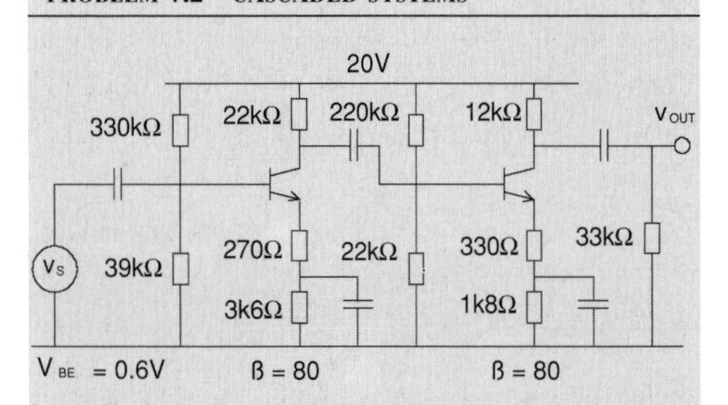

Figure 7.14

Find the voltage amplification and the current amplification for the above circuit.

Answer: $A_V = 531.5$, $A_i = 243$

DC conditions

Criteria

$10R_1$	vs	βR_E	Use
390 000		309 600	Thevenin
220 000		170 400	Thevenin

Thevenin resistance

$$R_{TH} = 39k\|330k = 34\,878\ \Omega$$

Thevenin voltage

$$V_{TH} = 20\frac{39k}{39k+330k} = 2.11\ V$$

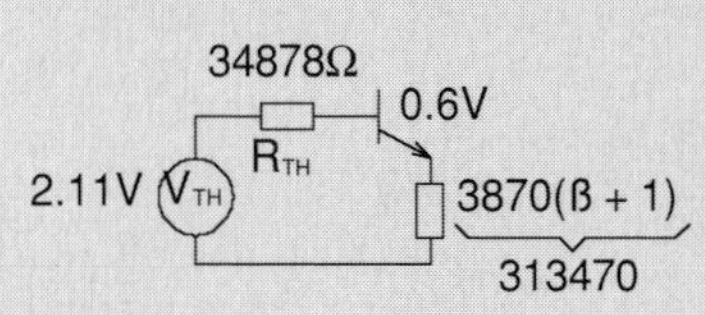

Figure 7.15

$$I_B = \frac{2.11-0.6}{34\,878+313\,470}$$

$$= \frac{1.51}{348\,348} = 4.33\ \mu A$$

$$I_E = 4.33 \times 10^{-6} \times 81 = 0.351\ mA$$

$$r_{e1} = 71.22\ \Omega$$

DC conditions , second stage

$$R_{TH} = 22k\|220k = 20\ k\Omega$$

$$V_{TH} = 20\frac{22k}{22k+220k} = 1.82\ V$$

$$I_B = \frac{1.82-0.6}{20000+2130(81)} = \frac{1.22}{192\,530} = 6.34\ \mu A$$

$$I_E = 6.34 \times 10^{-6} \times 81 = 0.51\ mA$$

$$r_{e2} = \frac{25}{0.51} = 49\ \Omega$$

AC conditions

Figure 7.16

$$A_{V1} = -\frac{R_C\|R_{IN2}}{r_{e1}+270}$$

$$= -\frac{22k\|12110}{71.22+270}$$

$$= -\frac{7811}{341.22} = -22.89$$

$$A_{V2} = -\frac{12k\|33k}{r_{e2} + 330}$$

$$= -\frac{8800}{379} = -23.22$$

Total voltage gain

$$A_V = 22.89 \times 23.22 = 531.5$$

In order to use the gain impedance formula, we need R_{IN1}:

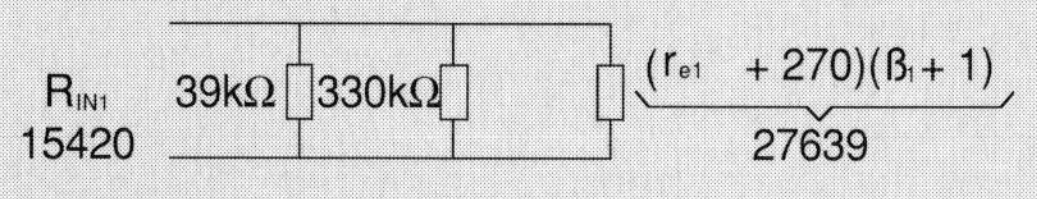

Figure 7.17

$$A_i = A_V \frac{R_{IN}}{R_L} = 531.5\frac{15\,420}{33\,000} = 248$$

Current gain calculation

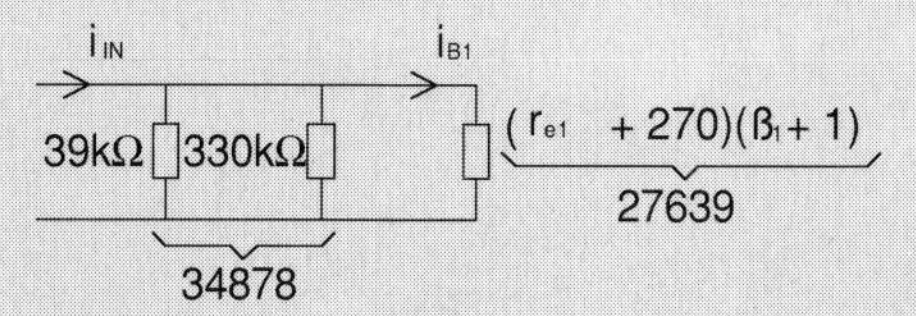

Figure 7.18

$$i_{B1} = i_{IN}\frac{34\,878}{34\,878 + 27\,639} = 0.56 i_{IN}$$

Output of first transistor

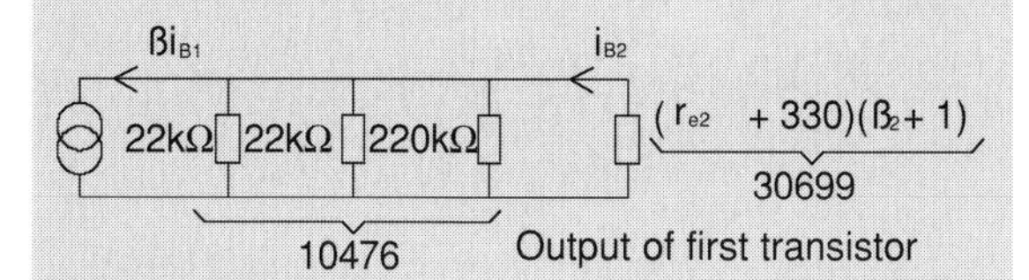

Figure 7.19

$$\beta i_{B1} = 80 \times 0.56 i_{IN} = 44.8 i_{IN}$$

$$i_{B2} = \beta i_{B1}\frac{10\,476}{10\,476 + 30\,699} = \beta i_{B1} \times 0.254$$

$$= 11.38 i_{IN}$$

Output of second transistor

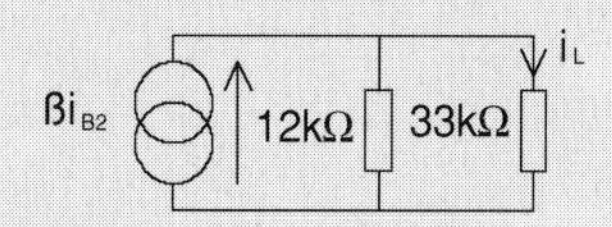

Figure 7.20

$$\beta i_{B2} = 80 \times 11.38 i_{IN} = 910.4 i_{IN}$$

$$i_L = \beta i_{B2}\frac{12k}{12k + 33k} = \beta i_{B2} \times 0.267$$

$$i_L = 0.267 \times 910.4 i_{IN} = 243 i_{IN}$$

PROBLEM 7.3 CASCADED SYSTEMS

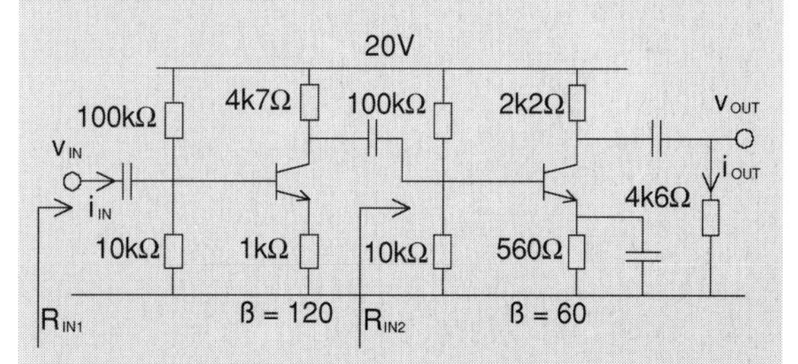

Figure 7.21

For the cascade amplifier, find the input resistances as shown, the voltage amplification and the current amplification.

> Answer: $R_{IN1} = 2082$, $R_{IN2} = 872$, $A_V = 3101$ and $A_i = 1369.5$

DC conditions

Criteria

$10R_1$	vs	βR_E	Use
100 000		120 000	Approx.

Approximate method

$$V_B = 20\frac{10}{10 + 100} = 1.82\text{ V}$$

$$V_E = 1.82 - 0.7 = 1.12\text{ V}$$

$$I_E = \frac{1.12}{1k} = 1.12\text{ mA}$$

$$r_{e1} = \frac{25}{1.12} = 22.32\ \Omega$$

Second stage

100 000 33 600 Thevenin

Thevenin method

$$V_{TH} = 20\frac{10}{10+100} = 1.82 \text{ V}$$

$$R_{TH} = 10k \| 100k = 9091\ \Omega$$

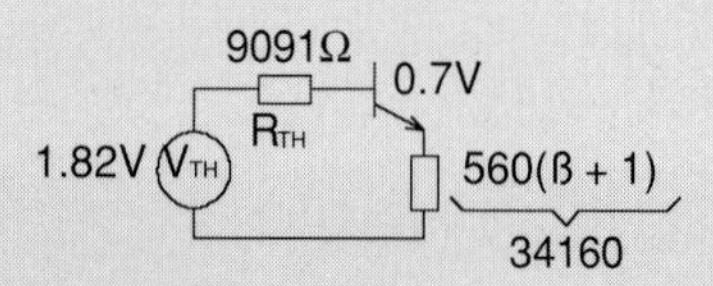

Figure 7.22

$$I_B = \frac{1.82-0.7}{9091+34160} = 25.89\ \mu\text{A}$$

$$I_E = I_B(\beta+1) = I_B \times 61 = 1.58 \text{ mA}$$

$$r_{e2} = \frac{25}{1.58} = 15.82\ \Omega$$

AC conditions

$$R_{IN1} = 10k \| 100k \| 22.32(121) = 2082\ \Omega$$

$$R_{IN2} = 10k \| 100k \| 15.82(61) = 872.41\ \Omega$$

$$A_{V1} = -\frac{R_C \| R_{IN2}}{r_{e1}}$$

$$= -\frac{4k7 \| 872.41}{22.32} = -32.967$$

$$A_{V2} = -\frac{R_C \| R_L}{r_{e2}}$$

$$= -\frac{2k2 \| 4k6}{15.82} = -94.073$$

Total voltage gain

$$A_V = 94.073 \times 32.967 = 3101$$

U sing the gain impedance formula

$$A_I = A_V \frac{R_{IN}}{R_L} = 3101\frac{2082}{4600} = 1403.5$$

Calculation of current gain

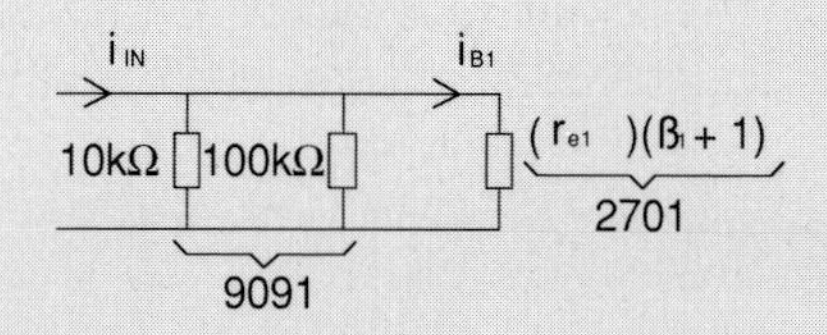

Figure 7.23

$$i_{B1} = i_{IN}\frac{9091}{9091+2701} = i_{IN} \times 0.771$$

Output of first transistor

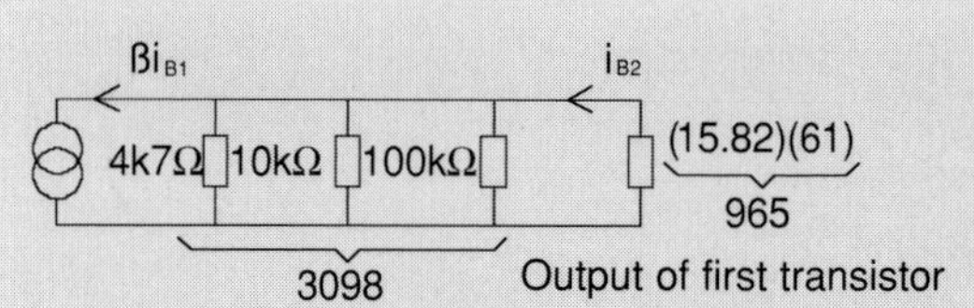

Figure 7.24

$$\beta i_{B1} = 120 i_{IN} \times 0.771 = 92.52 i_{IN}$$

$$i_{B2} = \beta i_{B1}\frac{3098}{3098+965} = 70.55 i_{IN}$$

Output side

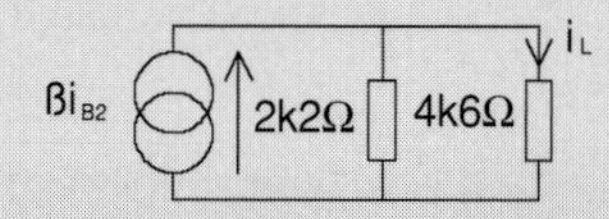

Figure 7.25

$$\beta i_{B2} = 70.55 i_{IN} \times 60 = 4233 i_{IN}$$

$$i_L = \beta i_{B2}\frac{2k2}{2k2+4k6} = 1369.5$$

PROBLEM 7.4 CASCADED SYSTEMS

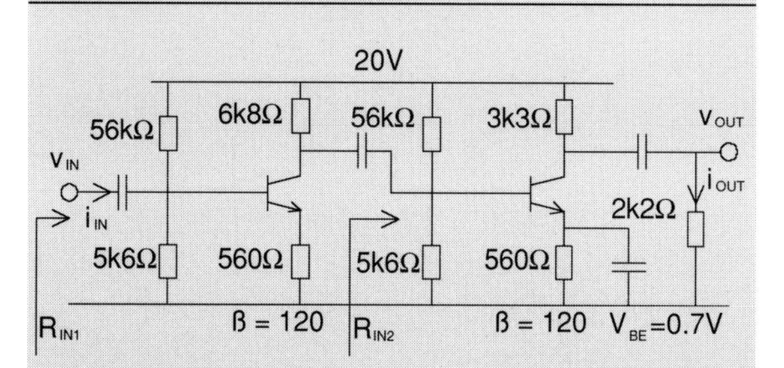

Figure 7.26

For the cascade amplifier, find the resistances as shown, the voltage amplification and the current amplification. Assume the temperature to be 20°C.

Answer: R_{IN1} = 1166, R_{IN2} = 1166, A_V = 8408 and A_i = 4383

DC conditions

Criteria

$10R_1$	vs	βR_E
56 000		67 000

$$V_B = 20\frac{5k6}{5k6 + 56k} = 1.82 \text{ V}$$

$$V_E = 1.82 - 0.7 = 1.12 \text{ V}$$

$$I_E = \frac{1.12}{560} = 2 \text{ mA}$$

$$r_{e1} = \frac{25}{2} = 12.5\ \Omega$$

The second amplifier has an identical circuit for DC conditions. Therefore

$$r_{e2} = 12.5\ \Omega$$

AC conditions

With both emitter resistors bypassed, we have

$$R_{IN1} = 5k6 \| 56k \| 12.5 \times 121 = 1166\ \Omega$$

$$R_{IN2} = 1166 \text{ (same circuit)}$$

Voltage amplification stage 1

$$A_{V1} = -\frac{6k8 \| R_{IN2}}{12.5} = -79.626$$

$$A_{V2} = -\frac{3k3 \| 2k2}{12.5} = -105.6$$

Total

$$A_V = 8408$$

Using the gain impedance formula

$$A_i = A_V\frac{R_{IN}}{R_L}$$

$$= 8408\frac{1166}{2k2} = 4456$$

Now the current gain (current division)

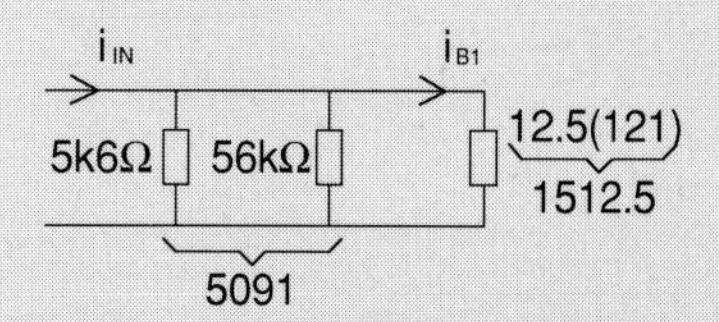

Figure 7.27

$$i_{B1} = i_{IN}\frac{5091}{5091 + 1512.5} = i_{IN} \times 0.771$$

Output first stage

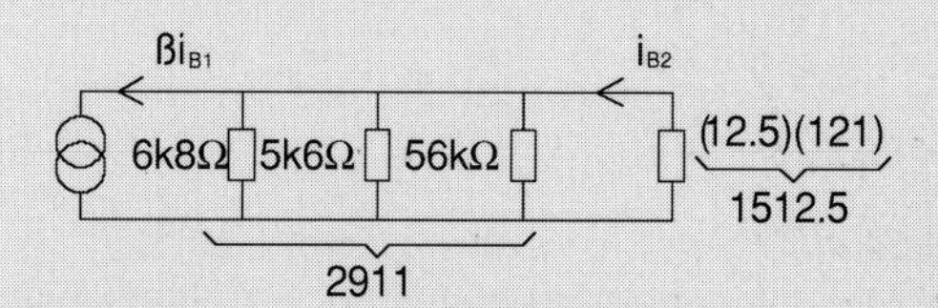

Figure 7.28

$$\beta i_{B1} = 120 i_{IN} \times 0.771 = 92.52 i_{IN}$$

$$i_{B2} = \beta i_{B1}\frac{2911}{2911 + 1512.5} = 60.88 i_{IN}$$

Output, second stage

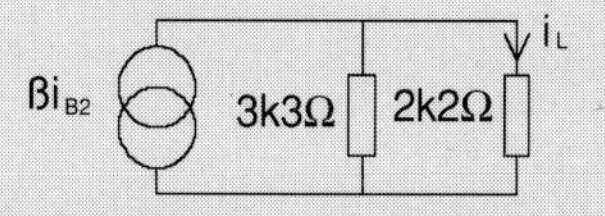

Figure 7.29

$$\beta i_{B2} = 60.88 i_{IN} \times 120 = 7305.6 i_{IN}$$

$$i_L = \beta i_{B2}\frac{3k3}{3k3 + 2k2} = 4383.36 i_{IN}$$

The gain impedance formula gave 4456, which is not too bad, considering all the rounding-off.

PROBLEM 7.5 CASCADED SYSTEMS

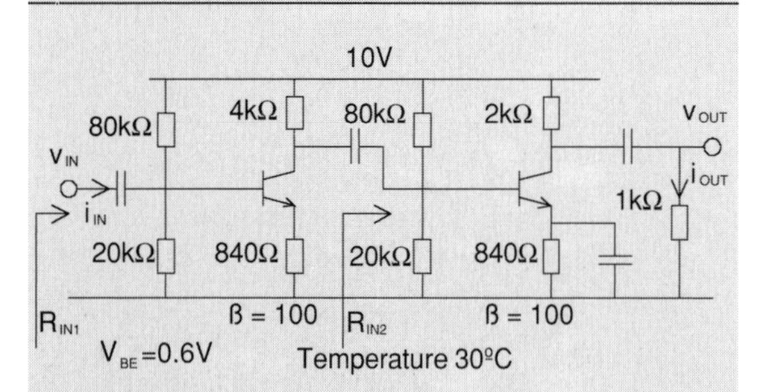

Figure 7.30

Find the midband voltage gain and current gain.

Answer: $A_V = 2286$, $A_i = 3759$

DC conditions

Criteria

$10R_1$	vs	βR_E	Use
200 000		84 000	Thevenin

Thevenin

$$V_{TH} = 10\frac{20k}{20k + 80k} = 2 \text{ V}$$

$$R_{TH} = 20k \| 80k = 16 \text{ k}\Omega$$

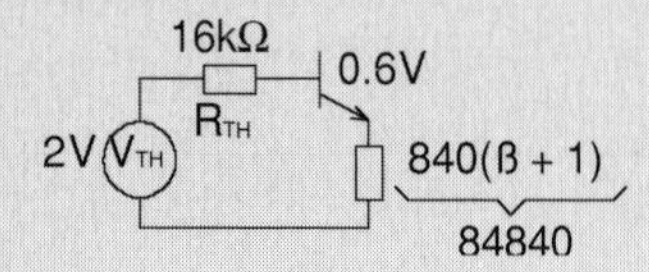

Figure 7.31

$$I_B = \frac{2 - 0.6}{16\,000 + 84\,840}$$

$$= \frac{1.4}{100\,840} = 13.88 \ \mu\text{A}$$

$$I_E = 13.88 \times 10^{-6} \times 101 = 1.4 \text{ mA}$$

$$r_e = \frac{26}{1.4} = 18.57 \ \Omega \text{ (for both ages)}$$

AC conditions – impedances

R IN1 1679 20kΩ 80kΩ (r e1)(ß1 + 1) 1876

This is also valid for R IN2

Figure 7.32

Both input circuits have the same value 1679 Ω.

Gain first stage

$$A_{V1} = -\frac{R_C \| R_{IN2}}{r_{e1}}$$

$$= -\frac{4k \| 1679}{18.57} = -63.68$$

Gain second stage

$$A_{V2} = -\frac{R_C \| R_L}{r_{e2}}$$

$$= -\frac{2k \| 1k}{18.57} = -35.9$$

Total

$$A_V = 63.68 \times 35.9 = 2286$$

$$A_i = A_V \frac{R_{IN}}{R_L} = 2286 \frac{1679}{1000} = 3838$$

Current gain

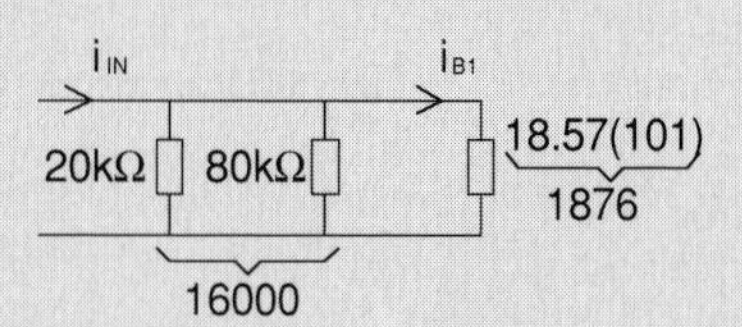

Figure 7.33

$$i_{B1} = i_{IN} \frac{16000}{16000 + 1876} = 0.895 i_{IN}$$

First stage output

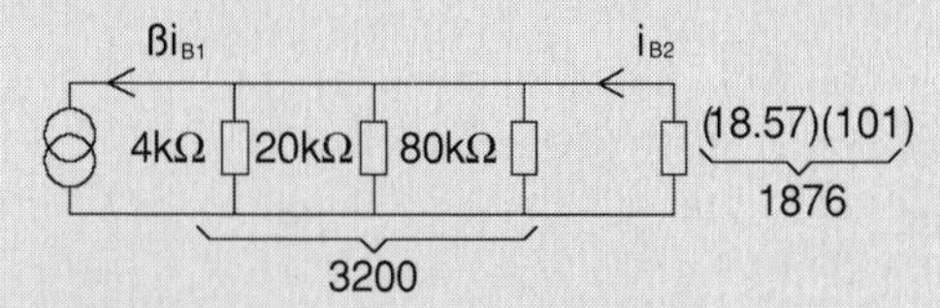

Figure 7.34

$$\beta i_{B1} = 100 \times 0.895 i_{IN} = 89.5 i_{IN}$$

$$i_{B2} = \beta i_{B1} \frac{3200}{3200 + 1876} = \beta i_{B1} \times 0.63$$

$$i_{B2} = 0.63 \times 89.5 i_{IN} = 56.385 i_{IN}$$

Output second stage

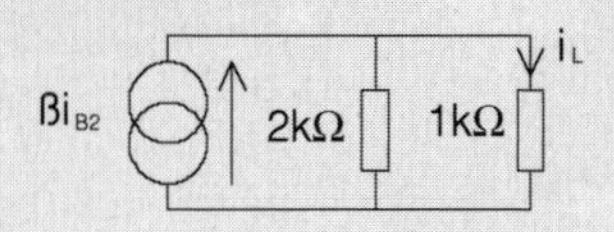

Figure 7.35

$$\beta i_{B2} = 100 \times 56.385 i_{IN} = 5638.5 i_{IN}$$

$$i_L = \beta i_{B2} \frac{2k}{1k + 2k} = \beta i_{B2} \frac{2}{3}$$

$$= 3759 i_{IN}$$

$$A_i = \frac{i_L}{i_{IN}} = 3759$$

PROBLEM 7.6 CASCADED SYSTEMS

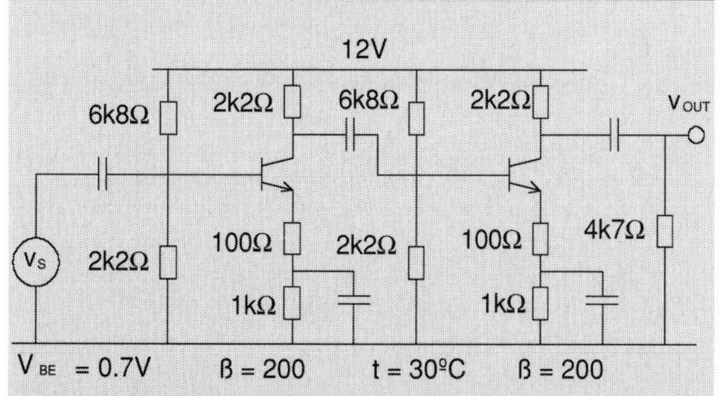

Figure 7.36

Find the overall voltage gain.

Answer: $A_V = 107$

Both stages are identical - approximate method

$$V_B = 12 \frac{2.2}{2.2 + 6.8} = 2.93 \text{ V}$$

$$V_E = 2.93 - 0.7 = 2.23 \text{ V}$$

$$I_E = \frac{2.23}{1100} = 2.03 \text{ mA}$$

$$r_e = \frac{26}{2.03} = 12.81\ \Omega = r_{e1} = r_{e2}$$

Voltage gain

$$\frac{v_{B2}}{v_{B1}} = -\frac{R_C || R_L}{r_{e1} + R_E}$$

R_L of the first stage is R_{IN} of the second stage

$$R_L = 2k2 || 6k8 || (r_{e2} + 100)(201) = 1549$$

$$\frac{v_{B2}}{v_{B1}} = -\frac{2k2 || 1549}{12.81 + 100} = -8.06$$

$$\frac{v_O}{v_{B2}} = -\frac{R_C || R_L}{r_{e2} + R_E} = -\frac{2k2 || 4k7}{112.81}$$

$$= -13.28$$

Total

$$A_V = 8.06 \times 13.28 = 107$$

PROBLEM 7.7 CASCADED SYSTEMS

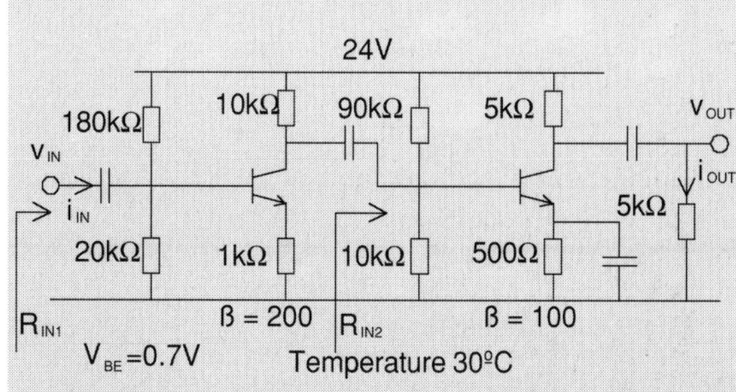

Figure 7.37

Find the current gain and the voltage gain of the amplifiers.

Answer: $A_i = 678.55$, $A_V = 208.25$

DC conditions - first stage - Thevenin

$$V_{TH} = 24 \frac{20}{20 + 180} = 2.4 \text{ V}$$

$$R_{TH} = 180k || 20k = 18 \text{ k}\Omega$$

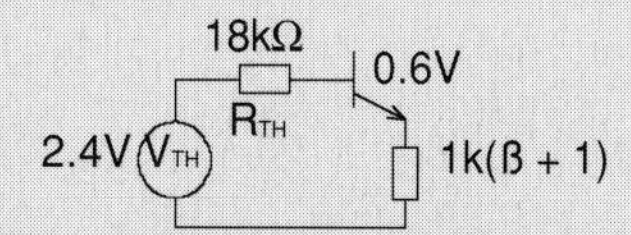

Figure 7.38

$$I_B = \frac{2.4 - 0.7}{1800 + 201\,000} = 7.763\ \mu\text{A}$$

$$I_{EQ} = 7.763 \times 10^{-6} \times 201 = 1.56 \text{ mA}$$

$$r_{e1} = \frac{26}{1.56} = 16.67\ \Omega$$

DC conditions – second stage – Thevenin

$$V_{TH} = 2.4 \text{ V} \qquad R_{TH} = 9 \text{ k}\Omega$$

$$I_B = \frac{2.4 - 0.7}{9k + 500(101)}$$

$$= \frac{1.7}{59\,500} = 28.57\ \mu\text{A}$$

$$I_{EQ} = 28.57 \times 10^{-6} \times 101 = 2.88\ \text{mA}$$

$$r_{e2} = \frac{26}{2.88} = 9.03\ \Omega$$

AC conditions – current gain

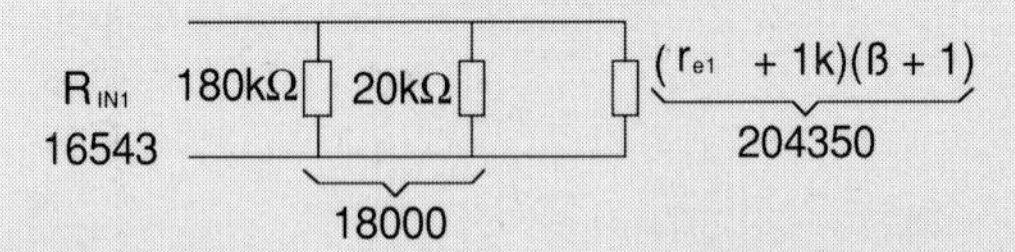

Figure 7.39

$$i_{B1} = i_{IN}\frac{18\,000}{18\,000 + 204\,350} = 0.081 i_{IN}$$

Output first stage

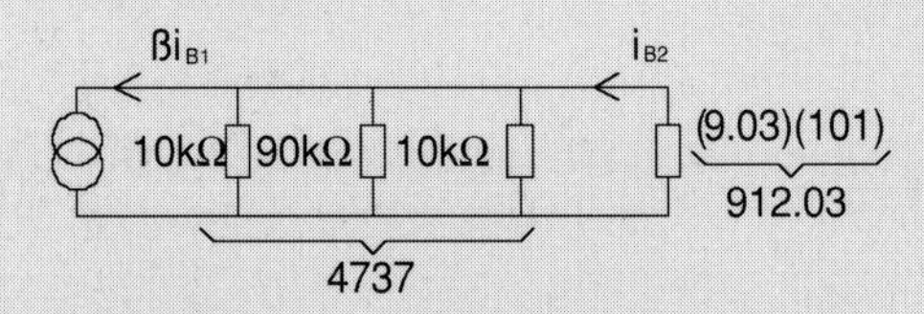

Figure 7.40

$$\beta i_{B1} = 200 \times 0.081 i_{IN} = 16.191 i_{IN}$$

$$i_{B2} = 16.191 i_{IN}\frac{48\,737}{48\,737 + 912.03} = 13.577 i_{IN}$$

$$i_L = \frac{\beta i_{B2}}{2} = \frac{100}{2} \times 13.577 i_{IN}$$

$$A_i = 678.85$$

Voltage gain

$$\frac{v_{B2}}{v_{B1}} = -\frac{R_C \| R_L}{r_{e1} + R_E}$$

R_L of first stage is R_{IN} of the second stage

$$R_L = 90\text{k} \| 10\text{k} \| 9.03 \times 101 = 828.11\ \Omega$$

$$\frac{V_{B2}}{V_{B1}} = -\frac{828.11 \| 10\text{k}}{1016.67} = -0.752\,2$$

$$\frac{v_O}{v_{B2}} = -\frac{R_C \| R_L}{r_{e2}}$$

$$= -\frac{5\text{k} \| 5\text{k}}{9.03} = -276.855$$

$$A_V = 0.752\,2 \times 276.855 = 208.25$$

PROBLEM 7.8 CASCADED SYSTEMS

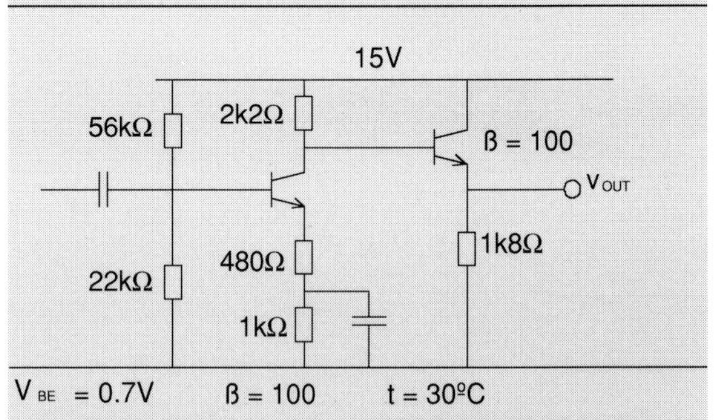

Figure 7.41

Find the voltage gain and the quiescent output voltage.

Answer: $A_V = -1.197$, $v_o = 9.548$ V

Thevenin

$$V_{TH} = 15\frac{22}{22 + 56} = 4.23\ \text{V}$$

$$R_{TH} = 22\text{k} \| 56\text{k} = 15\,795\ \Omega$$

$$I_B = \frac{4.23 - 0.7}{15\,795 + 1480 \times 101}$$

$$= \frac{3.53}{16\,5275} = 21.36\ \mu\text{A}$$

$$I_E = (\beta + 1)I_B = 101 \times 21.36 \times 10^{-6} = 2.16\ \text{mA}$$

$$r_{e1} = \frac{26}{2.16} = 12.04\ \Omega$$

We assume that there is no load from the last transistor. We find V_C on the first transistor

$$V_C = 15 - I_C \times 2\text{k}2 = 15 - 2.16 \times 2.2 = 10.248\ \text{V}$$

$$V_E = V_C - 0.7 = 10.248 - 0.7 = 9.548\ \text{V}$$

$$I_E = \frac{9.548}{1800} = 5.3\ \text{mA}$$

$$r_{e2} = \frac{26}{5.3} = 4.9\ \Omega$$

Second stage

$$R_{IN} = (\beta + 1)(r_{e2} + 1\text{k}8)$$

$$= 101 \times 1804.9$$

$$= 18\,2295\ \Omega$$

$$A_{V1} = -\frac{R_C \| R_{IN}}{r_{e2} + R_E}$$

$$= -\frac{2k2||18\,2295}{4.9 + 1800} = -\frac{2174}{1804.9} = -1.2$$

$$A_{V2} = \frac{R_E}{r_{e2} + R_E} = \frac{1800}{1804.9} = 0.997$$

Total

$$A_V = -0.977 \times 1.2 = -1.197$$

PROBLEM 7.9 CASCADED SYSTEMS

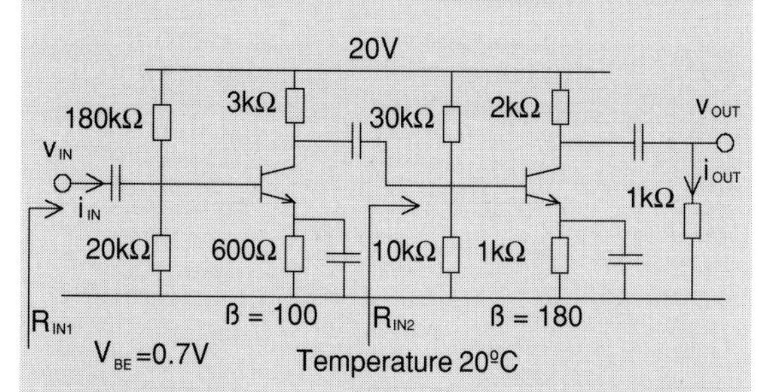

Figure 7.42

Find the voltage gain. Calculate the voltage output if the input is fed with a source of 2 mV peak to peak with an output resistance of 600 Ω.

> Answer: $A_V = 5450$, $v_{OUT} = 7.63$ V

First stage – Thevenin

$$V_{TH} = 20\frac{20}{180 + 20} = 2 \text{ V}$$

$$R_{TH} = 20k||180k = 18 \text{ k}\Omega$$

$$I_B = \frac{2 - 0.7}{18k + 600 \times 101}$$

$$= \frac{1.3}{78\,600} = 16.54\,\mu\text{A}$$

$$I_E = (\beta + 1)I_B = 101 \times 16.54 \times 10^{-6} = 1.67 \text{ mA}$$

$$r_{e1} = \frac{25}{1.67} = 14.97\ \Omega$$

Second stage – approximate method

$$V_B = 20\frac{10}{10 + 30} = 5 \text{ V}$$

$$V_E = 5 - 0.7 = 4.3 \text{ V}$$

$$I_E = \frac{4.3}{1k} = 4.3 \text{ mA}$$

$$r_{e2} = \frac{25}{4.3} = 5.81\ \Omega$$

Resistances

$$R_{IN1} = 20k||180k||14.97 \times 101 = 1395\ \Omega$$

$$R_{IN2} = 10k||30k||5.81 \times 181 = 923\ \Omega$$

Voltage gain

$$A_{V1} = -\frac{3k||R_{IN2}}{r_{e1}} = -\frac{3000||923}{14.97} = -47.15$$

$$A_{V2} = -\frac{2k||1k}{r_{e2}} = -\frac{666.66}{5.81} = -114.74$$

Total

$$A_V = 47.5 \times 114.74 = 5450$$

Feeding with 2 V into a load of 600 Ω

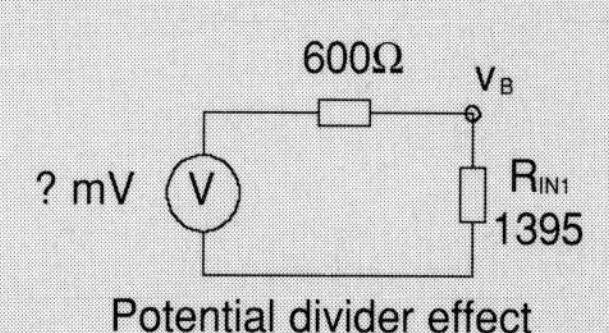

Figure 7.43

Potential divider effect

$$v_B = 2 \times 10^{-3}\frac{1395}{1395 + 600} = 1.4 \text{ mV}$$

$$v_{OUT} = 1.4 \times 10^{-3} \times 5450 = 7.63 \text{ V}$$

PROBLEM 7.10 CASCADED SYSTEMS

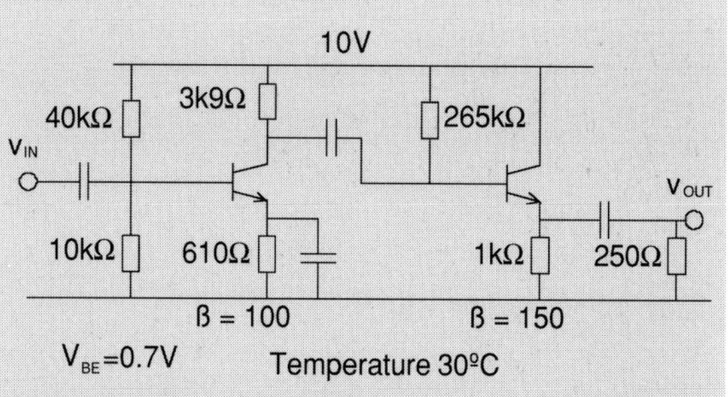

Figure 7.44

Find the voltage gain.

> Answer: $A_V = -261$

DC conditions – Thevenin method

$$V_{TH} = 10\frac{10}{10+40} = 2\text{ V}$$

$$R_{TH} = 10\text{k}\|40\text{k} = 8000\ \Omega$$

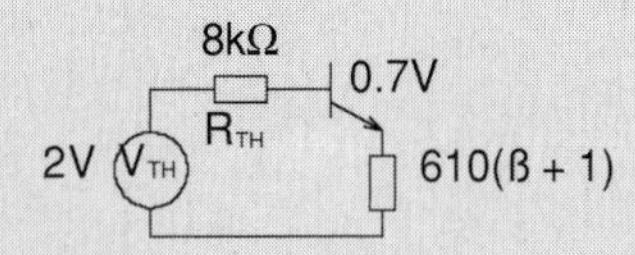

Figure 7.45

$$I_B = \frac{2-0.7}{8000+61\,610}$$

$$= \frac{1.3}{69\,610} = 18.68\ \mu\text{A}$$

$$I_E = (\beta+1)I_B = 101\times 18.68\times 10^{-6} = 1.89\text{ mA}$$

$$r_e = \frac{26}{1.89} = 13.76\ \Omega$$

Second stage

KVL

$$10 = I_B 265\text{k} + 0.7 + I_E 1\text{k}$$

$$9.3 = 265\,000 I_B + I_B(\beta+1)1\text{k}$$

$$9.3 = 265\,000 I_B + I_B 151\,000$$

$$9.3 = 416\,000 I_B$$

$$I_B = 22.36\ \mu\text{A}$$

$$I_E = (\beta+1)I_B = 151\times 22.36\times 10^{-6} = 3.38\text{ mA}$$

$$r_{e2} = \frac{26}{3.38} = 7.69\ \Omega$$

Second stage R_{IN}

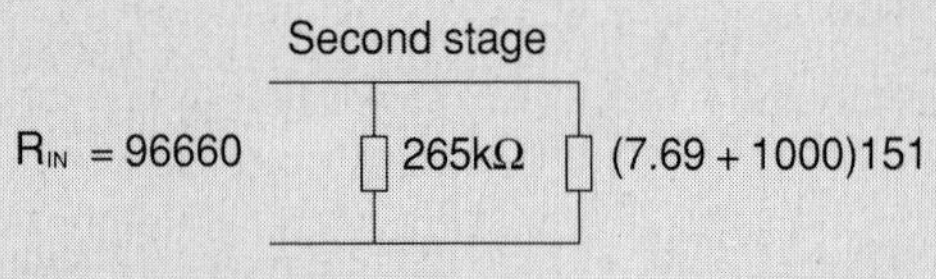

Figure 7.46

Voltage gain

$$A_{V1} = -\frac{3\text{k}9\|96\,660}{13.76} = -\frac{3749}{13.76} = -272$$

$$A_{V2} = \frac{1\text{k}\|250}{1\text{k}\|250+7.69} = \frac{200}{207.69} = 0.96$$

Total

$$A_V = 0.96\times(-272) = -261$$

PROBLEM 7.11 CASCADED SYSTEMS

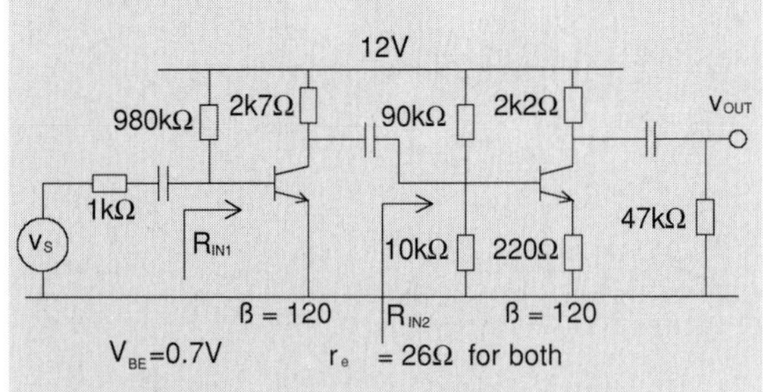

Figure 7.47

Find the voltage gain.

Answer: 345

As we have the values of r_e, we skip the DC conditions and we go directly to the AC conditions.

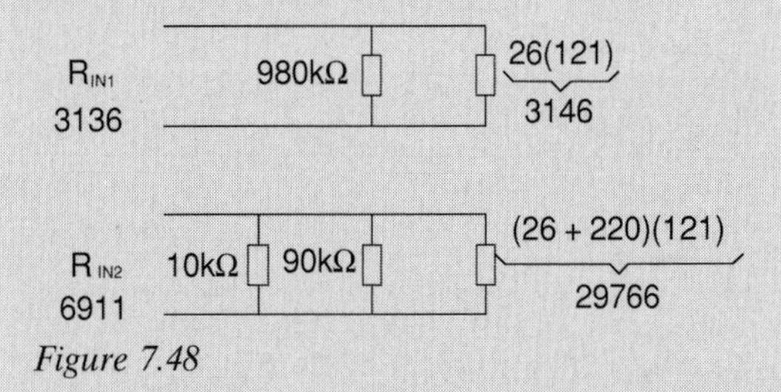

Figure 7.48

At the input there is a potential divider network

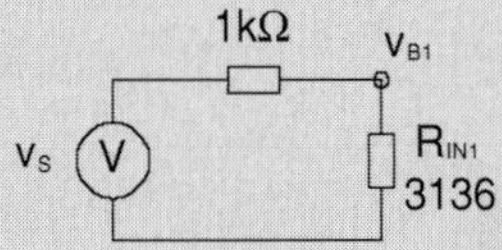

Figure 7.49

$$v_{B1} = v_S\frac{3136}{1\text{k}+3136} = 0.758V_S$$

Voltage gain

$$A_1 = -\frac{R_C\|R_L}{r_{e1}} = -\frac{2\text{k}2\|R_{IN2}}{r_{e1}}$$

$$= -\frac{2\text{k}7\|6911}{26} = -74.67$$

$$A_2 = -\frac{R_C\|R_L}{r_{e2}+R_E} = -\frac{2\text{k}2\|47\text{k}}{246}$$

$$= -\frac{1499}{246} = -6.09$$

Total

$$A_V = 0.758\times 74.67\times 6.09 = 345$$

PROBLEM 7.12 CASCADED SYSTEMS

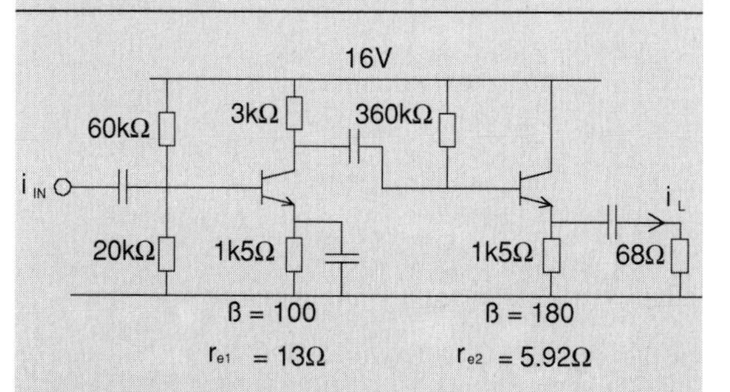

Figure 7.50

Find the current gain i_L/i_{IN}.

Answer: $A_i = -2997$

Input resistance, second and first stage.

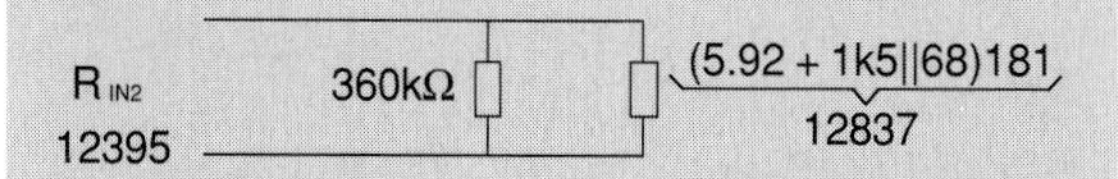

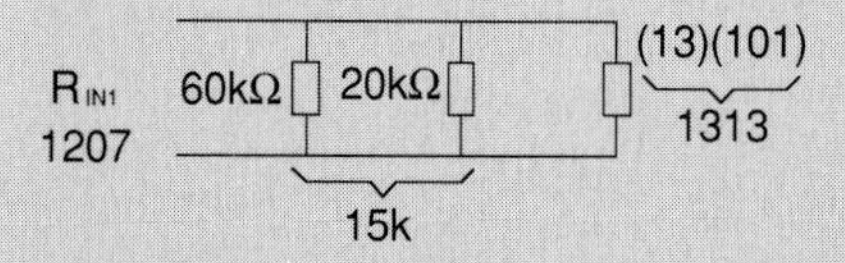

Figure 7.51

$$i_B = i_{IN}\frac{15\,000}{15\,000 + 1313} = i_{IN}\,0.92$$

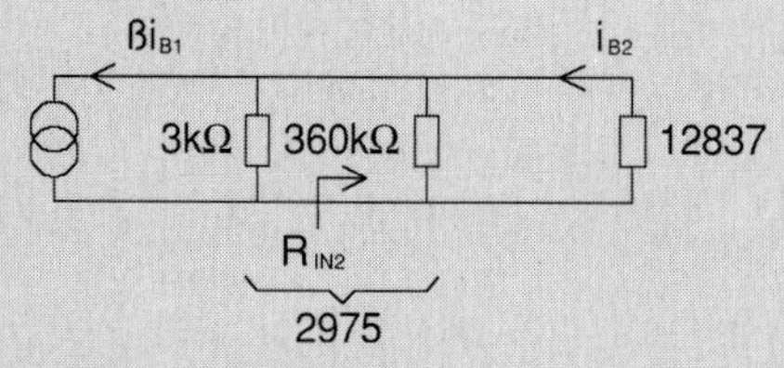

Figure 7.52

$$i_C = i_B\beta = 0.92 \times 100 i_{IN} = 92 i_{IN}$$

$$i_{B2} = 92 i_{IN}\frac{2975}{2975 + 12\,837} = 17.31 i_{IN}$$

$$i_{E2} = (\beta + 1) i_{B2} = 181 \times 17.31 i_{IN} = 3133 i_{IN}$$

$$i_L = -i_{E2}\frac{1500}{1500 + 68} = -3133 i_{IN}\frac{1500}{1568} = -2997 i_{IN}$$

PROBLEM 7.13 CASCADED SYSTEMS

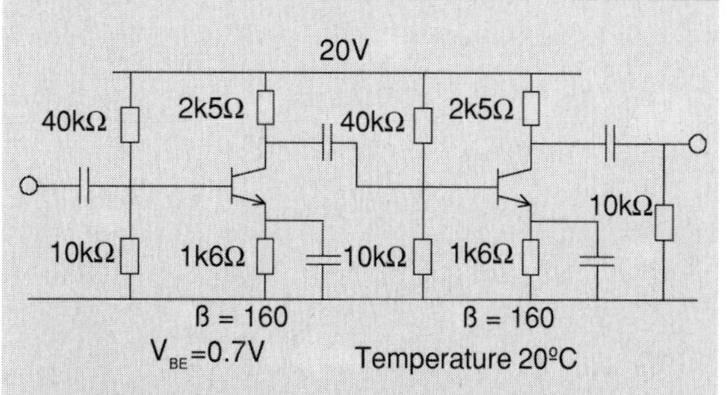

Figure 7.53

Find the overall voltage gain at midband frequencies.

Answer: $A_V = 2891$

First stage – DC conditions – approximate method

$$V_B = 20\frac{10}{10 + 40} = 2\text{ V}$$

$$V_E = 2 - 0.7 = 1.3\text{ V}$$

$$I_E = \frac{1.3}{1600} = 0.81\text{ V}$$

$$r_e = \frac{25}{0.81} = 30.86\ \Omega$$

First stage – AC model

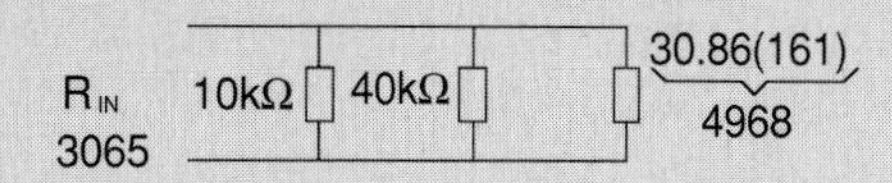

Figure 7.54

Second stage, the same as the first stage.

Voltage gain

$$\frac{v_{B2}}{v_{B1}} = -\frac{R_C \| R_L}{r_{e1}}$$

$$= -\frac{2\text{k}5 \| 3065}{30.86} = -44.62$$

$$\frac{v_O}{v_{B1}} = -\frac{R_C \| R_L}{r_{e2}}$$

$$= -\frac{2\text{k}5 \| 10\text{k}}{30.86} = -64.8$$

Total

$$A_V = 64.8 \times 44.62 = 2891$$

PROBLEM 7.14 CASCADED SYSTEMS

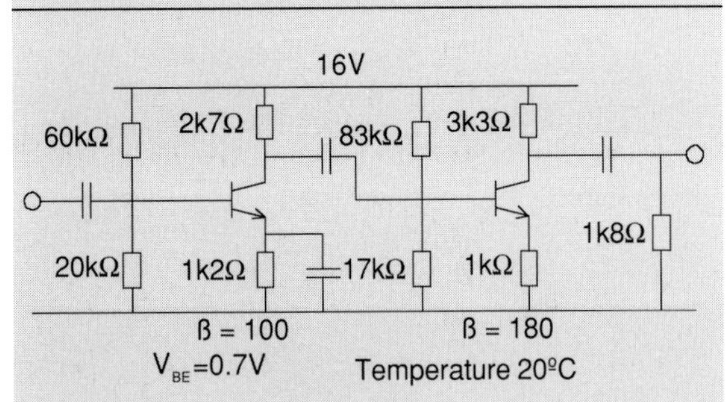

Figure 7.55

Find the current gain.

Answer: $A_i = 133$

First stage – Thevenin method

$$R_{TH} = 20\text{k}\|60\text{k} = 15\text{ k}\Omega$$

$$V_{TH} = 16\frac{20}{20+60} = 4\text{ V}$$

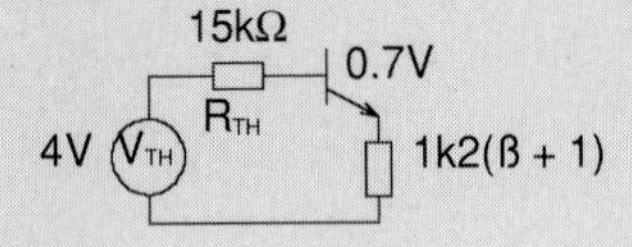

Figure 7.56

$$I_B = \frac{4-0.7}{15\,000 + 1200 \times 101}$$

$$= \frac{3.3}{136\,200} = 24.23\ \mu\text{A}$$

$$I_E = (\beta+1)I_B = 24.23 \times 10^{-6} \times 101 = 2.45\text{ mA}$$

$$r_{e1} = \frac{25}{2.45} = 10.2\ \Omega$$

AC model

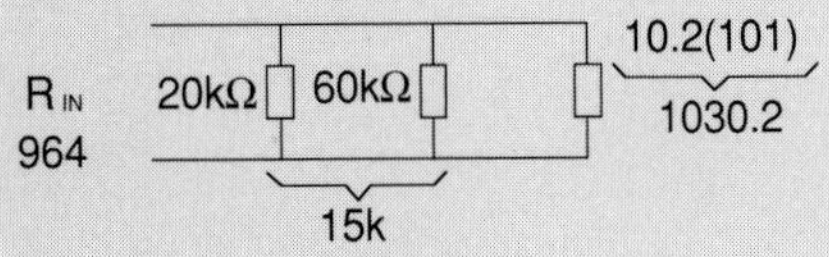

R IN 964 20kΩ 60kΩ 10.2(101) 1030.2 15k

Figure 7.57

Second stage – approximate method

$$V_B = 16\frac{17}{17+83} = 2.72\text{ V}$$

$$V_E = 2.72 - 0.7 = 2.02\text{ V}$$

$$I_E = \frac{2.02}{1\text{k}} = 2.02\text{ mA}$$

$$r_{e2} = \frac{25}{2.02} = 12.38\ \Omega$$

AC model – output of first transistor

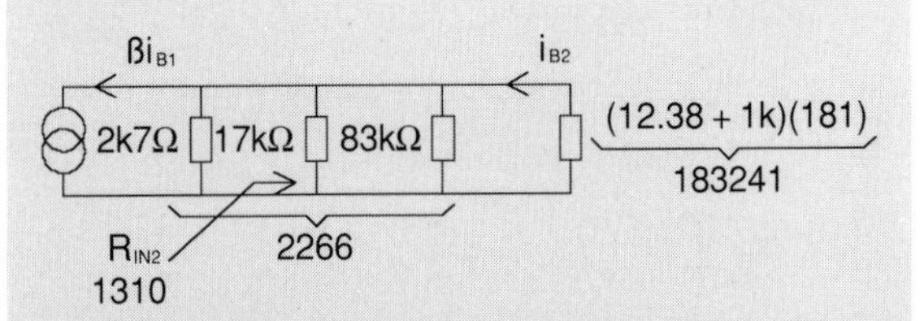

Figure 7.58

$$i_{B1} = i_{IN}\frac{15\,000}{15\,000 + 1030.2} = 0.936 i_{IN}$$

$$i_{C1} = \beta i_{B1} = 100 \times 0.936 i_{IN} = 93.6 i_{IN}$$

$$i_{B2} = i_{C1}\frac{2266}{2266 + 18\,3241} = i_{C1} \times 0.012\,2$$

$$i_{B2} = 0.012\,2 \times 93.6 i_{IN} = 1.142 i_{IN}$$

$$i_{C2} = \beta i_{B2} = 180 \times 1.142 i_{IN}$$

$$i_L = i_{C2}\frac{3.3}{3.3+1.8} = i_{C2} \times 0.647$$

$$i_L = 0.647 \times 205.56 i_{IN} = 133 i_{IN}$$

$$A_i = 133$$

PROBLEM 7.15 CASCADED SYSTEMS

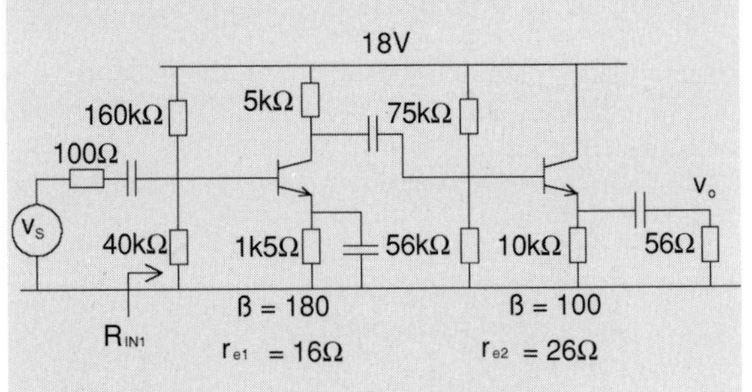

Figure 7.59

Find the total voltage gain.

Answer: $A_V = -112$

$$R_{IN1} = 40k||160k||16(181) = 2656\ \Omega$$

Load to transistor 1

$$R_{L1} = 5k||56k||75k||(26 + 10k||56)101 = 2750\ \Omega$$

Voltage gain

$$A_1 = -\frac{R_{L1}}{r_{e1}} = -\frac{2750}{16} = -171.86$$

$$A_2 = \frac{10k||56}{10k||56 + r_{e2}}$$

$$= \frac{55.69}{55.69 + 26} = 0.68$$

Potential divider at input

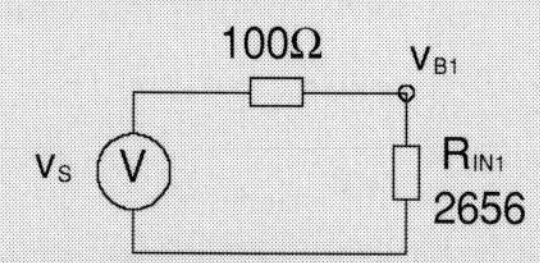

Figure 7.60

$$\frac{v_{B1}}{v_S} = \frac{2656}{2656 + 100} = 0.96$$

Total gain

$$A_V = -0.96 \times 171.86 \times 0.68 = -112$$

PROBLEM 7.16 CASCADED SYSTEMS

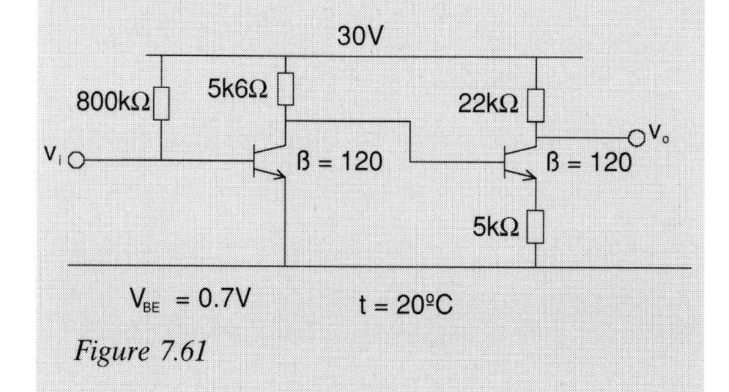

Figure 7.61

Find the overall voltage gain.

Answer: $A_V = 4349$

First transistor circuit – KVL

$$30 - 0.7 = I_{B1} 800k$$

$$I_{B1} = \frac{29.3}{800\,000}$$

$$I_{C1} = \beta I_{B1} = \frac{29.3 \times 120}{800\,000} = 4.395\text{ mA}$$

$$I_{E1} = 4.43\text{ mA}$$

We assume no loading from the second stage

$$V_{C1} = 30 - I_{C1} \times 5k6$$

$$= 30 - 4.395 \times 5.6 = 5.39\text{ V}$$

Second stage

$$V_{E2} = V_{C1} - V_{BE} = 5.39 - 0.7 = 4.69\text{ V}$$

$$I_{E2} = \frac{4.69}{5000} = 0.938\text{ mA}$$

$$r_{e2} = \frac{25}{0.938} = 26.65\ \Omega$$

$$r_{e1} = \frac{25}{4.43} = 5.64\ \Omega$$

Voltage gain

$$A_1 = -\frac{5600}{5.64} = -993$$

$$A_2 = -\frac{22\,000}{r_{e2} + R_E}$$

$$= -\frac{22\,000}{5026.65} = -4.38$$

Total

$$A_V = 4.38 \times 993 = 4349$$

PROBLEM 7.17 CASCADED SYSTEMS

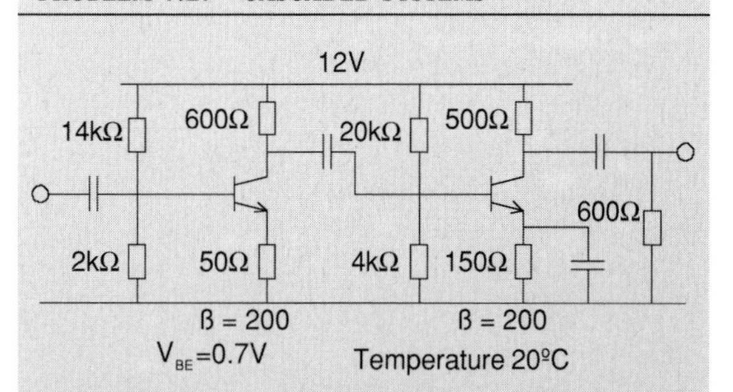

Figure 7.62

Find the current amplification.

Answer: $A_i = 1716$

AC model – input side

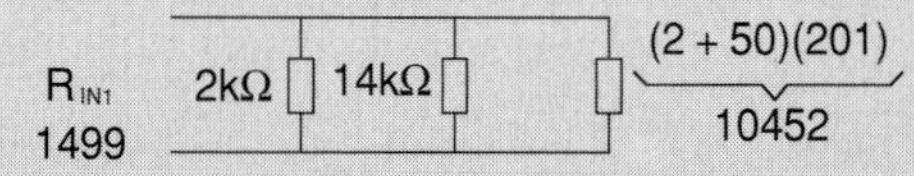

R_IN1 1499 | 2kΩ | 14kΩ | (2 + 50)(201) = 10452

Figure 7.63

$$i_{B1} = i_{IN}\frac{1750}{1750+10\,452}$$

$$= i_{IN}\frac{1750}{12\,202} = 0.143 i_{IN}$$

$$I_{C1} = \beta i_{B1} = 200 \times 0.143 i_{IN} = 28.6 i_{IN}$$

Transistor 1, output

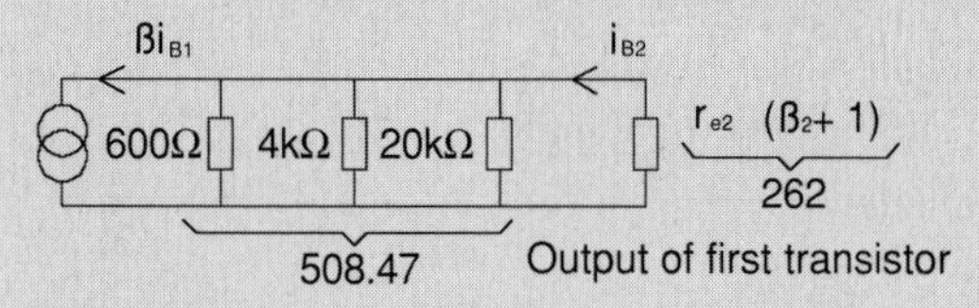

Figure 7.64

$$i_{B2} = i_{C1}\frac{508.47}{508.47+262} = i_{C1} \times 0.66$$

$$= 28.6 \times 0.66 \times i_{IN} = 18.876 i_{IN}$$

$$i_{C2} = \beta i_{B2} = 200 \times 18.876 i_{IN} = 3775 i_{IN}$$

$$i_L = 3775 i_{IN}\frac{500}{500+600} = 1716 i_{IN}$$

$$A_i = \frac{i_L}{i_{IN}} = 1716$$

PROBLEM 7.18 CASCADED SYSTEMS

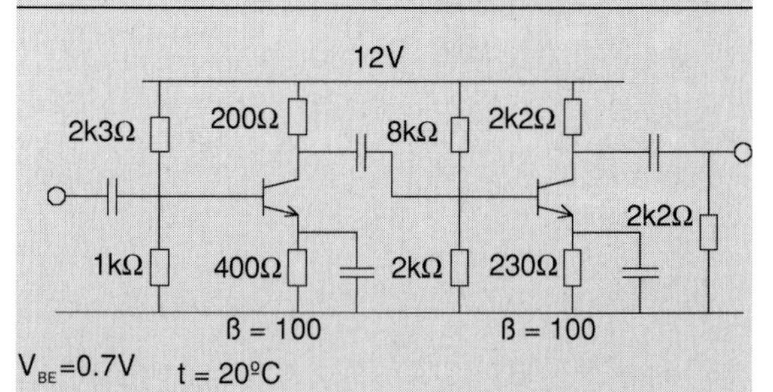

Figure 7.65

Find the voltage gain. $r_{e1} = r_{e2} = 3.4\ \Omega$

Answer: 11 184

Voltage gain

$$\frac{V_{B2}}{V_{B1}} = -\frac{R_C \| R_L}{r_{e1}}$$

Load resistance for first stage

$$R_L = 2k \| 8k \| 3.4(101) = 283\ \Omega$$

$$\frac{V_{B2}}{V_{B1}} = -\frac{200 \| 283}{3.4}$$

$$= \frac{117.53}{3.4} = -34.57$$

$$\frac{V_O}{V_{B2}} = -\frac{R_C \| R_L}{r_{e2}}$$

$$= -\frac{2k2 \| 2k2}{3.4} = -323.53$$

Total

$$A_V = 34.57 \times 323.53 = 11\,184$$

PROBLEM 7.19 CASCADED SYSTEMS

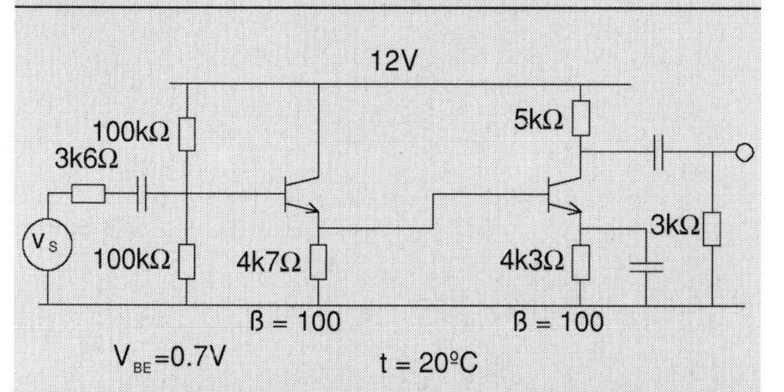

Figure 7.66

Find the overall voltage gain.

Answer: $A_V = -70.3$

DC conditions

$$V_B = 12\frac{100}{100+100} = 6 \text{ V}$$

$$V_{E1} = 6 - 0.7 = 5.3 \text{ V}$$

$$I_{E1} = \frac{5.3}{4k7} = 1.13 \text{ mA}$$

$$r_{e1} = \frac{25}{1.13} = 22.12\ \Omega$$

$$V_{E2} = V_{E1} - V_{BE} = 5.3 - 0.7 = 4.6 \text{ V}$$

$$I_{E2} = 1.07 \text{ mA}$$

$$r_{e2} = \frac{25}{1.07} = 23.36\ \Omega$$

AC model

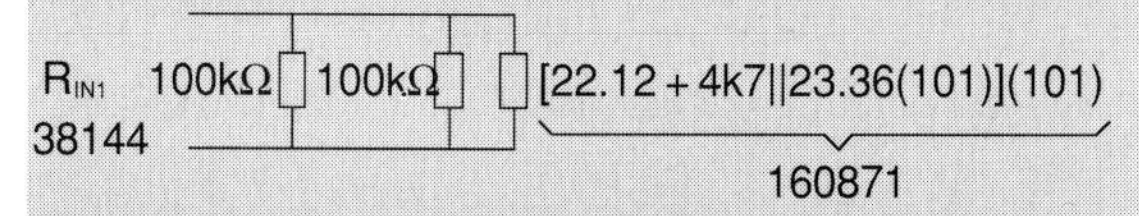

Figure 7.67

R_{E1} is in parallel with the reflected value of r_{e2}

Potential divider at input

$$\frac{v_{B1}}{v_S} = \frac{38\,144}{38\,144 + 3600}$$

$$= \frac{38\,144}{41\,744} = 0.91$$

$$\frac{v_{B2}}{v_{B1}} = \frac{R_{E1} || r_{e2}(\beta + 1)}{R_{E1} || r_{e2}(\beta + 1) + r_{e1}}$$

$$= \frac{1571}{1593} = 0.99$$

$$\frac{V_O}{V_{B2}} = -\frac{R_C || R_L}{r_{e2}} = -\frac{5k||3k}{23.36} = -80.26$$

Total

$$A_V = -0.91 \times 0.99 \times 80.26 = -70.3$$

PROBLEM 7.20 CASCADED SYSTEMS

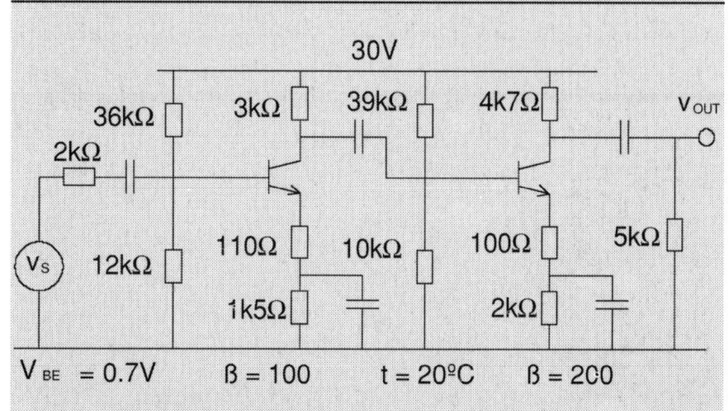

Figure 7.68

Find the voltage gain.

Answer: $A_V = 273.32$

First stage – DC conditions – Thevenin

$$V_{TH} = 30\frac{12}{12+36} = 7.5 \text{ V}$$

$$R_{TH} = 12k||36k = 9\ k\Omega$$

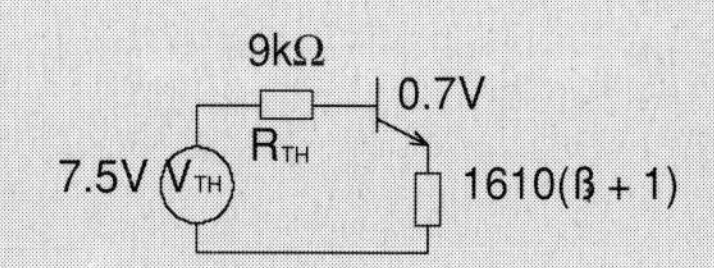

Figure 7.69

$$I_B = \frac{7.5 - 0.7}{9000 + 162\,610}$$

$$= \frac{6.8}{171\,610} = 39.62\ \mu A$$

$$I_{E1} = 39.62 \times 10^{-6} \times 101 = 4\ mA$$

$$r_{e1} = \frac{25}{4} = 6.25\ \Omega$$

Second stage – DC conditions – approximate method

$$V_B = 30\frac{10}{10+39} = 6.12\ V$$

$$V_E = 6.12 - 0.7 = 5.42\ V$$

$$I_E = \frac{5.42}{2100} = 2.58\ mA$$

$$r_{e2} = \frac{25}{2.58} = 9.69\ \Omega$$

AC model

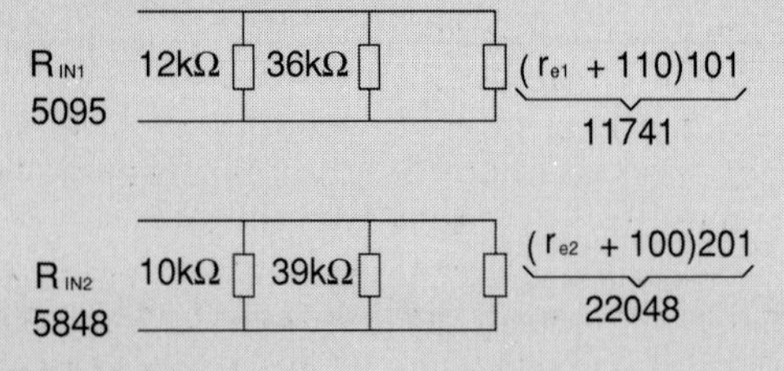

Figure 7.70

Potential divider at input

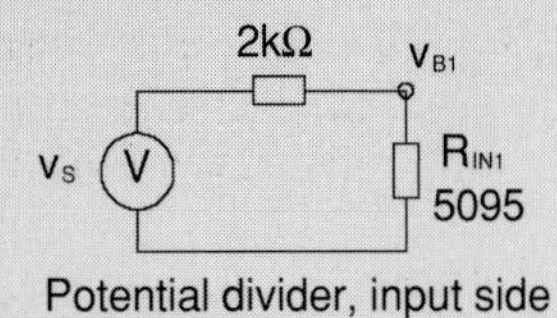

Figure 7.71

$$\frac{V_{B1}}{V_S} = \frac{5095}{5095 + 200} = \frac{5095}{7095} = 0.718$$

$$A_i = -\frac{R_C || R_{IN2}}{r_{e1} + R_{E1}}$$

$$= -\frac{3k||5831}{116.25} = -17.04$$

$$A_2 = -\frac{R_C || R_L}{r_{e2} + R_{E2}}$$

$$= -\frac{4k7||5k}{109.69} = -22.09$$

Total

$$A_V = 0.718 \times 17.04 \times 22.09 = 270.26$$

8

Bode plots

Bode plots are a simple method of graphically evaluating the frequency response of a circuit. The plot is done on log–log paper and in this way a compression of the frequency is effected. The gain is plotted on one graph and there is a different graph to plot the phase angle. If you join the gain and the phase angle in one graph, you obtain the Nyquist plot. The general arrangement is shown in Figure 8.1. We have a network in the middle. A sweep generator on the input of the network and an oscilloscope or a frequency analyser in the output.

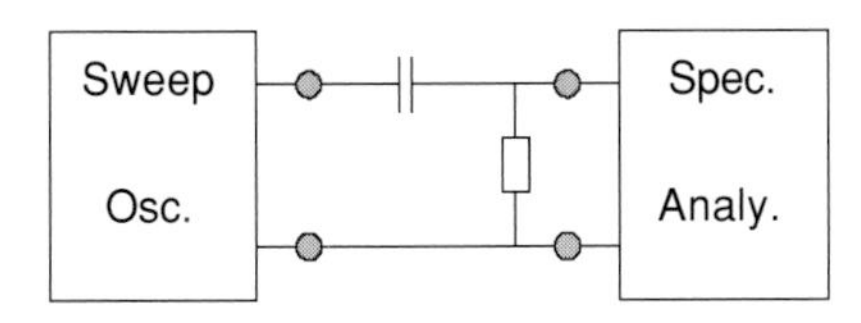

Figure 8.1

With the help of this simple circuit we can define the transfer function. The transfer function is the ratio between the output and the input. In this simple example the circuit is a potential divider and this will be shown in the transfer function.

$$\frac{v_o}{v_i} = \frac{R}{R + \dfrac{1}{j\omega C}} = \frac{j\omega CR}{j\omega CR + 1}$$

Using the complex frequency operator $s = \sigma + j\omega$ we can rewrite

$$\frac{v_o}{v_i} = \frac{R}{R + \dfrac{1}{sC}} = \frac{sCR}{sCR + 1}$$

In general the transfer function of any passive network will be of the form

$$F(s) = K\frac{(s - Z_1)(s - Z_2)\ldots(s - Z_n)}{(s - P_1)(s - P_2)\ldots(s - P_m)}$$

With reference to the simple quadratic equation we can say that if X_1 and X_2 are the solutions of a quadratic equation, then $(X - X_1)(X - X_2) = 0$ is the equation.
Example

$$X^2 - 7X + 12 = 0$$

$$(X - 3)(X - 4) = 0$$

3 and 4 are the solutions of the equation. Similarly, in the general equation, Z is a zero value, that is to say, a value that will make the top term equal to zero.

P is a pole, that is to say a value that will make the denominator equal to zero. Zeroes and poles are the solution of the transfer function and it can be plotted in a system of coordinates.

In frequency response we are interested in the magnitude of the function and in the phase angle at a given frequency. The magnitude is given by

$$|F(s)| = K\frac{|s - Z_1||s - Z_2|\ldots}{|s - P_1||s - P_2|\ldots}$$

The value in-between the bars is the absolute value. Because the function is factorised and in polar form, the angle is simply found

$$\angle F(s) = \frac{\alpha + \beta + \cdots}{\rho + \sigma + \cdots}$$

$$= \alpha + \beta - \rho - \sigma \pm \cdots$$

In order to systematically plot these factors as Bode first suggested, we use the general equation to see how many different forms we will encounter. These are:

$F(s) = K$

$F(s) = Ks$ type A

$F(s) = K\dfrac{1}{s}$ type B

$F(s) = K(s - Z)$ type C

$$F(s) = \frac{K1}{(s-P)} \qquad \text{type D}$$

The first form $F(s) = K$, is only a constant. As such it is not affected by the frequency. That is to say, its value will not change with a change in the frequency. It will, however, affect the magnitude of the function. It can be argued whether or not this is a Bode form. We don't mind, but we will concentrate on the four types of Bode forms, types A to D. The four different types are shown in Figure 8.2.

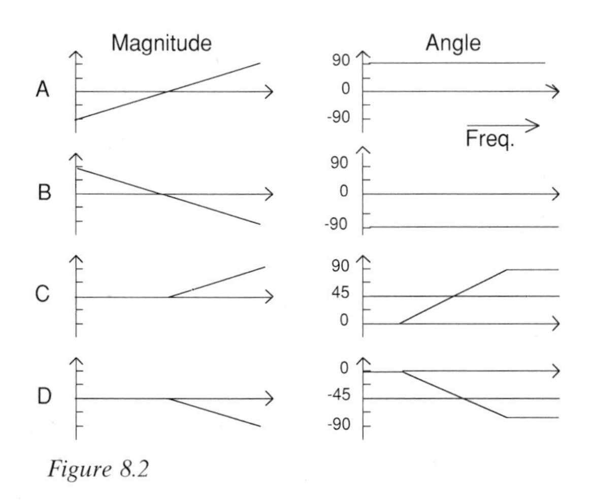

Figure 8.2

We have replaced the complex frequency operator $s = \sigma + j\omega$ by just $j\omega$ as we are mainly interested in the sinusoidal oscillations. The magnitude is the absolute value of the function and j will not form an important part of the plot. For the angle representation the j is very important. In type A, the angle is always 90° for any value of ω.

For type B, the angle is always –90°. In order to see this we go back to the system of coordinates in Figure 8.3. The real values go on the X axis and the imaginary values go on the Y axis. The angles are measured from the real axis in an anticlockwise direction.

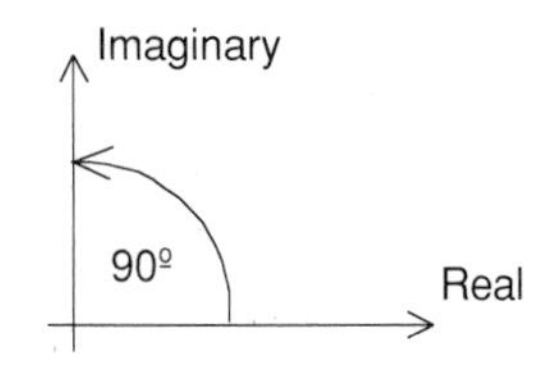

Figure 8.3

In type B

$$\frac{1}{j\omega} = \frac{j}{jj\omega} = -\frac{j}{\omega}$$

So far, we have seen type A and type B. They are both straightforward. The magnitude plot is a straight line going up for one, down for the other and crossing the zero line at one point.

Slope

An interesting part of this representation is that the slope of the line in type A is 20 dB per decade, whereas in type B it is –20 dB per decade. Some authors refer to it as 6 dB per octave.

Let us see where these values come from. The representation of Bode plots is done on log–log paper. That is to say, we represent the frequency in a log scale and the magnitude in decibels.

An octave from music is the distance between two notes where one is twice the frequency of the other

$$\begin{aligned} A &= 20 \log 2 \\ &= 20 \times 0.30103 \\ &= 6.0206 \text{ dB} \end{aligned}$$

6 dB per octave is an approximation.

A decade is a factor of 10 up or down in the frequency scale

$$\begin{aligned} A &= 20 \log 10 \\ &= 20 \times 1 \\ &= 20 \text{ dB} \end{aligned}$$

20 dB per decade is an exact value.

Using a slope of 20 dB per decade means that we can easily construct the plots. All we need to know is the point where they cross the ω axis. We can then draw the straight line up for type A, or down for type B, with a slope of ±20 dB per decade.

As we saw in Figure 8.2, type C and type D have a bent line. The point where the line bends is called the corner frequency or break frequency.

As you probably suspect by now, type C and type D are very much related. We will only consider type C in detail, but by extension you will find type D.

Type C

We would start with something like:

$j25\omega + 75$

with suitable factorising and normalising we obtain the basic form which is:

$H(\omega) = j\omega + 1$

The absolute value of this function is:

As $\omega \to 0$, $|H(\omega)| \to 1$ $\qquad A \to 0$ db

As $\omega \to \infty$, $|H(\omega)| \to \omega$ $\qquad A \to \infty$

Another way of explaining this is the Bode way. We distinguish three cases:

a) $|j\omega| \gg 1$. If this is the case we can ignore the 1 as it would be insignificant compared to the ω term. The value of magnitude A in decibels will be:

$A = 20\log \omega$ dB

b) $|j\omega| << 1$. In this case the term $j\omega$ can be ignored and the 1 will be the significant part of the function.

$A = 20\log 1 = 0$ dB

c) The other alternative is

$|j\omega| = 1 \qquad \omega = 1$

The imaginary part is equal to the real part. The line changes direction and we have the corner, or break frequency.

For the magnitude plot of the type C function, all we need to do is to find the corner frequency. The left-hand side of this is a straight line at 0 dB level. To the right of this we have a straight line at 20 dB per decade.

On type C functions the angle plot is more complicated, but we can come to an agreement that will make the plotting straightforward. First of all, at the corner frequency the real part is equal to the imaginary part. This means that the angle of the combined vector is 45° as seen in Figure 8.4.

In Figure 8.5, if ω increases, the real part will still be 1, as before, but the imaginary part will grow. The angle that was 45° will be larger. As ω increases, the angle will become larger and larger, but the limit will be 90°.

Conversely, as we see in Figure 8.5, if ω decreases, the real part will still be 1 as before, but the imaginary part part will become smaller. The angle that was 45° at the corner frequency, will now be smaller.

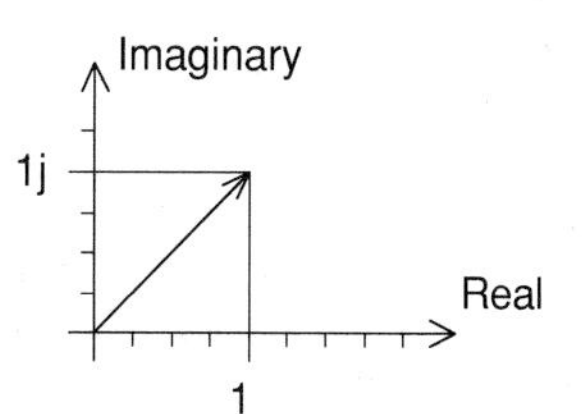

Figure 8.4

lower freq.	freq. f	higher freq.
Angle near 0	45º exact	Angle near 90º
Say 0º		Say 90º

Figure 8.5

As ω decreases the angle will become smaller and smaller, but it will never be smaller than the limit which is 0°. The agreement that is necessary here to simplify matters is to define:

angle = 90° at 10 times the corner frequency
angle = 0° at 0.1 times the corner frequency

We can now see the three alternatives for the type C function. You can work out the type D which is very similar, but with a negative angle.

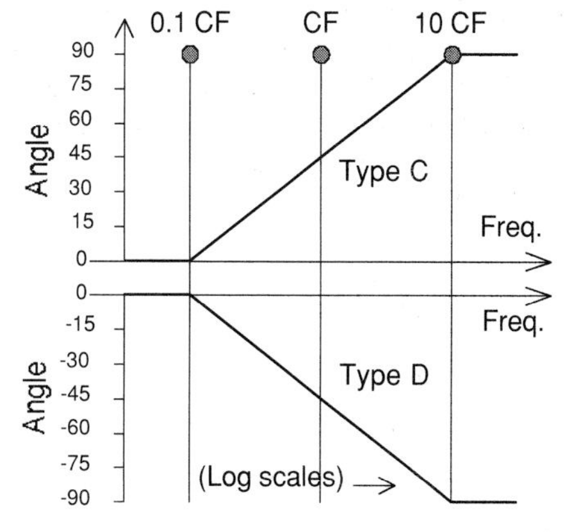

Figure 8.6

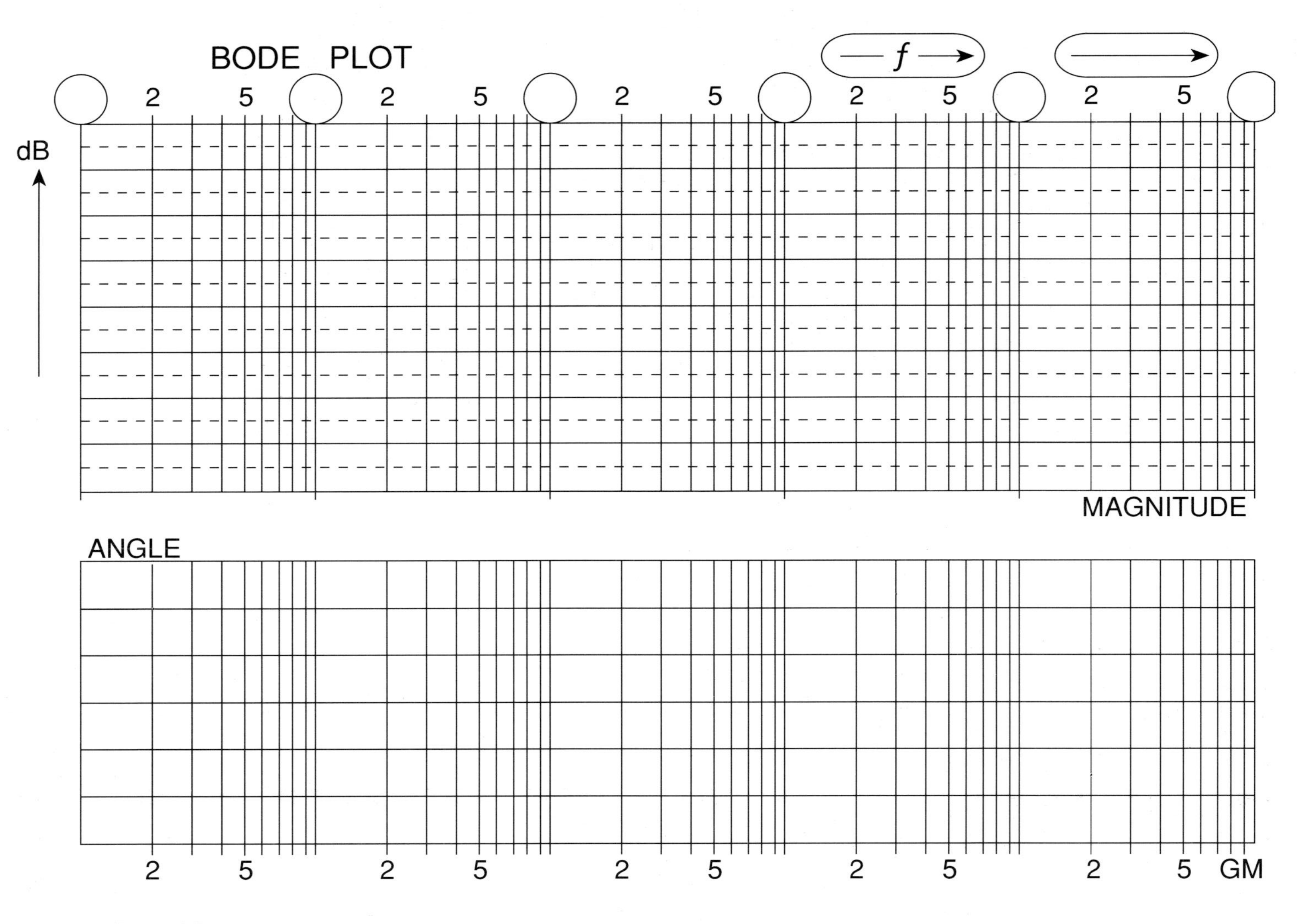

Figure 8.7

Just to clarify matters at this stage we show Figure 8.6. We can see the angle plot for a type C and type D. CF stands for the corner frequency.

We are almost ready to start solving problems with Bode plots, but before we do so we must add one more comment. As you saw at the beginning, the general transfer function will be several factors that multiply or divide. Every one of these factors will be either type A, B, C or D. As it is in logarithmic form, these will add individually. Example

$$\log 12 = \log (6 \times 2) = \log 6 + \log 2$$
$$= 1.079\,18 = 0.778\,15 + 0.301\,03$$

This logarithmic transformation of products into sums means that you can do the individual components separately and join them in a graphical sum to obtain the total response. The same technique can be applied to angles. You plot individual angles of each factor and add them (taking care of the sign) together.

Figure 8.7 shows a full page Bode form to be used in problem solving. Copy as required. There is a choice of whether you use *f* in cycles per second or ω in radians per second (delete one of them when you have decided). The scales in decibels are for you to adjust trying to accommodate the response in the space provided. The same is valid for the angle scale that you can adjust to suit your problem.

PROBLEM 8.1 BODE

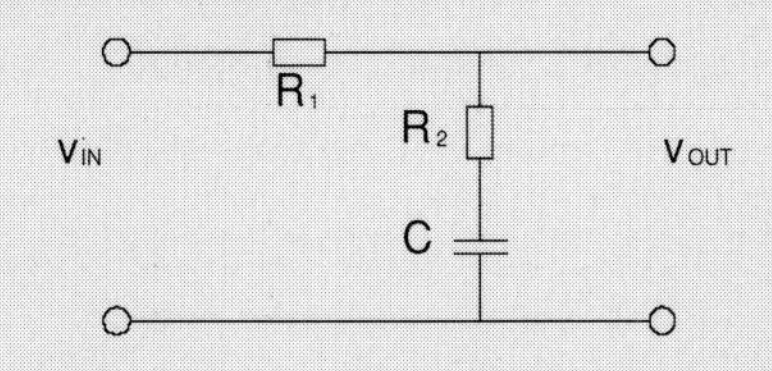

Figure 8.8

Find the transfer function for this circuit.

Answer: Type C and type D

$$\frac{v_o}{v_i} = \frac{R_2 + \dfrac{1}{j\omega C}}{R_2 + \dfrac{1}{j\omega C} + R_1} \qquad | \times j\omega C$$

$$= \frac{j\omega C R_2 + 1}{j\omega C(R_1 + R_2) + 1}$$

This is a type C and a type D

PROBLEM 8.2 BODE

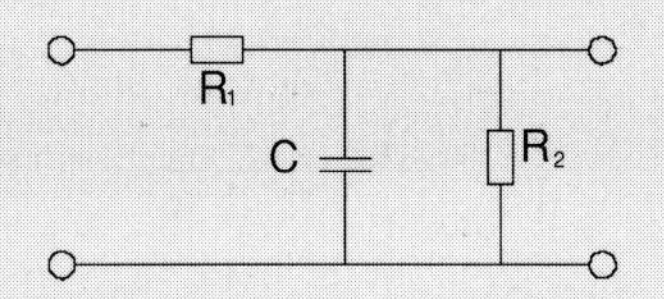

Figure 8.9

Find the transfer function for this circuit.

Answer: $\dfrac{v_o}{v_s} = \dfrac{R_2}{R_1 + R_2} \times \dfrac{1}{1 + \dfrac{R_1 R_2}{R_1 + R_2} j\omega C}$

$$\frac{v_o}{v_s} = \frac{\dfrac{R_2 \times \dfrac{1}{j\omega C}}{R_2 + \dfrac{1}{j\omega C}}}{\dfrac{R_2 \times \dfrac{1}{j\omega C}}{R_2 + \dfrac{1}{j\omega C}} + R_1} \qquad | \times \left(R_2 + \frac{1}{j\omega C} \right)$$

$$= \frac{R_2 \times \dfrac{1}{j\omega C}}{R_2 \times \dfrac{1}{j\omega C} + R_1 \left(R_2 + \dfrac{1}{j\omega C} \right)} \qquad | \times j\omega C$$

$$\frac{v_o}{v_s} = \frac{R_2}{R_2 + R_1 R_2 j\omega C + R_1}$$

$$= \frac{R_2}{R_1 + R_2 + R_1 R_2 j\omega C}$$

We factorise by $R_1 + R_2$

$$\frac{v_o}{v_s} = \frac{R_2}{R_1 + R_2} \frac{1}{\left[1 + \dfrac{R_1 R_2}{R_1 + R_2} j\omega C \right]}$$

The first part is a constant and the second part is a Bode type D ,

PROBLEM 8.3 BODE

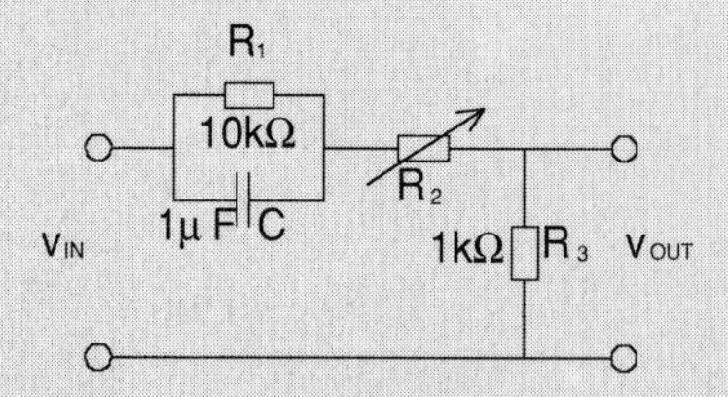

Figure 8.10

Find the transfer function and the value of R_2 that will produce a corner frequency at ω = 500 rad/s. What is the frequency of the other break?

Answer: R_2 = 1500 Ω, corner frequency 100 rad/s

$$H(s)=\frac{R_3}{\dfrac{R_1\dfrac{1}{sC}}{R_1+\dfrac{1}{sC}}+R_2+R_3} \qquad |\times sC$$

$$=\frac{R_3}{\dfrac{R_1}{sCR_1+1}+R_2+R_3} \qquad |\times(sCR_1+1)$$

$$H(s)=\frac{R_3(sCR_1+1)}{R_1+sCR_1R_2+R_2+sCR_1R_3+R_3}$$

$$=\frac{R_3sCR_1+1}{R_1+R_2+R_3+s(CR_1R_2+CR_1R_3)}$$

Factorising

$$H(s)=\frac{R_3(sCR_1+1)}{(R_1+R_2+R_3)\left(1+s\dfrac{CR_1R_2+CR_1R_3}{R_1+R_2+R_3}\right)}$$

One type C and one type D functions.

Corner frequency

$$\omega_C=\frac{1}{CR_1}=\frac{1}{10^{-6}\times10^{-4}}=100\text{ rad / s}$$

$$\omega_D=500=\frac{R_1+R_2+R_3}{CR_1R_2+CR_1R_3}$$

$$R_2=1500\ \Omega$$

PROBLEM 8.4 BODE

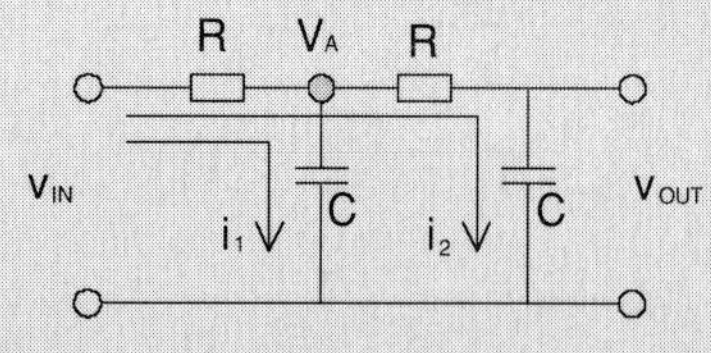

Figure 8.11

Find the transfer function.

Answer: $\dfrac{v_o}{v_i}=\dfrac{1}{1-\omega^2C^2R^2+3j\omega CR}$

$$v_A=v_o+i_2R$$

$$i_2=\frac{v_o}{\dfrac{1}{j\omega C}}=v_oj\omega C$$

$$i_1=\frac{v_A}{\dfrac{1}{j\omega C}}=v_Aj\omega C$$

$$v_i=v_A+(i_1+i_2)R$$

Replace i_1+i_2

$$v_i=v_A+v_Aj\omega CR+v_oj\omega CR$$
$$=v_A(1+j\omega CR)+v_oj\omega CR$$

Replace v_A

$$v_i=(v_o+i_2R)(1+j\omega CR)+v_oj\omega CR$$
$$=(v_o+v_oj\omega CR)(1+j\omega CR)+v_oj\omega CR$$
$$=v_o[(1+j\omega CR)^2+j\omega CR]$$
$$=v_o(1+2j\omega CR-\omega^2C^2R^2+j\omega CR)$$
$$=v_o(1+3j\omega CR-\omega^2C^2R^2)$$

$$\frac{v_o}{v_i}=\frac{1}{1-\omega^2C^2R^2+3j\omega CR}$$

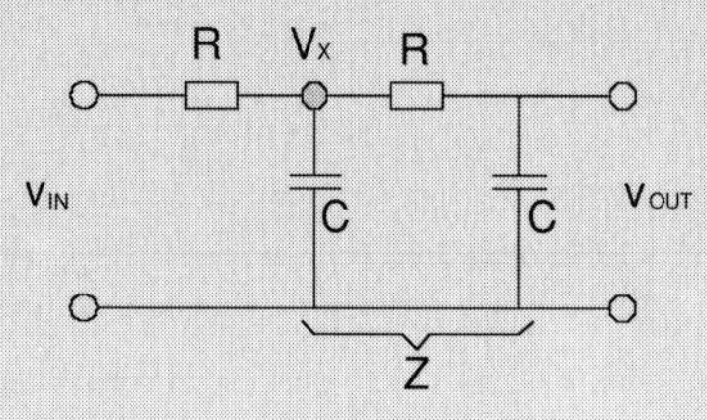

Figure 8.12

Find the transfer function. Alternative method (using s operator).

$$v_o = v_x \frac{\frac{1}{sC}}{R + \frac{1}{sC}} = v_x \frac{1}{sRC + 1}$$

$$v_x = v_o(sRC + 1)$$

Impedance Z

$$Z = \frac{\frac{1}{sC}\left(R + \frac{1}{sC}\right)}{\frac{1}{sC} + R + \frac{1}{sC}} \qquad | \times sC$$

$$= \frac{R + \frac{1}{sC}}{sRC + 2}$$

Potential divider

$$\frac{v_x}{v_i} = \frac{\frac{R + \frac{1}{sC}}{2 + sCR}}{\frac{R + \frac{1}{sC}}{2 + sCR} + R} \qquad | \times (2 + sCR)$$

$$\frac{v_x}{v_i} = \frac{R + \frac{1}{sC}}{R + \frac{1}{sC} + 2R + sR^2C} \qquad | \times sC$$

$$= \frac{sRC + 1}{sRC + 1 + 2sRC + s^2R^2C^2}$$

Replace v_x

$$\frac{v_o(sRC + 1)}{v_i} = \frac{sRC + 1}{1 + 3sRC + s^2R^2C^2}$$

$$\frac{v_o}{v_i} = \frac{1}{1 + 3sRC + s^2R^2C^2}$$

PROBLEM 8.5 BODE

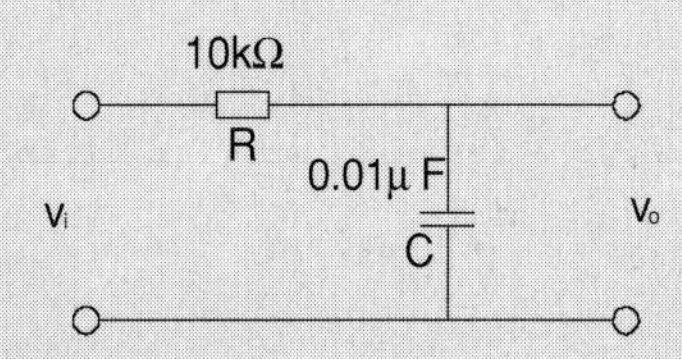

Figure 8.13

Draw the Bode plot for the above circuit.

Answer: See Figure 8.14.

Transfer function $\frac{v_o}{v_i}$

$$\frac{v_o}{v_i} = \frac{\frac{1}{j\omega C}}{\frac{1}{j\omega C} + R} \qquad | \times j\omega C$$

$$= \frac{1}{1 + j\omega CR}$$

Type D function. Corner frequency

$$\omega CR = 1$$

$$\omega = \frac{1}{CR}$$

$$\omega = \frac{1}{0.01 \times 10^{-6} \times 10^4} = 10\,000 \text{ rad/s}$$

$$f = \frac{10\,000}{2\pi} = 1592 \text{ Hz}$$

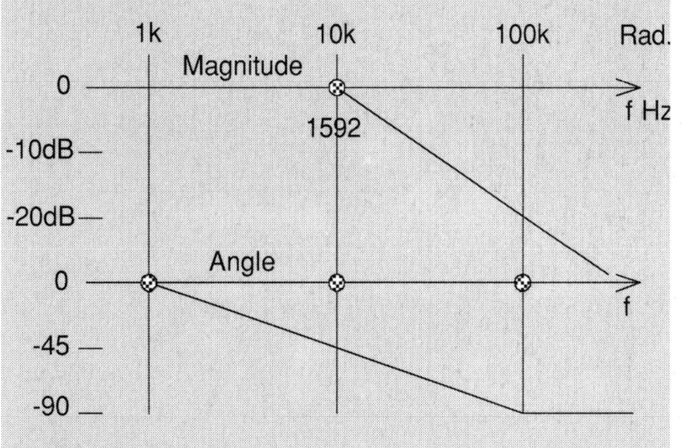

Figure 8.14

PROBLEM 8.6 BODE

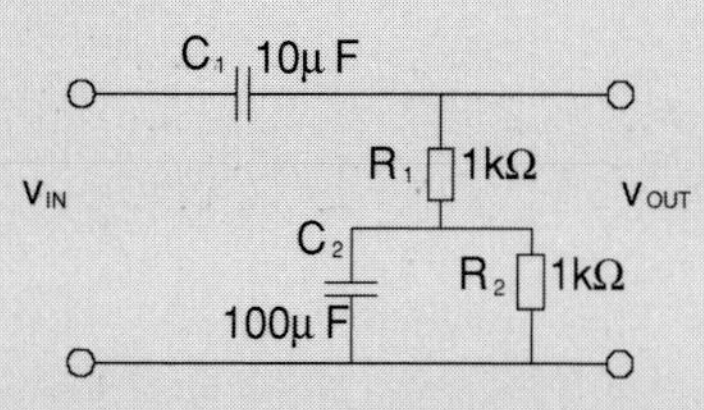

Figure 8.15

Draw the Bode plot and check the values of magnitude and phase angle for 5 and 50 rad/s.

Answer: –20.9 dB, 72.41°, –7.22 dB, 54.15°

$$\frac{v_o}{v_i}=\frac{R_1+\dfrac{\frac{R_2}{sC_2}}{R_2+\frac{1}{sC_2}}}{\dfrac{1}{sC_1}+R_1+\dfrac{\frac{R_2}{sC_2}}{R_2+\frac{1}{sC_2}}}\quad |\times\left(R_2+\frac{1}{sC_2}\right)$$

$$\frac{v_o}{v_i}=\frac{R_1\left(R_2+\frac{1}{sC_2}\right)+\frac{R_2}{sC_2}}{\frac{R_2}{sC_1}+R_1R_2+\frac{1}{s^2C_1C_2}+\frac{R_1}{sC_2}+\frac{R_2}{sC_2}}\quad |\times s^2C_1C_2$$

$$\frac{v_o}{v_i}=\frac{s^2R_1R_2C_1C_2+sR_1C_1+sR_2C_1}{sR_2C_2+s_2R_1R_2C_1C_2+1+sC_1R_1+sR_2C_1}$$

$$\frac{v_o}{v_i}=\frac{s^2 10^6 10^{-5}10^{-4}+s(10^{-2}+10^{-2})}{s^2 10^{-3}+s(10^{-1}+10^{-2}+10^{-2})+1}$$

$$=\frac{s^2 10^{-3}+2s\times10^{-2}}{s^2 10^{-3}+s(0.12)+1}\quad |\times 10^3$$

$$=\frac{s^2+20s}{s^2+120s+1000}$$

$$=\frac{s(s+20)}{(s+9)(s+111)}\ \text{(approximately)}$$

Normalising (achieved by factorising)

$$H(s)=\frac{20}{9\times111}\times s\frac{\frac{s}{20}+1}{\left(\frac{s}{9}+1\right)\left(\frac{s}{111}+1\right)}$$

$K=0.02$ $\log K=-1.69$

$20\log K=-33.9$ dB

The Bode plot of this function is seen in Figure 8.16.

Magnitude for 5 rad / s

$$\frac{5\times\sqrt{5^2+20^2}}{\sqrt{5^2+9^2}\sqrt{5^2+111^2}}$$

$$=\frac{5\times20.62}{10.3\times111.11}=0.09$$

$\log\ 0.09=-1.045$

$20\log\ 0.09=-20.9$ dB

Angle

$$\frac{\angle90\angle14.04}{\angle29.05\angle2.58}=72.41^\circ$$

Magnitude for 50 rad / s

$$\frac{50\times\sqrt{50^2+20^2}}{\sqrt{50^2+9^2}\sqrt{50^2+111^2}}$$

$$=\frac{50\times53.85}{50.8\times121.74}=0.4354$$

$\log 0.4354=-0.36111$

$20\log 0.4354=-7.22$ dB

Angle

$$\frac{\angle90\angle68.2}{\angle79.8\angle24.25}=54.15^\circ$$

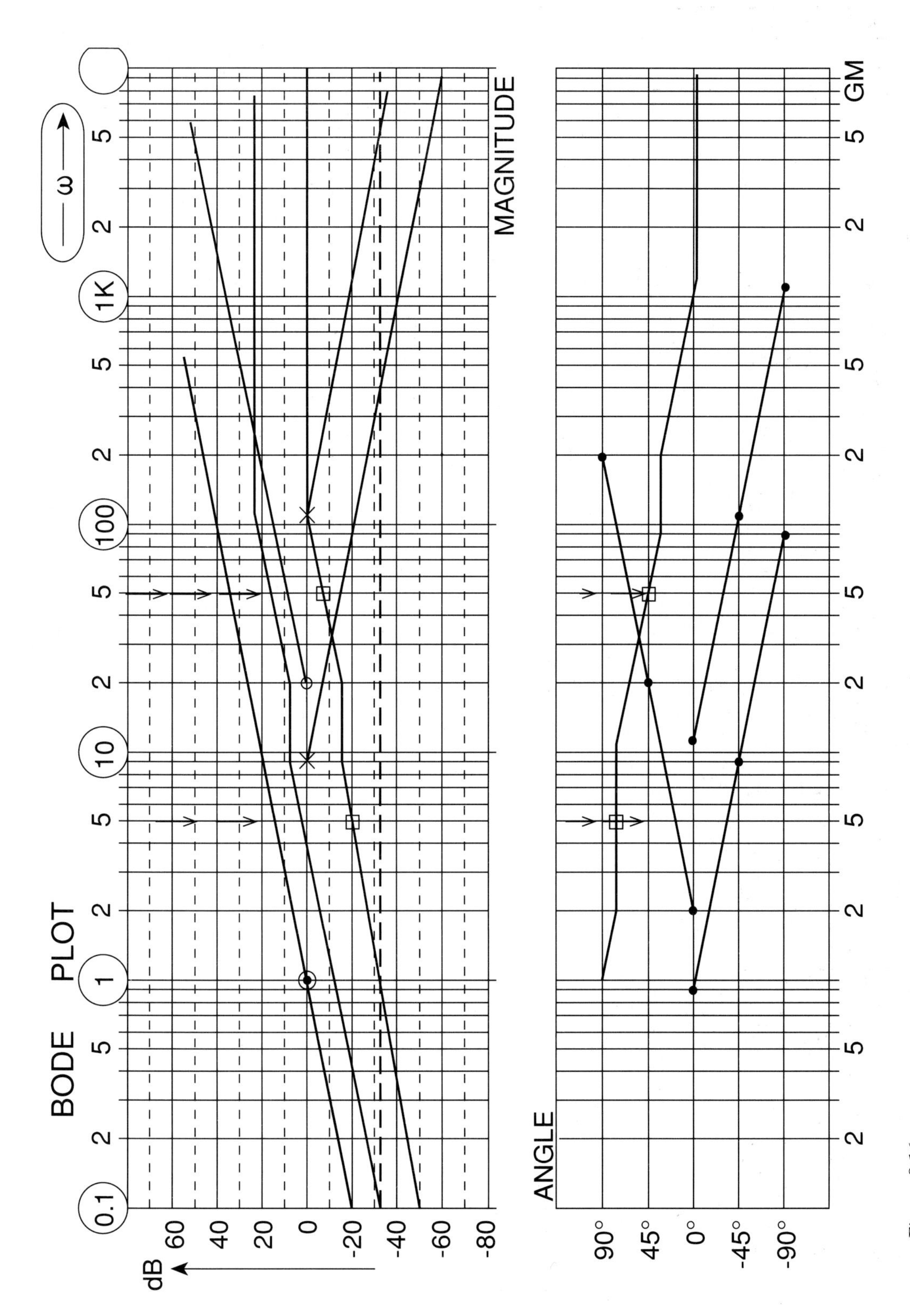
BODE PLOT
ω
0.1
2
5
1
2
5
10
2
5
100
2
5
1K
2
5
dB
60
40
20
0
-20
-40
-60
-80
MAGNITUDE
ANGLE
90°
45°
0°
-45°
-90°
GM

Figure 8.16

PROBLEM 8.7 BODE

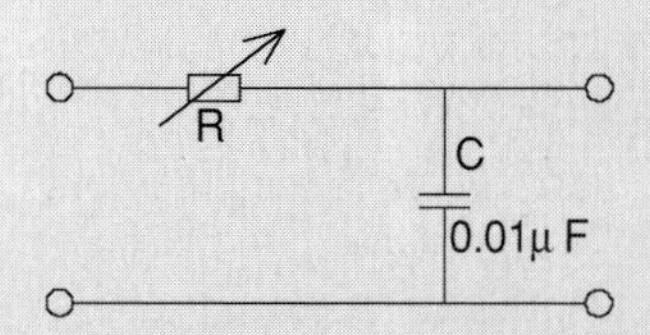

Figure 8.17

Find the transfer function and express it in the standard Bode form. Find the limits of a resistor that will produce a corner frequency from 1000 rad/s to 10000 rad/s.
Draw the Bode plot for R = 10 kΩ for magnitude and angle and obtain the value of magnitude and angle for ω = 5000 rad/s

> Answer: $\dfrac{1}{1+j\omega CR}$, R = 10 kΩ to 100 kΩ, magnitude 0 dB, angle −30°

$$\frac{v_o}{v_i}=\frac{\dfrac{1}{j\omega C}}{\dfrac{1}{j\omega C}+R}=\frac{1}{1+j\omega CR}$$

Only one type D

Corner frequency (when imaginary part is equal to the real part)

$$\omega CR = 1$$

At $\omega = 1000$ $\quad R=\dfrac{1}{\omega C}=\dfrac{1}{10^3\times 0.01\times 10^{-6}}=10^5$

At $\omega = 10\,000$ $\quad R=\dfrac{1}{\omega C}=\dfrac{1}{10^4\times 0.01\times 10^{-6}}=10^4$

From the Bode plot (Figure 8.18)

PROBLEM 8.8 BODE

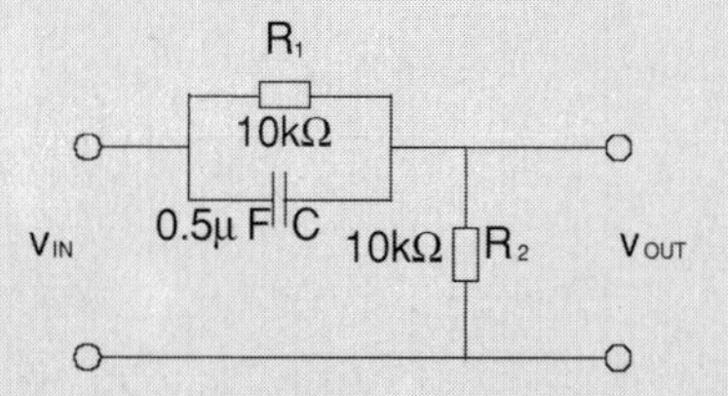

Figure 8.19

Draw the Bode plot and find, from the plot, the maximum magnitude in dB and the range of frequencies at which the phase angle is greater than zero.

> Answer: See Figure 8.20, maximum magnitude –1 dB, angle 20 to 4000 rad/s

$$\frac{v_o}{v_i}=\frac{R_2}{R_2+\dfrac{R_1\dfrac{1}{j\omega C}}{R_1+\dfrac{1}{j\omega C}}}\qquad |\times\left(R_1+\frac{1}{j\omega C}\right)$$

$$=\frac{R_1R_2+\dfrac{R_2}{j\omega C}}{R_1R_2+\dfrac{R_2}{j\omega C}+\dfrac{R_1}{j\omega C}}\qquad |\times j\omega C$$

$$\frac{v_o}{v_i}=\frac{R_1R_2 j\omega C+R_2}{R_1R_2 j\omega C+R_2+R_1}$$

We factorise to obtain the form $(j\omega K + 1)$

$$\frac{v_o}{v_i}=\frac{R_2}{R_1+R_2}\,\frac{j\omega CR_1+1}{\dfrac{R_1R_2}{R_1+R_2}j\omega C+1}$$

We replace values

$$\frac{v_o}{v_i}=\frac{1}{2}\,\frac{j\omega 0.5\times 10^{-2}+1}{j\omega 2.5\times 10^{-3}+1}$$

We have a constant, a type C and a type D function.

$$20\log\frac{1}{2}=-6.02\text{ dB}$$

$$\omega_C = 200\text{ rad / s}\qquad\qquad \omega_D = 400\text{ rad / s}$$

From the plot in Figure 8.20, the maximum magnitude is −1 dB. The angle is greater than zero from 20 to 4000 rad / s

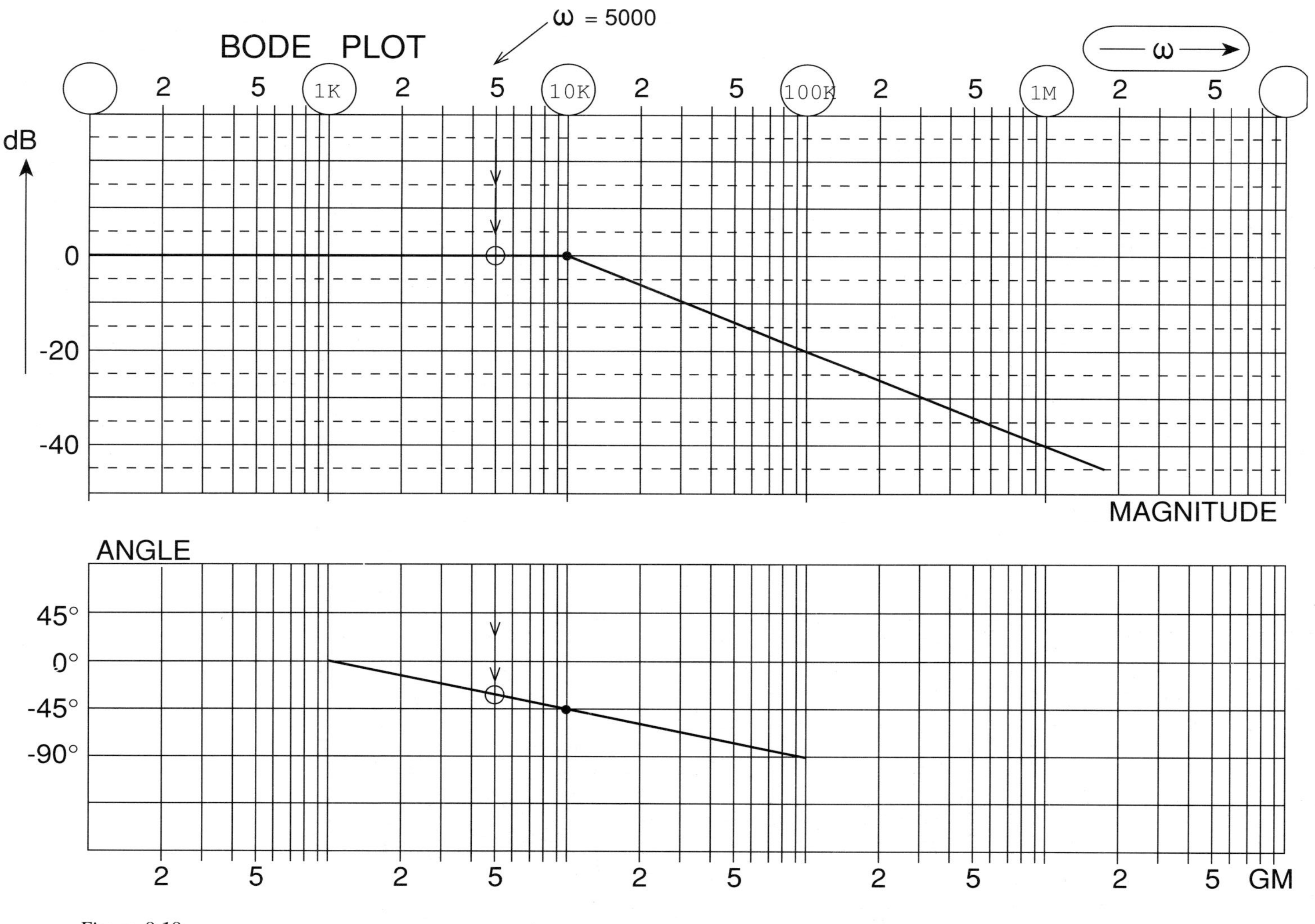
BODE PLOT
ω = 5000
ω
2
5
1K
2
5
10K
2
5
100K
2
5
1M
2
5
dB
0
-20
-40
MAGNITUDE
ANGLE
45°
0°
-45°
-90°
2
5
2
5
2
5
2
5
2
5
GM

Figure 8.18

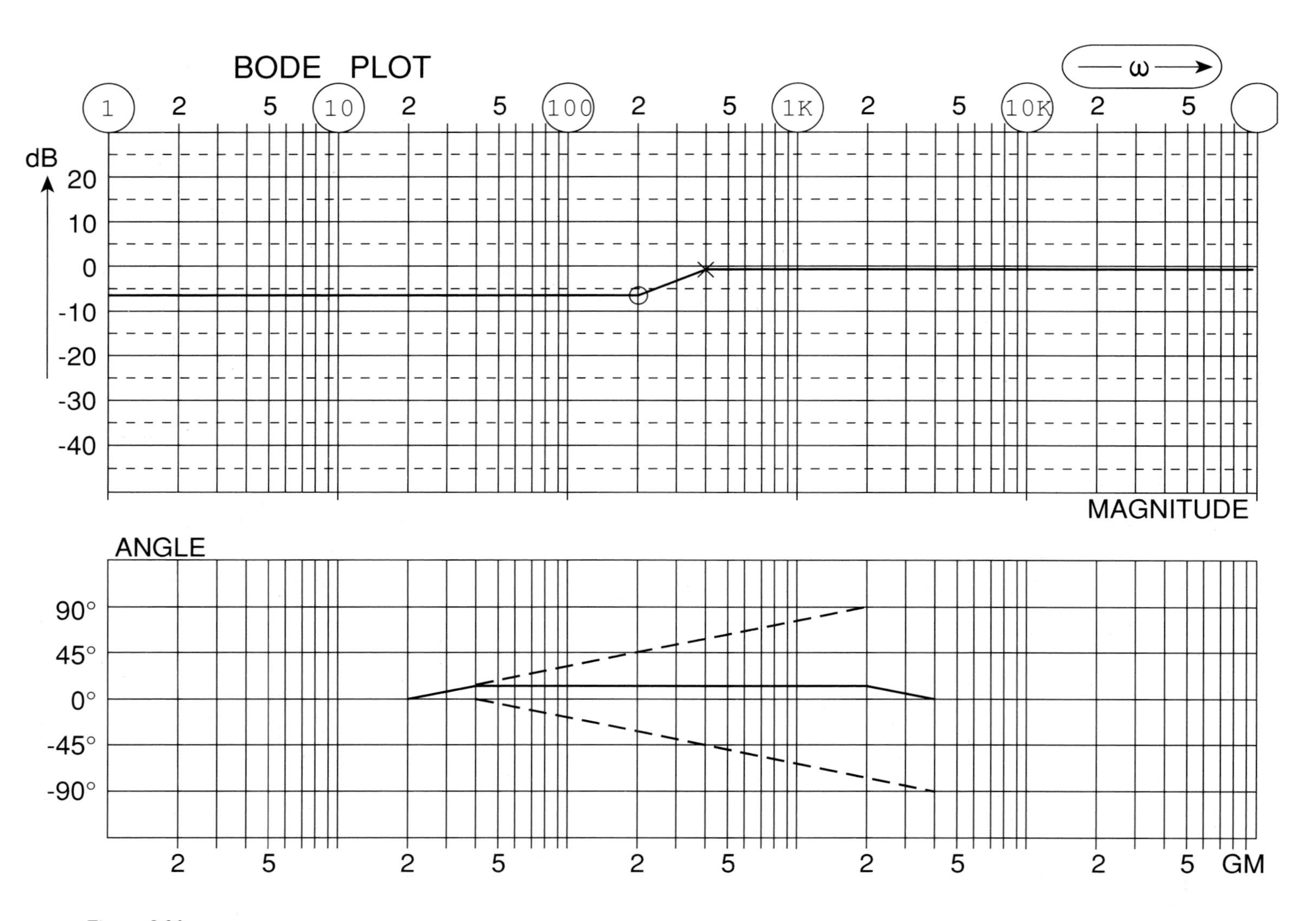
BODE PLOT
ω
1
10
100
1K
10K
2
5
dB
20
10
0
-10
-20
-30
-40
MAGNITUDE
ANGLE
90°
45°
0°
-45°
-90°
GM

Figure 8.20

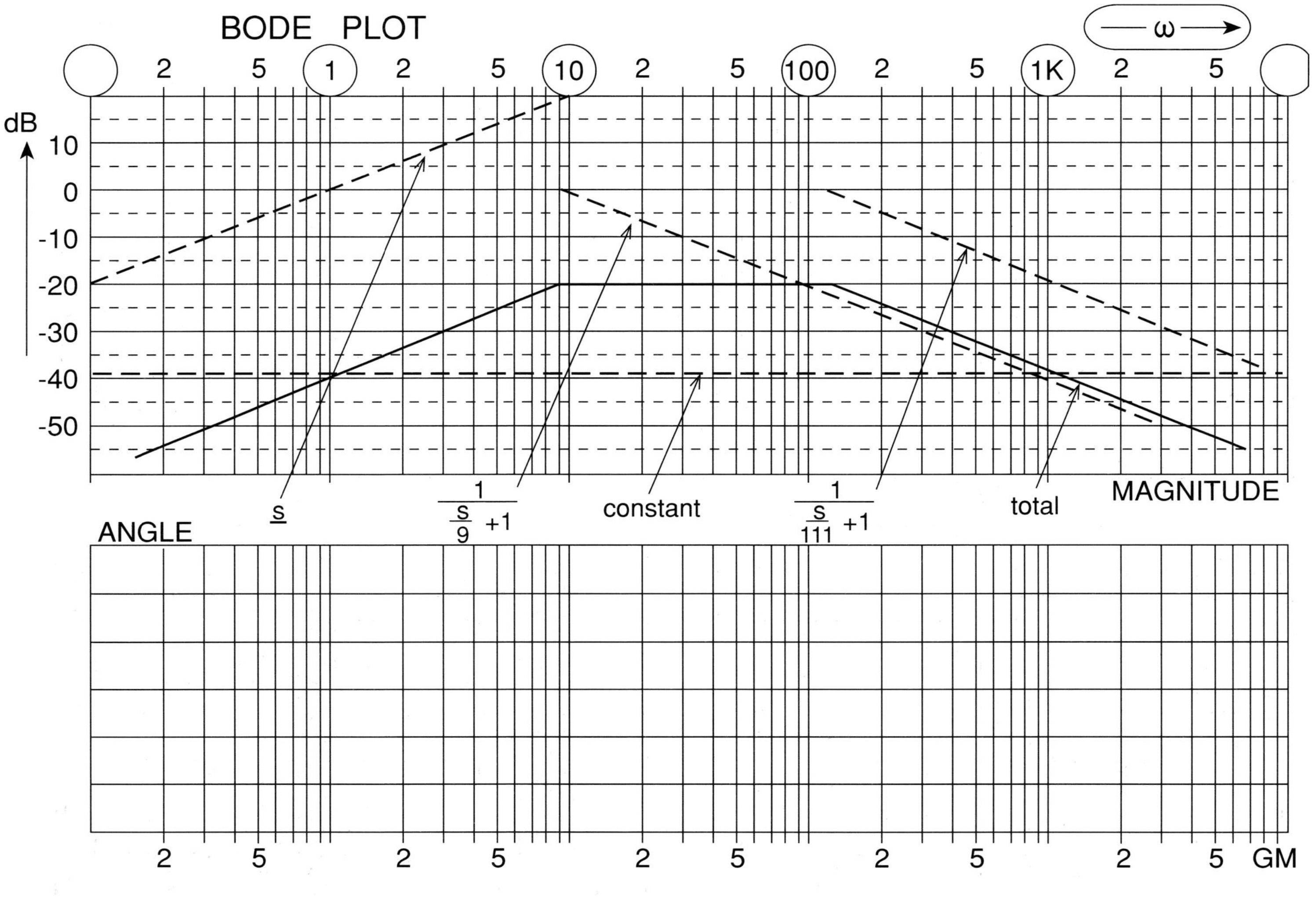
BODE PLOT
ω
1
10
100
1K
dB
10
0
-10
-20
-30
-40
-50
s
$\frac{1}{\frac{s}{9}+1}$
constant
$\frac{1}{\frac{s}{111}+1}$
total
MAGNITUDE
ANGLE
GM

Figure 8.22

PROBLEM 8.9 BODE

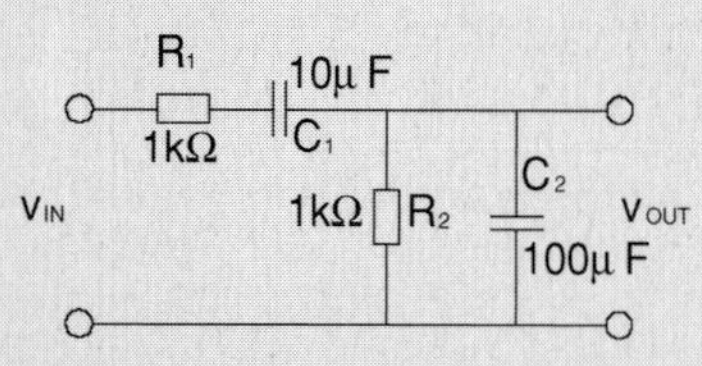

Figure 8.21

Find the transfer function and draw the Bode plot for magnitude only.

> Answer: See Figure 8.22,
> $$H(s)=\frac{10s}{(s+9)(s+111)}$$

Transfer function

$$H(s)=\frac{v_o}{v_i}=\frac{\dfrac{R_2\dfrac{1}{sC_2}}{R_2+\dfrac{1}{sC_2}}}{\dfrac{R_2\dfrac{1}{sC_2}}{R_2+\dfrac{1}{sC_2}}+R_1+\dfrac{1}{sC_1}} \qquad |\times sC_2$$

$$\frac{v_o}{v_i}=\frac{\dfrac{R_2}{sC_2R_2+1}}{\dfrac{R_2}{sC_2R_2+1}+R_1+\dfrac{1}{sC_1}} \qquad |\times (sC_2R_2+1)$$

$$\frac{v_o}{v_i}=\frac{R_2}{R_2+sR_1R_2C_2+R_1+\dfrac{R_2C_2}{C_1}+\dfrac{1}{sC_1}} \qquad |\times sC_1$$

$$=\frac{sC_1R_2}{sC_1\left(R_1+R_2+\dfrac{R_2C_2}{C_1}\right)+s^2R_1R_2C_1C_2+1}$$

$$H(s)=\frac{sC_1R_2}{s^2R_1R_2C_1C_2+s(R_1C_1+R_2C_1+R_2C_2)+1}$$

Factorising

$$=\frac{C_1R_2}{R_1R_2C_1C_2}\times\frac{s}{s^2+s\dfrac{R_1C_1+R_2C_1+R_2C_2}{R_1R_2C_1C_2}+\dfrac{1}{R_1R_2C_1C_2}}$$

Replacing values

$$H(s)=\frac{10s}{s^2+120s+1000}$$

The quadratic for $s^2+120s+1000$

gives $s=-60\pm 51$

$s_1=-9 \quad s_2=-111$

$$H(s)=\frac{10s}{(s+9)(s+111)}$$

$$=\frac{10}{9\times 1111}\times\frac{s}{\dfrac{s}{9}+1}\,\frac{s}{111+1}$$

Constant 0.010 01

20 log 0.010 01 = –39.99 dB

PROBLEM 8.10 BODE

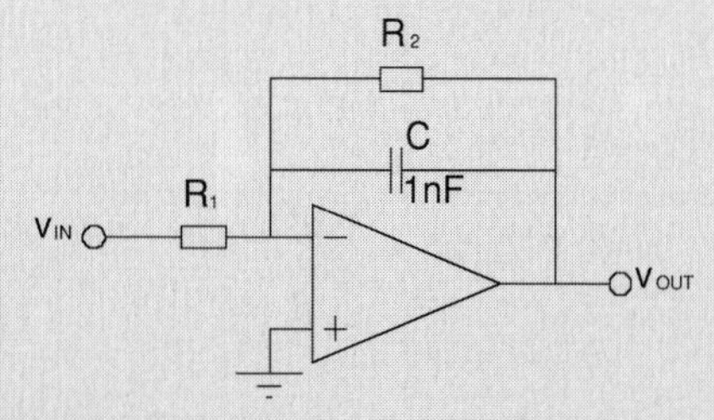

Figure 8.23

Find R_1 and R_2 such that the transfer function has a corner frequency at $\omega = 10^4$ rad/s and the value of the transfer function is –2.5 at $\omega = 0$.

> Answer: R_1 = 40 kΩ, $R_2 = 10^5$ Ω

Equating currents (v_- is the virtual earth)

$$\frac{v_i}{R_1}=-v_oY=-v_o\left(\frac{1}{R_2}+j\omega C\right)$$

Transfer function

$$\frac{v_o}{v_i}=-\frac{1}{R_1\left(\dfrac{1}{R_2}+j\omega C\right)} \qquad |\times R_2$$

$$=-\frac{R_2}{R_1}\frac{1}{1+j\omega R_2C}$$

For $\omega = 0$

$$\frac{v_o}{v_i}=-2.5=-\frac{R_2}{R_1} \qquad R_2=2.5R_1$$

Corner frequency type D when

$$\omega R_2 C = 1$$

$$R_2 = \frac{1}{\omega C} = \frac{1}{10^4 \times 10^{-9}} = 10^5$$

$$R_1 = \frac{R_2}{2.5} = \frac{10^5}{2.5} = 40\ \text{k}\Omega$$

PROBLEM 8.11 BODE

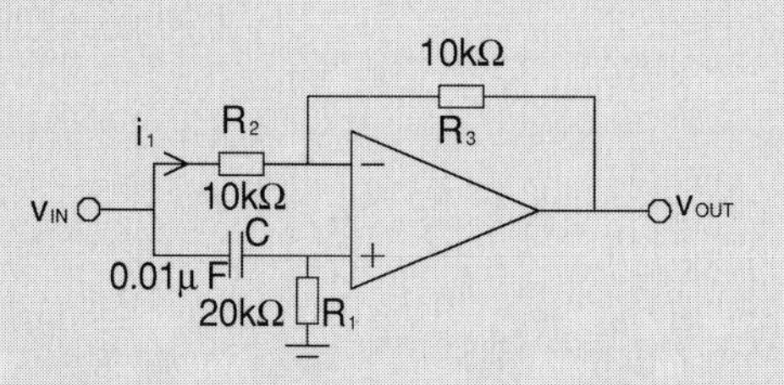

Figure 8.24

Find the phase and magnitude of the transfer function at v = 5000 rad/s

> Answer: magnitude = 1, angle = 90°

$$v_+ = v_i \frac{R_1}{R_1 + \dfrac{1}{sC}} = v_i \frac{sCR_1}{sCR_1 + 1}$$

$$v_+ = v_-$$

$$i_1 = \frac{v_i - v_+}{R_2} = \frac{v_i}{R_2} \times \left(1 - \frac{sCR_1}{sCR_1 + 1}\right)$$

$$= \frac{v_i}{R_2} \frac{1}{sCR_1 + 1}$$

$$v_o = v_+ - i_1 R_3$$

$$= v_i \frac{sCR_1}{sCR_1 + 1} - v_i \frac{R_3}{R_2} \frac{1}{sCR_1 + 1}$$

But $R_2 = R_3$

$$v_o = v_i \frac{sCR_1}{sCR_1 + 1} - \frac{1}{sCR_1 + 1}$$

$$= v_i \frac{sCR_1 - 1}{sCR_1 + 1}$$

Transfer function

$$\frac{v_o}{v_i} = \frac{sCR_1 - 1}{sCR_1 + 1}$$

Replacing values and making $s = j\omega$, with ω = 5000 rad / s

$$\frac{v_o}{v_i} = \frac{j5000 \times 0.01 \times 10^{-6} \times 20\,000 - 1}{j5000 \times 0.01 \times 10^{-6} \times 20\,000 + 1}$$

$$\frac{v_o}{v_i} = \frac{j - 1}{j + 1}$$

Magnitude = 1

Angle

$$\frac{j - 1}{j + 1} \frac{-j + 1}{-j + 1}$$

$$= \frac{1 + j + j - 1}{2} = j \quad \text{Therefore angle } 90°$$

PROBLEM 8.12 BODE

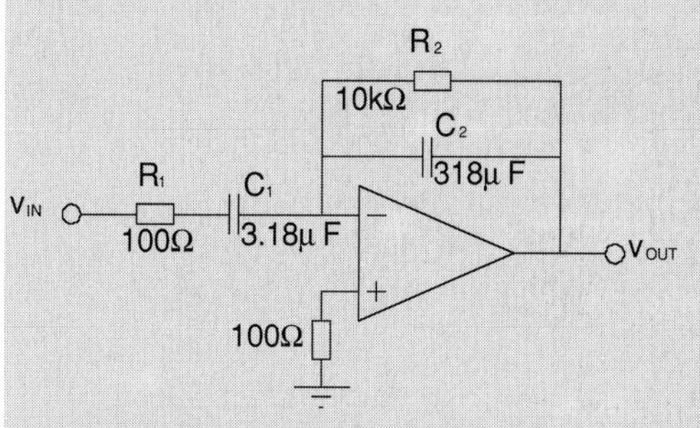

Figure 8.25

Draw the Bode plot for this band pass filter and find the two corner frequencies.

> Answer: See Figure 8.26, 0.05 Hz and 500 Hz

$$\frac{v_o}{v_i} = -\frac{\dfrac{R_2\dfrac{1}{sC_2}}{R_2 + \dfrac{1}{sC_2}}}{R_1 + \dfrac{1}{sC_1}} \qquad | \times sC_2$$

$$= -\frac{\dfrac{R_2}{sC_2R_2 + 1}}{\dfrac{sC_1R_1 + 1}{sC_1}} \qquad | \times sC_1$$

$$\frac{v_o}{v_i} = -\frac{sC_1R_2}{(sC_2R_2 + 1)(sC_1R_1 + 1)}$$

This represents: a constant, a type A and two type D functions. The constant and the type A can be combined into just one as required.

NOTE: The minus sign represents a shift of 180°. As the angle plot is not required in this case, the minus sign has no significance in this problem.

Replacing values

$$\frac{v_o}{v_i} = \frac{3.18 \times 10^{-2} s}{(3.18s + 1)(3.18 \times 10^{-4} s + 1)}$$

$$f_1 = \frac{1}{2\pi 3.18} = 0.05 \text{ Hz}$$

$$f_2 = \frac{1}{2\pi 3.18 \times 10^{-4}} = 500 \text{ Hz}$$

Crossing type A

$$f_3 = \frac{1}{2\pi C_1 R_2} = \frac{1}{2\pi 3.18 \times 10^{-6} \times 10^4} = 5 \text{ Hz}$$

The Bode plot is seen in Figure 8.26.

PROBLEM 8.13 BODE

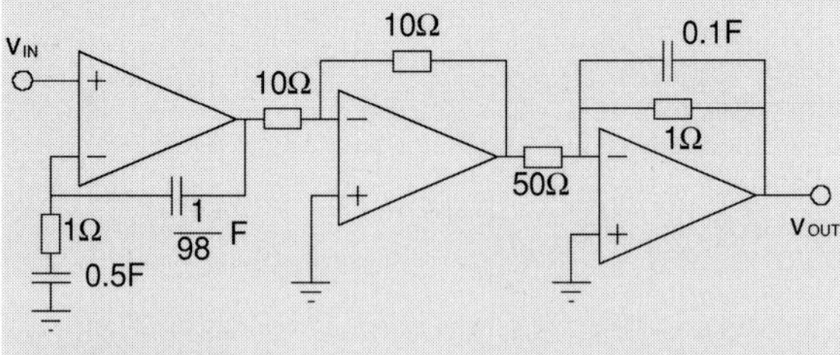

Figure 8.27

Find the transfer function and draw the Bode plot.

Answer: See Figure 8.28,

$$H(s) = \frac{s + 100}{5(s + 2)(s + 10)}$$

We can use the complex frequency operator $s = \sigma + j\omega$. The transfer function is made up of three parts.

$$T_1 = 1 + \frac{Z_2}{Z_1} = 1 + \frac{\dfrac{98}{s}}{1 + \dfrac{2}{s}}$$

$$= 1 + \frac{98}{s + 2} = \frac{s + 100}{s + 2}$$

$$T_2 = -1$$

$$T_3 = -\frac{Z_2}{Z_1} = -\frac{\dfrac{1 \times \dfrac{10}{5}}{1 + \dfrac{10}{5}}}{50}$$

$$T_3 = -\frac{\dfrac{10}{s + 10}}{50} = -\frac{1}{5(s + 10)}$$

$$H(s) = \frac{s + 100}{5(s + 2)(s + 10)}$$

Change into Bode form, ready for plotting

$$H(s) = \frac{\dfrac{s}{100} + 1}{\left(\dfrac{s}{2} + 1\right)\left(\dfrac{s}{10} + 1\right)}$$

Constant

K = 1 $\qquad$ 20 log K = 0 dB

This gives one type C and two type D. See the plot in Figure 8.28.

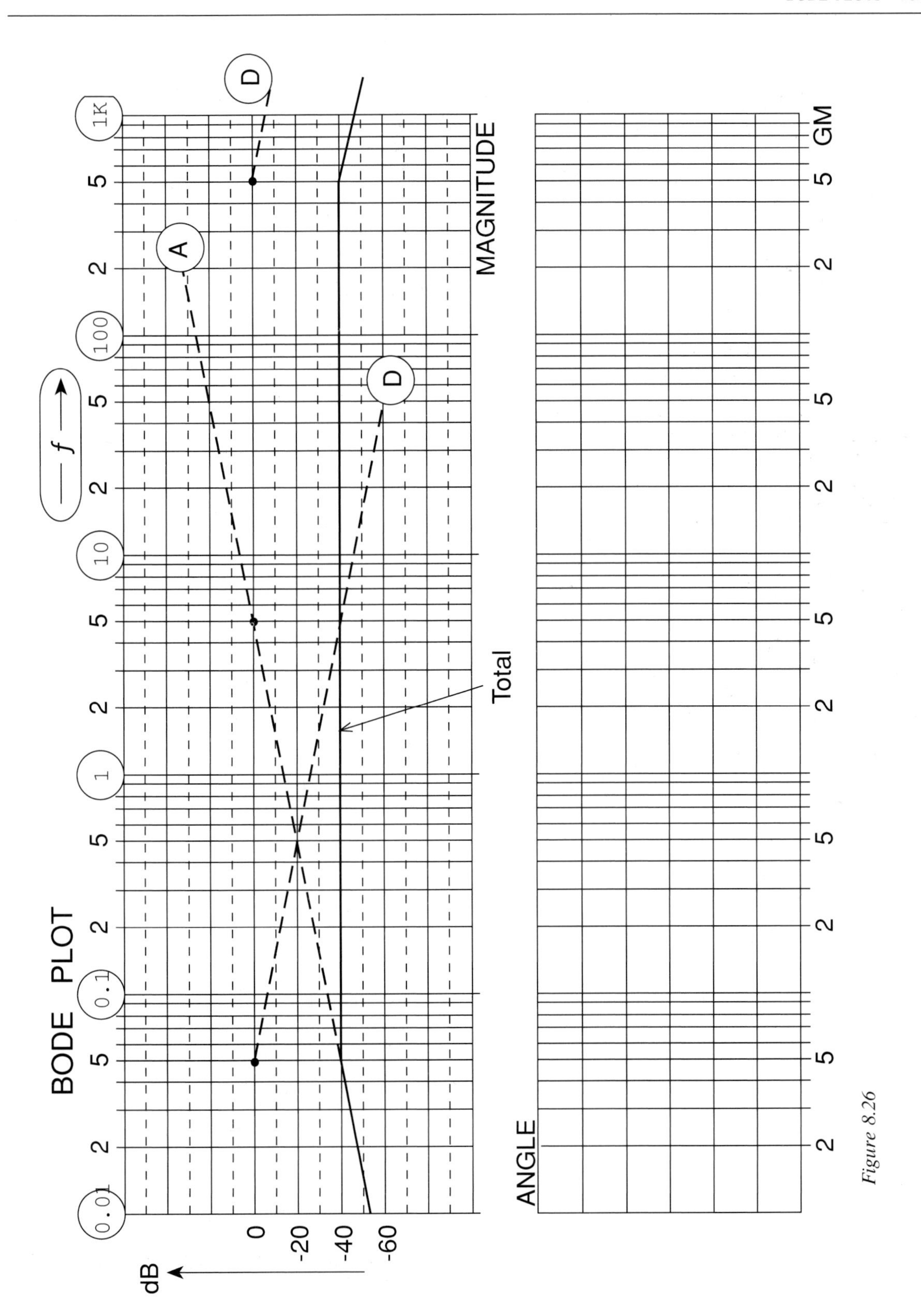
BODE PLOT
MAGNITUDE
ANGLE
GM
Total
f
dB
A
D
D
0.01
0.1
1
10
100
1K
0
-20
-40
-60

Figure 8.26

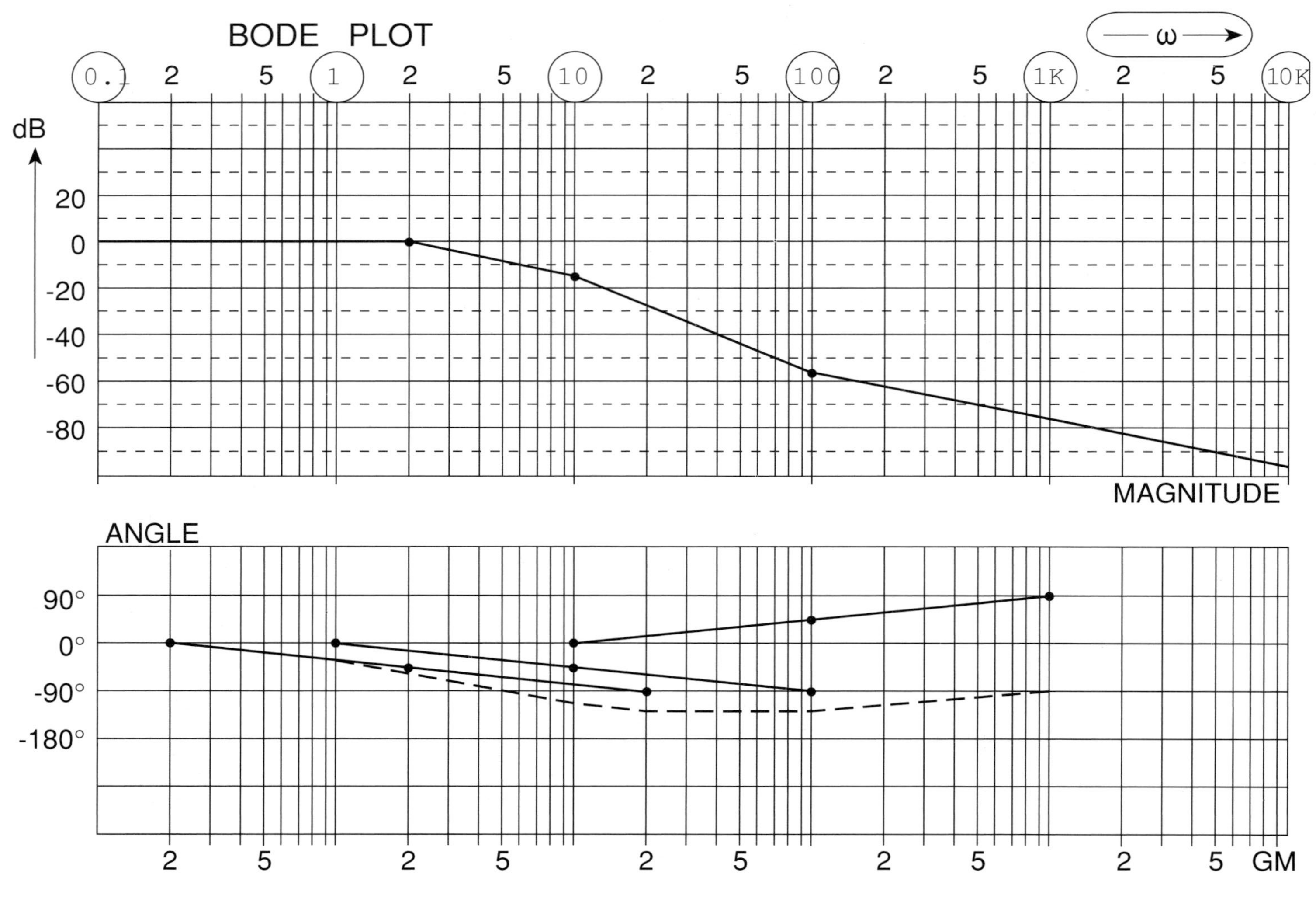
BODE PLOT
ω
0.1
1
10
100
1K
10K
2
5
dB
20
0
-20
-40
-60
-80
MAGNITUDE
ANGLE
90°
0°
-90°
-180°
GM

Figure 8.28

PROBLEM 8.14 BODE

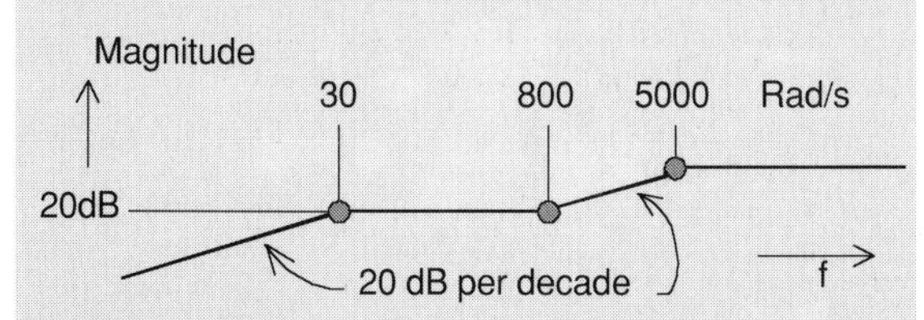

Figure 8.29

Find the transfer function, including the amplification constant and express it in the standard Bode form.

Answer: $H(j\omega)=$

$$62.5\ j\omega\frac{j\omega+800}{(j\omega+30)(j\omega+5000)}$$

This type of problem is a reverse Bode. The function is given. We have to find the transfer function.

We try to find the original components of the transfer function. It is possible that you might find more than one possibility. This is possible, although you will find that they will all end up in the same transfer function. If you imagine a base level, a type A, two type D and a type C as shown will produce the desired response.

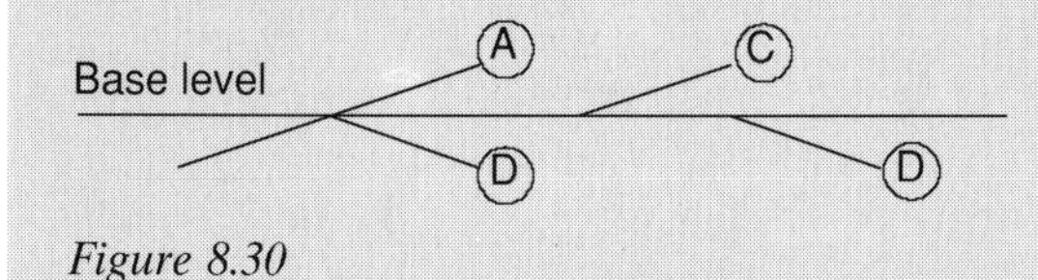

Figure 8.30

The base level is at 20 dB and this will give us the constant.

$20 \log K = 20$ dB

$\log K = 1$

$K = 10$

The type A will be of the type $j\omega K$ (a different K). The crossing of the base level is at 30 rad/s. We need a value of $j\omega/30$ because at $\omega = 30$ the value is 1 which in dB is 0.

The type C has a corner frequency at 800 rad/s. The factor will be

$$\frac{j\omega}{800}+1$$

We can now write the transfer function

$$H(j\omega)=\frac{10\times\frac{j\omega}{30}\times\frac{j\omega}{800}+1}{\left(\frac{j\omega}{30}+1\right)\left(\frac{j\omega}{5000}+1\right)}$$

We amplify by 800, 5000 and 30 to remove the denominators

$$H(j\omega)=\frac{10\times5000\times j\omega(j\omega+800)}{800(j\omega+30)(j\omega+5000)}$$

finally

$$H(j\omega)=\frac{6.25\ j\omega(j\omega+800)}{(j\omega+30)(j\omega+5000)}$$

PROBLEM 8.15 BODE

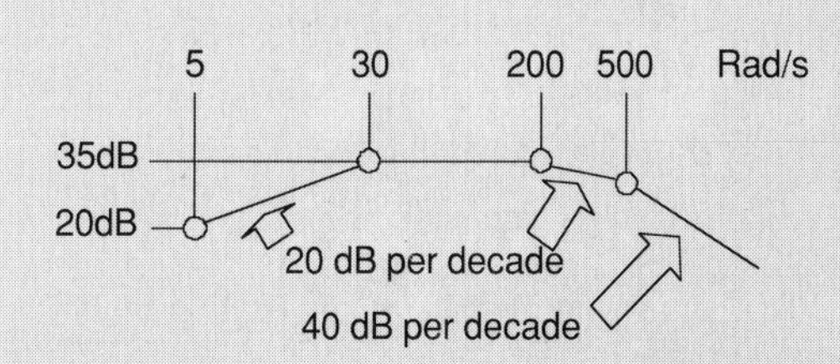

Figure 8.31

Given this response, find the transfer function and verify the value of magnitude at $\omega = 100$ rad/s.

Answer: $H(j\omega)=\dfrac{6\times10^6(j\omega+5)}{(j\omega+30)(j\omega+200)(j\omega+500)}$

Magnitude $= 34.06$ dB

The diagram has a zero at ω = 5 and three poles at ω = 30, 200 and 500 rad/s.

Normalised branches will be

$$\frac{\frac{j\omega}{5}+1}{\left(\frac{j\omega}{30}+1\right)\left(\frac{j\omega}{200}+1\right)\left(\frac{j\omega}{500}+1\right)}$$

The response has been 'lifted' by 20 dB

$$20 \log K = 20$$

$$K = 10$$

The transfer function is

$$\frac{10\times\left(\frac{j\omega}{5}+1\right)}{\left(\frac{j\omega}{30}+1\right)\left(\frac{j\omega}{200}+1\right)\left(\frac{j\omega}{500}+1\right)}$$

Rearranging it

$$\frac{10\times 30\times 200\times 500\times(j\omega+5)}{5(j\omega+30)(j\omega+200)(j\omega+500)}$$

$$H(j\omega)=\frac{6\times 10^6(j\omega+5)}{(j\omega+30)(j\omega+200)(j\omega+500)}$$

Exact response for $\omega = 100$ rad / s

$$H(j100)=\frac{6\times 10^6\sqrt{100^2+5^2}}{\sqrt{(100^2+30^2)(100^2+200^2)(100^2+500^2)}}$$

$$H(j100)=50.468$$

$$=34.06 \text{ dB}$$

PROBLEM 8.16 BODE

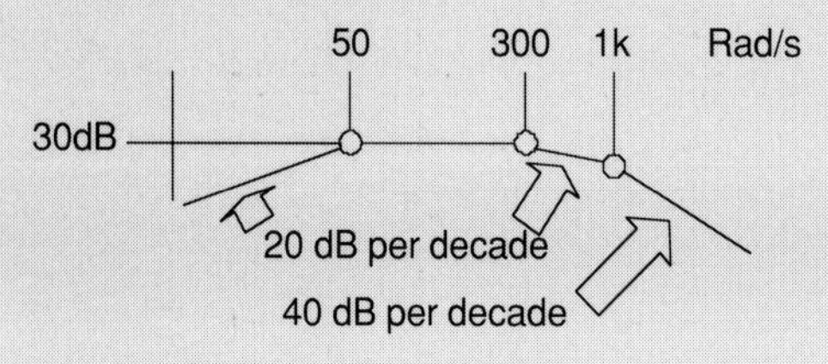

Figure 8.32

Given the response shown, find the transfer function and verify the values by calculating the magnitude and angle for 10, 125 and 2000 rad/s.

Answer: $\frac{94.869\times 10^5 j\omega}{(j\omega+50)(j\omega+300)(j\omega+1000)}$

15.48 dB, 76.21°, 28.59 dB, − 7.9°, 6.433 dB, − 143°

Cons tan t

$$30 \text{ dB} = 20 \log K$$

$$\log K = \frac{30}{20} = 1.5$$

$$K = 31.623$$

Transfer function

One type A and 3 type D.

$$H(j\omega)=\frac{31.623\frac{j\omega}{50}}{\left(\frac{j\omega}{50}+1\right)\left(\frac{j\omega}{300}+1\right)\left(\frac{j\omega}{1000}+1\right)}$$

$$=\frac{31.623\frac{j\omega}{50}\times 50\times 300\times 1000}{\left(\frac{j\omega}{50}+1\right)\left(\frac{j\omega}{300}+1\right)\left(\frac{j\omega}{1000}+1\right)\times 50\times 300\times 100}$$

$$H(j\omega)=\frac{31.623\times 300\times 1000\, j\omega}{(j\omega+50)(j\omega+300)(j\omega+1000)}$$

$$H(j\omega)=\frac{94.869\times 10^5 j\omega}{(j\omega+50)(j\omega+300)(j\omega+1000)}$$

The Bode plot can be seen in Figure 8.33.

We need to obtain the values for the three frequencies requested.

For $\omega = 10$ rad / s

$$M=\frac{94.869\times 10^5\times 10}{\sqrt{(10^2+50^2)}\sqrt{(10^2+300^2)}\sqrt{(10^2+1000^2)}}$$

$$=\frac{94.869\times 10^6}{51\times 300.16\times 1000.05}$$

$$=6.197$$

$$\log 6.197 = 0.792$$

$$20 \log 6.197 = 15.84 \text{ dB}$$

$$\text{Angle} = \frac{\angle 90}{\angle 11.31\angle 1.91\angle 0.57} = 76.21^\circ$$

For $\omega = 125$ rad / s

$$M=\frac{94.869\times 10^5\times 125}{\sqrt{(125^2+50^2)}\sqrt{(125^2+300^2)}\sqrt{(125^2+1000^2)}}$$

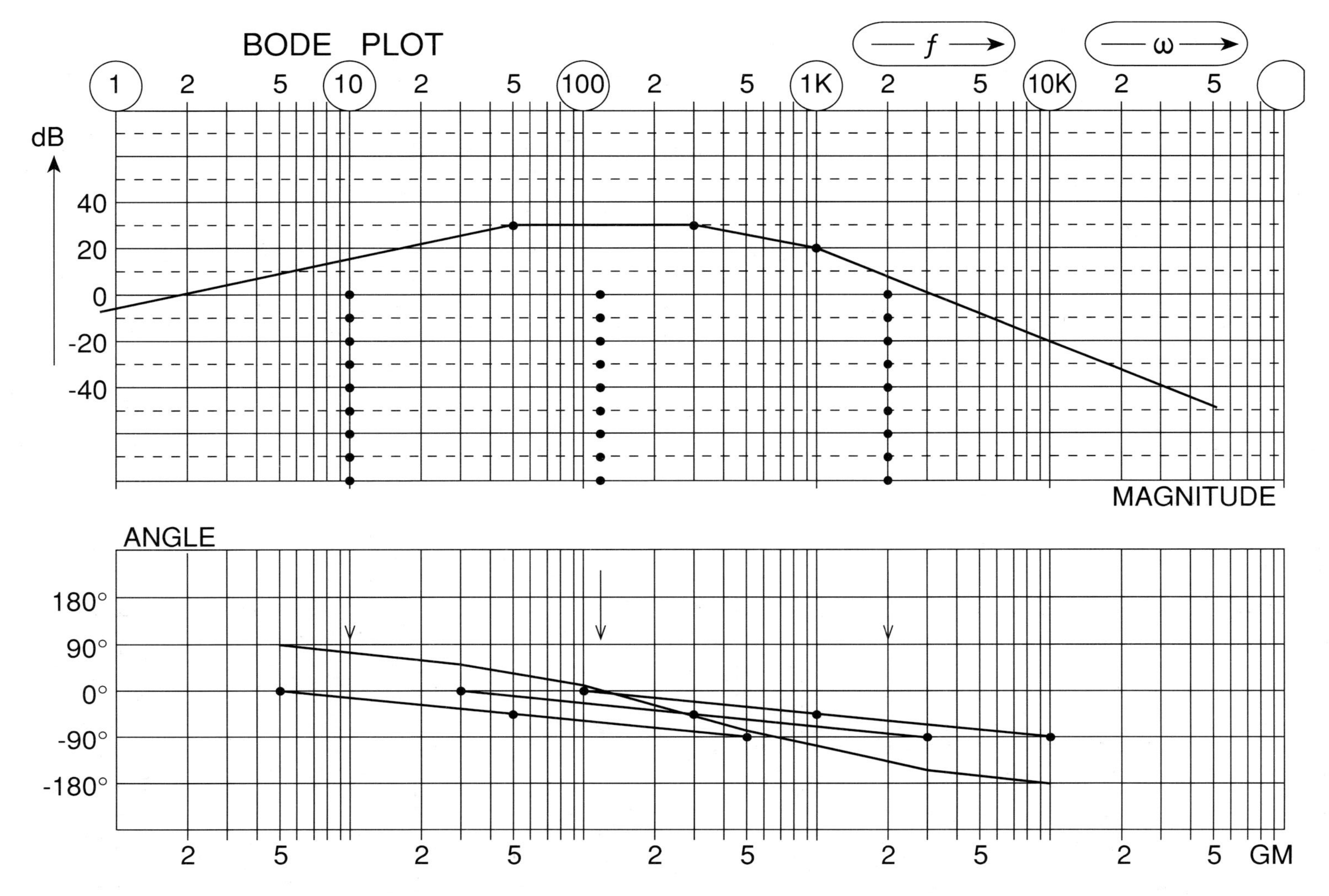
BODE PLOT
f
ω
1
2
5
10
2
5
100
2
5
1K
2
5
10K
2
5
dB
40
20
0
-20
-40
MAGNITUDE
ANGLE
180°
90°
0°
-90°
-180°
2
5
2
5
2
5
2
5
2
5 GM

Figure 8.33

$$M = \frac{9\,486\,900 \times 125}{134.629 \times 325 \times 1007.78}$$

$$= 26.893$$

$\log 26.893 = 1.4296$

$20 \log 26.893 = 28.59$ dB

$$\text{Angle} = \frac{\angle 90}{\angle 68.2 \angle 22.62 \angle 7.125} = -7.9^\circ$$

For $\omega = 2000$ rad / s

$$M = \frac{9\,486\,900 \times 2000}{\sqrt{(2000^2 + 50^2)}\sqrt{(2000^2 + 300^2)}\sqrt{(2000^2 + 1000^2)}}$$

$$= \frac{9\,486\,900 \times 2000}{2000.6 \times 2022.37 \times 2236} = 2.097$$

$\log 2.097 = 0.322$

$20 \log 2.097 = 6.433$ dB

$$\text{Angle} = \frac{\angle 90}{\angle 88.57 \angle 81.47 \angle 63.43} = -143^\circ$$

Bearing in mind that Bode plots are only approximations, the results are reasonably good.

PROBLEM 8.17 BODE

Given

$$H(s) = \frac{10(s+100)}{(s+0.1)(s+1)(s+10)}$$

Draw the Bode plot and find the gain and phase angle for 30 rad/s

> Answer: magnitude = –30 dB, angle = –225°, see Figure 8.34.

We transform the function to standard Bode form for plotting .

$$H(s) = \frac{10(s+100) \times 0.1 \times 10 \times 100}{(s+0.1)(s+1)(s+10) \times 0.1 \times 10 \times 100}$$

$$H(s) = 1000 \frac{\left(\frac{s}{100} + 1\right)}{\left(\frac{s}{0.1} + 1\right)(s+1)\left(\frac{s}{10} + 1\right)}$$

Constant

$20 \log 1000 = 60$ dB

Zeroes at $\omega = 100$ (type C)

Poles (type D) at

$\omega = 0.1 \quad \omega = 1 \quad$ and 10 rad / s

We can now draw the Bode plot of Figure 8.34.

From the Bode plot for 30 rad / s the magnitude is – 30 dB and the phase angle is – 225°.

Check the magnitude

$$\frac{10\sqrt{(30^2 + 100^2)}}{\sqrt{(30^2 + 0.1^2)}\sqrt{(30^2 + 1^2)}\sqrt{(30^2 + 10^2)}}$$

$$= 0.036\,66$$

which is – 28.7 dB

Check the angle:

$$\frac{\angle 16.7}{\angle 89.8 \angle 88.1 \angle 71.6} = -232.8^\circ$$

which compares very well with the plot.

PROBLEM 8.18 BODE

For the following expression draw the Bode plot and give the exact value of magnitude and angle for a frequency of 100 Hz.

$$H(j\omega) = -\frac{100}{\left(1 - \frac{j10}{f}\right)\left(1 - \frac{j100}{f}\right)\left(1 + \frac{jf}{10^4}\right)}$$

> Answer: See Figure 8.36, magnitude = 36.94 dB, angle = 225°

We examine the denominator

What is

$$\frac{1}{\left(1 - \frac{j10}{f}\right)}$$

in terms of type A, B, C or D?

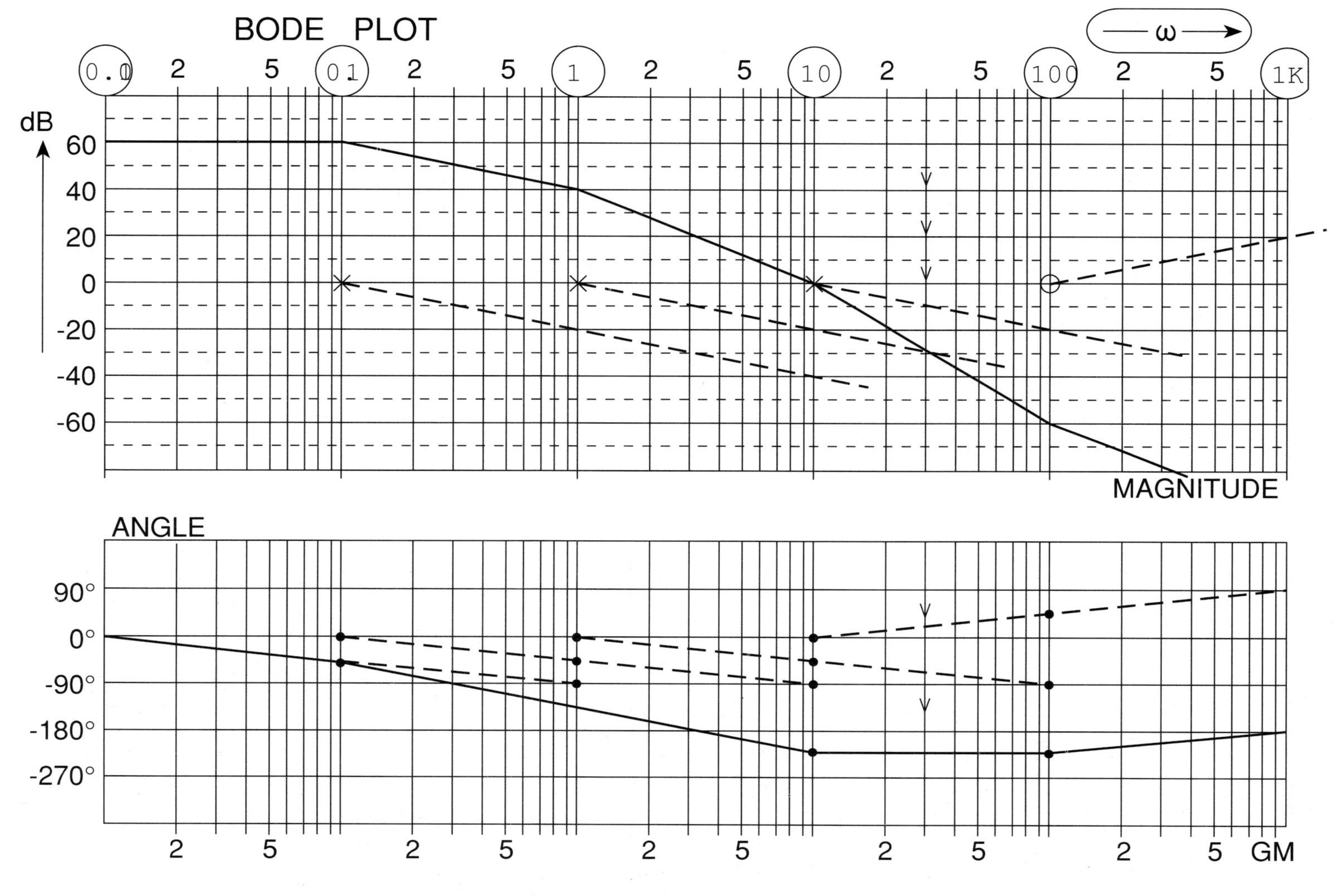
BODE PLOT
ω
0.1
1
10
100
1K
dB
60
40
20
0
-20
-40
-60
MAGNITUDE
ANGLE
90°
0°
-90°
-180°
-270°
GM

Figure 8.34

Amplify by j

$$\frac{1}{\left(1-\frac{jj10}{jf}\right)}=\frac{1}{1+\frac{10}{jf}} \qquad |\times jf$$

$$=\frac{jf}{jf+10} \qquad |\times \frac{1}{10}$$

$$=\frac{\frac{jf}{10}}{\frac{jf}{10+1}}$$

This is a type A and a type D as seen in Figure 8.35

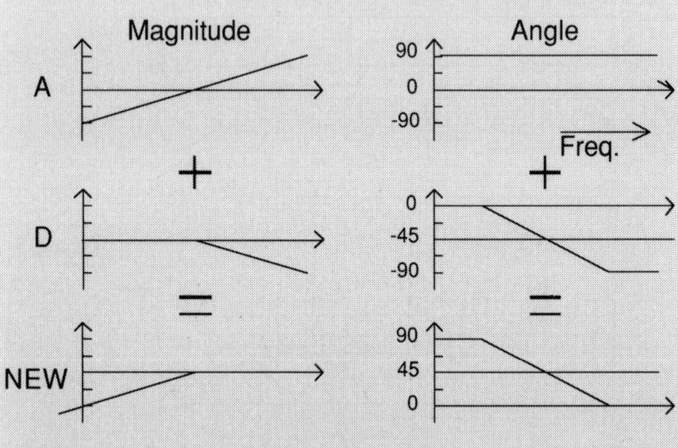

Figure 8.35

We can now draw the plot of Figure 8.36

Constant

$\log 100 = 2$

$20 \log 100 = 40$ dB

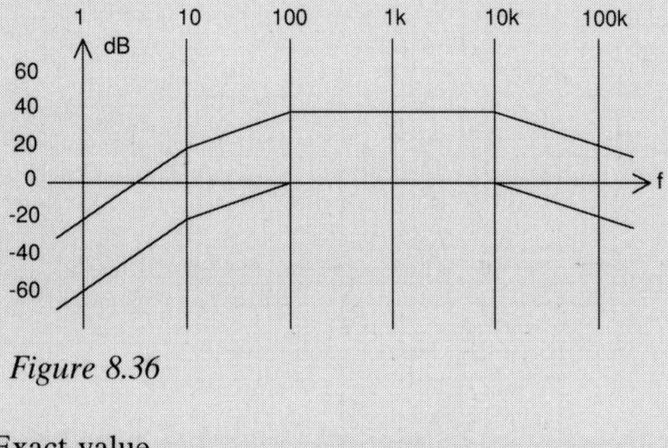

Figure 8.36

Exact value

$$|H|=\frac{100}{\sqrt{(1+0.01)}\sqrt{(1+1)}\sqrt{(1+0.000\,1)}}$$

$$=\frac{100}{1.005\times 1.414\times 1}=70.37$$

$|H| = 20 \log 70.37 = 36.94$ dB

$$\text{Angle}=\frac{\angle 180}{\angle -5.7\angle -45\angle 5.7}=225^{\circ}$$

PROBLEM 8.19 BODE

Given

$$A(s)=\frac{1000}{\left(1+\frac{s}{10^4}\right)\left(1+\frac{s}{10^5}\right)^2}$$

draw the Bode plot, find the frequency at which the phase shift is 180° and the critical value of β.

Answer: $\omega = 10^5$, $\beta = 10^{-2}$

There are three type D functions ready for plotting with corner frequencies at 10 krad/s and two at 100 krad/s.

Constant

$20 \log 1000 = 60$ dB

The Bode plot is shown in Figure 8.37.
The critical value is found from the plot at $\omega = 10^5$ rad/s with the magnitude being 40 dB.

$A = 40$ dB

$= 20 \log A \qquad A = 100$

The critical value of β is when $A\beta = 1$. Therefore $\beta = 10^{-2}$.

By calculation (at $\omega = 10^5$)

$$A=\frac{1000}{\sqrt{(1+10^2)}\sqrt{(1+1)}\sqrt{(1+1)}}=49.75$$

$A = 20 \log 49.75 = 34$ dB

For this value

$$\beta=\frac{1}{49.75}=0.02$$

NOTE: The Bode plot is an approximation. The real output is rounded at the corner and this should be taken into account in the interpretation of the results.

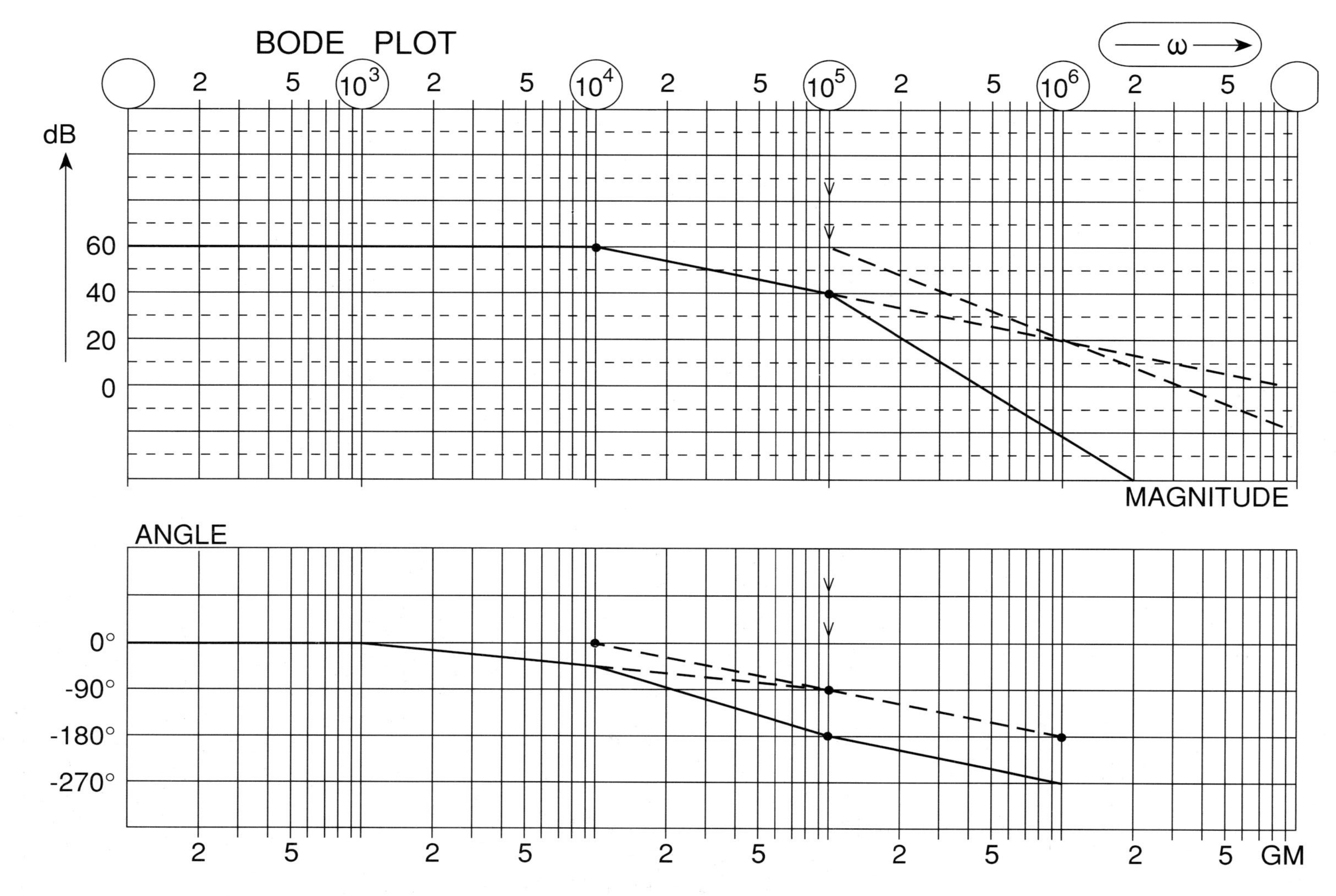
BODE PLOT
ω
2
5
10³
2
5
10⁴
2
5
10⁵
2
5
10⁶
2
5
dB
60
40
20
0
MAGNITUDE
ANGLE
0°
-90°
-180°
-270°
GM

Figure 8.37

PROBLEM 8.20 BODE

Given $\beta = 10^{-4}$ and the function

$$A = \frac{10^5}{\left(1+\frac{jf}{10^4}\right)\left(1+\frac{jf}{10^5}\right)\left(1+\frac{jf}{3\times10^6}\right)\left(1+\frac{jf}{8\times10^6}\right)}$$

Draw the Bode plot and find the phase and magnitude margins.

Answer: See Figure 8.38, phase margin 48°, gain margin 35 dB

ω and f readily can be exchanged as they are different units like miles and kilometres.

Here we have four type D functions

Constant

$$k = 10^5$$

$$20 \log k = 100 \text{ dB}$$

The value of interest from the plot is

$$20 \log \frac{1}{\beta} = 80 \text{ dB}$$

The difference between the magnitude and the 80 dB line is

$$20 \log A - 20 \log \frac{1}{\beta}$$

$$= 20 \log A + 20 \log \beta$$

$$= 20 \log A\beta$$

When this difference is zero in dB the value of $A\beta = 1$. The phase margin, therefore, can be seen at this frequency. When we move to 180° phase, we can see the magnitude margin. From the plot the phase margin ≈ 48° and the gain margin ≈ 35 dB.

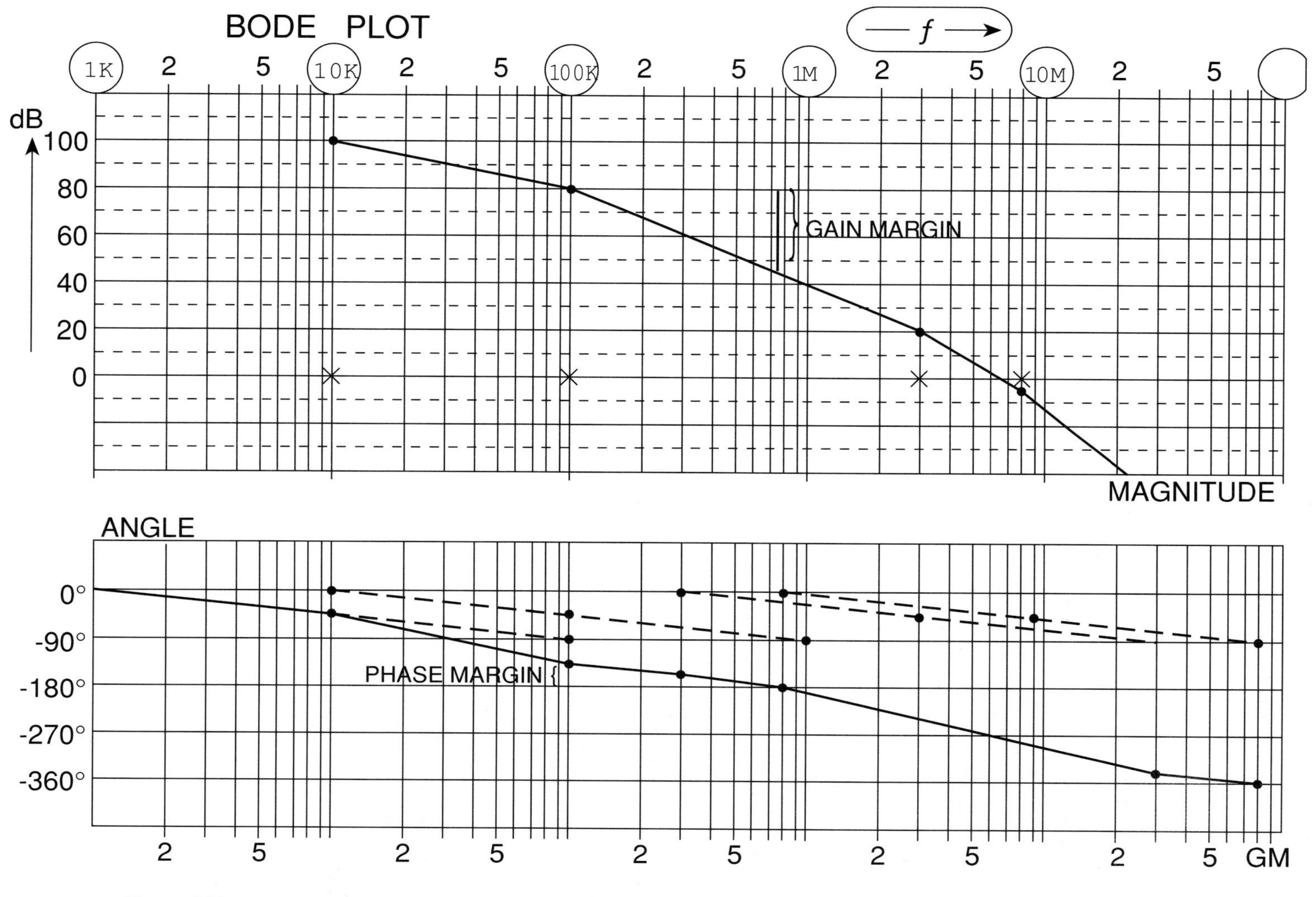
BODE PLOT
f
1K
10K
100K
1M
10M
dB
100
80
60
40
20
0
GAIN MARGIN
MAGNITUDE
ANGLE
0°
-90°
-180°
-270°
-360°
PHASE MARGIN
GM

Figure 8.38

9

Effect of capacitors

It is customary when introducing electronics to students to assume that there are no effects from the capacitors and that β is constant. Sooner or later, however, you will come to know that β drops with increasing frequency and that you reach a point where the transistor ceases to amplify. There are also parasitic capacitors which spoil the operation of the transistor at high frequency. However, we are now concentrating at the lower end of the frequency spectrum. We are not talking about parasitic capacitances, but about capacitors that we have purposely put in the circuit to block the DC supply or to manipulate the feedback.

Figure 9.1, shows the change in output of a transistor against the variation of frequency. This

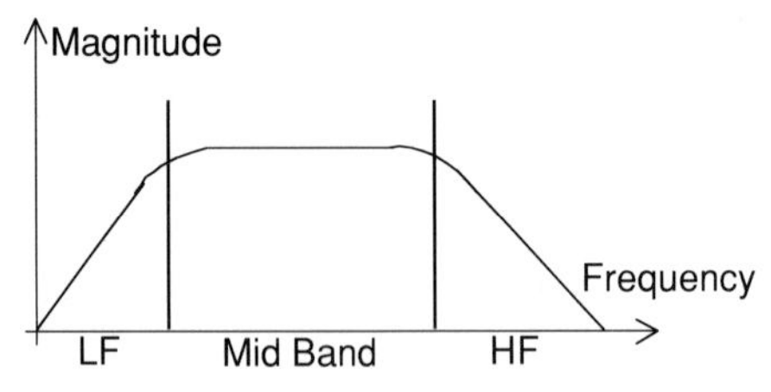

Figure 9.1

shows a uniform output at mid frequency, whereas there is decay both at high frequency and low frequency.

Input capacitor C_1

We start with the general circuit shown in Figure 9.2. We first look at capacitor C_1. It is best to

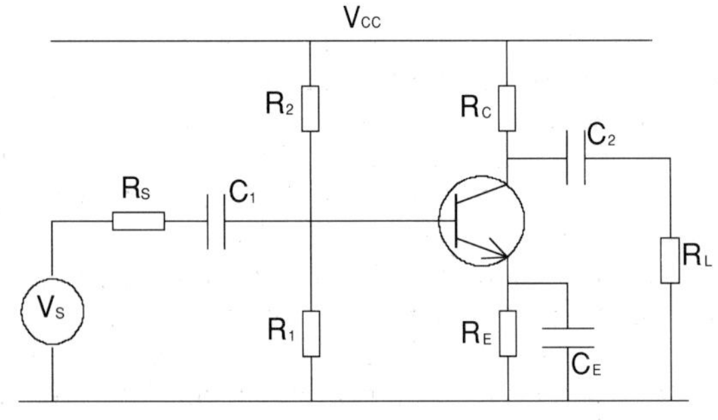

Figure 9.2

draw the equivalent circuit using our knowledge of transistor modelling.

The point to notice here is that it would be too difficult to consider all the capacitors in one go. So what we do is to consider each capacitor at a time, assuming that the others are not affecting the circuit. So when we calculate the effect of C_1, C_2 and C_E are considered to be very low impedances.

.

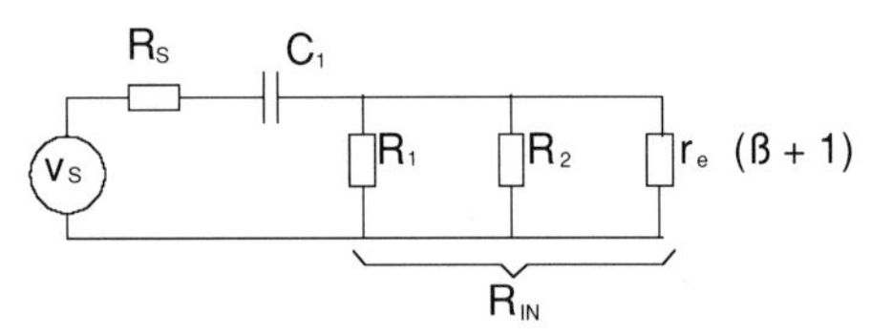

Figure 9.3

Let us look at Figure 9.3. This is the equivalent circuit. At the corner frequency

$$R = X_C \quad R = R_S + R_{IN}$$

$$R = X_C = \frac{1}{\omega C_1} = \frac{1}{2\pi f C_1}$$

$$\boxed{f = \frac{1}{2\pi R C_1}}$$

Remember that R in this case is $R_S + R_{IN}$ and that

$$R_{IN} = R_1 \parallel R_2 \parallel r_e\,(\beta + 1)$$

Output capacitor C_2

The equivalent circuit associated with C_2 is shown in Figure 9.4.

Note that R_E or C_E do not appear anywhere. This is in line with what we have said before. That is, that when we calculate the effect of one capacitor, in this case C_2, we assume that the others are working normally on very low impedances. This means that C_E is conducting

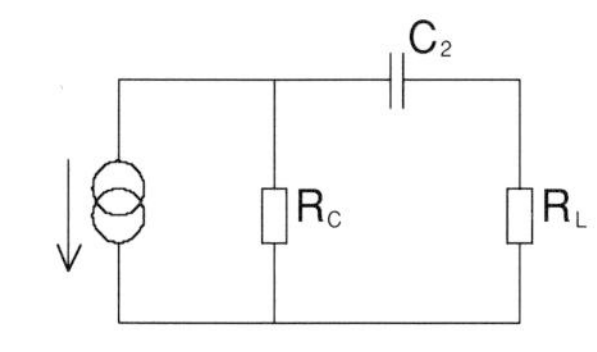

Figure 9.4

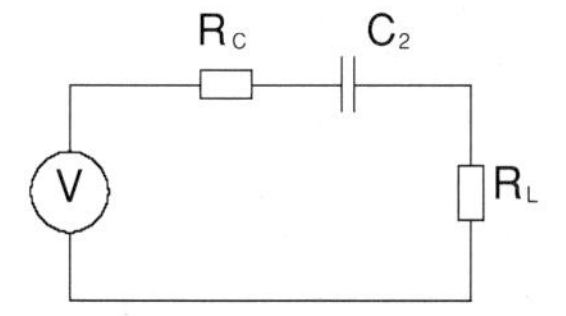

Figure 9.5

and bypassing R_E. So the emitter of the transistor is at ground potential.

Many students have difficulty in knowing whether R_C and R_L are in series or parallel for the charging or discharging of the capacitor. In order to resolve this dilema, we use a Norton to Thevenin conversion. The result is shown in Figure 9.5.

We can now see clearly that they are in series.

$$R_C + R_L = X_C = \frac{1}{2\pi f C_2}$$

$$\boxed{f = \frac{1}{2\pi(R_C + R_L)C_2} \quad \text{Hz}}$$

Bypass capacitor C_E

This circuit is more complicated than the two previous ones. We are now in the emitter circuit and this circuit is linked to the base circuit.

We draw the equivalent circuit starting from the capacitor C_E as shown in Figure 9.6.

We can use a Thevenin to Norton conversion to get the circuit shown in Figure 9.7. In this circuit the left-hand side is in the base circuit. On the right-hand side we have i_E which is $(\beta + 1)$ times i_B. We can compensate for this, assuming that i_E flows everywhere, if we modify the values of R_S, R_1 and R_2. In this way we arrive at the circuit of Figure 9.8.

$$R = X_C = \frac{1}{\omega C_3} = \frac{1}{2\pi f C_E}$$

$$\boxed{f = \frac{1}{2\pi R C_E}}$$

R in this case is

$$R = R_E || \left(r_e + \frac{R_1 || R_2 || R_S}{\beta + 1} \right)$$

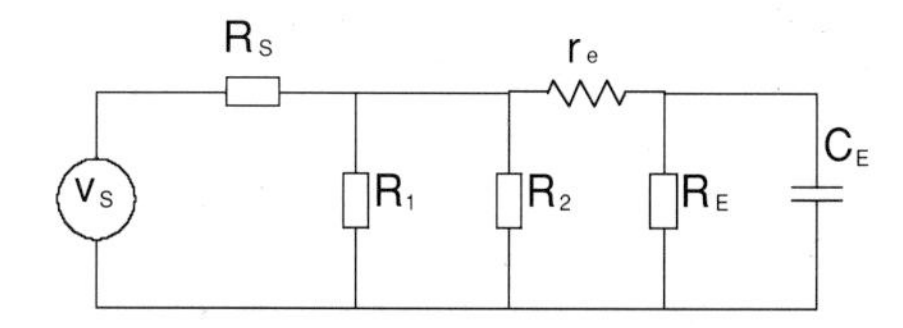

Figure 9.6

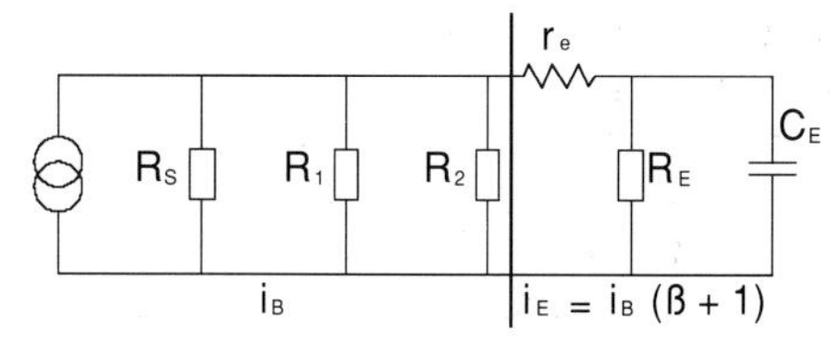

Figure 9.7

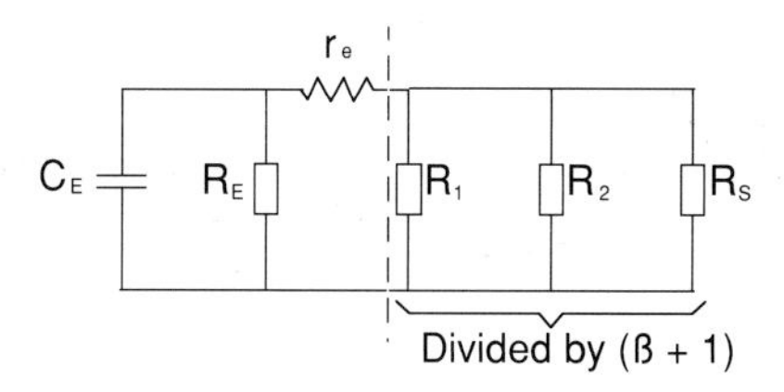

Figure 9.8

Cascaded 3 dB points

We wish to calculate the corner frequencies due to capacitors C_1, C_2 and C_3 as indicated in Figure 9.9.

We start by looking at capacitor C_1 which forms part of the input side of the first stage of the amplification. We draw the equivalent circuit under AC small signals conditions as we see in Figure 9.10. In problems of this nature we will be looking for the corner frequency for a given capacitance or finding the value of a capacitance to achieve the corner frequency at a given point. We are trying to define the edge of a passband, therefore we are defining the bandwidth of the midband, at the lower end.

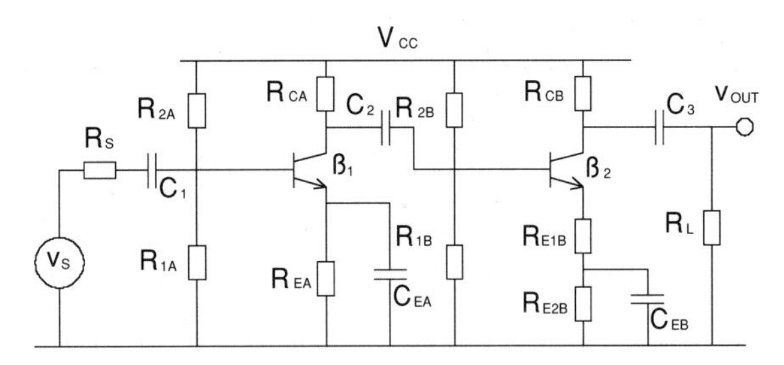

Figure 9.9

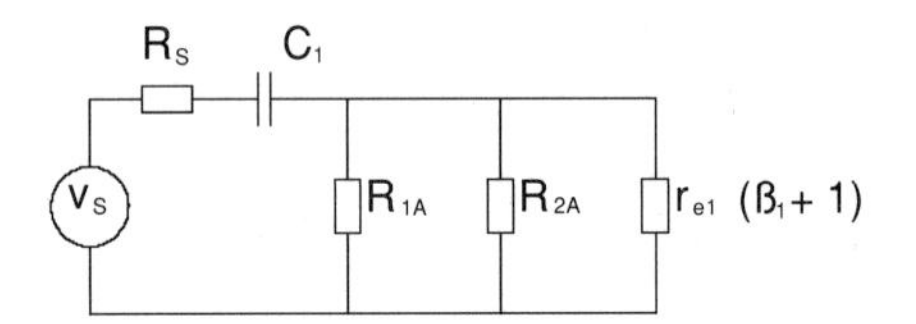

Figure 9.10

For the corner frequency we have

$$R = X_C$$

R is the total resistance of the circuit 'seen' by the capacitor. In this case the resistance is

$$R = R_S + R_{1A} || R_{2A} || r_{e1}(\beta_1 + 1)$$

$$\boxed{f = \frac{1}{2\pi RC}}$$

Effect of C_2

Now we move onto the effect of the second capacitor which is located between the stages. As is normal in a cascaded system, the load of the first stage is the input resistance of the second stage. This can be seen in Figure 9.11, where we show the equivalent circuit around C_2.

We can convert from Norton to Thevenin on the left of Figure 9.11, to obtain Figure 9.12, where it is easier to see the circuit around capacitor C_2.

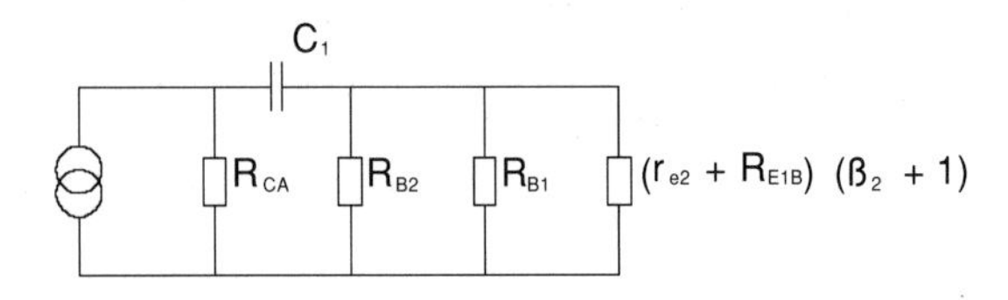

Figure 9.11

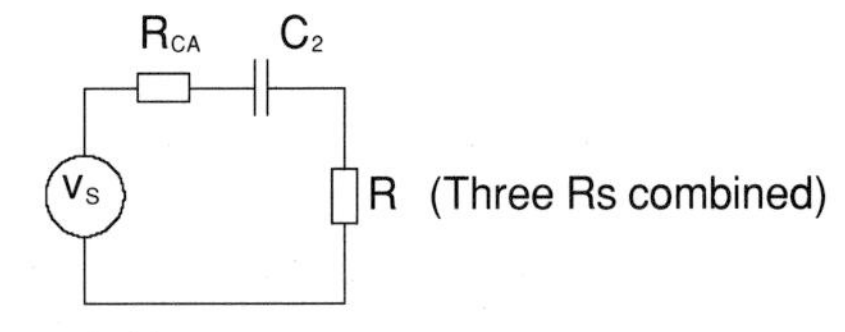

Figure 9.12

The resistance of the circuit is

$$R = R_{CA} + R_{B2} \,||\, R_{B1} \,||\, (r_{e2} + R_{E1B})(\beta 2 + 1)$$

$$\boxed{f = \frac{1}{2\pi(R_{CA} + R)C_2}}$$

Cascaded C_3

We now move to examine C_3 in the cascaded system shown in Figure 9.9. This should be the same as the output capacitor of a single stage. However, in this case the emitter resistor is not

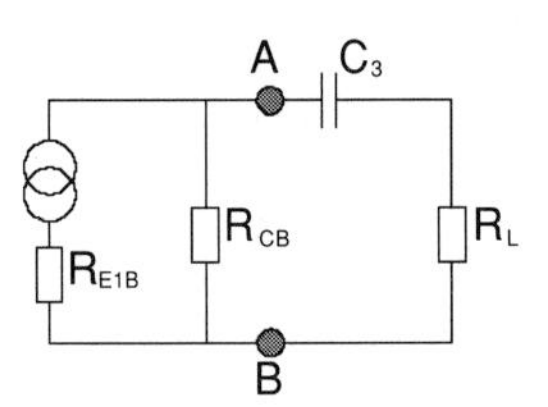

Figure 9.13

fully bypassed and R_{E1B} will appear in the equivalent circuit. The equivalent circuit appears in Figure 9.13. In this case C_3 sees a resistance of R_L in series with R_{CB}.

In order to see this we transform the left of AB into a Thevenin equivalent. The resulting circuit is shown in Figure 9.14.

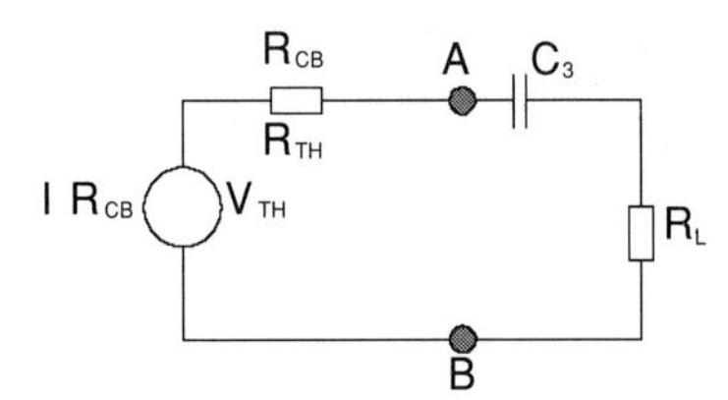

Figure 9.14

$V_{TH} = IR_{CB}$

$R_{TH} = R_{CB}$ (current source open circuit)

In this case the resistor R_{E1B} is not seen by capacitor C_3. The total resistance of the circuit is R_{CB} in series with R_L.

$$f = \frac{1}{2\pi RC_3} = \frac{1}{2\pi(R_{CB}+R_L)C_3}$$

PROBLEM 9.1 EFFECT OF CAPACITORS

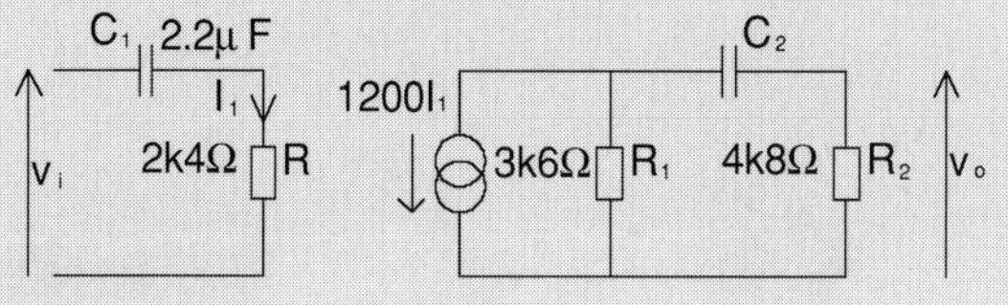

Figure 9.15

Find the break frequency due to C_1 and the value of C_2 to give a break frequency of 70 Hz.

Answer: 30 Hz, 0.27 µF

Capacitor C_1

$$\omega = \frac{1}{RC} \quad f = \frac{1}{2\pi RC}$$

$$f = \frac{1}{2\pi \times 2400 \times 2.2 \times 10^{-6}}$$

$$f = 30 \text{ Hz}$$

Capacitor C_2

$$\omega = \frac{1}{C_2(R_1 + R_2)}$$

$$C_2 = \frac{1}{2\pi f(R_1 + R_2)}$$

$$= \frac{1}{2\pi 70 \times 8.4 \times 10^3}$$

$$= 0.27\ \mu\text{F}$$

PROBLEM 9.2 EFFECT OF CAPACITORS

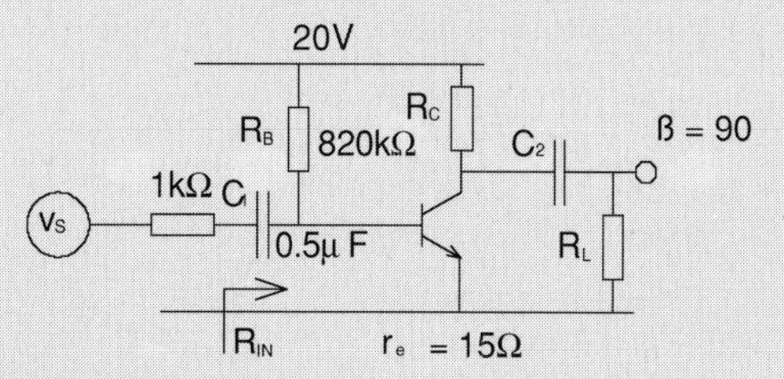

Figure 9.16

Find the corner frequency due to C_1.

Answer: 134.7 Hz

AC model

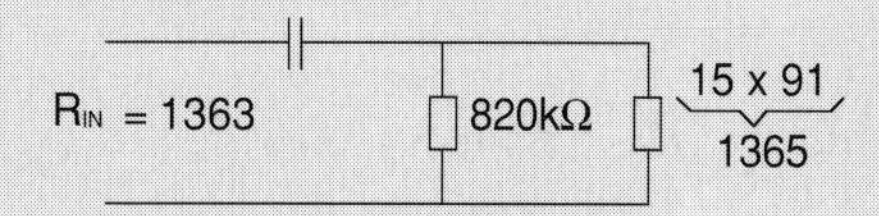

Figure 9.17

Frequency

$$f = \frac{1}{2\pi(1\text{k} + 1363)C_1}$$

$$f = \frac{1}{2\pi \times 2363 \times 0.5 \times 10^{-6}}$$

$$f = 134.7 \text{ Hz}$$

PROBLEM 9.3 EFFECT OF CAPACITORS

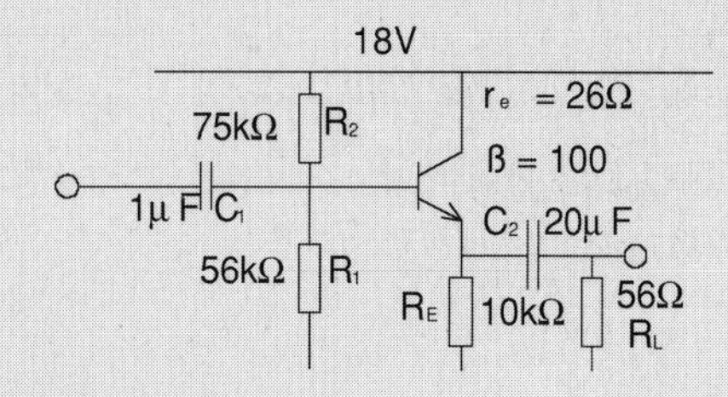

Figure 9.18

Find the corner frequency due to C_2.

Answer: 23.97 Hz

AC model

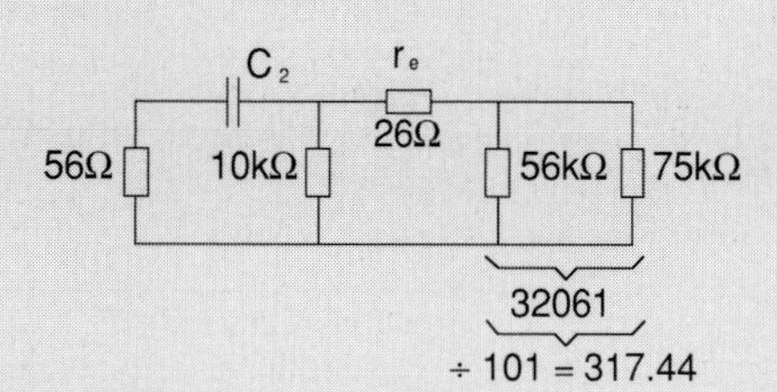

Figure 9.19

$$R = 56 + 10\text{k}||(26 + 317.44)$$
$$= 56 + 10\text{k}||343.44$$
$$= 332.04$$

$$f = \frac{1}{2\pi RC}$$
$$= \frac{1}{2\pi \times 332.04 \times 20 \times 10^{-6}}$$
$$= 23.97 \text{ Hz}$$

PROBLEM 9.4 EFFECT OF CAPACITORS

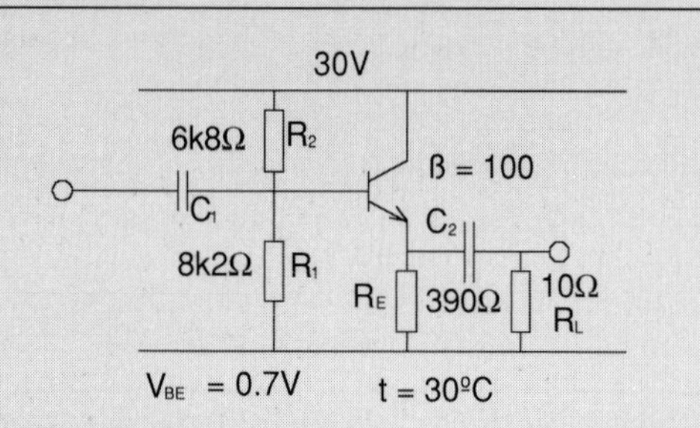

Figure 9.20

Use the approximate method to evaluate the DC conditions and calculate the value of C_2 to produce a corner frequency of 120 Hz.

Answer: 30 μF

DC conditions

$$V_B = 30\frac{8.2}{8.2 + 6.8} = 16.4 \text{ V}$$

$$V_E = 16.4 - 0.7 = 15.7 \text{ V}$$

$$I_E = \frac{15.7}{390} = 40.26 \text{ mA}$$

$$r_e = \frac{26}{40.26} = 0.646\ \Omega$$

AC model

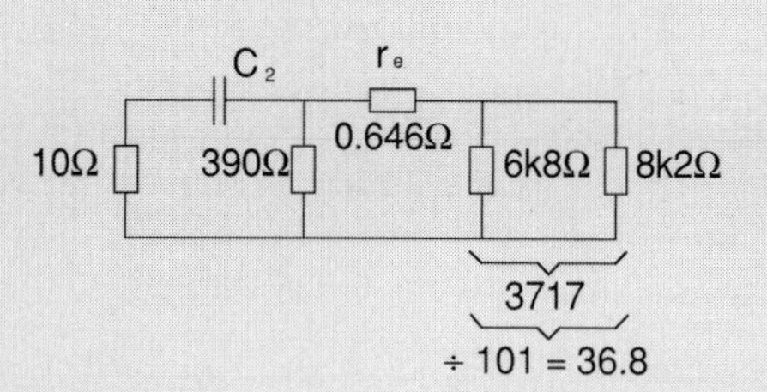

Figure 9.21

$$R = 10 + 390||(0.646 + 36.8)$$
$$= 10 + 34.17 = 44.17$$

$$C_2 = \frac{1}{2\pi \times 120 \times 44.17} = 30\ \mu\text{F}$$

PROBLEM 9.5 EFFECT OF CAPACITORS

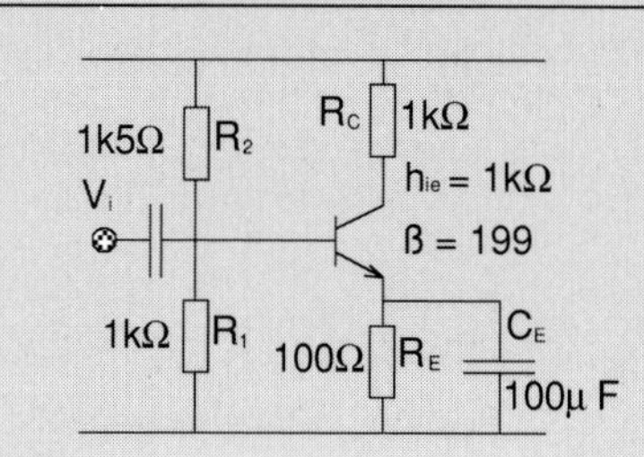

Figure 9.22

Find the corner frequency due to C_E.

Answer: 214.78 Hz

Resistance

$$r_e = \frac{h_{ie}}{\beta + 1}$$

$$= \frac{1k}{200} = 5\ \Omega$$

AC model

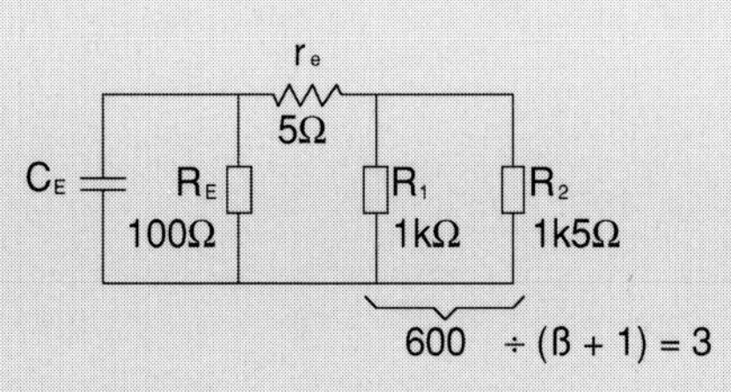

Figure 9.23

$$R = 100 \| (5+3)$$

$$= 100 \| 8 = 7.41\ \Omega$$

$$f = \frac{1}{2\pi \times 7.41 \times 100 \times 10^{-6}}$$

$$f = 214.78\ \text{Hz}$$

Approximate method

$$V_B = 9\frac{2.2}{2.2 + 6.8} = 2.2\ \text{V}$$

$$V_E = 2.2 - 0.7 = 1.5\ \text{V}$$

$$I_E = \frac{1.5}{1k5} = 1\ \text{mA}$$

$$r_e = \frac{26}{1} = 26\ \Omega$$

AC model

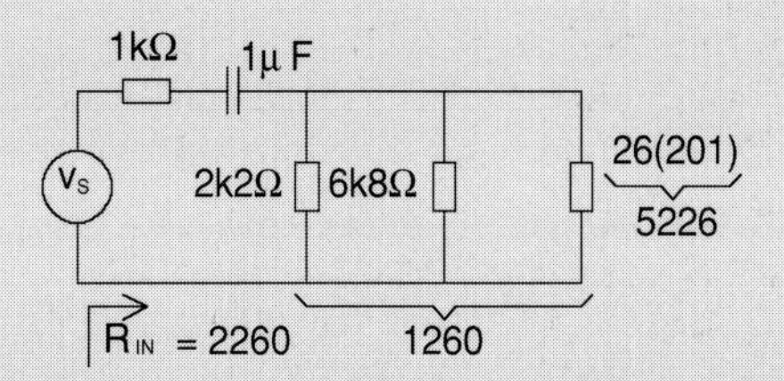

Figure 9.25

$$\omega = \frac{1}{RC} = \frac{1}{2260 \times 1 \times 10^{-6}} = 442.48\ \text{rad/s}$$

$$f = \frac{\omega}{2\pi} = 70.42\ \text{Hz}$$

PROBLEM 9.6 EFFECT OF CAPACITORS

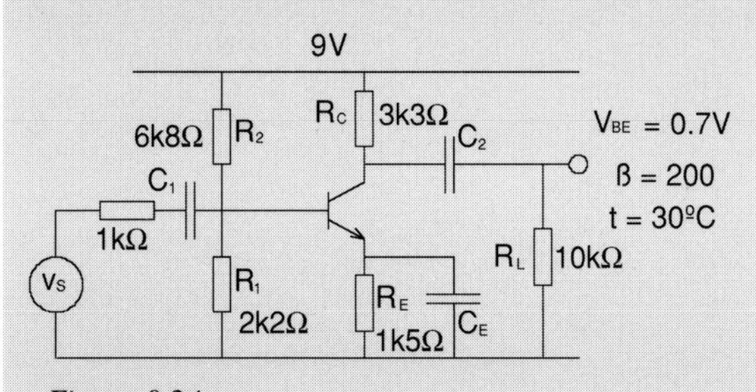

Figure 9.24

Find the lower 3 dB frequency if C_1 is 1 μF.

Answer: $f = 70.42$ Hz

PROBLEM 9.7 EFFECT OF CAPACITORS

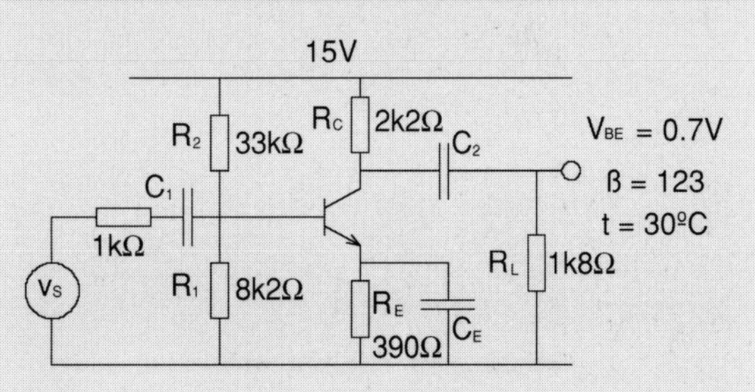

Figure 9.26

Find the value of C_E that will produce a corner frequency at 40 Hz.

Answer: $C_E = 341$ μF

DC conditions

82k vs 47k use Thevenin

Thevenin

$$V_{TH} = 15\frac{8.2}{8.2+33} = 2.99 \text{ V}$$

$$R_{TH} = 8k2 || 33k = 6568\ \Omega$$

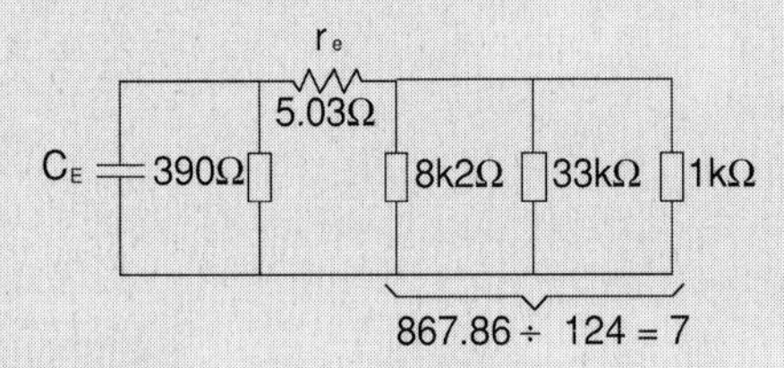

Figure 9.27

$$I_B = \frac{2.99-0.7}{6568+48\,360} = \frac{2.29}{54928}$$

$$I_E = \frac{2.29\times 124}{54\,928} = 5.17 \text{ mA}$$

$$r_e = \frac{26}{5.17} = 5.03\ \Omega$$

AC model

Figure 9.28

$$R = 390||(5.03+7) = 11.67\ \Omega$$

$$C_E = \frac{1}{2\pi \times 40 \times 11.67} = 341\ \mu\text{F}$$

PROBLEM 9.8 EFFECT OF CAPACITORS

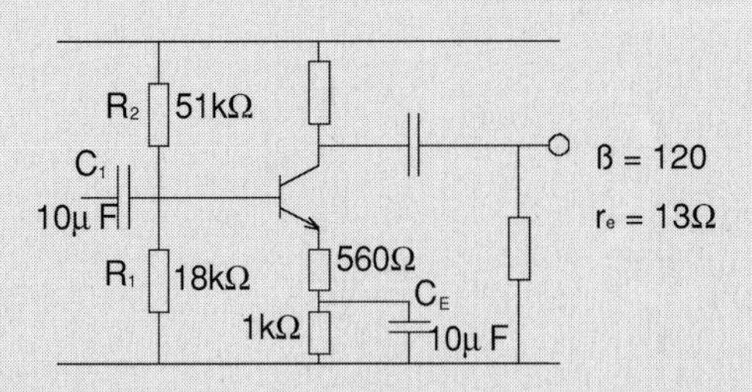

Figure 9.29

Find the corner frequencies due to C_1 and C_E.

Answer: 1.43 Hz, 39.22 Hz

AC model

$$R_{IN} = 18k||51k||(13+560)121$$
$$= 18k||51k||69\,333$$
$$= 111\,62.39\ \Omega$$

$$f = \frac{1}{2\pi \times 11162.39 \times 10 \times 10^{-6}} = 1.43 \text{ Hz}$$

Emitter capacitor

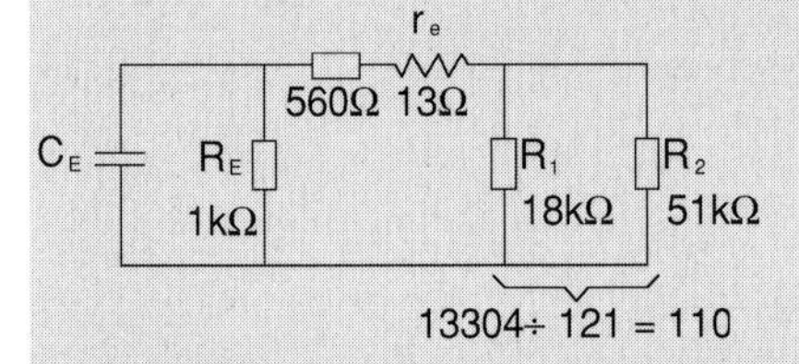

Figure 9.30

$$R = 1k||(560+13+110)$$
$$= 1k||683$$
$$= 405.82$$

$$f = \frac{1}{2\pi \times 405.82 \times 10 \times 10^{-6}} = 39.22 \text{ Hz}$$

PROBLEM 9.9 EFFECT OF CAPACITORS

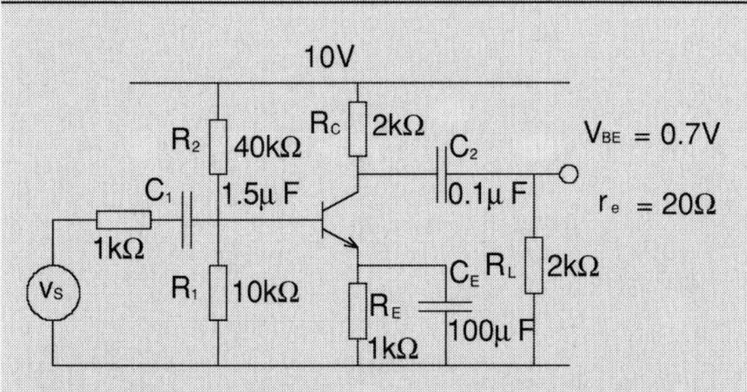

Figure 9.31

Find the corner frequencies due to C_1, C_2 and C_E.

Answer: 28.93 Hz, 398 Hz and 66.75 Hz

AC model C_1. See Figure 9.32.

$$f = \frac{1}{2\pi RC} = \frac{1}{2\pi \times 3667 \times 1.5 \times 10^{-6}} = 29.93 \text{ Hz}$$

C_2

$$f = \frac{1}{2\pi(R_C + R_L)C}$$

$$= \frac{1}{2\pi \times 4000 \times 0.1 \times 10^{-6}} = 398 \text{ Hz}$$

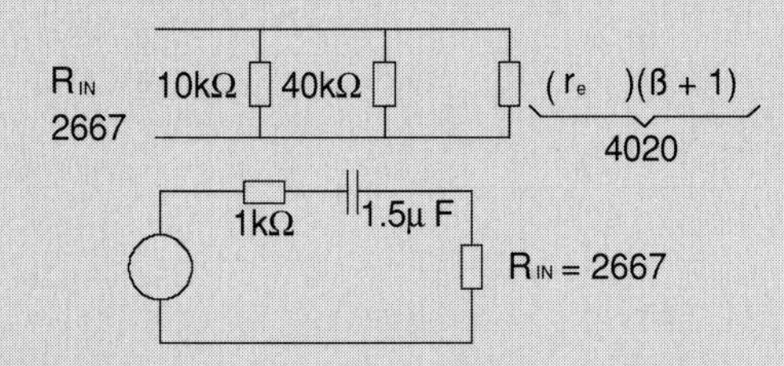

Figure 9.32

C_E

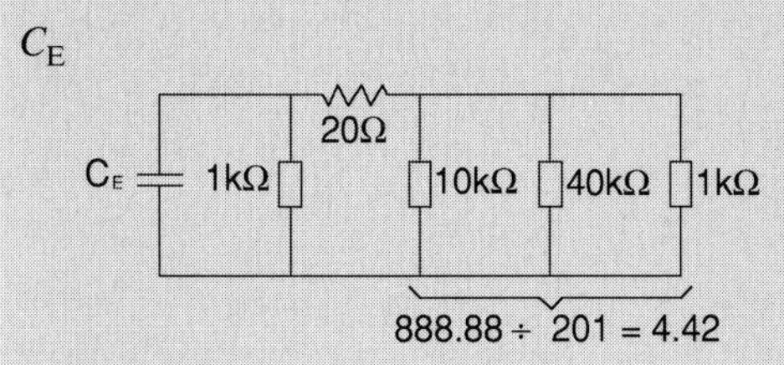

Figure 9.33

$$R = 1000 \| (20 + 4.42) = 23.84$$

$$f = \frac{1}{2\pi RC} = \frac{1}{2\pi \times 23.84 \times 100 \times 10^{-6}} = 66.75 \text{ Hz}$$

PROBLEM 9.10 EFFECT OF CAPACITORS

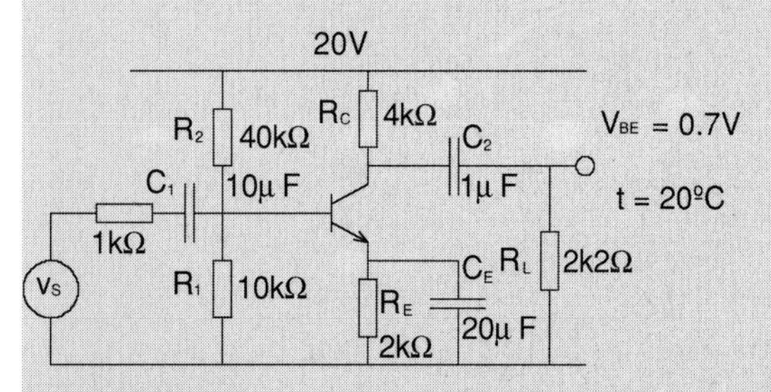

Figure 9.34

Find the lowest corner frequency due to C_1, C_2 or C_E.

Answer: 6.97 Hz (due to C_1)

DC conditions

$$V_B = 20\frac{10}{40 + 10} = 4 \text{ V}$$

$$V_E = 4 - 0.7 = 3.3 \text{ V}$$

$$I_E = \frac{3.3}{2\text{k}} = 1.65 \text{ mA}$$

$$r_e = \frac{25}{1.65} = 15.15 \ \Omega$$

C_1 – AC model

$$R_{IN} = 10\text{k} \| 40\text{k} \| r_e(\beta + 1)$$

$$= 10\text{k} \| 40\text{k} \| 15.15 \times 101 = 1284 \ \Omega$$

$$f = \frac{1}{2\pi RC_1} \quad (R = 1284 + 1\text{k})$$

$$= \frac{1}{2\pi \times 2284 \times 10 \times 10^{-6}} = 6.97 \text{ Hz}$$

C_2

$$f = \frac{1}{2\pi RC_2} \quad (R = R_C + R_L = 6\text{k}2)$$

Bypass capacitor

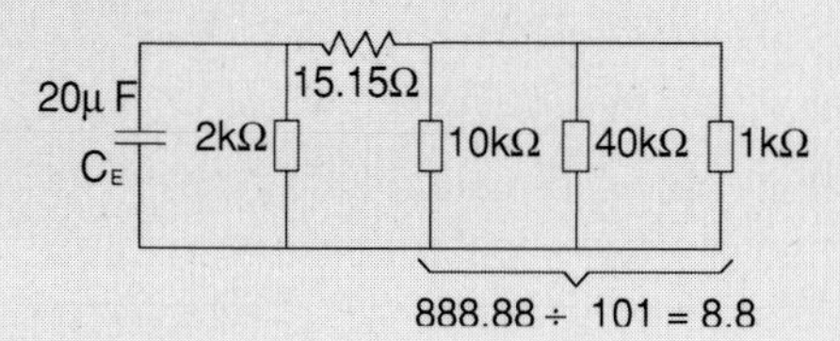

Figure 9.35

$$R = R_E || (r_e + 8.8)$$
$$= 2k||23.95 = 23.67\ \Omega$$
$$f = \frac{1}{2\pi RC} = \frac{1}{2\pi \times 23.67 \times 20 \times 10^{-6}}$$
$$= 336\ \text{Hz}$$

The lowest corner frequency is due to C_1 and is 6.97 Hz.

PROBLEM 9.11 EFFECT OF CAPACITORS

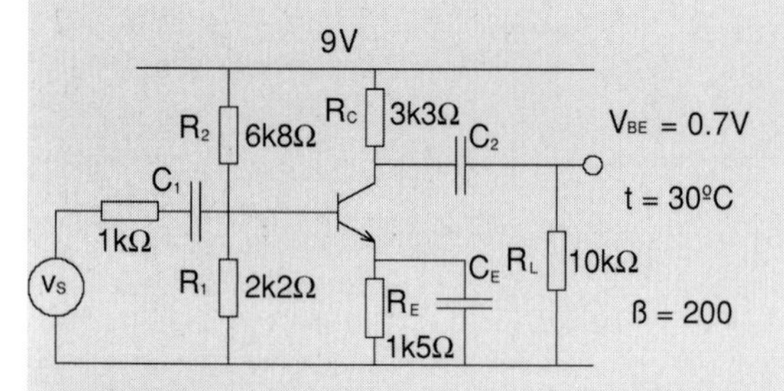

Figure 9.36

Find the value of C_E that will produce a corner frequency of 50 Hz and the value of C_2 that will produce a corner frequency of 90 Hz.

Answer: C_E = 111.5 μF, C_2 = 0.133 μF

DC conditions

$$V_B = 9\frac{2.2}{2.2 + 6.8} = 2.2\ \text{V}$$
$$V_E = 2.2 - 0.7 = 1.5\ \text{V}$$
$$I_E = \frac{1.5}{1k5} = 1\ \text{mA}$$
$$r_e = \frac{26}{1} = 26\ \Omega$$

AC model – bypass capacitor

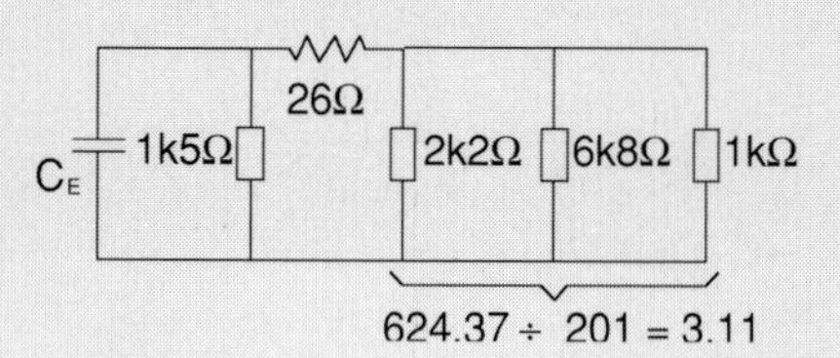

Figure 9.37

$$R = 1k5 || (26 + 3.11) = 28.55\ \Omega$$
$$f = \frac{1}{2\pi RC_E} \qquad C_E = \frac{1}{2\pi Rf}$$
$$C_E = \frac{1}{2\pi \times 28.55 \times 50} = 111.5\ \mu\text{F}$$

Output capacitor C_2

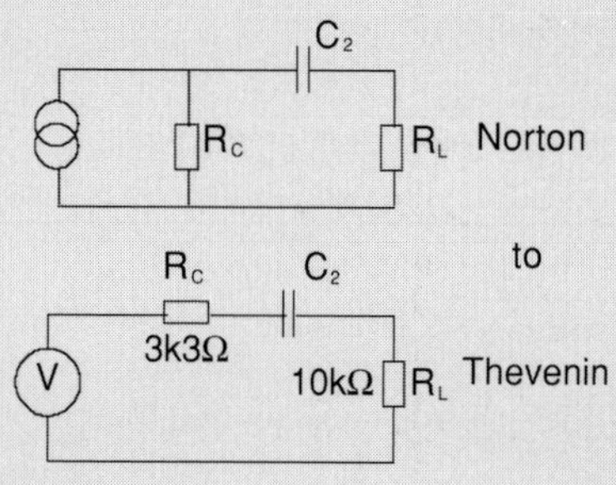

Figure 9.38

$$f = \frac{1}{2\pi(R_C + R_L)C_2}$$
$$C_2 = \frac{1}{2\pi \times 13\,300 \times 90}$$
$$= 0.133\ \mu\text{F}$$

PROBLEM 9.12 EFFECT OF CAPACITORS

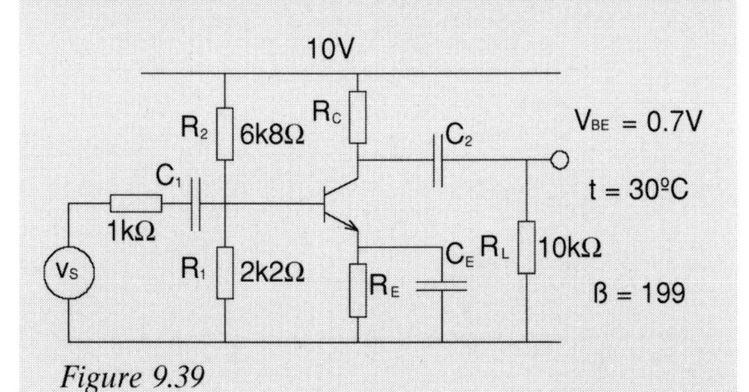

Figure 9.39

Find the 3 dB frequency due to C_1 if C_1 is 0.772 µF.

Answer: 100 Hz (approx.)

Approximate method

$$V_B = 10\frac{2k2}{9k} = 2.444 \text{ V}$$

$$V_E = 2.444 - 0.7 = 1.744 \text{ V}$$

$$I_E = \frac{1.744}{1k} = 1.744 \text{ mA}$$

$$r_e = \frac{25}{1.744} = 14.33\ \Omega$$

AC conditions

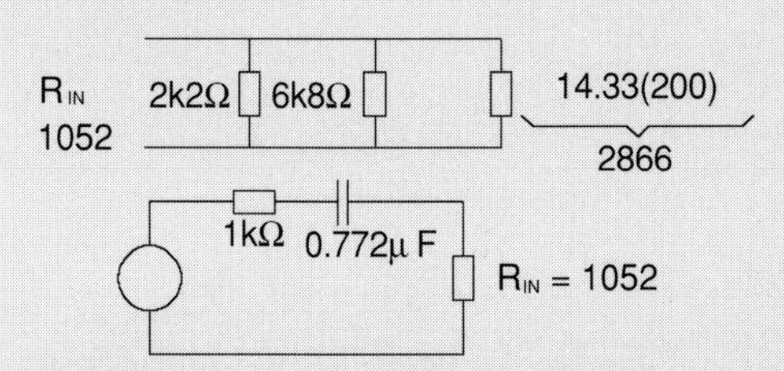

Figure 9.40

$$\omega = \frac{1}{RC} \qquad f = \frac{1}{2\pi RC}$$

$$f = \frac{1}{2\pi \times 2052 \times 0.772 \times 10^{-6}}$$

$$= 100.46 \text{ Hz}$$

PROBLEM 9.13 EFFECT OF CAPACITORS

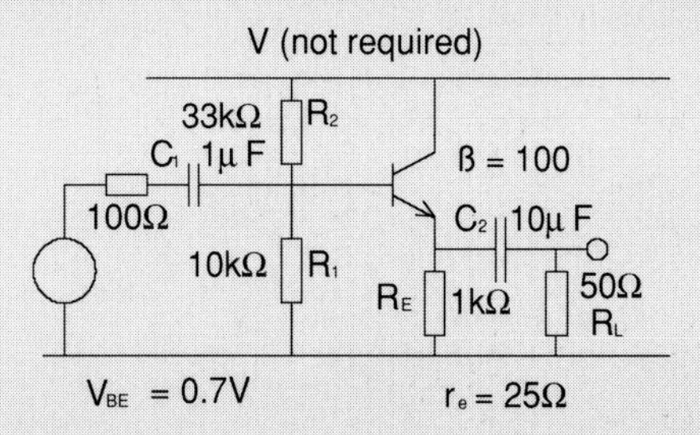

Figure 9.41

Find the lower corner frequencies.

Answer: 36.87 Hz and 211.3 Hz

AC model – C_1

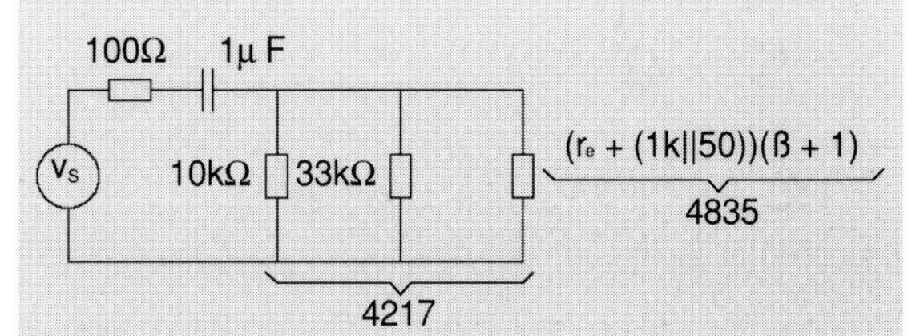

Figure 9.42

$$f = \frac{1}{2\pi(100 + 4217)10^{-6}} = 36.87 \text{ Hz}$$

Capacitor C_2

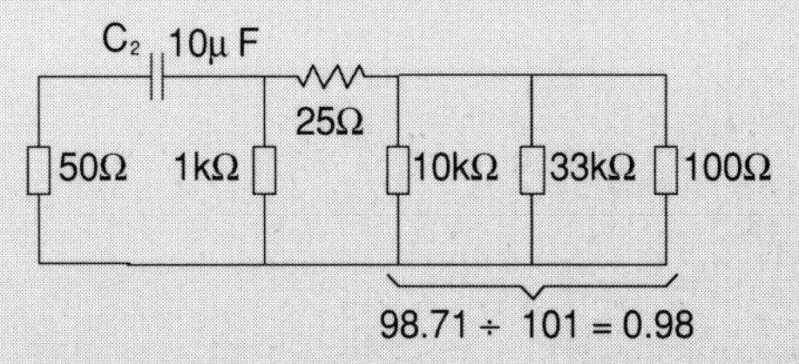

Figure 9.43

$$R = 50 + 1k\|(25 + 0.98)$$

$$= 50 + 25.32$$

$$= 75.32\ \Omega$$

$$f = \frac{1}{2\pi \times 75.32 \times 10 \times 10^{-6}} = 211.3 \text{ Hz}$$

PROBLEM 9.14 EFFECT OF CAPACITORS

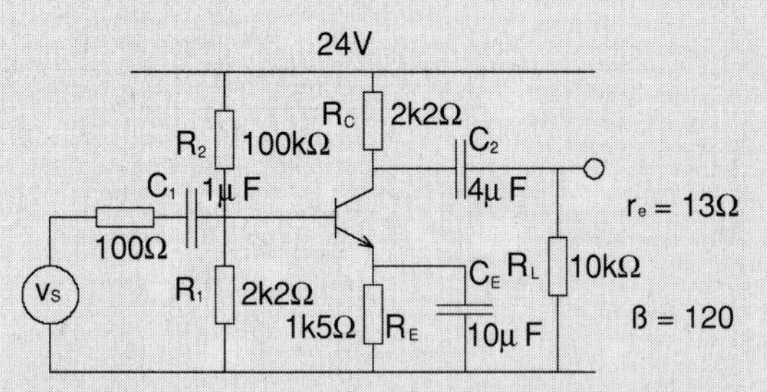

Figure 9.44

Find the corner frequencies due to C_1, C_2 and C_E.

> Answer: 102.88 Hz, 10.25 Hz and 1162.56 Hz

AC model

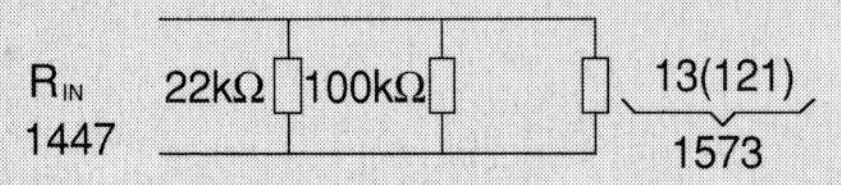

RIN 1447 22kΩ 100kΩ 13(121) 1573

Figure 9.45

Capacitor C_1

$$f = \frac{1}{2\pi(100 + 1447)C_1} = 102.88 \text{ Hz}$$

Capacitor C_2

$$R = R_C + R_L = 2k2 + 10k = 12k2$$

$$f = \frac{1}{2\pi \times 12.2 \times 10^3 C_2}$$

$$= \frac{1}{2\pi \times 12.2 \times 4 \times 10^{-3}} = 10.25 \text{ Hz}$$

Bypass capacitor

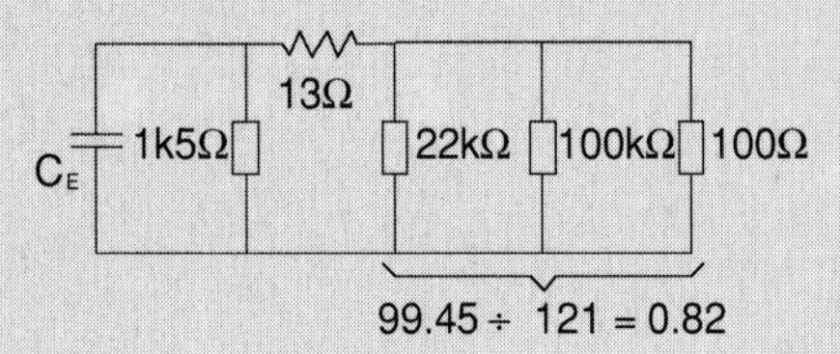

Figure 9.46

$$R = 1k5 \| 13.82 = 13.69$$

$$f = \frac{1}{2\pi \times 13.69 \times 10 \times 10^{-6}}$$

$$= 1162.56 \text{ Hz}$$

PROBLEM 9.15 EFFECT OF CAPACITORS

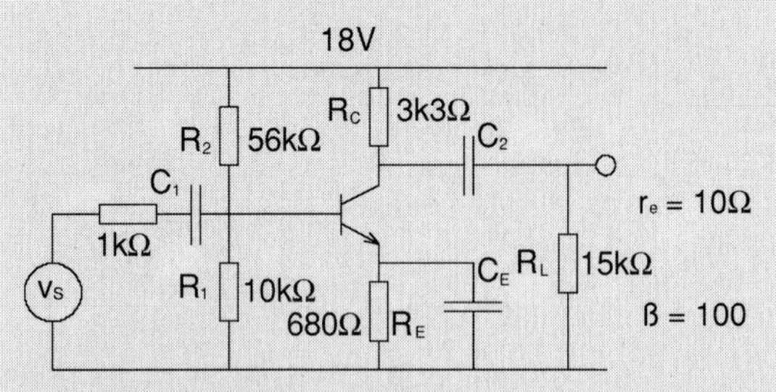

Figure 9.47

Find the values of C_1, C_2 and C_E so that the three lower corner frequencies are at 150 Hz.

> Answer: C_1 = 0.557 μF, C_2 = 0.058 μF and C_E = 57.82 μF

AC model

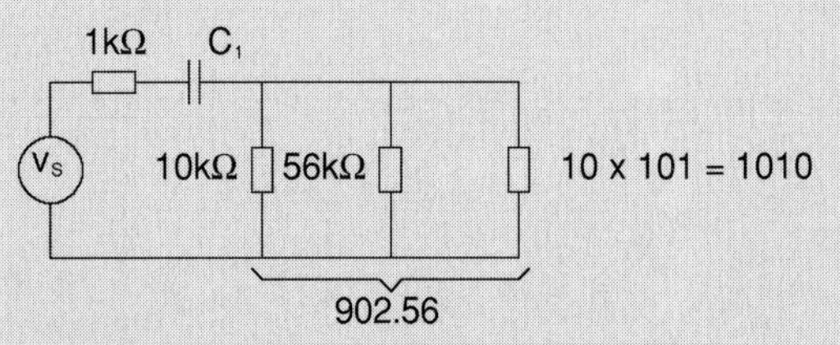

Figure 9.48

Capacitor C_1

$$C_1 = \frac{1}{2\pi \times 150 \times 1902.56} = 0.557\,\mu\text{F}$$

Capacitor C_2

$$C_2 = \frac{1}{2\pi \times 150 \times 18\,300} = 0.058\,\mu\text{F}$$

Bypass capacitor C_E

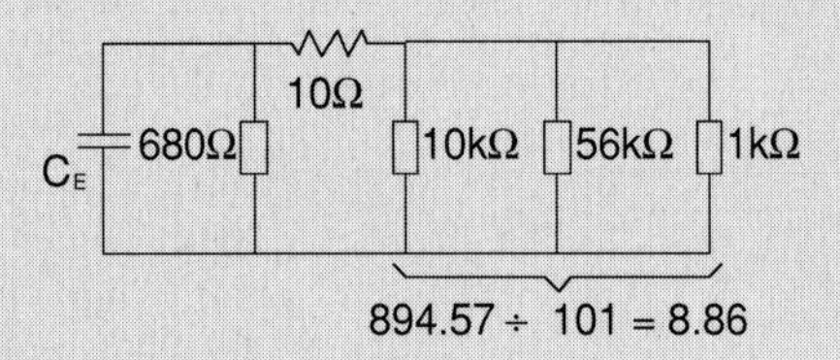

Figure 9.49

$$R = 680 \| 18.86 = 18.35\ \Omega$$

$$C_E = \frac{1}{2\pi \times 150 \times 18.35} = 57.82\,\mu\text{F}$$

PROBLEM 9.16 EFFECT OF CAPACITORS

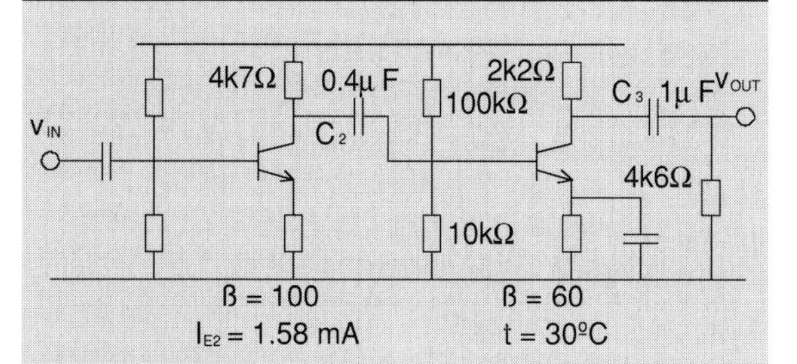

Figure 9.50

Find the corner frequencies due to C_2 and C_3.

Answer: 71 Hz and 23.4 Hz

Resistance

$$r_{e2} = \frac{26}{1.58} = 16.46\ \Omega$$

AC model – C_2

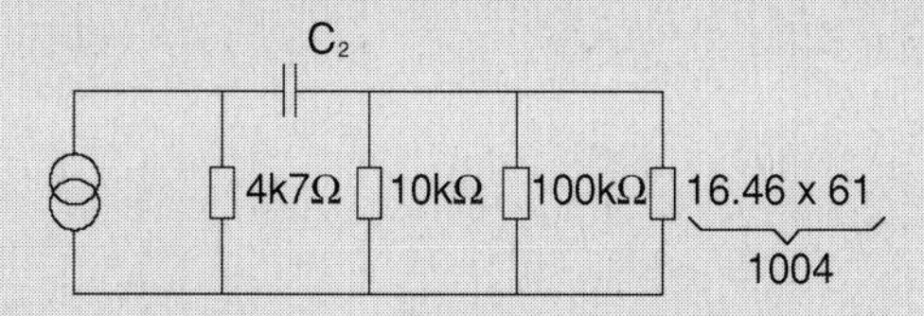

Figure 9.51

$$R = 4k7 + 10k||100k||1004$$

$$= 4700 + 904 = 5604\ \Omega$$

$$f = \frac{1}{2\pi RC}$$

$$= \frac{1}{2\pi \times 5604 \times 0.4 \times 10^{-6}}$$

$$= 71\text{ Hz}$$

Capacitor C_3

$$f = \frac{1}{2\pi(2k2 + 46k) \times 10^{-6}}$$

$$= \frac{1}{2\pi \times 6800 \times 10^{-6}}$$

$$= 23.4\text{ Hz}$$

PROBLEM 9.17 EFFECT OF CAPACITORS

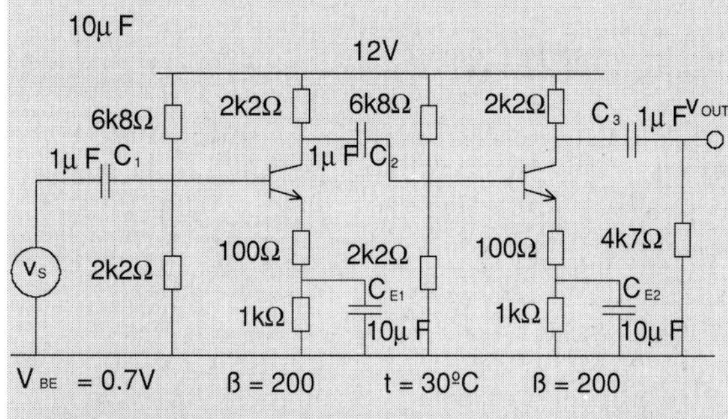

Figure 9.52

Find the corner frequencies due to C_1 and C_2.

Answer: 102.75 Hz and 42.45 Hz

Approximate method

$$V_B = 12\frac{2.2}{2.2 + 6.8} = 2.93\text{ V}$$

$$V_E = 2.93 - 0.7 = 2.23\text{ V}$$

$$I_E = \frac{2.23}{1100} = 2.03\text{ mA}$$

$$r_e = \frac{26}{2.03} = 12.81\ \Omega$$

AC model – C_1

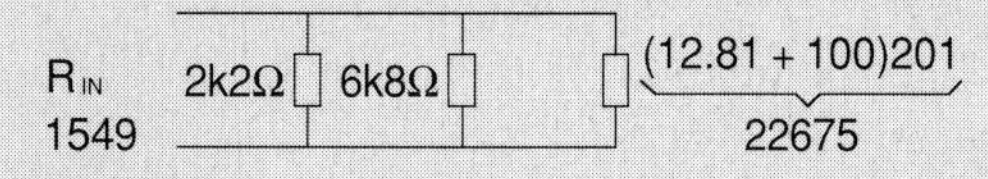

Figure 9.53

C_1 frequency

$$f = \frac{1}{2\pi \times 1549 \times 10^{-6}} = 102.75\text{ Hz}$$

C_2 frequency

$$R_{IN1} = R_{IN2} = 1549\ \Omega$$

$$f = \frac{1}{2\pi(1549 + 2k2) \times 10^{-6}}$$

$$f = 42.45\text{ Hz}$$

PROBLEM 9.18 EFFECT OF CAPACITORS

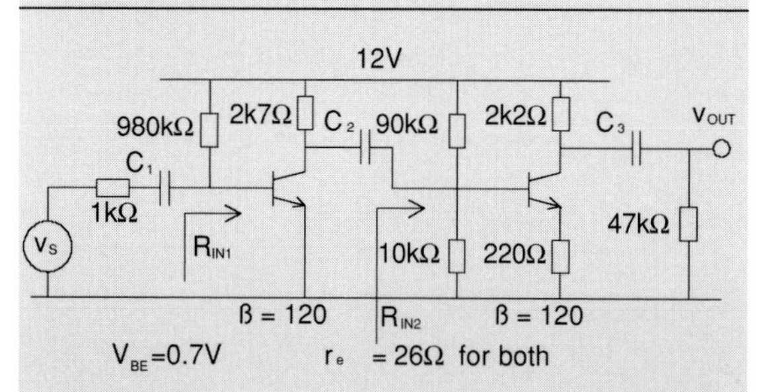

Figure 9.54

Find the values of C_1, C_2 and C_3 to produce corner frequencies at 50, 100 and 150 Hz.

> Answer: $C_1 = 0.77$ μF, $C_2 = 0.165$ μF and $C_3 = 0.022$ μF

AC model

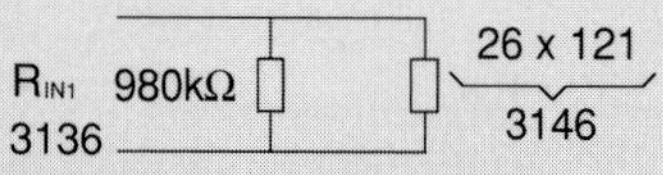

R_IN1 3136 — 980kΩ ∥ 26 x 121 = 3146

Figure 9.55

$$C_1 = \frac{1}{2\pi fR}$$

$$= \frac{1}{2\pi \times 50 \times (3136 + 1\text{k})}$$

$$= 0.77\ \mu\text{F}$$

Capacitor C_2

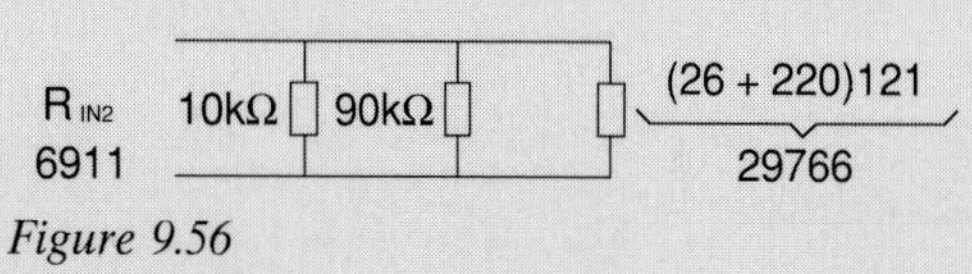

R_IN2 6911 — 10kΩ ∥ 90kΩ ∥ (26 + 220)121 = 29766

Figure 9.56

$$C_2 = \frac{1}{2\pi fR}$$

$$= \frac{1}{2\pi \times 100 \times (2\text{k}7 + 6911)}$$

$$= 0.165\ \mu\text{F}$$

Capacitor C_3

$$C_3 = \frac{1}{2\pi fR}$$

$$= \frac{1}{2\pi \times 150 \times (2\text{k}2 + 47\text{k})}$$

$$= 0.022\ \mu\text{F}$$

PROBLEM 9.19 EFFECT OF CAPACITORS

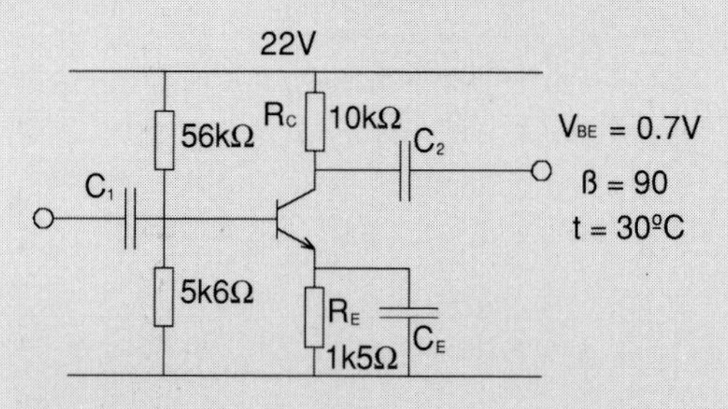

Figure 9.57

Find the value of C_E to produce a corner frequency at 100 Hz.

> Answer: 19.58 μF

DC conditions – approximate method

$$V_B = 22\frac{5.6}{5.6 + 56} = 2\ \text{V}$$

$$V_E = 2 - 0.7 = 1.3\ \text{V}$$

$$I_E = \frac{1.3}{1\text{k}5} = 0.866\ \text{mA}$$

$$r_e = \frac{26}{0.866} = 30\ \Omega$$

AC model

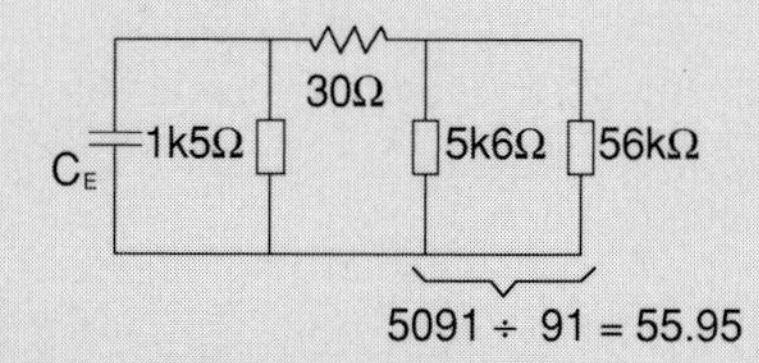

Figure 9.58

$$R = 1\text{k}5 \| (30 + 55.95) = 81.29\ \Omega$$

$$C_E = \frac{1}{2\pi fR}$$

$$= \frac{1}{2\pi \times 100 \times 81.29}$$

$$= 19.58\ \mu\text{F}$$

PROBLEM 9.20 EFFECT OF CAPACITORS

15V
33kΩ
Rc
2k7Ω
VBE = 0.7V
C1
ß = 100
t = 20ºC
12kΩ
RE
1k5Ω

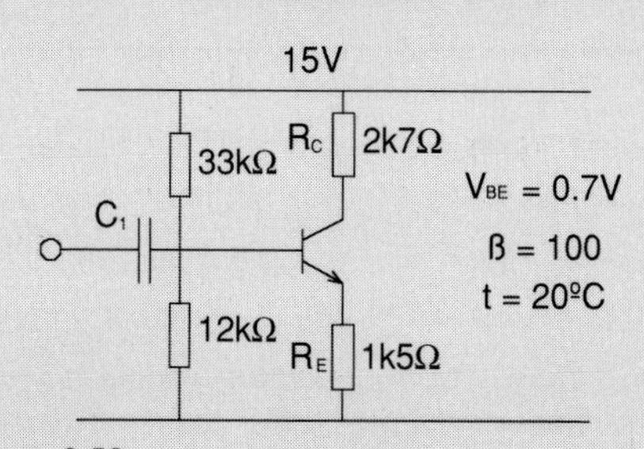

Figure 9.59

Find the value of C_1 that will produce a corner frequency at 77 Hz.

Answer: 0.248 μF

DC conditions – approximate method

$$V_B = 15\frac{12k}{45k} = 4\ V$$

$$V_E = 4 - 0.7 = 3.3\ V$$

$$I_E = \frac{3.3}{1k5} = 2.2\ mA$$

$$r_e = \frac{25}{2.2} = 11.36\ \Omega$$

AC conditions

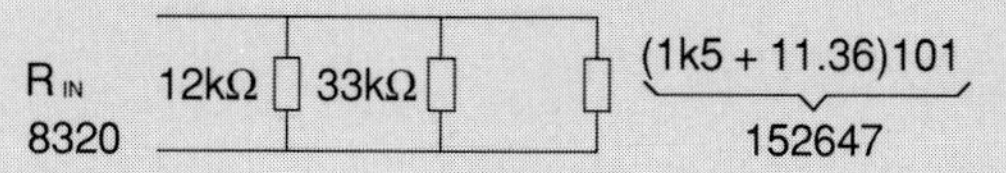

RIN
8320
12kΩ
33kΩ
(1k5 + 11.36)101
152647

Figure 9.60

$$C_1 = \frac{1}{f2\pi R}$$

$$= \frac{1}{77 \times 2\pi \times 8320} = 0.248\ \mu F$$

10

High frequency

We will now concentrate on the gain of a transistor amplifier at high frequency. The calculations are complicated and would normally be done with the help of a computer. We are going to do it with a calculator.

It requires skill to carry out the gain evaluation by hand, as it would be done under examination conditions. You really need to concentrate, and dominate all the theorems and all the electronic laws. You need to have a knowledge of transistor modelling and high frequency performance. You also need to follow a procedure, and the following is recommended:

1) Evaluate the DC conditions. Obtain the quiescent conditions and the value of r_e
2) Evaluate the gain at midband
3) Draw the equivalent circuit for high frequency operation
4) Simplify the circuit as much as possible
5) Apply Miller's theorem. You will end up with two circuits instead of one.
6) Use Thevenin–Norton conversions to reach the simplest possible circuits at input and output sides
7) Make the final calculations in these circuits

There will be several variations depending on the type of load, the type of emitter resistor bypass, the type of source impedance, the type of biasing, but on the whole the above procedure can be used as a guide.

A typical circuit is shown in Figure 10.1. The problem here is to find the voltage gain at a given high frequency.

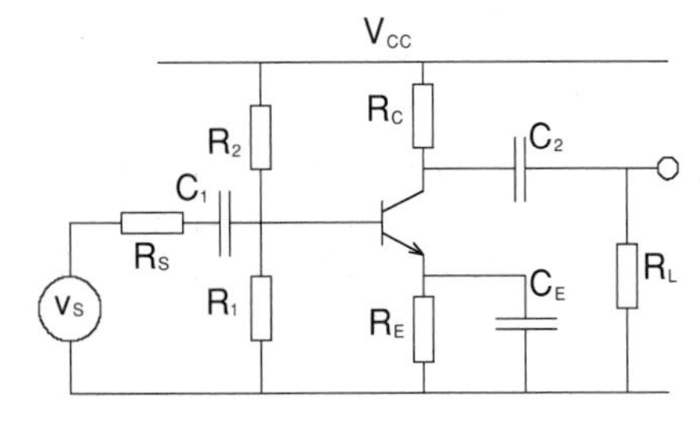

Figure 10.1

We will need more information to be able to complete the problem. The extra information required could be:

$R_{CB'}$ $R_{B'C}$

R_{CE} β

$C_{B'C}$ f_T

and the temperature.

1) DC conditions

In the circuit above, we have potential divider biasing. We have a choice of two methods to evaluate the DC conditions: the approximate method, or the Thevenin method. We will be able to decide which method we should use by using the criteria:

$10R_1$ against βR_E

We will find the value of I_B, I_C, I_E and knowing the temperature we can decide on the two options that we have for r_e:

$$r_e = \frac{25}{I_E} \quad \text{at } 20^\circ\text{C}$$

$$r_e = \frac{26}{I_E} \quad \text{at } 30^\circ\text{C}$$

The value of I_E is entered in mA

2) Midband gain

We now need the model of the transistor for small signal conditions. We see this in Figure 10.2. This circuit is similar to the one used previously for midband gain, with the exception of the

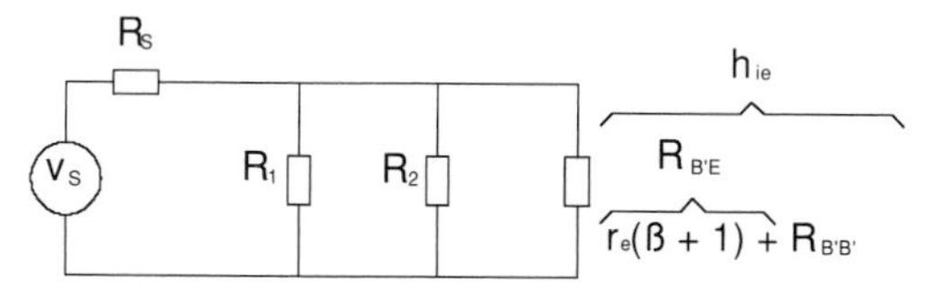

Figure 10.2

right-most resistor which carries i_B. R_E is bypassed by C_E and it doesn't appear anywhere. $R_{B'E}$ is the value of little r_e, reflected to the input side (i.e., multiplied by $(\beta + 1)$). We also need to include $R_{BB'}$ and this will be part of the information given at the beginning of the problem. All the above will represent the value of h_{ie} (the input hybrid parameter under common emitter configuration).

All the resistors form a potential divider, where the base voltage will be a proportion of the voltage source, given by

$$v_B = v_S \frac{R_1 || R_2 || h_{ie}}{R_S + R_1 || R_2 || h_{ie}}$$

We can now evaluate the midband gain as

$$\left| \frac{v_O}{v_B} \right| = \frac{R_C || R_L}{r_e}$$

and

$$\left| \frac{v_O}{v_S} \right| = \frac{v_O}{v_B} \frac{v_B}{v_S}$$

3) HF equivalent circuit

The HF circuit was developed in the modelling, (Chapter 4) and based on that we have the circuit of Figure 10.3. This will be the basic circuit used and the information given as part of the problem will allow us to do some simplifications at this stage.

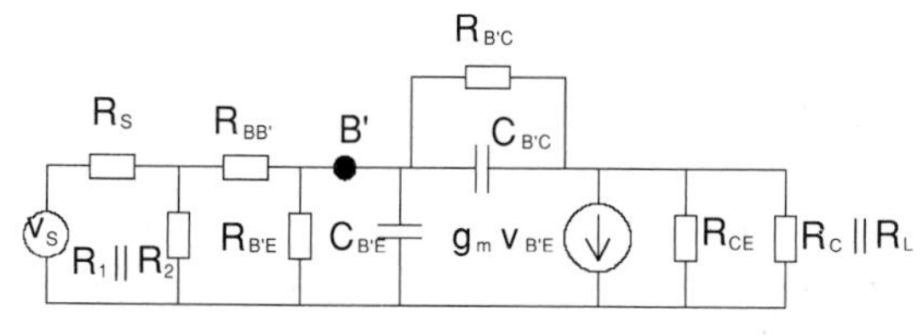

Figure 10.3

4) Simplifications to the circuit

The value of $R_{B'C}$ is usually very high (2 MΩ) and as it is much larger than $R_{B'E}$, $R_{B'C}$ can be disregarded. The next consideration is the source resistance. If this is small we can assume $R_S = 0$ and in that case we don't need to bother with $R_1 \,||\, R_2$. If the source resistance is not negligible, then we can use Thevenin–Norton conversions to simplify the circuit at the left of *B'*. This being a purely resistive circuit will present no difficulty with the Thevenin–Norton conversions.

Miller's theorem could have been applied earlier. We only need to know the midband amplification to apply the theorem, but we have decided to leave Miller's theorem for step 5.

5) Miller's theorem

When we apply Miller's theorem, we divide the circuit in two. The appearance of the circuit after applying Miller's theorem is shown in Figure 10.4.

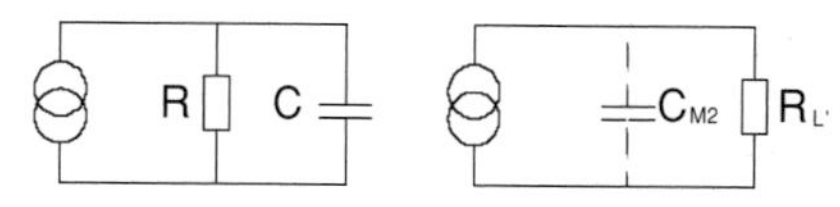

Figure 10.4

The current generator on the left, will be the result of making Thevenin–Norton transformations from left to right. We started with a source v_S and we end up with a current source as the last resistor is in parallel with the capacitor. The simplest we can therefore obtain is a Norton equivalent with the Norton resistor included in the resistor *R*. We will see more of this in step 6.

The capacitor *C* on the circuit in the left is the result of the parallel combination of

$$C = C_{B'E} + C_{M1}$$

$$\boxed{C_{M1} = C_{B'C}(1 - A_V)}$$

Sometimes the value of $C_{B'E}$ is not known, although we will know the value of the transition frequency. In this case, the following relation can be used

$$\boxed{f_T = \frac{1}{(r_e 2\pi C_T)}}$$

C_T (total capacitance) is defined as:

$$C_T = C_{B'E} + C_{B'C}$$

In the circuit on the right, the capacitor has been shown in dotted lines. This capacitor corresponds to C_{M2}, the Miller's theorem capacitor.

$$C_{M2} = C_{B'C}\left(1 - \frac{1}{A_V}\right) \approx C_{B'C}$$

Due to the fact that the impedance of the capacitor is usually much greater than R_L', then the capacitor can be neglected, leaving only R_L' in the output side of the transistor. R_L' will be a combination of resistors including

The transistor output resistor
The collector resistance
The load resistance

6) Thevenin–Norton conversions

It is sometimes possible, according to the circuit, to apply Miller's theorem, divide the circuit in two and still have some reductions to do.

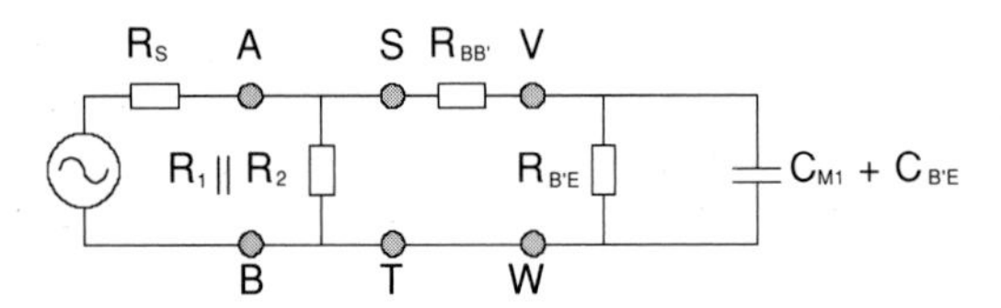

Figure 10.5

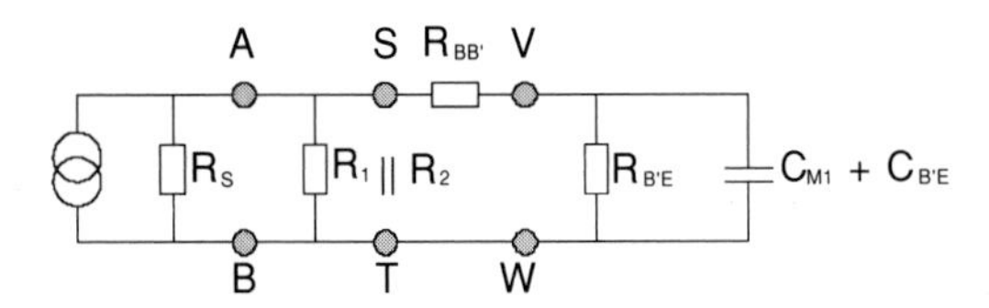

Figure 10.6

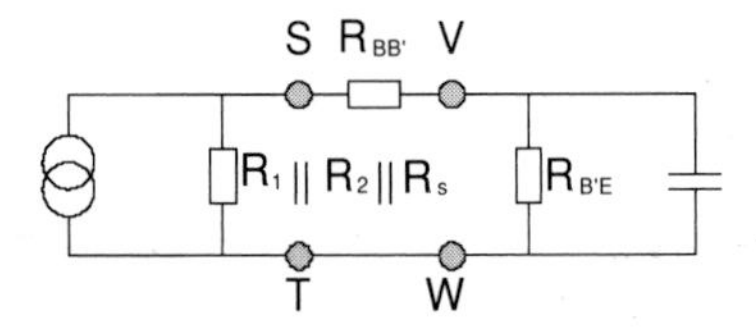

Figure 10.7

We could have the situation of Figure 10.5. This circuit can be reduced by changing the circuit on the left of AB to a Norton equivalent as we see in Figure 10.6. The resistors at the left of ST can be joined together. As a result we get the circuit shown in Figure 10.7. The circuit at the left of ST can be converted from a Norton to a Thevenin. We would continue reducing the circuit from left to right. This procedure is quite simple and avoids having to manipulate complex numbers if we get the capacitor involved in the calculations.

7) Final calculations

We will finally arrive at two circuits like the one shown in Figure 10.8. We have the current source

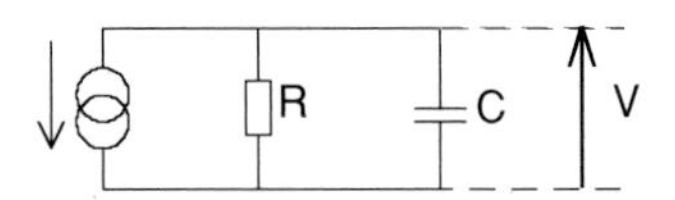

Figure 10.8

and only two components R and C. We want to find the voltage across the RC components. This is simply:

$$V = IZ = \frac{I}{Y}$$

In the first instance we use the impedance, in the second we use the formula with the admittance. The second alternative seems best in this case.

There is one more simplification which is to use the absolute value of the admittance as we are only pursuing the gain of the transistor and the phase angle is of no significance in this case.

$$|Y| = \sqrt{\left(\frac{1}{R}\right)^2 + (2\pi fC)^2}$$

So we have

$$V = \frac{I}{|Y|}$$

The voltage calculated in this way will be $v_{B'E}$.

On the output side we will have a circuit as shown in Figure 10.9. The output voltage is given by Ohm's law:

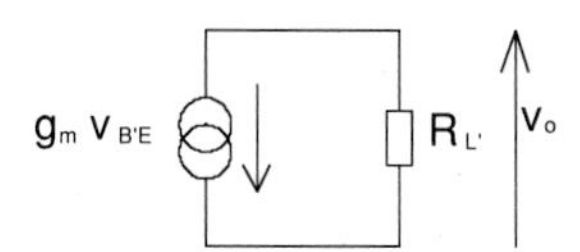

Figure 10.9

$$v_o = -g_m v_{B'E} R_L'$$

As $v_{B'E}$ is a function of the input source voltage this expression will give us the overall gain, which is what we have been looking for.

One question to ask here is: where does $g_m v_{B'E}$ come from? We will see this.

Transconductance

The transconductance is the ratio of output current to input voltage with the output voltage constant.
Trans, means from input to output and conductance implies the reciprocal of resistance and it is measured in siemens.

$$g_m = \frac{i_C}{v_{B'E}} \quad \text{at } v_{CE} \text{ constant}$$

We are now familiar with the emitter resistance r_e

$$r_e = \frac{V_T}{I_E}$$

V_T is the constant which includes the absolute temperature and is referred as the diode thermal voltage. We defined this as

$$r_e = \frac{25}{I_E} \quad \text{at } 20^\circ\text{C}$$

$$r_e = \frac{26}{I_E} \quad \text{at } 30^\circ\text{C}$$

The transconductance is related to r_e in the following way. If the transistor is working in the active region, that is to say it is not in the saturated region, then

$$g_m = \frac{I_C}{V_T}$$

Therefore we can use approximately:

$$g_m = \frac{1}{r_e}$$

and this will be the way in which we will calculate the transconductance.

Current source g_m $v_{B'E}$

In order to see this more clearly we start with the relation in the transistor:

$$i_C = \beta i_B$$

In the high frequency model (Figure 10.10) we define the point *B'* which is an imaginary contact inside the transistor, and not accessible from the outside.

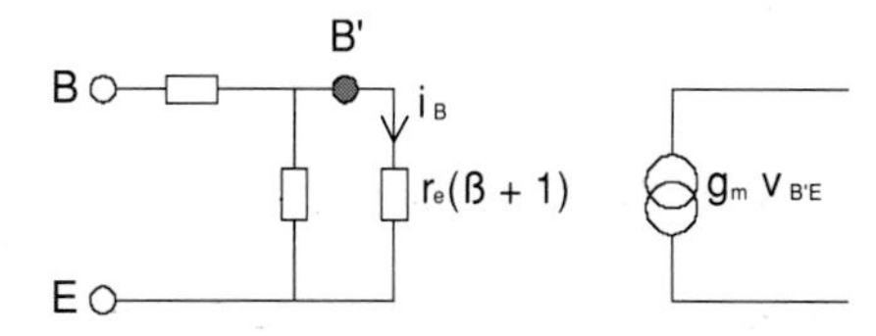

Figure 10.10

Through this point B' is where i_B circulates and that resistor is r_e. Remember that R_E is bypassed. Normally r_e carries i_E which is $(\beta + 1)$ times i_B. This is why we use the reflected value of r_e and simply i_B.

The voltage is

$$v_{B'E} = i_B r_e (\beta + 1)$$

so

$$i_B = \frac{v_{B'E}}{r_e(\beta + 1)}$$

On the output side

$$i_C = \beta i_B$$

We replace i_B, then

$$i_C = \frac{\beta v_{B'E}}{r_e(\beta + 1)}$$

As β is usually big we can compromise and simplify β with (β + 1).
So

$$i_C = \frac{v_{B'E}}{r_e}$$

As we are interested in the voltage gain, we start with a voltage source, $v_{B'E}$ is a function of the voltage source. That is why the current source $g_m v_{B'E}$ (which is the same as $v_{B'E}/r_e$), is also a function of the voltage source. The expression

$$v_o = - g_m v_{B'E} R_L'$$

therefore, contains the overall voltage gain.

The minus sign is to satisfy the current direction chosen in Figure 10.9.

All we wanted to do in this last section was to explain where $g_m v_{B'E}$ came from and we hope that we have done so.

We can now continue with the problems.

PROBLEM 10.1 HIGH FREQUENCY

A transistor has a quiescent current of 0.2 mA, f_T = 3 GHz, and $C_{B'C}$ = 0.1 pF. The temperature is 20° Find the value of $C_{B'C}$.

Answer: $C_{B'E} = 0.32$ pF

$$r_e = \frac{25}{0.2} = 125\ \Omega$$

$$f_T = \frac{1}{2\pi r_e (C_{B'C} + C_{B'E})}$$

$$C_T = \frac{1}{2\pi r_e f_T}$$

$$= \frac{1}{2\pi \times 125 \times 3 \times 10^9} = 0.42 \text{ pF}$$

$$C_{B'C} + C_{B'E} = 0.42 \times 10^{-12}$$

$$0.1 \text{ pF} + C_{B'E} = 0.42 \times 10^{-12}$$

$$C_{B'E} = 0.32 \text{ pF}$$

PROBLEM 10.2 HIGH FREQUENCY

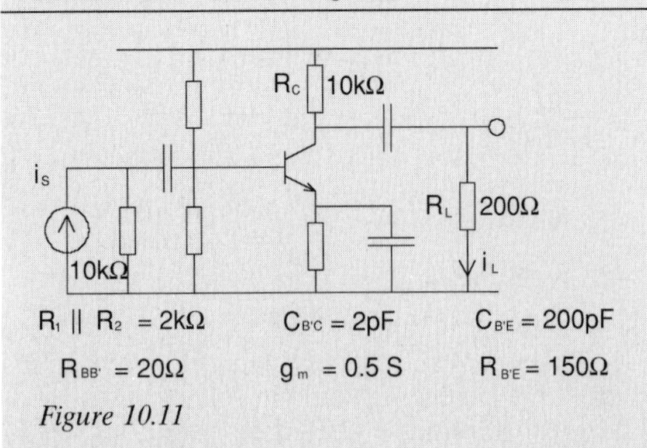

Figure 10.11

Find the corner frequency f_H.

Answer: 2.59 MHz

Midband gain

$$A_V = -\frac{R_C || R_L}{r_e} = -98$$

Miller's theorem

$$C_{M1} = C_{B'C}(1 - A_V)$$

$$= 2 \times 99 = 198 \text{ pF}$$

Resistance

$$R = 10\text{k} || R_1 || R_2 || (150 + 20)$$

$$= 10\text{k} || 2\text{k} || 170 = 154.26\ \Omega$$

Capacitance

$$C = C_{B'E} + C_{M1}$$

$$= 200 + 198 = 398 \text{ pF}$$

$$f_H = \frac{1}{2\pi RC}$$

$$= \frac{1}{2\pi \times 154.26 \times 398 \times 10^{-12}}$$

$$= 2.59 \text{ MHz}$$

PROBLEM 10.3 HIGH FREQUENCY

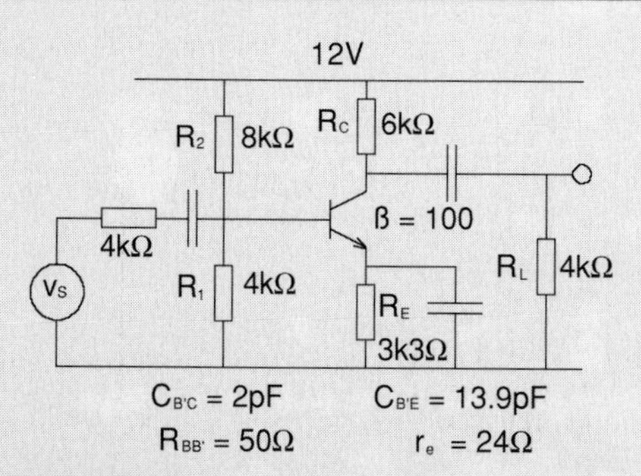

Figure 10.12

Find the total input capacitance of the transistor and the corner frequency f_H, due to the capacitance.

Answer: 215.9 pF and 764.84 kHz

Midband gain (to use in Miller' s theorem)

$$A_V = -\frac{R_C \| R_L}{r_e} = -\frac{6k\|4k}{24} = -100$$

$$C_{M1} = C_{B'C}(1 - A_V) = 2(101) = 202 \text{ pF}$$

Input capacitance

$$C_{M1} + C_{B'C} = 202 + 13.9 = 215.9 \text{ pF}$$

Equivalent resistance

$$R = 4k\|4k\|8k\|r_e(\beta + 1)$$
$$= 4k\|4k\|8k\|2424 = 963.82$$

$$f_H = \frac{1}{2\pi RC} = \frac{1}{2\pi \times 963.82 \times 215.9 \times 10^{-12}}$$
$$= 764.84 \text{ kHz}$$

PROBLEM 10.4 HIGH FREQUENCY

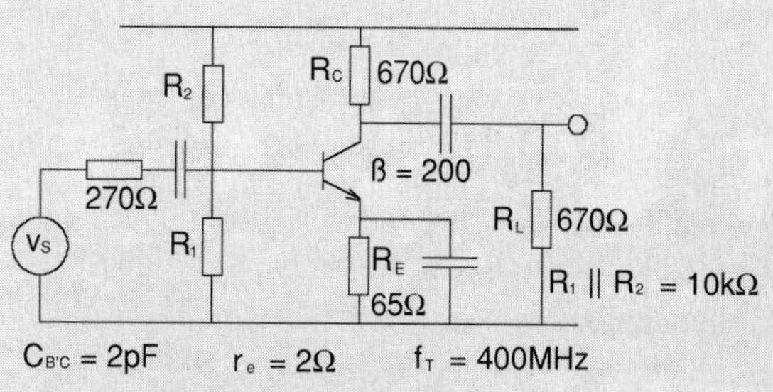

Figure 10.13

Find the high frequency corner frequency.

Answer: 1.874 MHz

Midband gain

$$A_V = -\frac{R_C \| R_L}{r_e}$$
$$= -\frac{335}{2} = -167.5$$

Capacitor $C_{B'E}$ (from transition frequency)

$$C_{B'C} + C_{B'E} = \frac{1}{2\pi r_e f_T}$$
$$= \frac{1}{2\pi 2 \times 2 \times 400 \times 10^6} = 199 \times 10^{-12}$$

$$2 + C_{B'E} = 199 \text{ pF} \qquad C_{B'E} = 197 \text{ pF}$$

Miller's theorem

$$C_{M1} = C_{B'C}(1 - A_V)$$
$$= 2(168.5) = 337 \text{ pF}$$

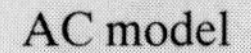

AC model

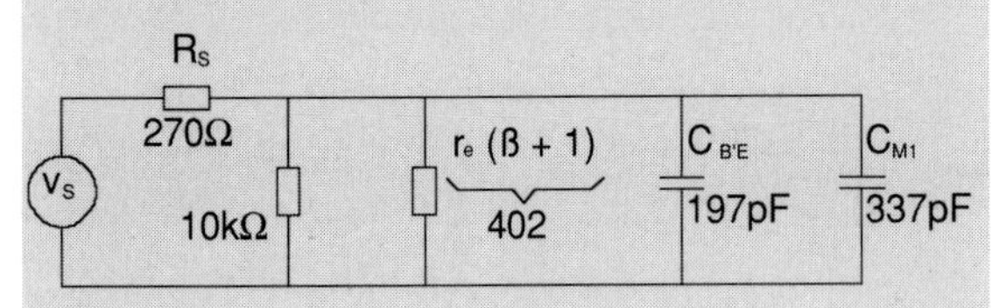

Figure 10.14

Equivalent resistance

$$R = 270 \| 10k \| 402 = 159\ \Omega$$

Equivalent capacitance

$$C = 197 + 337 = 534\ \text{pF}$$

$$f_H = \frac{1}{2\pi RC}$$

$$= \frac{1}{2\pi \times 159 \times 534 \times 10^{-12}}$$

$$= 1.874\ \text{MHz}$$

PROBLEM 10.5 HIGH FREQUENCY

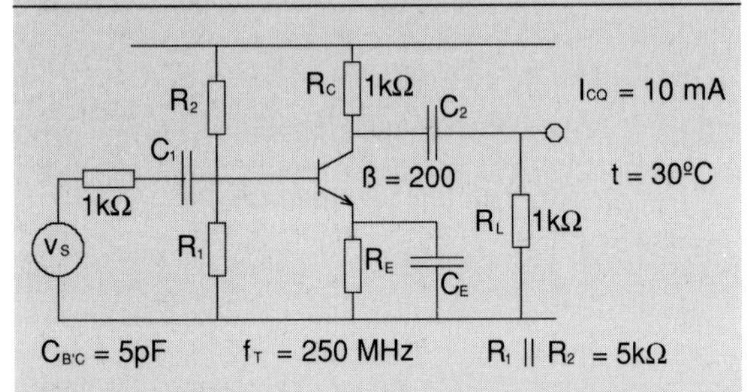

Figure 10.15

Find the high frequency and corner frequency.

Answer: $f_H = 411$ kHz

Note: C_1, C_2 and C_E are related to the low frequency cut-off. Now we are concerned with high frequency cut-off. The capacitors do not appear in the diagram.

AC model

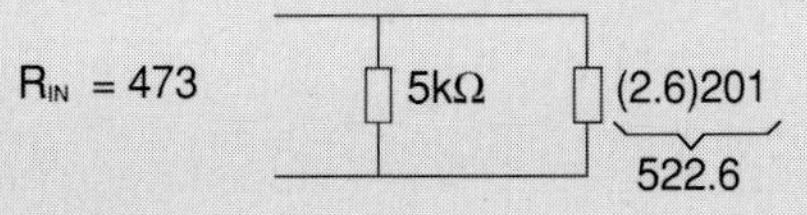

RIN = 473   5kΩ   (2.6)201 = 522.6

Figure 10.16

$$r_e = \frac{26}{10} = 2.6\ \Omega$$

Midband gain used to be in Miller's formula

$$A_V = -\frac{R_C \| R_L}{r_e} = -\frac{500}{2.6} = -192.3$$

Transition frequency

$$C_{B'C} + C_{B'E} = \frac{1}{2\pi r_e f_T}$$

$$= \frac{1}{2\pi \times 2.6 \times 250 \times 10^6} = 245\ \text{pF}$$

$$5 + C_{B'E} = 245\ \text{pF}$$

$$C_{B'E} = 240\ \text{pF}$$

Miller's theorem ($C_{B'C}$ split into C_{M1} and C_{M2})

$$C_{M1} = C_{B'C}(1 - A_V)$$

$$= 5(193.3) = 966.5\ \text{pF}$$

AC model

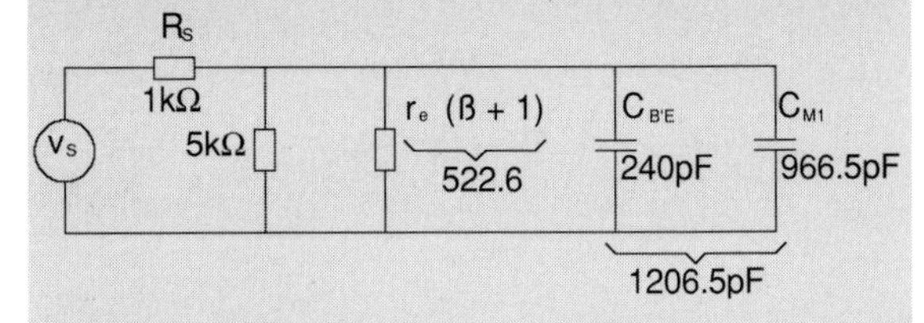

Figure 10.17

Equivalent resistance

$$R = 1k \| 5k \| 522.6 = 321\ \Omega$$

$$f_H = \frac{1}{2\pi RC}$$

$$= \frac{1}{2\pi \times 321 \times 1206.5 \times 10^{-12}}$$

$$= 411\ \text{KHz}$$

PROBLEM 10.6 HIGH FREQUENCY

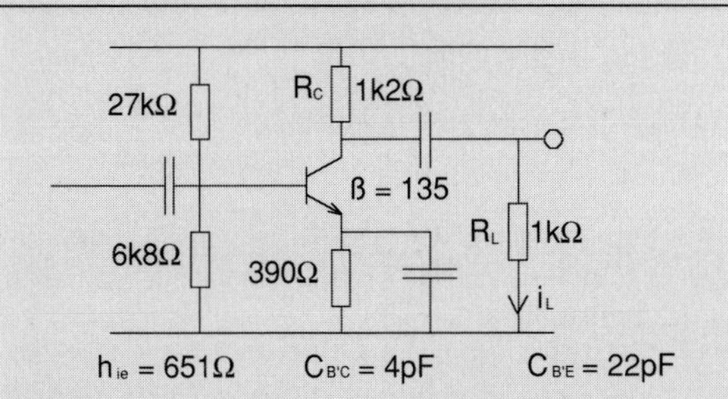

Figure 10.18

Find f_H.

> Answer: 569 kHz

We can calculate r_e from h_{ie}.

$$h_{ie} = (\beta + 1)r_e$$

$$r_e = \frac{651}{136} = 4.79\ \Omega$$

Midband gain

$$A_V = -\frac{R_C \| R_L}{r_e} = -\frac{1k2 \| 1k}{4.79} = -113.87$$

Miller's theorem

$$C_{M1} = C_{B'C}(1 - A_V)$$
$$= 4(114.87) = 459.48 \text{ pF}$$

AC model

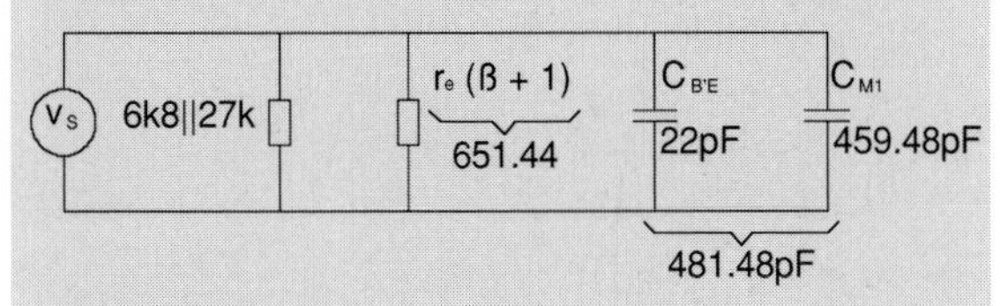

Figure 10.19

$$R = 6k8 \| 27k \| 651.44 = 581.68\ \Omega$$

$$C = 22 + 459.48 = 481.48 \text{ pF}$$

$$f_H = \frac{1}{2\pi RC} = \frac{1}{2\pi \times 581 \times 481.48 \times 10^{-12}} = 569 \text{ kHz}$$

PROBLEM 10.7 HIGH FREQUENCY

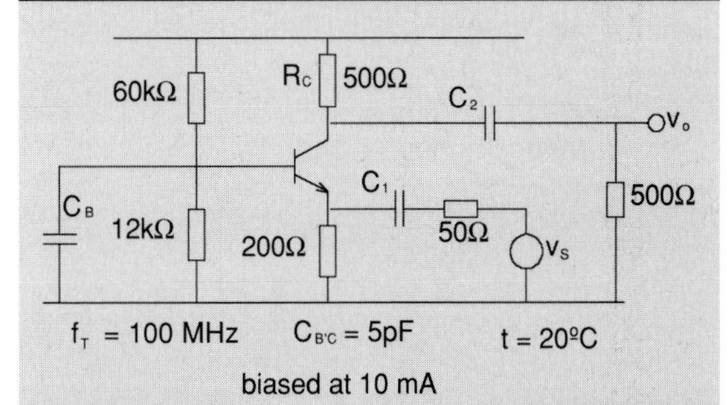

Figure 10.20

Find the input and output corner frequencies on the high frequency side.

> Answer: 107 MHz and 127.32 MHz

Note: This is a common base amplifier. The input is in phase with the output. The Miller effect does not apply as it would need a negative gain!

$$r_e = \frac{25}{10} = 2.5\ \Omega$$

$$C_{B'C} + C_{B'E} = \frac{1}{2\pi r_e f_T}$$

$$= \frac{1}{2\pi \times 25 \times 10^8} = 636.6 \text{ pF}$$

$$C_{B'E} = 636.6 - 5 = 631.6 \text{ pF}$$

Output side. C_B takes the base to ground

$$C = 5 \text{ pF}$$

$$R = 500 \| 500 = 250\ \Omega$$

$$f_{OH} = \frac{1}{2\pi RC} = \frac{1}{2\pi \times 250 \times 5 \times 10^{-12}} = 127.32 \text{ MHz}$$

Input side

$$R = r_e \| R_E \| R_S = 2.5 \| 200 \| 50 = 2.35\ \Omega$$

$$f_{IH} = \frac{1}{2\pi RC} = \frac{1}{2\pi \times 2.35 \times 631.6 \times 10^{-12}}$$

$$f_{IH} = 107 \text{ MHz}$$

PROBLEM 10.8 HIGH FREQUENCY

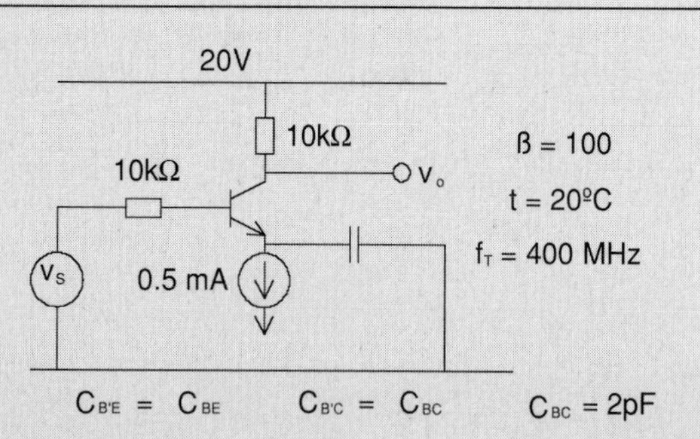

Figure 10.21

Find the midband gain and the upper corner frequency.

Answer: −67.11 and 116.28 kHz

$$r_e = \frac{25}{0.5} = 50\ \Omega$$

Transistor gain

$$A_V = -\frac{R_C}{r_e} = -\frac{10k}{50} = -200$$

Overall gain – potential divider

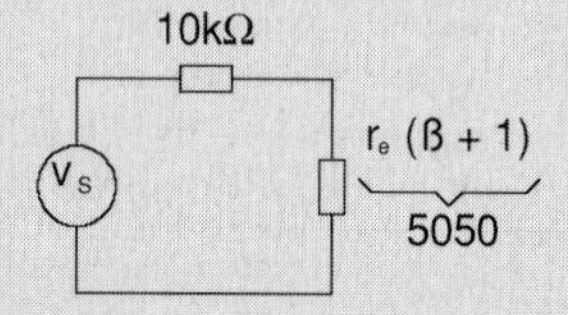

Figure 10.22

$$\frac{v_B}{v_S} = \frac{5050}{10k + 5050} = 0.335\,5$$

Voltage gain

$$A_V = \frac{5050}{10k + 5050} \times (-200) = -67.11$$

Corner frequency

$$C_{B'C} + C_{B'E} = \frac{1}{2\pi r_e f_T} = \frac{1}{2\pi \times 50 \times 400 \times 10^6}$$

$$= 7.96 \text{ pF}$$

$$C_{B'E} = 7.96 - 2 = 5.96 \text{ pF}$$

Upper corner frequency

$$R = 10k\|5050 = 3355\ \Omega$$

$$C_{M1} = 2(1 + 200) = 402 \text{ pF}$$

$$C = C_{B'E} + C_{M1} = 5.96 + 402 = 407.96 \text{ pF}$$

$$f = \frac{1}{2\pi RC} = \frac{1}{2\pi \times 3355 \times 407.96 \times 10^{-12}}$$

$$f = 116.28 \text{ kHz}$$

PROBLEM 10.9 HIGH FREQUENCY

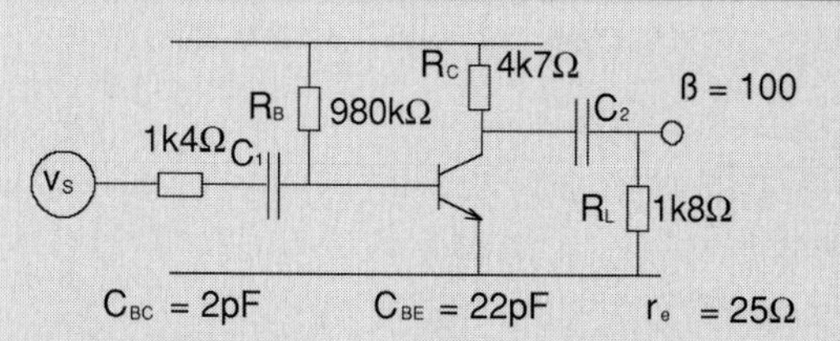

Figure 10.23

Find the input upper corner frequency.

Answer: 1.38 MHz

Midband gain (transistor only)

$$A_V = -\frac{4k7\|1k8}{25} = -\frac{1302}{25} = -52.06$$

Amplifier model

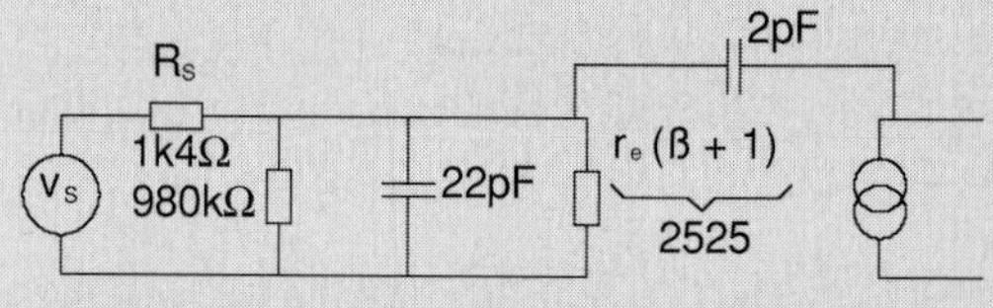

Figure 10.24

$$C_{M1} = 2(1 + 52.06) = 106.12 \text{ pF}$$

Input capacitance

$$C = C_{M1} + 22 \text{ pF} = 128.12 \text{ pF}$$

Input resistance

$$R = 1k4\|980k\|2525 = 900\ \Omega$$

$$f_H = \frac{1}{2\pi RC} = \frac{1}{2\pi \times 900 \times 128.12 \times 10^{-12}}$$

$$f_H = 1.38 \text{ MHz}$$

PROBLEM 10.10 HIGH FREQUENCY

A transistor has a quiescent current of 1 mA. The temperature is 30°. Additional data: $C_{B'C}$ = 1 pF, $C_{B'E}$ = 10 pF, β = 150.
Find f_T and f_β.

Answer: f_T = 556.485 MHz, f_β = 3.71 MHz

$$r_e = \frac{26}{1} = 26\ \Omega$$

$$\omega_T = \frac{1}{r_e(C_{B'E} + C_{B'C})}$$

$$= \frac{1}{26(10+1)\times 10^{-12}} = 3.496\,5\ \text{G rad/s}$$

$$f_T = \frac{\omega_T}{2\pi} = 556.485\ \text{MHz}$$

$$\omega_T = \beta_o \omega_\beta$$

$$f_T = \beta_o \times f_\beta$$

$$f_\beta = \frac{f_T}{\beta} = \frac{556.485 \times 10^6}{150}$$

$$= 3.71\ \text{MHz}$$

PROBLEM 10.11 HIGH FREQUENCY

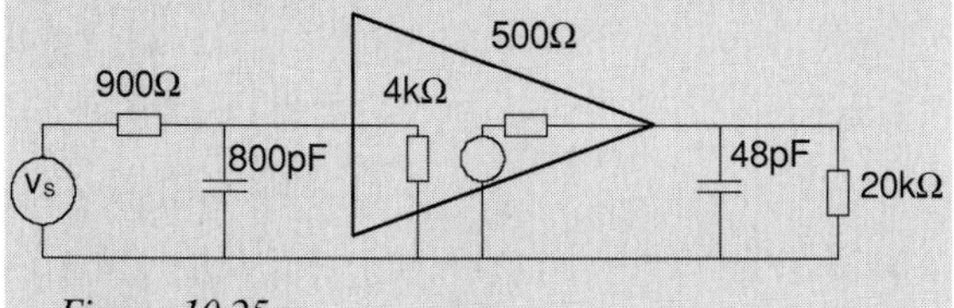

Figure 10.25

Find the upper break frequencies.

Answer: 271 kHz, 6.8 MHz

Input side – equivalent resistance

$$R = 900 \| 4\text{k} = 734.69\ \Omega$$

$$f = \frac{1}{2\pi RC} = \frac{1}{2\pi \times 734.69 \times 800 \times 10^{-12}}$$

$$= 271\ \text{kHz}$$

Output side – equivalent resistance

$$R = 500 \| 20\text{k} = 488\ \Omega$$

$$f = \frac{1}{2\pi RC} = \frac{1}{2\pi \times 488 \times 48 \times 10^{-12}}$$

$$= 6.8\ \text{MHz}$$

PROBLEM 10.12 HIGH FREQUENCY

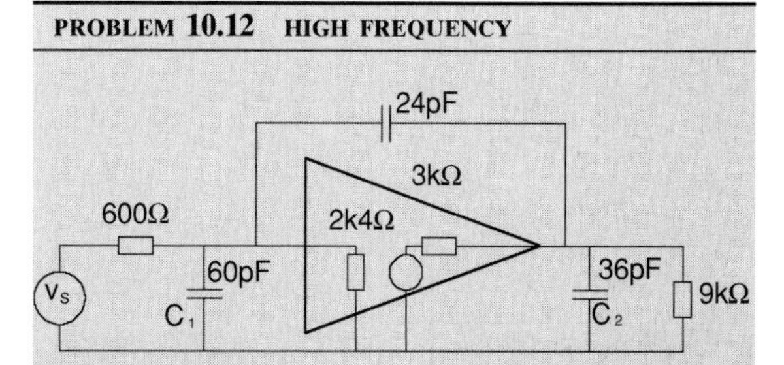

Figure 10.26

The amplifier has a gain of –150. Use Miller's theorem to calculate the input and output upper corner frequencies.

Answer: 90 kHz, 1.179 MHz

Miller effect on 24 pF capacitor

$$C_{M1} = 24(1+150) = 3624 \text{ pF}$$

$$C_{M2} \approx 24 \text{ pF}$$

Input resistance

$$R = 600 \| 2\text{k}4 = 480 \ \Omega$$

Input capacitance

$$C = 3624 + 60 = 3684 \text{ pF}$$

$$f_{1H} = \frac{1}{2\pi RC} = \frac{1}{2\pi \times 480 \times 3864 \times 10^{-12}}$$

$$f_{1H} = 90 \text{ kHz}$$

Output side

$$R = 3\text{k} \| 9\text{k} = 2250 \ \Omega$$

$$C = C_{M2} + C_2 = 24 + 36 = 60 \text{ pF}$$

$$f_{2H} = \frac{1}{2\pi RC} = \frac{1}{2\pi \times 2250 \times 60 \times 10^{-12}}$$

$$f_{2H} = 1.179 \text{ MHz}$$

PROBLEM 10.13 HIGH FREQUENCY

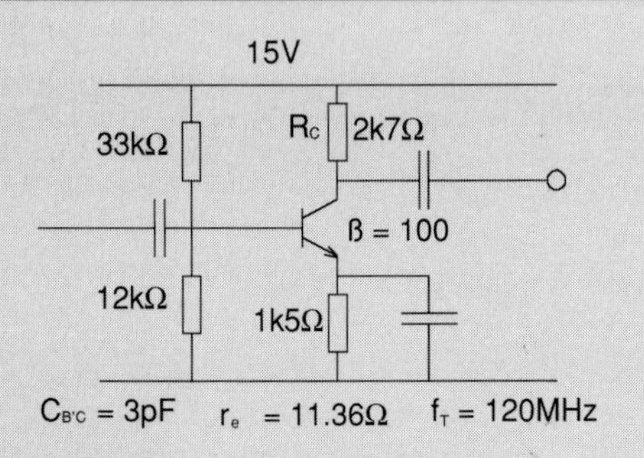

Figure 10.27

Find the voltage gain at 2.5 MHz.

Answer: 205

High frequency model.

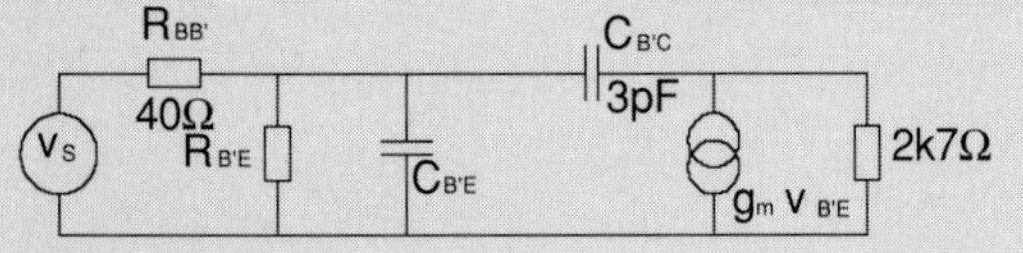

Figure 10.28

$$R_{B'E} = r_e(\beta + 1) = 11.36 \times 101 = 1147.36 \ \Omega$$

Miller

$$C_{M1} = C_{B'C}(1 + g_m R_L)$$

$$= C_{B'C}\left(1 + \frac{R_L}{r_e}\right)$$

$$= 3 \times (1 + 237.68) = 716 \text{ pF}$$

Transition frequency

$$f_T = \frac{1}{2\pi r_e (C_{B'E} + C_{B'C})}$$

$$C_{B'C} + C_{B'E} = \frac{1}{2\pi r_e f_T} = \frac{1}{2\pi \times 11.36 \times 120 \times 10^6}$$

$$= 116.75 \text{ pF}$$

$$C_{B'E} = 116.75 - 3 = 113.75 \text{ pF}$$

$$C = C_{M1} + C_{B'E} = 716 + 113.75 = 829.75 \text{ pF}$$

Equivalent circuit after Thevenin to Norton conversion

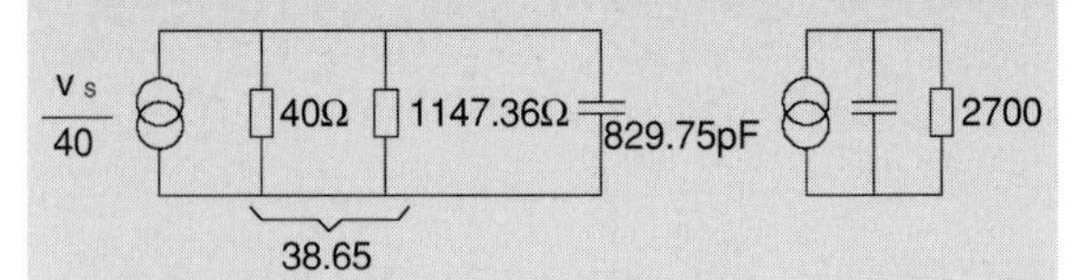

Figure 10.29

Absolute value of admittance in input side

$$|Y| = \sqrt{\left(\left(\frac{1}{38.65}\right)^2 + (2\pi \times 2.5 \times 10^6 \times 829.75 \times 10^{-12})^2\right)}$$

$$= \sqrt{\left(669.4 \times 10^{-6} + 169.88 \times 10^{-6}\right)}$$

$$= \sqrt{\left(839.28 \times 10^{-6}\right)} = 28.97 \text{ mS}$$

$$v_{B'E} = \frac{v_S}{40} \frac{1}{28.97 \times 10^{-3}} = v_S \times 0.863$$

$(|i_C| = g_m v_{B'E}$ and $v_o = |i_C| R_L)$

$$|v_o| = g_m v_{B'E} R_L$$

$$= \frac{1}{11.36} \times v_S \times 0.863 \times 2700$$

$$\left|\frac{v_o}{v_S}\right| = 205$$

PROBLEM 10.14 HIGH FREQUENCY

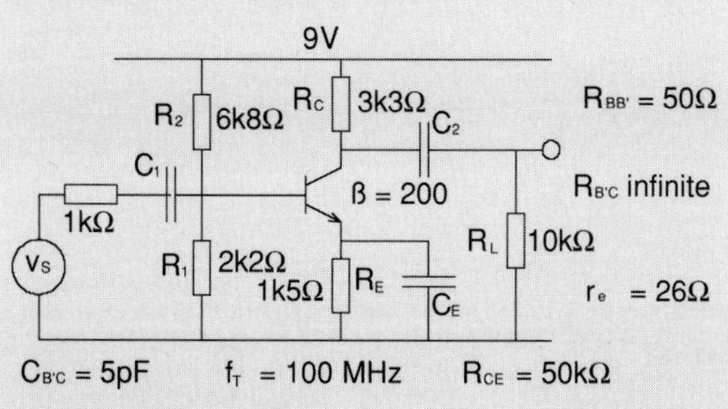

Figure 10.30

Find the voltage gain at 2 MHz. For this assume that R_s is reduced to zero.

Answer: 585.76

High frequency model (R_s = 0 therefore $R_1 \| R_2$ not required)

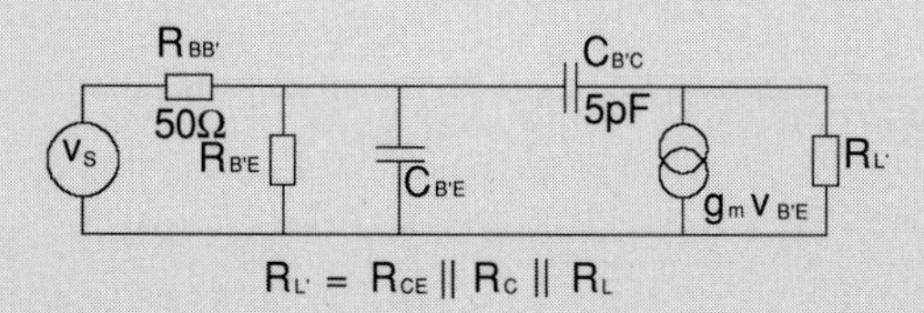

Figure 10.31

$$R_{B'E} = r_e(\beta + 1) = 26 \times 201 = 5226\ \Omega$$

$$R_{L'} = R_{CE} \| R_C \| R_L = 50\text{k} \| 3\text{k}3 \| 10\text{k} = 2364\ \Omega$$

Miller

$$C_{M1} = C_{B'C}(1 + g_m R_{L'})$$

$$= 5\left(1 + \frac{2364}{26}\right) = 459.61\ \text{pF}$$

$$f_T = \frac{1}{2\pi r_e (C_{B'E} + C_{B'C})}$$

$$C_{B'E} + C_{B'C} = \frac{1}{2\pi r_e f_T} = \frac{1}{2\pi \times 26 \times 10^8} = 61.21\ \text{pF}$$

$$C_{B'E} = 61.21 - 5 = 56.21\ \text{pF}$$

$$C = C_{M1} + C_{B'E} = 459.61 + 56.21 = 515.82\ \text{pF}$$

Equivalent circuit after Thevenin to Norton conversion

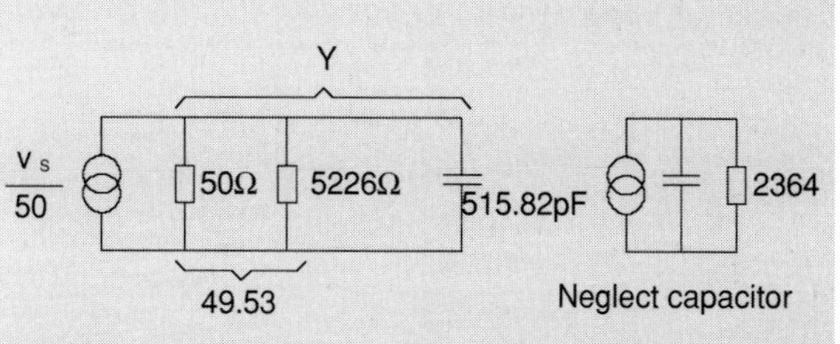

Figure 10.32

$$|Y| = \sqrt{\left(\frac{1}{49.53}\right)^2 + (2\pi \times 2 \times 10^6 \times 515.82 \times 10^{-12})^2}$$

$$|Y| = 21.205 \times 10^{-3}\ \text{S}$$

Voltages

$$v_{B'E} = \frac{\frac{v_S}{50}}{21.205 \times 10^{-3}} = v_S \times 0.9432$$

$$(|i_C| = g_m v_{B'E} \text{ and } v_o = i_C R_{L'})$$

$$|v_o| = g_m v_{B'E} R_{L'}$$

$$= \frac{1}{26} \times v_S \times 0.9432 \times 2364$$

$$= 85.76 v_S$$

Therefore

$$\left|\frac{v_o}{v_S}\right| = 85.76$$

PROBLEM 10.15 HIGH FREQUENCY

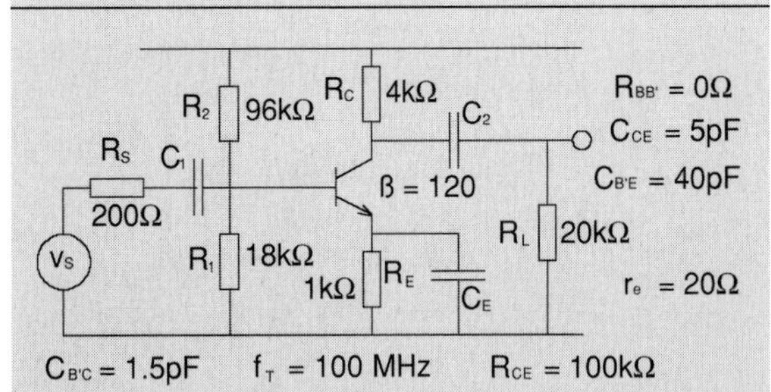

Figure 10.33

Find f_β and high frequency cut-off due to base and collector circuits.

Answer: 3.093 MHz, 6.366 MHz and 0.833 MHz

Midband gain

$$A_V = -\frac{R_C || R_L || R_{CE}}{r_e}$$

$$= -\frac{4k||20k||100k}{20} = -161.29$$

AC model

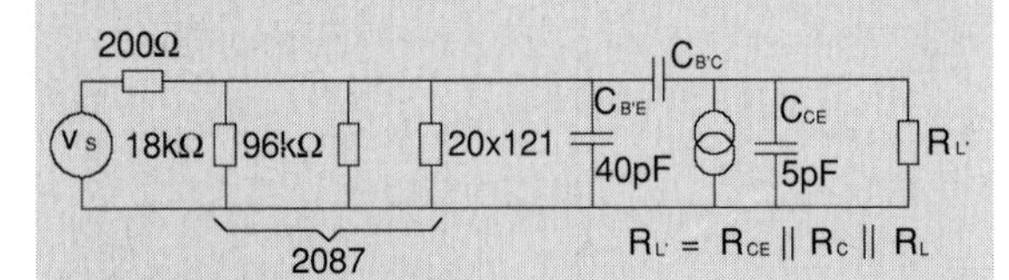

Figure 10.34

Miller's theorem

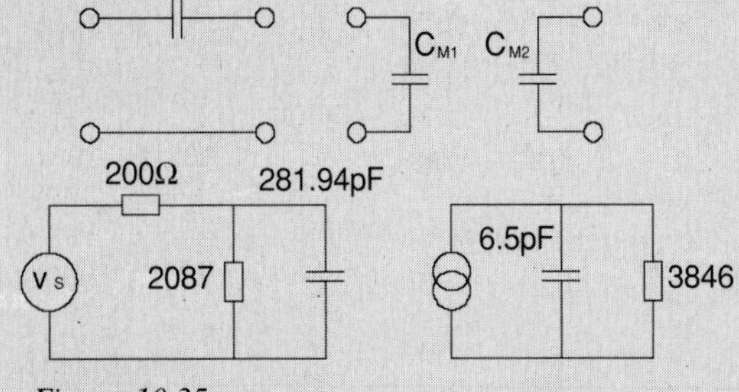

Figure 10.35

$$C_{m1} = 1.5(1 + 161.29) = 241.94 \text{ pF}$$

$$C_{m2} \approx 1.5 \text{ pF}$$

Thevenin to Norton

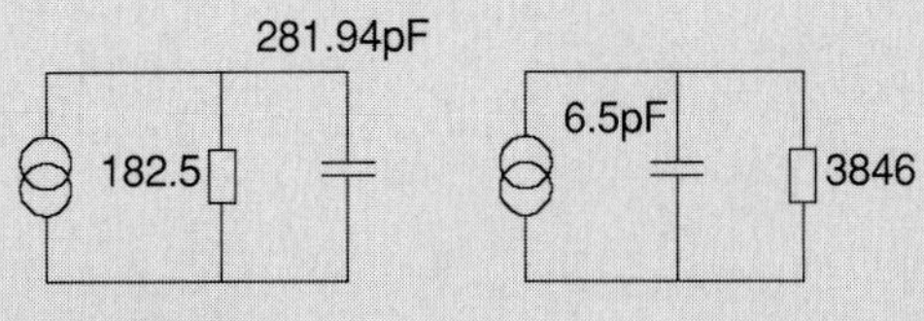

Figure 10.36

$$f_{1H} = \frac{1}{2\pi RC} = \frac{1}{2\pi \times 182.5 \times 281.94 \times 10^{-12}}$$

$$= 3.093 \text{ MHz}$$

$$f_{2H} = \frac{1}{2\pi RC} = \frac{1}{2\pi \times 3846 \times 6.5 \times 10^{-12}}$$

$$= 6.366 \text{ MHz}$$

Bandwidth f_β

$$f_\beta = \frac{f_T}{\beta} = \frac{100}{120} = 0.833 \text{ MHz}$$

PROBLEM 10.16 HIGH FREQUENCY

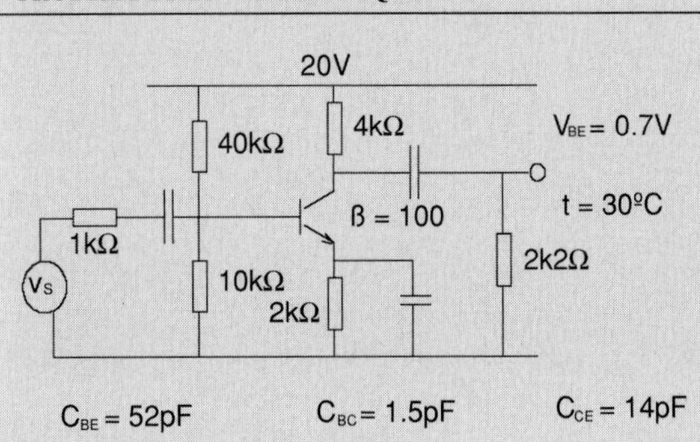

Figure 10.37

Find the corner frequencies at input and output and f_β.

Answer: 1.479 MHz, 7.236 MHz and 1.888 MHz

Approximate method

$$V_B = 20\frac{10}{10 + 40} = 4 \text{ V}$$

$$V_E = 4 - 0.7 = 3.3 \text{ V}$$

$$I_E = \frac{3.3}{2k} = 1.65 \text{ mA}$$

$$r_e = \frac{26}{1.65} = 15.76\ \Omega$$

Transistor gain (not overall gain!)

$$A_V = -\frac{R_C \| R_L}{r_e} = -\frac{4k\|2k2}{15.76}$$

$$= -90.06$$

AC model

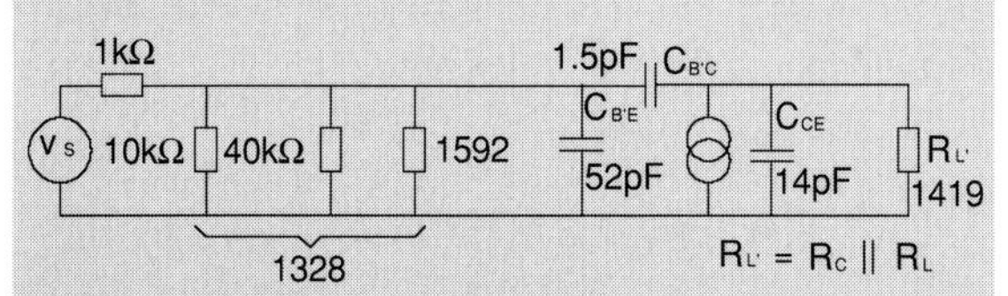

Figure 10.38

Miller's theorem

$$C_{M1} = 1.5(1+90.06) = 136.59 \text{ pF}$$

$$C_{M2} \approx 1.5 \text{ pF}$$

New circuit

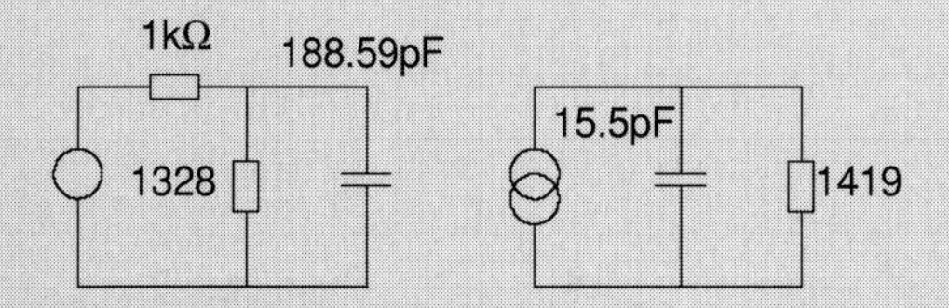

Figure 10.39

Input

$$f_{1H} = \frac{1}{2\pi RC} = \frac{1}{2\pi(1k\|1328)\times 188.59\times 10^{-12}}$$

$$= \frac{1}{2\pi\times 570.45\times 188.59\times 10^{-12}} = 1.479 \text{ MHz}$$

Output

$$f_{2H} = \frac{1}{2\pi RC} = \frac{1}{2\pi\times 1419\times 15.5\times 10^{-12}}$$

$$= 7.236 \text{ MHz}$$

$$f_\beta = f_T\frac{1}{\beta}$$

$$= \frac{1}{\beta}\frac{1}{2\pi r_e(C_{B'C}+C_{B'E})}$$

As we don' t have values for either $C_{B'C}$ or $C_{B'E}$, we assume that they are the same as C_{BC} and C_{BE}.

$$f_\beta = \frac{1}{\beta}\frac{1}{2\pi\times 15.76\times 53.5\times 10^{-12}}$$

$$= 1.888 \text{ MHz}$$

and

$$f_T = f_\beta\beta = 1.888\times 10^6\times 100 = 188.8 \text{ MHz}$$

PROBLEM 10.17 HIGH FREQUENCY

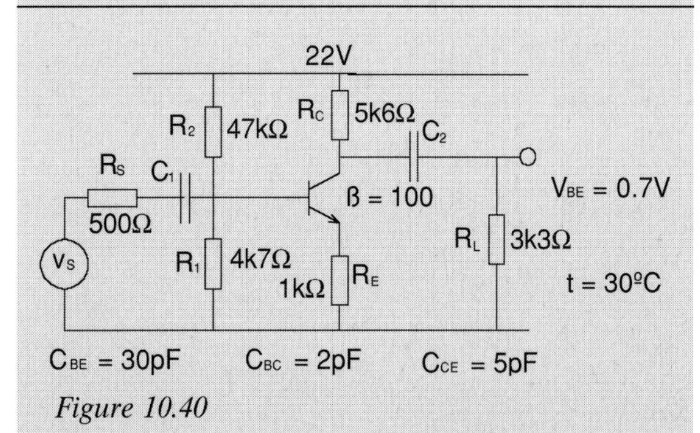

Figure 10.40

Find the input lower and higher corner frequency. $C_1 = 1\ \mu\text{F}$

Answer: 34.58 Hz and 9.898 MHz

DC conditions – approximate method

$$V_B = 22\frac{4.7}{4.7+47} = 22\frac{4.7}{51.5} = 2 \text{ V}$$

$$V_E = 2 - 0.7 = 1.3 \text{ V}$$

$$I_E = \frac{1.3}{1k} = 1.3 \text{ mA}$$

$$r_e = \frac{26}{1.3} = 20\ \Omega$$

Lower corner frequency

$$R = 500 + 4k7\|47k\|(101)(20+1k)$$

$$= 500 + 4k7\|47k\|103020 = 4602\ \Omega$$

$$f_{1L} = \frac{1}{2\pi RC} = \frac{1}{2\pi\times 4602\times 10^{-6}}$$

$$= 34.58 \text{ MHz}$$

Voltage gain

$$A_V = -\frac{R_C\|R_L}{r_e + R_E} = -\frac{5k6\|3k3}{1020}$$

$$= -\frac{2076.4}{1020} = -2.04$$

AC model

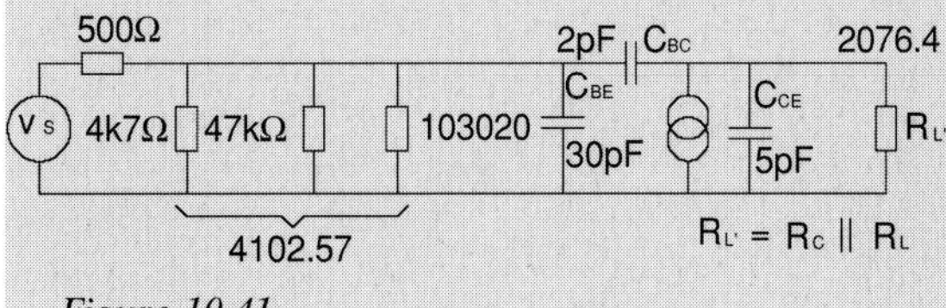

Figure 10.41

Miller's theorem

$$C_{M1} = 2(1 + 2.04) = 6.08 \text{ pF}$$

$$C_{M2} = 2\left(1 + \frac{1}{2.04}\right) = 2.98 \text{ pF}$$

$$R_{IN} = 500 \| 4102.57 = 445.68\ \Omega$$

$$C_{IN} = C_{BE} + C_{M1} = 30 + 6.08 = 36.08 \text{ pF}$$

$$f_{1H} = \frac{1}{2\pi RC} = \frac{1}{2\pi \times 445.68 \times 36.08 \times 10^{-12}}$$

$$= 9.898 \text{ MHz}$$

PROBLEM 10.18 HIGH FREQUENCY

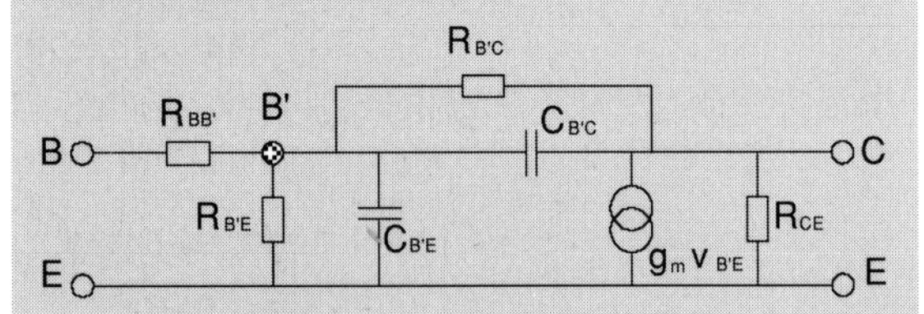

Figure 10.42

Determine the hybrid π model parameters for a transistor operating with a quiescent current of 2.6 mA, at a temperature of 300°K and with the following parameters.

$h_{ir} = 2k2\ \Omega$	$h_{fe} = 200$
$h_{re} = 10^{-4}$	$h_{oe} = 20\ \mu s$
$f_f = 300$ MHz	$C_{B'C} = 4$ pF

Answer:	$R_{BB'} = 190\ \Omega$	$R_{B'E} = 2010\ \Omega$
	$R_{B'C} = 20.097\ \Omega$	$R_{CE} = 50\ k\Omega$
	$C_{B'E} = 49$ pF	$C_{B'C} = 4$ pF

Little r_e

$$r_e = \frac{26}{2.6} = 10\ \Omega$$

$$g_m = \frac{1}{r_e} = 0.1 \text{ S}$$

$$R_{CE} = \frac{1}{h_{oe}} = \frac{1}{20 \times 10^{-6}} = 50 \text{ k}\Omega$$

$$h_{ie} = B_{BB'} + r_e(1 + h_{fe}) \quad (R_{B'E} = r_e(1 + h_{fe}))$$

$$2200 = R_{BB'} + 10(1 + 200)$$

$$2200 = R_{BB'} + 2010$$

$$R_{BB'} = 190\ \Omega \qquad R_{B'E} = 2010$$

$$h_{re} = \frac{R_{B'E}}{R_{B'E} + R_{B'C}} \qquad \text{(potential divider)}$$

$$10^{-4} = \frac{2010}{2010 + R_{B'C}}$$

$$R_{B'C} = 20.097 \text{ M}\Omega$$

Capacitances

$$f_T = \frac{g_m}{2\pi(C_{B'E} + C_{B'C})}$$

$$300 \text{ MHz} = \frac{1}{10 \times 2\pi \times C_T}$$

$$C_T = \frac{1}{300 \times 10^6 \times 10 \times 2\pi}$$

$$= \frac{1000 \times 10^{-12}}{3 \times 2 \times \pi} = 53 \text{ pF}$$

$$C_{B'C} = 4 \text{ pF}$$

$$C_{B'E} = 53 - 4 = 49 \text{ pF}$$

PROBLEM 10.19 HIGH FREQUENCY

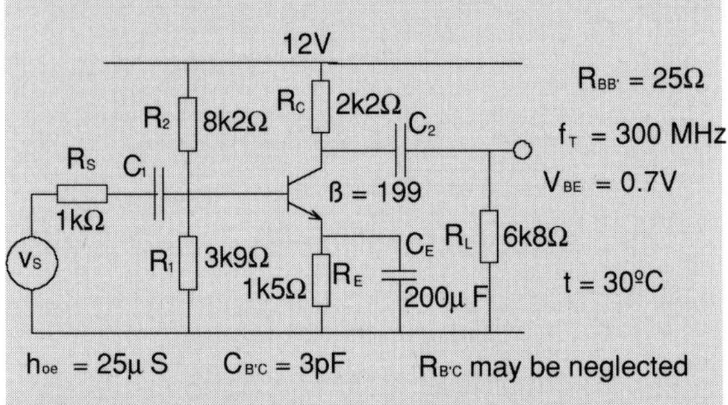

Figure 10.43

Find the voltage gain at 10 MHz.

Additional data:

$\beta = 199$	$h_{oe} = 25\ \mu s$
$R_{BB'} = 25\ \Omega$	$C_{B'C} = 3$ pF
$f_T = 300$ MHz	$R_{B'C}$ may be neglected
$V_{BE} = 0.7$ V	t = 30°C

Answer: 4.4

DC conditions – (Q values)

$$V_B = 12\frac{3.9}{3.9+8.2} = 3.87\ \text{V}$$

$$V_E = 3.87 - 0.7 = 3.17\ \text{V}$$

$$I_E = \frac{3.17}{1k5} = 2.11\ \text{mA}$$

$$r_e = \frac{26}{2.11} = 12.3\ \Omega$$

AC model

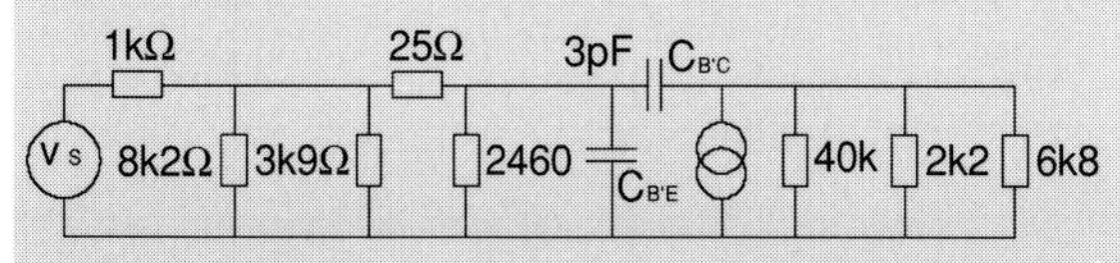

Figure 10.44

$$C_{B'E} = \frac{1}{2\pi f_T r_e} - C_{B'C}$$

$$= \frac{1}{2\pi \times 300 \times 10^6 \times 12.3} - 3\ \text{pF} = 40.13$$

Miller's theorem

$$C_{M1} = 3 \times (1 + g_m R_{L'})$$

$$R_{L'} = 40\text{k}||2\text{k}2||6\text{k}8 = 1596\ \Omega$$

$$C_{M1} = 3 \times 10^{-12}\left(1 + \frac{1596}{12.3}\right) = 392.3\ \text{pF}$$

$$C = C_{M1} + C_{B'E} = 392.3 + 40.13 = 432.43\ \text{pF}$$

New circuit

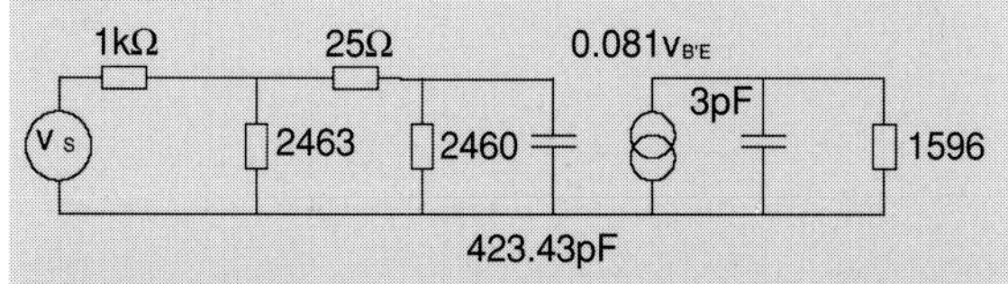

Figure 10.45

Thevenin to Norton

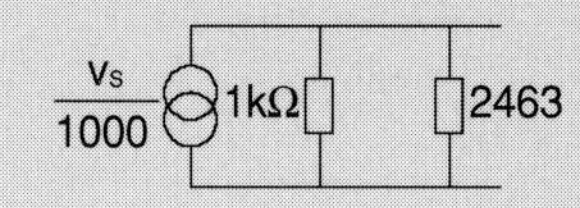

Figure 10.46

Norton to Thevenin

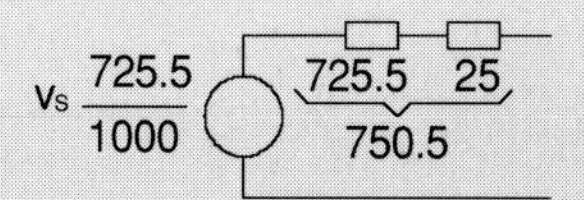

Figure 10.47

Thevenin to Norton

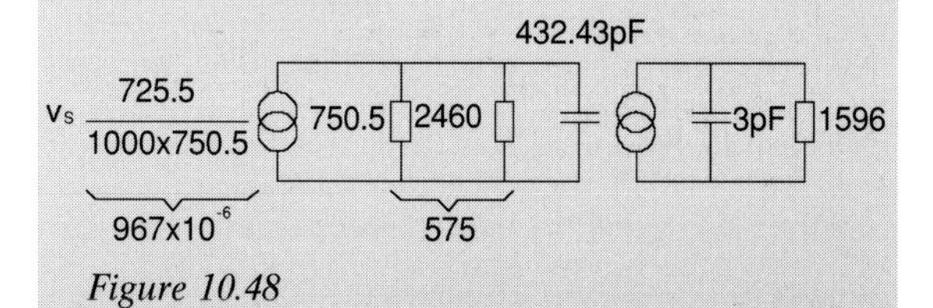

Figure 10.48

Magnitude of Y_1 at 10 MHz

$$|Y_1| = \sqrt{\left(\frac{1}{575}\right)^2 + (2\pi \times 10^7 \times 432.43 \times 10^{-12})^2}$$

$$|Y_1| = \sqrt{3.62 \times 10^{-6} + 738.23 \times 10^{-6}}$$

$$= \sqrt{741.25 \times 10^{-6}} = 27.226 \times 10^{-3}\ \text{S}$$

$$|v_{B'E}| = \frac{967 \times 10^{-6}}{27.226 \times 10^{-3}} = 35.52 \times 10^{-3} v_s$$

Magnitude of Y_2 at 10 MHz

$$|Y_2| = \sqrt{\left(\frac{1}{1596}\right)^2 + (2\pi \times 10^7 \times 3 \times 10^{-12})^2}$$

$$= \sqrt{0.392 \times 10^{-6} + 0.0355 \times 10^{-6}}$$

$$= \sqrt{0.4275 \times 10^{-6}} = 653.83 \times 10^{-6}\ \text{S}$$

Therefore

$$\frac{v_o}{v_s} = \frac{35.53 \times 10^{-3} \times 0.081}{653.83 \times 10^{-6}}$$

$$= 4.4$$

If the 3 pF in the output circuit is ignored the result is 4.59 instead of 4.4. This approximation can be made without affecting the result too much.

PROBLEM 10.20 HIGH FREQUENCY

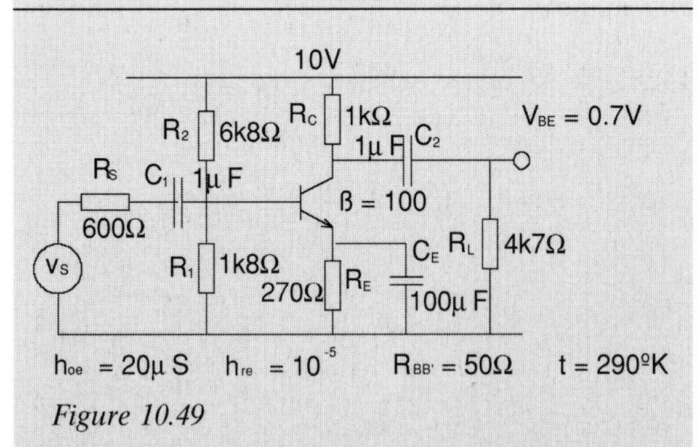

Figure 10.49

Find r_e, the voltage gain at midband frequency (5 kHz) and the voltage gain ay 5 MHz assuming f_T = 250 MHz and $C_{B'C}$ = 4 pF.

Answer: 4.85 Ω, –59.86 and 9.78

DC conditions – approximate method

$$V_B = 10\frac{1.8}{1.8+6.8} = 2.093 \text{ V}$$

$$I_{EQ} = \frac{V_B - V_{BE}}{R_E} = \frac{2.093-0.7}{270} = 5.159 \text{ mA}$$

$$r_e = \frac{25}{5.159} = 4.85\ \Omega$$

$$h_{ie} = R_{BB'} + r_e(1+h_{fe})$$

$$= 50 + 4.85(1+100) = 540\ \Omega$$

Voltage gain midband – AC conditions

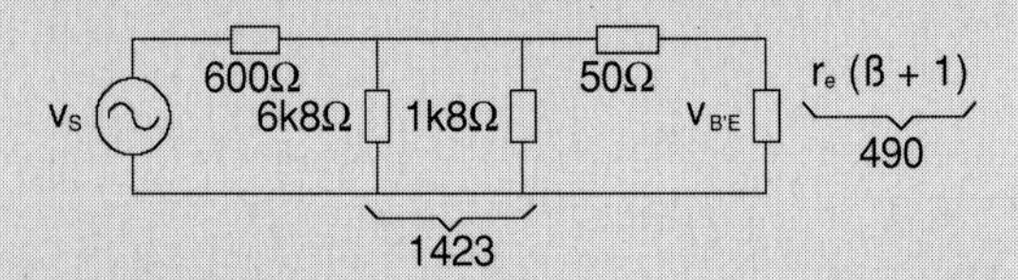

Figure 10.50

Thevenin to Norton

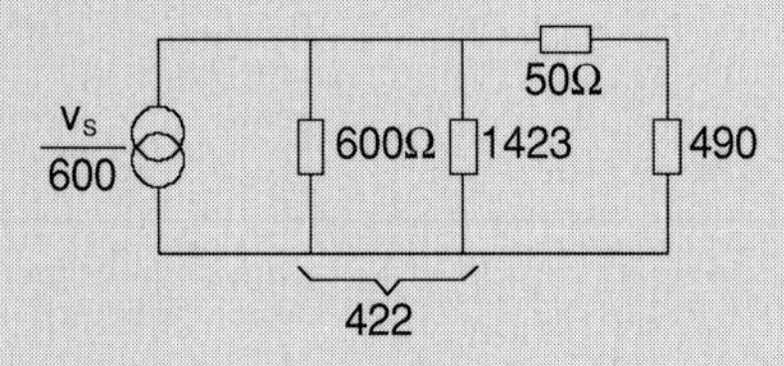

Figure 10.51

Norton to Thevenin

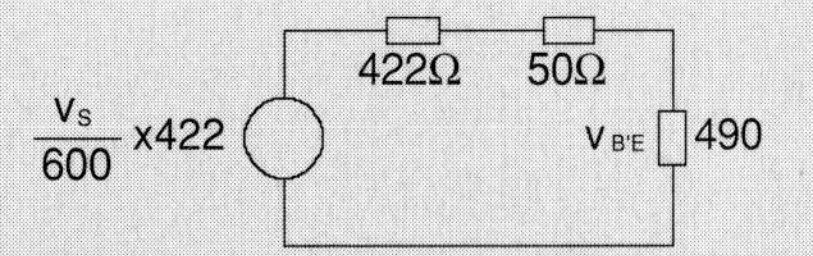

Figure 10.52

$$v_{B'E} = \frac{v_S}{600} \times 422 \frac{490}{490+50+422}$$

$$= 0.358 v_S$$

Output side

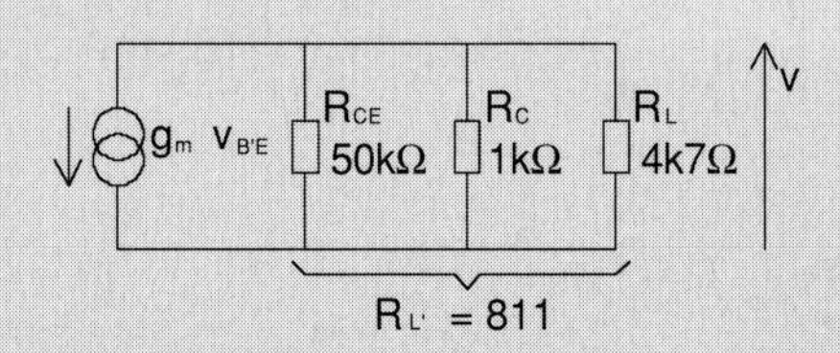

Figure 10.53

$$R_{CE} = \frac{1}{h_{oe}} = \frac{10^6}{20} = 50 \text{ k}\Omega$$

Voltage amplification

$$v_{B'E} = 0.358 v_S$$

$$g_m v_{B'E} = \frac{0.358 v_S}{4.85}$$

$$v_{OUT} = -g_m v_{B'E} R_L = -v_S \frac{0.358}{4.85} \times 811$$

$$= -59.86 v_S$$

$$\frac{v_{OUT}}{v_S} = -59.86$$

Voltage gain at 5 MHz

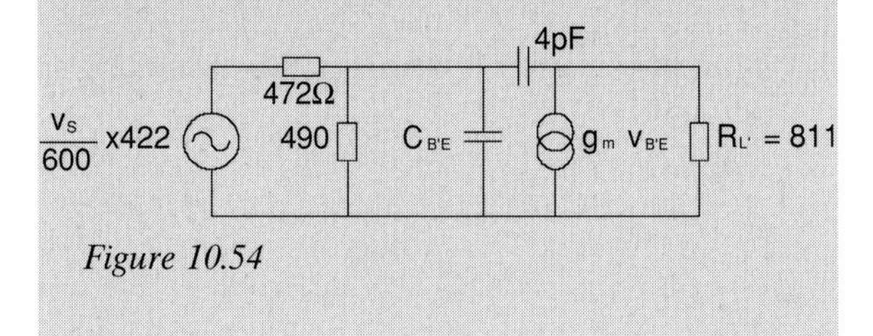

Figure 10.54

$C_{B'E}$ is unknown, but we know f_T.

The voltage gain from $v_{B'E}$ to $-g_m v_{B'E} R_L$, is $-g_m R_L$

$$f_T = \frac{1}{2\pi r_e (C_{B'E} + C_{B'C})}$$

$$C = \frac{1}{2\pi \times 4.85 \times 250 \times 10^6} = 131 \text{ pF}$$

$$C_{B'E} = 131 - 4 = 127 \text{ pF}$$

Miller's theorem

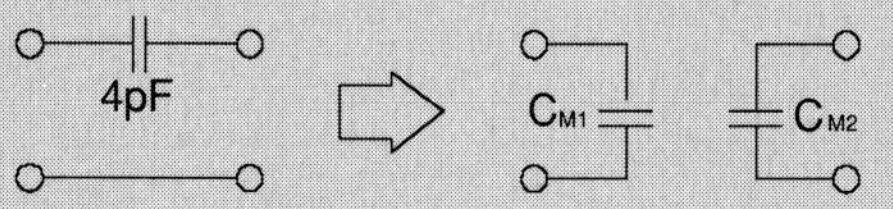

Figure 10.55

$$C_{M2} \approx 4 \text{ pF} \qquad C_{M1} = 4\,(1 - \text{gain})$$

$$\text{gain} = -g_m R_L = -\frac{811}{4.85} = -167.22$$

$$C_{M1} = 4 \times 168.22 = 673 \text{ pF}$$

We now divide the circuit and we change the circuit from Thevenin to Norton.

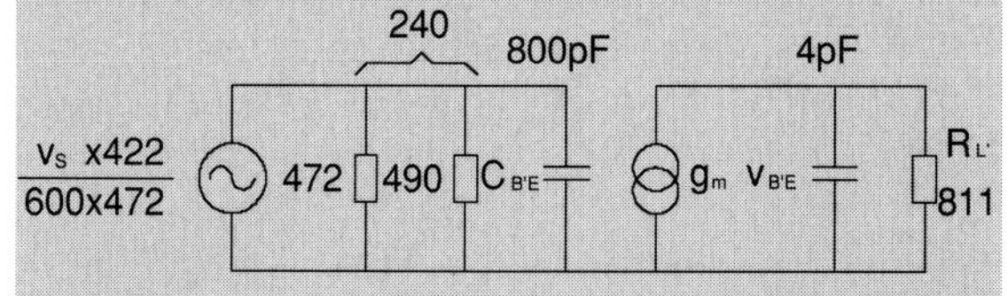

Figure 10.56

We work out the magnitude of Y at 5 MHz

$$Y = \frac{1}{R} + j\omega C$$

$$|Y| = \sqrt{\left(\frac{1}{R}\right)^2 + (\omega C)^2}$$

$$= \sqrt{17.36 \times 10^{-6} + (2\pi 5 \times 10^6 \times 800 \times 10^{-12})^2}$$

$$= 25.48 \times 10^{-3} \text{ S (siemens)}$$

$$v_{B'E} = \frac{v_S}{600} \frac{422}{472} \frac{1}{|Y|}$$

$$\frac{v_{B'E}}{v_S} = \frac{422}{600} \frac{1}{472} \frac{10^3}{25.48} = 58.48 \times 10^{-3}$$

Output side

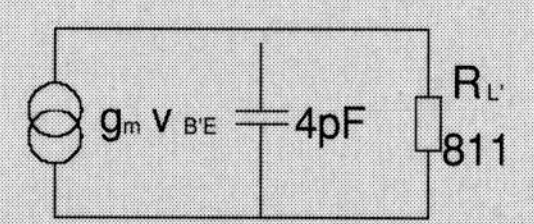

Figure 10.57

Ignoring the capacitor 4 pF

$$v_{OUT} = v_S \times \frac{58.48}{4.85} \times 811 \times 10^{-3}$$

$$\left|\frac{v_{OUT}}{v_S}\right| = 9.78$$

11

Operational amplifiers: general

Properties

a) Very high open loop voltage gain which is about 100 000 for DC and low frequency AC. The gain decreases as the frequency increases.
b) Very high input impedance, which is about 10^8 to 10^{10} Ω, so that the current taken from the source is minute. The input voltage is passed to the Op-Amp with little loss.
c) Very low output impedance, which is around 100 Ω which results in an efficient transfer of input voltage to any load greater than a few kΩs.

Physical details

With reference to the popular 741 Op-Amp we can say that it has one output and two inputs. The non-inverting input is marked +, and the inverting input is marked – as shown in Figure 11.1. Its

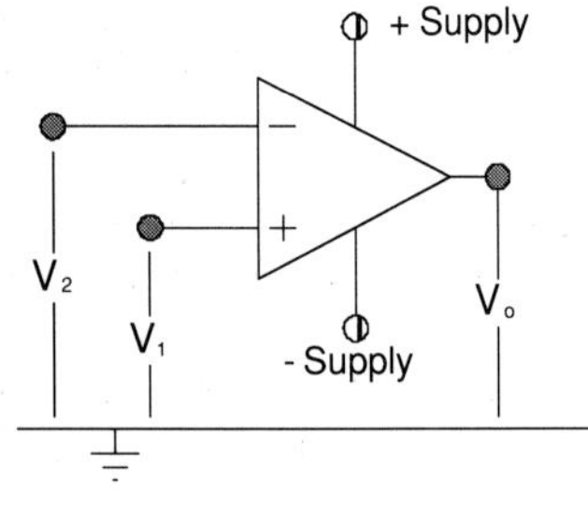

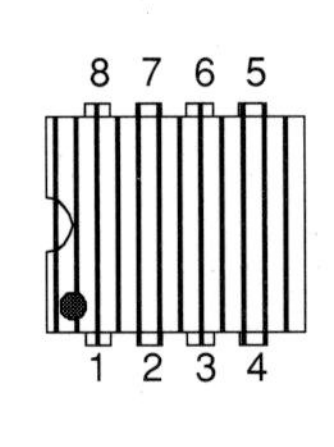

Figure 11.1

operation is convenient, from a dual balanced power supply giving equal positive and negative voltages V_S in the range of ±5 V to ±15 V. The 0 V is the centre point of the power supply and is common to the input and output circuits and is usually the voltage reference point.

Two important consequences

- The difference of voltage at the input is approximately zero.
- The current flowing into or out of the input is approximately zero.

These two consequences are very important and will be used as a technique to solve most of the Op-Amp problems that will follow.

The first consequence can be stated as $V_+ = V_-$. This is sensible due to the large amplification of the Op-Amp. If we have an output of 5 V and the amplification is 100 000, it means that the input must be 0.000 05 V. This quantity being so small can be approximated to 0 V. This is indeed an approximation, but a very useful one.

The second consequence can be stated as $I_+ = I_- = 0$. That is to say, there are no currents going in or out of the Op-Amp inputs. This is indeed an approximation, but a very useful one.

Inverting amplifier

The inverting Op-Amp is shown in Figure 11.2. The input voltage is applied to resistor R_1 to the inverting input V_-. The output will have a – sign,

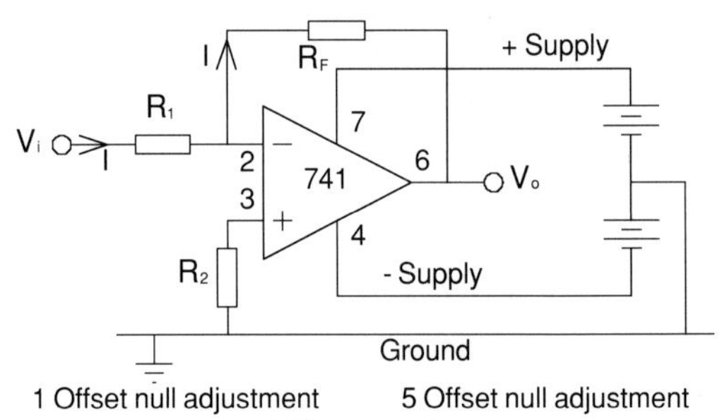

Figure 11.2

that is to say it will be in antiphase to the input. The non-inverting input is connected to 0 V.

This configuration is a closed loop, the output is connected to the input V_- through the resistor R_F which feeds back some of the output and is called the negative feedback resistor.

As we mentioned before, V_- is at the same potential as V_+ and is called the virtual earth. Under these conditions, the input voltage is applied to resistor R_1 and the voltage across R_F is the ouput voltage V_o.

When V_i is positive, the current I flows through R_1 and as we said before, due to the high impedance at the Op-Amp input we assume that all the current goes through R_F. This is a very useful approximation.
So

$$I = \frac{V_i}{R_1} = -\frac{V_o}{R_F}$$

The closed loop gain is

$$A = \frac{V_o}{V_i} = -\frac{R_F}{R_1}$$

This brings us the normal formula

$$\boxed{V_o = -V_i \frac{R_F}{R_1}}$$

Another circuit of the inverting Op-Amp, where you can see the potential divider effect better is shown in Figure 11.3.

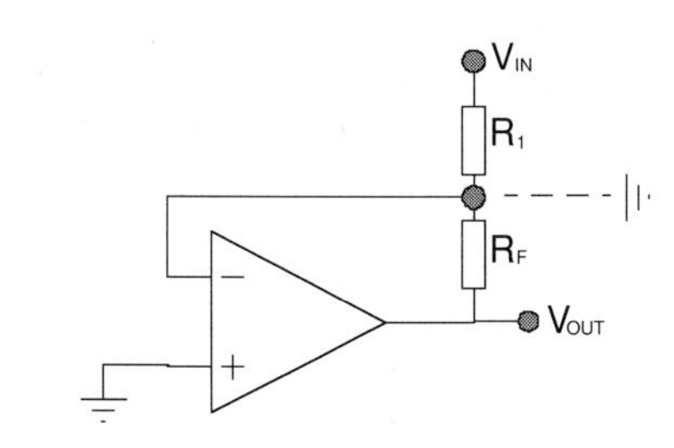

Figure 11.3

Offset voltage

In practical terms, when there is no signal present, there may be a slight quiescent voltage present. This is called the differential input offset voltage. Although this input signal could only have a value of around 1 mV, due to the high amplification of the Op-Amp, it would be unacceptable in most cases.

The 741 Op-Amp has a correction circuit between pin 1 and 5, the offset null adjustment. A potentiometer of 10 KΩ can be connected to pins 1 and 5 with the centre point connected to the negative of the supply. The potentiometer is adjusted to give zero output when the input is zero volts.

For AC operation, a capacitor at the output removes any DC voltage arising from the input offset.

Output current

A typical Op-Amp can supply a maximum output current of 5 mA, therefore the minimum load that can be fed is roughly 2 kΩ, on an operation of ±10 V. If a greater input current is required, an emitter follower output stage or an Op-Amp current booster can be used.

Input bias current

Although we have approximated the input currents of the Op-Amp to zero and accepted this as an approximation, the Op-Amp will not work if there isn't a current flowing into or out of the inputs of the amplifier. The input currents are called the input bias currents and are of the order of 0.1 μA in the case of the 741. As the Op-Amp is symmetrical, the current is normally balanced out.

In Figure 11.2 we introduced R_2 as an input balancing resistor. It will not affect the formula for the inverting Op-Amp, but it will improve the performance. R_2 is calculated to be equivalent to the load on the $V-$ input, that is to say, R_1 and R_F in parallel. If the source impedance is significant, this is added to R_1.

Negative feedback

Op-Amps can use negative feedback obtained by feeding back some of the ouput to the inverting input. The part of the output fed back to the input produces a voltage at the output that opposes the one from which it is taken, thereby reducing the new output of the amplifier. The resulting closed loop gain A is then less than the open loop gain A_o. In compensation, a wider range of values of voltage can be applied to the input for amplification. The opposite would be the result if the feedback is applied to the non-inverting input (positive feedback). Negative feedback also gives greater stability, less distortion and increased bandwidth.

Non-inverting amplifier

The non-inverting Op-Amp is shown in Figure 11.4. The formula for the output of this amplifier is:

$$V_o = V_i\left(1+\frac{R_2}{R_1}\right)$$

This circuit cannot be used as an attenuator, unlike the inverting amplifier. The input impedance is very high, approximately equal to the Op-Amp input impedance.

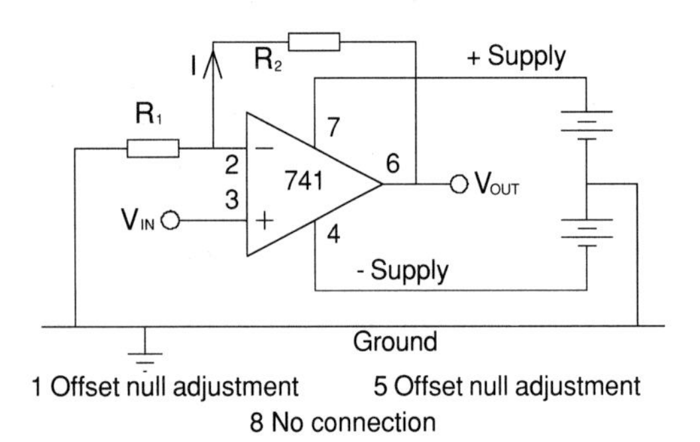

Figure 11.4

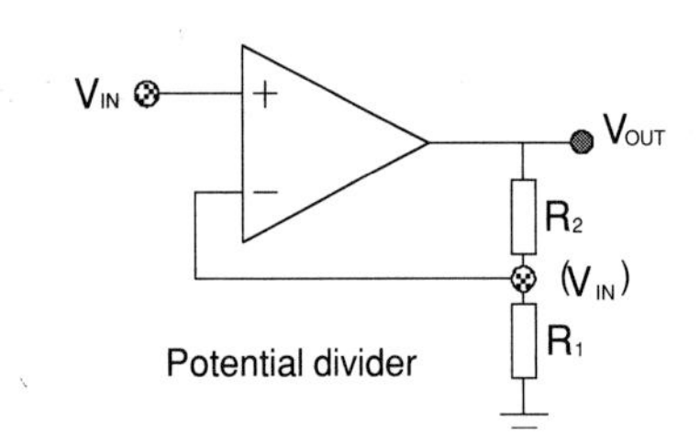

Figure 11.5

Figure 11.5 shows the circuit redrawn to show the potential divider effect of the non-inverting Op-Amp.

Differential Op-Amp

The circuit shown in Figure 11.6, shows a differential Op-Amp with inputs V_1 and V_2. These correspond to the – and + inputs of the Op-Amp. Transistors T_1 and T_2 form the differential part of the circuit and, as shown, both inputs are identical. One output will be inverted and the other non–inverted, but that part of the circuit is not shown.

In a simpler form the emitters of T_1 and T_2 are joined together and through a common resistor R_E go to the negative rail. In Figure 11.6, however, we use a constant current generator to supply a constant current to transistors T_1 and T_2. The constant current circuit is formed by transistor T_3 which also has temperature compensation by the addition of D_1. As the temperature varies T_3 and D_2 vary in the same amount. They can be hand picked to behave as equally as possible.

V_{R1} is a potentiometer to compensate for the manufacturing differences between T_1 and T_2.

An Op-Amp like the 741 will have a differential input like the one in Figure 11.6, plus several other stages to amplify the signals to the required

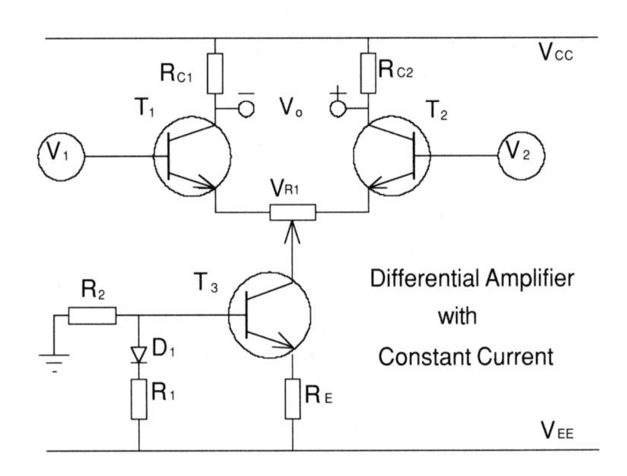

Figure 11.6

level, to shift the level of DC to allow an equal swing of the output, to bias the different transistors, to protect the output circuits, etc.

In the circuit of Figure 11.6, V_1 is compared with V_2. If both inputs are equal, both transistors will be conducting equally and the voltage difference at the output should be zero.

As soon as one voltage is larger than the other, one transistor will conduct more than the other and as there is a constant current source, as one transistor takes more current the other will have to take less. The larger output then, will prevail.

Positive feedback

The principle of positive feedback can be understood with reference to the sequence of three sketches shown in Figures 11.7–11.9.

The first, Figure 11.7, shows an Op-Amp with an input to the inverting input. The positive feedback is done by R_2 which feeds a fraction of the output to the non–inverting input. The fraction is $R_1/(R_1 + R_2)$.

We assume an ideal case in Figure 11.7. There is no input, i.e. 0 V input. We also show a 0 V output. This is only theoretically possible because in reality, any difference in the input will produce a small output. This small output will feedback a signal which will increase the previous difference and the output will continue to grow, the feedback will become larger until the Op-Amp is saturated.

Normally the difference between V_+ and V_- is very small. Due to the large amplification of the Op-Amp a very small input is sufficient to produce the required output. When the Op-Amp reaches saturation, this is not the case, as we will see.

Figure 11.8 shows the first case. We know that with zero input and zero output the Op–Amp with positive feedback is unstable. We assume here that the first difference was positive. As we described earlier, the feedback will help the signal and the output will continue to grow until it reaches the maximum. This maximum is a little less than the rails. The circuit settles at 14 V output, 7 V input which is the positive feedback from the potential divider. In order to change the output from positive to negative, the input has to grow to over 7 V.

Figure 11.9, shows the second case, when without an input the tiny difference at the input goes negative. As we already know, the Op-Amp with positive feedback is unstable and it will immediately reach saturation, this time negative saturation.

The circuit settles with an output of –14 V. The negative rail is –15 V. The feedback is –7 V. At the input we have 7 V with the polarity as shown. The V_- side is higher and the output will be inverted, i.e., negative. In order to change the ouput from negative to positive the input voltage V_i will have to be lower than –7 V. All the Op-Amp can do is to change from positive to negative saturation. It behaves in a digital manner with two states. An interesting analogy is the common light switch which rocks from one position to the other and it is unstable in the middle. If you want to change it from one position to the other you have to overcome the latch, but once it goes, it goes. The analogy is a good one.

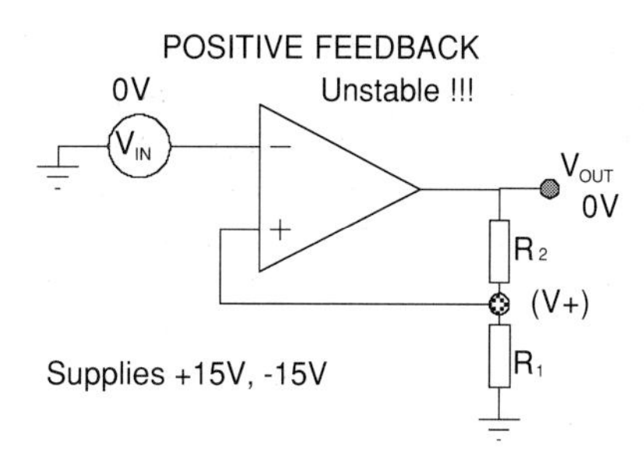

Figure 11.7

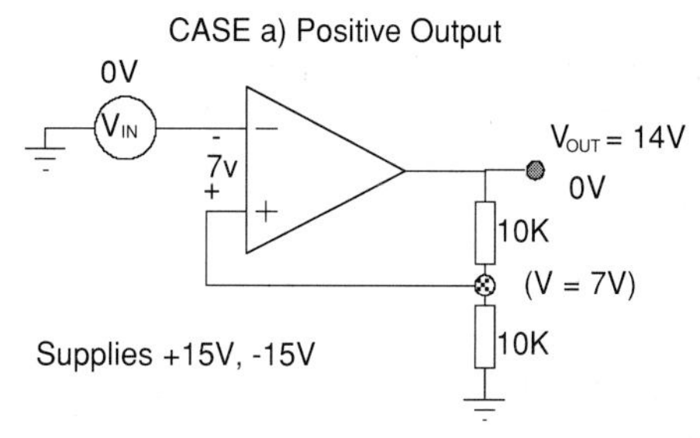

Figure 11.8

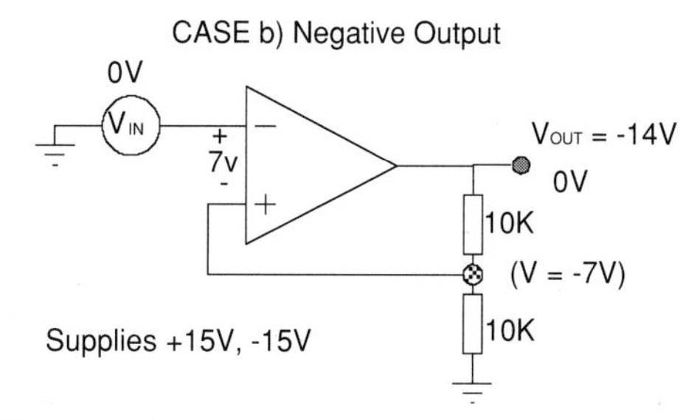

Figure 11.9

PROBLEM 11.1 OP-AMP GENERAL

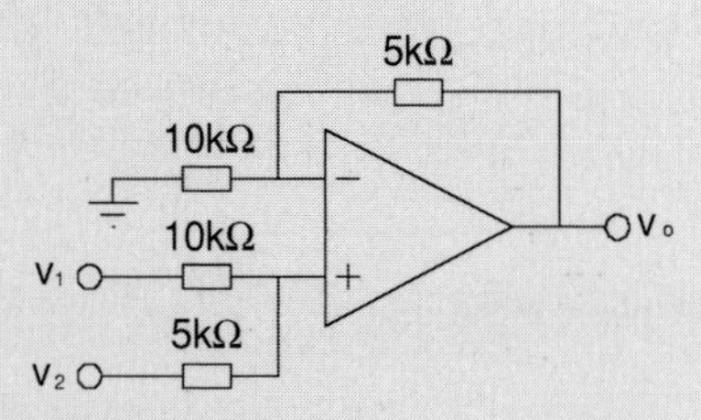

Figure 11.10

Find v_o in terms of v_1 and v_2.

$$\text{Answer: } v_0 = \frac{v_1}{2} + v_2$$

Superposition. Set $v_1 = 0$ and calculate v_0 due to v_2 only.

$$v_+ = v_2 \frac{10}{15}$$

$$v_- = v_+ = v_2 \frac{2}{3}$$

$$i = \frac{v_-}{10\text{k}} = \frac{v_2}{10\text{k}} \frac{2}{3}$$

$$v_{01} = v_- + i \times 5\text{k}$$

$$= \left(v_2 \frac{2}{3}\right) + \left(\frac{v_2}{10\text{k}} \frac{2}{3} \times 5\text{k}\right) = v_2$$

Alternatively we apply the formula

$$v_{01} = v_-\left(1 + \frac{5\text{k}}{10\text{k}}\right) = 1.5 v_-$$

$$= 1.5\, v_2 \frac{2}{3} = v_2$$

Set $v_2 = 0$. Calculate v_0 due to v_1 alone (second part of superposition)

$$v_+ = v_1 \frac{5}{15}$$

$$v_- = v_+ = v_1 \frac{1}{3}$$

$$v_{02} = v_-\left(1 + \frac{5\text{k}}{10\text{k}}\right) = v_1\left(\frac{1}{3}\right) \times 1.5 = v_1\left(\frac{1}{2}\right)$$

Total solution

$$v_0 = v_{01} + v_{02} = v_2 + \frac{v_1}{2}$$

PROBLEM 11.2 OP-AMP GENERAL

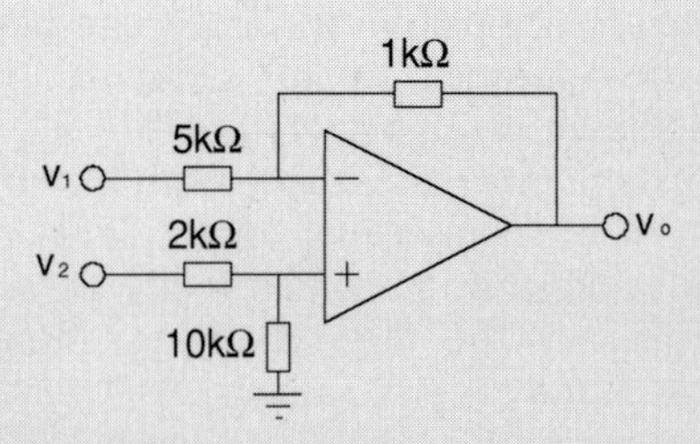

Figure 11.11

Find v_0 in terms of v_1 and v_2.

$$\text{Answer: } v_0 = -\frac{v_1}{5} + v_2$$

Superposition

a) set $v_1 = 0$ b) set $v_2 = 0$

Set $v_1 = 0$

$$v_+ = v_2 \frac{10}{12}$$

$$v_+ = v_-$$

$$v_{oa} = v_-\left(1 + \frac{1\text{k}}{5\text{k}}\right)$$

$$= v_2 \frac{10}{12} \times \frac{6}{5} = v_2$$

Set $v_2 = 0$

$$v_2 = 0 \text{ and } v_+ = 0$$

$$v_{ob} = -v_1 \frac{1\text{k}}{5\text{k}} = -\frac{v_1}{5}$$

Total solution

$$v_o = v_2 - \frac{v_1}{5}$$

PROBLEM 11.3 OP-AMP GENERAL

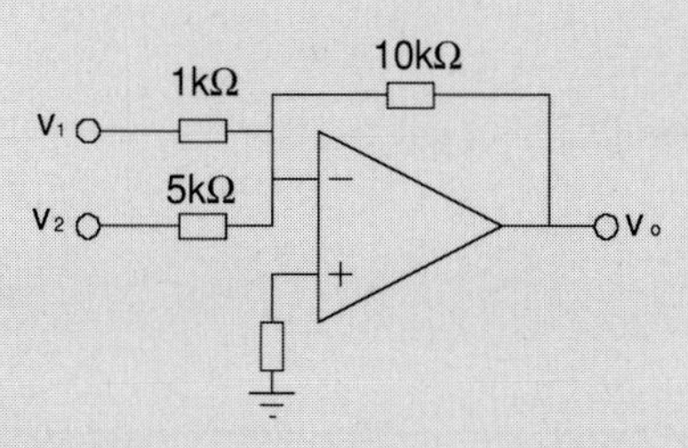

Figure 11.12

Find v_0 in terms of v_1 and v_2.

Answer: $-10v_1 - 2v_2$

Note that the inputs on the inverting side appear in the result with a minus sign, whereas inputs to the non-inverting side appear in the result with a plus sign (as expected).

Superposition

$v_+ = v_- = 0$ (virtual earth)

a) Set $v_1 = 0$. The 1 kΩ branch has 0 V at both sides. There is no current flow through the 1 kΩ resistor in this case.

$$v_{oa} = -v_2 \frac{10k}{5k} = -2v_2$$

b) Set $v_2 = 0$. The 5k branch is now dead

$$v_{ob} = -v_1 \frac{10k}{1k} = -10v_1$$

Solution

$$v_o = v_{oa} + v_{ob} = -2v_2 - 10v_1$$

PROBLEM 11.4 OP-AMP GENERAL

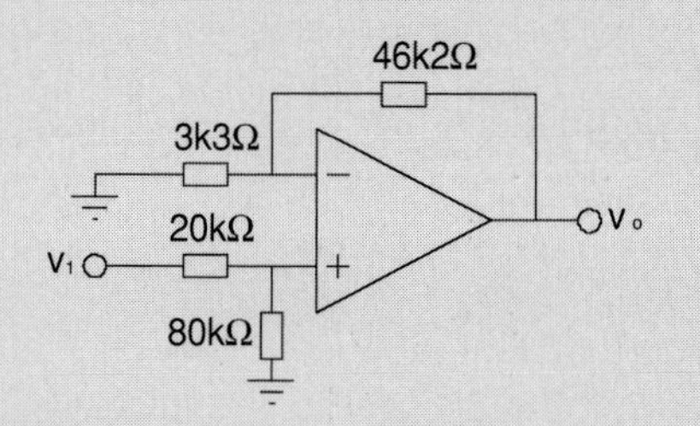

Figure 11.13

v_1 is equal to 750 mV. Find v_0.

Answer: 9 V

$$v_+ = 750 \times 10^{-3} \times \frac{80k}{100k}$$

$$= 600 \text{ mV} = v_-$$

$$v_o = 600 \times 10^{-3}\left(1 + \frac{46k2}{3k3}\right)$$

$$= 600 \times 10^{-3}(1 + 14)$$

$$= 600 \times 10^{-3} \times 15 = 9 \text{ V}$$

PROBLEM 11.5 OP-AMP GENERAL

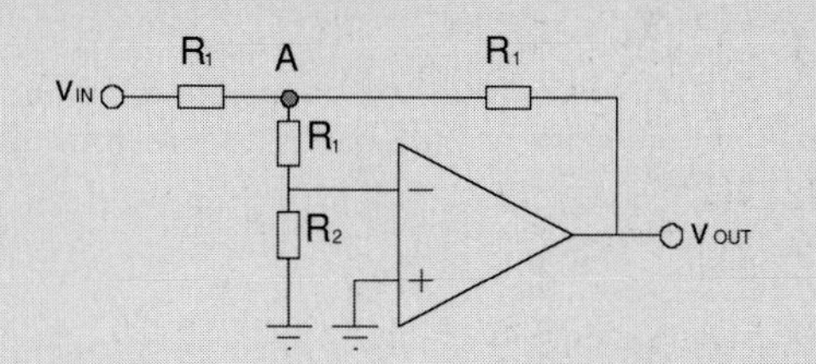

Figure 11.14

Find the ratio $\frac{v_o}{v_i}$ assuming no saturation.

Answer: −1

This is a simple, but puzzling problem.

At v_- we have virtual earth. R_2 has no current. The resistor R_1 next to R_2, has no current either. The current has nowhere to go. At point A, therefore, we also have a virtual earth. The current going from v_i to v_o will be

$$\frac{v_i - v_A}{R_1} = \frac{v_A - v_o}{R_1}$$

but

$$v_A = 0$$

$$\frac{v_i}{R_1} = -\frac{v_o}{R_1}$$

$$\frac{v_o}{v_i} = -1$$

PROBLEM 11.6 OP-AMP GENERAL

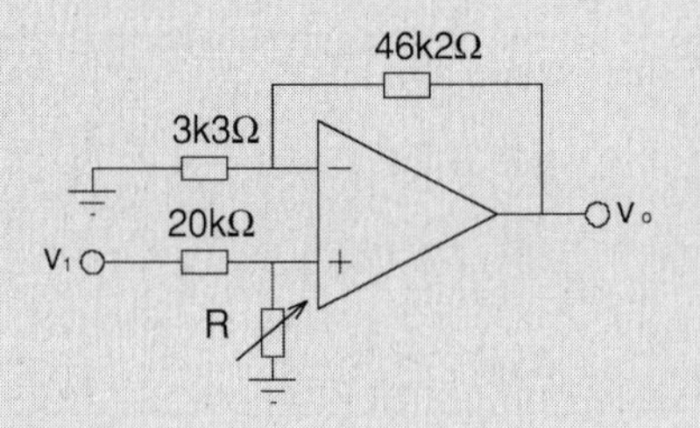

Figure 11.15

v_1 is equal to 750 mV. The Op-Amp saturates at ± 11 V. The variable resistor is increased. What is the minimum value of the resistor within saturation.

Answer: 880 kΩ

When the resistor is increased, the output will go to +11 V.

$$v_o = 11 \text{ V}$$

$$v_- = 11\frac{3k3}{46k2 + 3k3}$$

$$v_+ = \frac{750}{1000} \times \frac{R}{20\,000 + R}$$

$$v_+ = v_-$$

$$\frac{750}{1000} \times \frac{R}{20\,000 + R} = 11\frac{3.3}{49.5}$$

$$\frac{R}{20\,000 + R} = \frac{11 \times 3.3 \times 1000}{750 \times 49.5}$$

$$= 0.977\,777\,7$$

$$R = 19\,555.555\,5 + 0.977\,777\,R$$

$$0.0222\,222\,2\,R = 19\,555.5555$$

$$R = 880 \text{ k}\Omega$$

PROBLEM 11.7 OP-AMP GENERAL

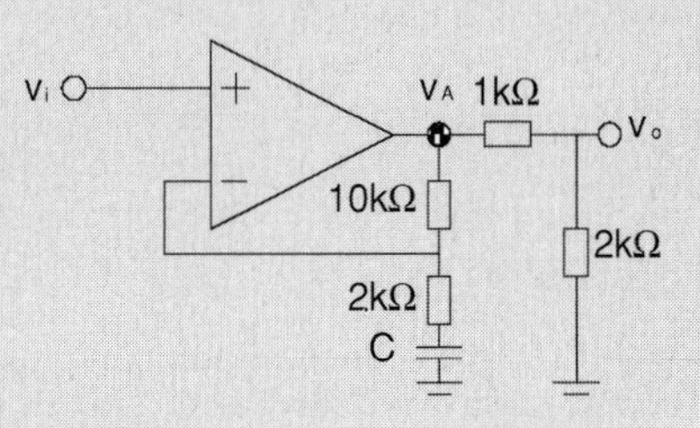

Figure 11.16

Calculate v_o if

a) v_i = 0.75 V

b) v_i = 75 mV sine wave at a frequency where C becomes a short circuit.

Answer: (a) 0.5 V and (b) 300 mV

a) v_i = 0.75 V

The capacitor isolates the DC. There is no current through the 10k resistor.

$$v_+ = v_- = v_i$$

The output is a potential divider.

$$v_o = v_A\frac{2k}{1k + 2k} = v_A\frac{2}{3}$$

$$v_o = 0.75\frac{2}{3} = 0.5 \text{ V}$$

b) C is a short circuit under AC conditions

$$v_A = 75 \times 10^{-3}\left(1 + \frac{10k}{2k}\right) = 0.45 \text{ V}$$

$$v_o = v_A\frac{2k}{1k + 2k} = 0.45 \times \frac{2}{3} = 0.3 \text{ V}$$

PROBLEM 11.8 OP-AMP GENERAL

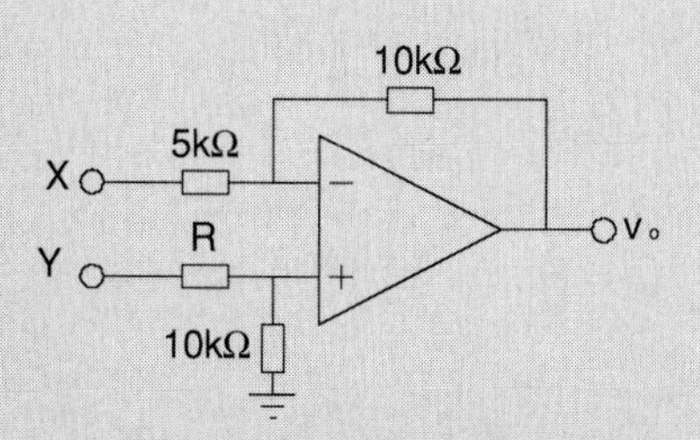

Figure 11.17

Find R such that $v_o = \frac{Y}{3} + 2X$

Answer: $R = 80$ kΩ

Superposition

a) $Y = 0$

$$v_{oa} = -X\frac{10k}{5k} = -2X$$

b) $X = 0$

$$v_+ = Y\frac{10k}{R + 10k}$$

$$v_{ob} = v_+\left(1 + \frac{10k}{5k}\right)$$

$$= Y\frac{10k}{R + 10k}\frac{15k}{5k}$$

$$= 3Y\frac{10k}{R + 10k}$$

This must be equal to $\frac{Y}{3}$

$$\frac{Y}{3} = 3Y\frac{10k}{R + 10k} \qquad \text{(Find } R\text{)}$$

$$10k + R = 90k$$

$$R = 80 \text{ k}\Omega$$

PROBLEM 11.9 OP-AMP GENERAL

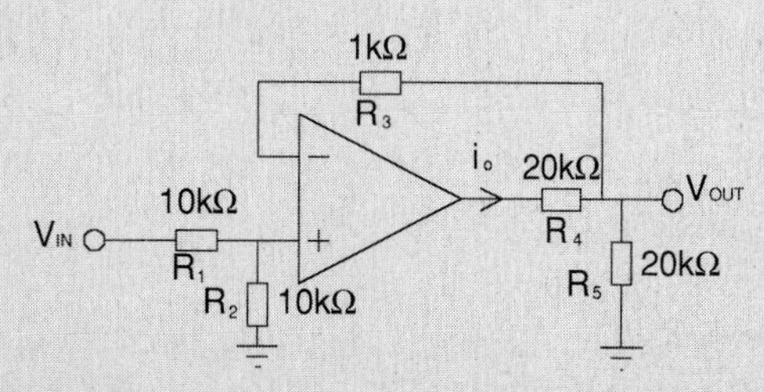

Figure 11.18

The input is set at 40 mV.
Find the value of i_o. If R_4 is increased state whether the output v_o will increase, decrease or remains the same.

Answer: $i_o = 1$ μA. The same.

R_3 is not conducting. This means that $v_o = v_-$. We also have $v_+ = v_-$.

$$v_+ = 40 \times 10^{-3} \times \frac{10}{10 + 10} = 20 \text{ mV}$$

$$v_o = 20 \text{ mV}$$

$$i_o = \frac{v_o}{R_5} = \frac{20 \text{ mV}}{20k} = 1\,\mu\text{A}$$

If R_4 is increased, the voltage on R_4 will increase, but v_o will still be given by $v_o = v_- = v_+$. We assume that the Op-Amp has not reached saturation.

PROBLEM 11.10 OP-AMP GENERAL

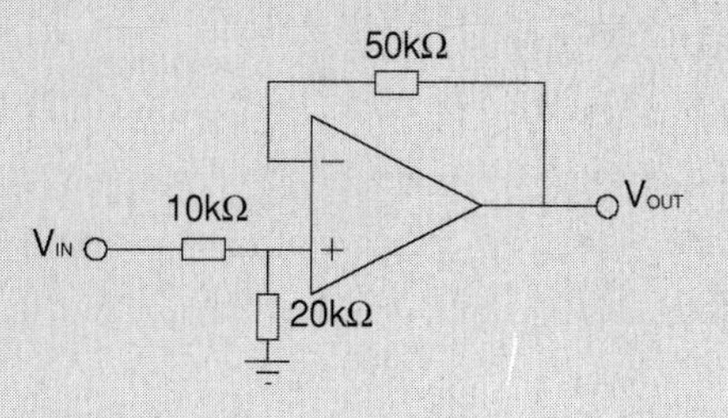

Figure 11.19

v_i is 75 mV. Find v_o.

Answer: 50 mV

There is no current through the 50 kΩ resistor.

$v_o = v_-$

At the input end

$v_+ = v_-$ and then $v_+ = v_o$

v_+ is part of a potential divider

$$v_+ = v_i \frac{20k}{10k + 20k} = v_i \frac{2}{3}$$

$$v_o = v_i \frac{2}{3} = 75 \frac{2}{3} = 50 \text{ mV}$$

PROBLEM 11.11 OP-AMP GENERAL

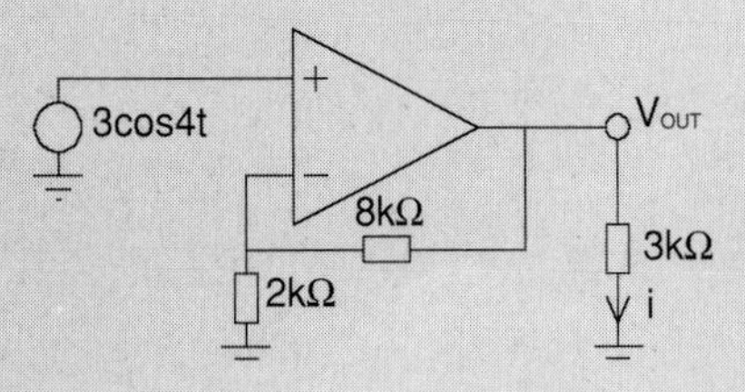

Figure 11.20

Find i.

Answer: 5cos 4t mA

This is a non-inverting Op-Amp

$$v_o = 3\cos 4t\left(1 + \frac{8k}{2k}\right)$$

$$= 3\cos 4t\,(1 + 4)$$

$$= 15\cos 4t$$

$$i = \frac{v_o}{3k} = \frac{15\cos 4t}{3k}$$

$$= 5\cos 4t \text{ mA}$$

PROBLEM 11.12 OP-AMP GENERAL

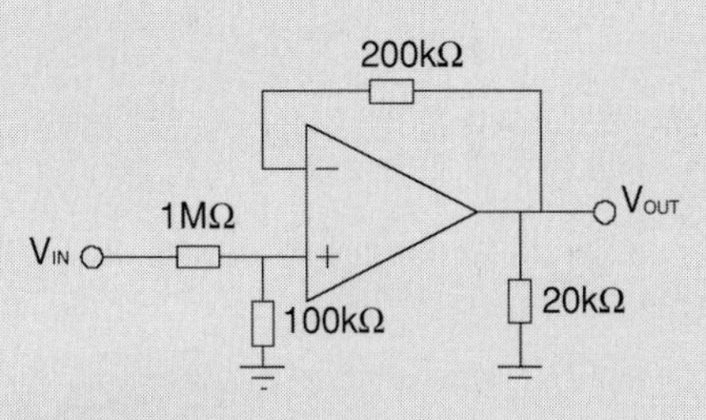

Figure 11.21

Find v_o in terms of v_i.

Answer: $v_o = v_i \frac{1}{11}$

This is an interesting problem which shows how Op-Amps work. The 200 kΩ resistor has no current.

$v_o = v_-$

Potential divider

$$v_+ = v_i \frac{100k}{1M + 100k} = \frac{v_i}{11}$$

But $v_+ = v_-$, so

$$v_o = v_- = v_+ = \frac{v_i}{11}$$

PROBLEM 11.13 OP-AMP GENERAL

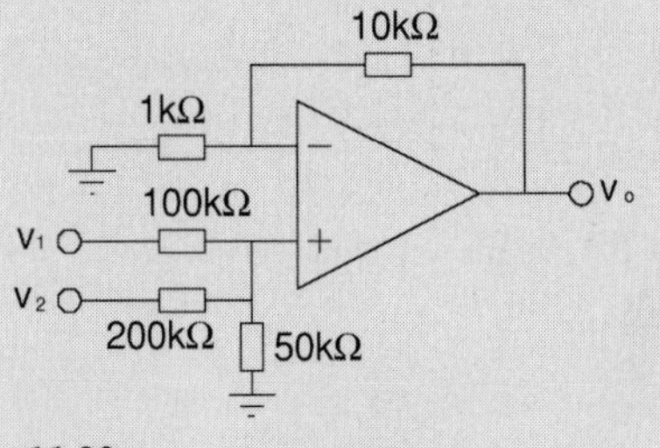

Figure 11.22

Find v_o.

Answer: $v_o = \frac{11}{7}(2v_1 + v_2)$

Superposition

a) $v_2 = 0$

$50\text{k}\|200\text{k} = 40\ \text{k}\Omega$

$$v_+ = v_1 \frac{40}{40+100} = v_1 \frac{2}{7}$$

b) $v_1 = 0$

$50\text{k}\|100\text{k} = 33\,333.33\ \Omega$

$$v_+ = v_2 \frac{33\,333.33}{200\text{k} + 33\,333.33} = \frac{v_2}{7}$$

Amplification

$$1 + \frac{10\text{k}}{1\text{k}} = 11$$

$$v_o = \frac{11}{7}(2v_1 + v_2)$$

PROBLEM 11.14 OP-AMP GENERAL

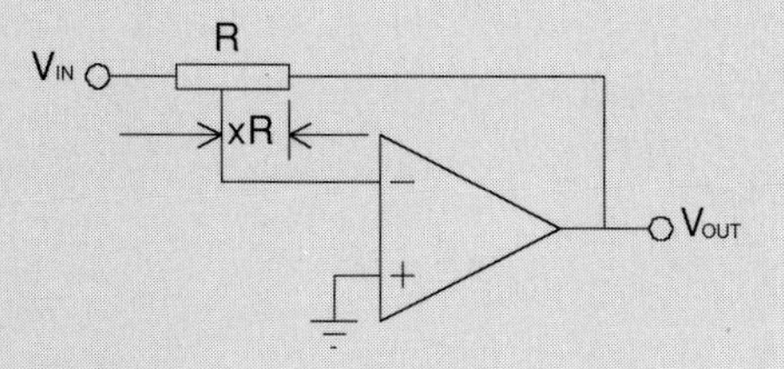

Figure 11.23

Calculate $\frac{v_o}{v_i}$ as a function of x.

Answer: $-\frac{x}{1-x}$

Inverting the Op - Amp

$$v_o = -v_i \frac{xR}{R - xR}$$

$$\frac{v_o}{v_i} = -\frac{xR}{R - xR}$$

$$\frac{v_o}{v_i} = -\frac{x}{1-x}$$

PROBLEM 11.15 OP-AMP GENERAL

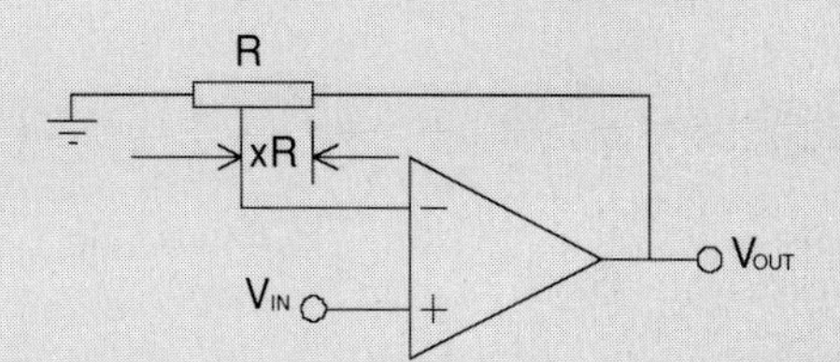

Figure 11.24

Calculate $\frac{v_o}{v_i}$ in function of x.

Answer: $\frac{1}{1-x}$

Non - inverting Op - Amp

$$v_o = v_i\left(1 + \frac{R_F}{R_1}\right)$$

$$= v_i\left(1 + \frac{xR}{R - xR}\right)$$

$$= v_i \frac{R - xR + xR}{R - xR}$$

$$= v_i \frac{R}{R - xR}$$

$$= v_i \frac{1}{1-x}$$

PROBLEM 11.16 OP-AMP GENERAL

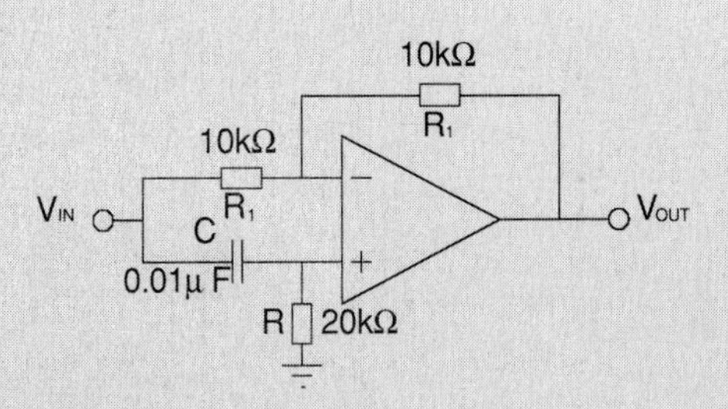

Figure 11.25

Find the magnitude and phase angle for ω = 5000 rad/s.

Answer: magnitude = 1, angle = 90°

Superposition. Although there is only one input, we can separate them into lower and higher branches.

a) Lower branch equal to zero.

$v_{oa} = -v_i$ inverter with gain 1 to 1.

b) Higher branch equal zero.

$$v_{ob} = v_+\left(1+\frac{10k}{10k}\right) = 2v_+$$

$$v_+ = v_i\frac{R}{R+\dfrac{1}{j\omega C}} \qquad |\times j\omega C$$

$$v_+ = \frac{v_i j\omega CR}{j\omega CR+1}$$

$$v_{ob} = \frac{2v_i j\omega CR}{j\omega CR+1}$$

$$v_o = v_{oa} + v_{ob} = \frac{2v_i j\omega CR}{j\omega CR+1} - v_i$$

$$\frac{v_o}{v_i} = \frac{2j\omega CR}{j\omega CR+1} - 1 = \frac{j\omega CR-1}{j\omega CR+1}$$

For ω = 5000 rad/s

$$\frac{v_o}{v_i} = \frac{j5000\times0.01\times10^{-6}\times20\,000-1}{j\omega5000\times0.01\times10^{-6}\times20\,000+1}$$

$$\frac{v_o}{v_i} = \frac{j-1}{j+1} = j$$

Magnitude = 1, angle 90°

PROBLEM 11.17 OP-AMP GENERAL

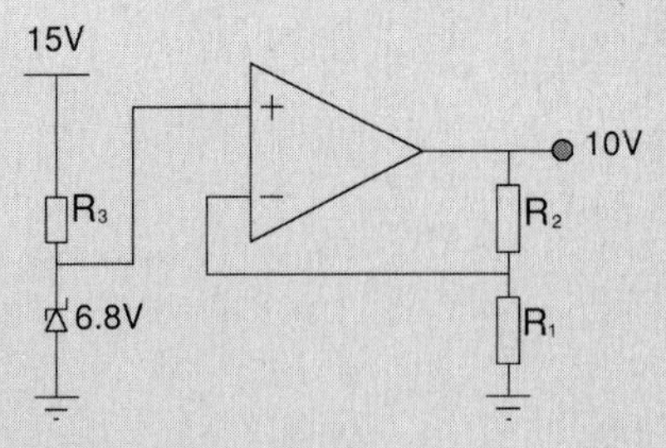

Figure 11.26

The Zener diode current is 1 mA. The current through the resistive network is 0.1 mA.
Find R_1, R_2 and R_3 to produce an output of 10 V.

Answer: 68 kΩ, 32 kΩ and 8k2 Ω

At Zener

$$15 - 6.8 = 8.2 \text{ V}$$

$$R_3 = \frac{8.2}{10^{-3}} = 8\text{k}2\ \Omega$$

$$\frac{v_o}{v_i} = \frac{10}{6.8} = 1+\frac{R_2}{R_1} = \frac{R_1+R_2}{R_1}$$

At output

$$(R_1+R_2)\times0.1\times10^{-3} = 10$$

$$R_1+R_2 = 10^5$$

$$\frac{10}{6.8} = \frac{R_1+R_2}{R_1} = \frac{10^5}{R_1}$$

$$R_1 = 10^5\frac{6.8}{10} = 68\text{ k}\Omega$$

$$R_2 = 10^5 - 68\text{k} = 32\text{ k}\Omega$$

PROBLEM 11.18 OP-AMP GENERAL

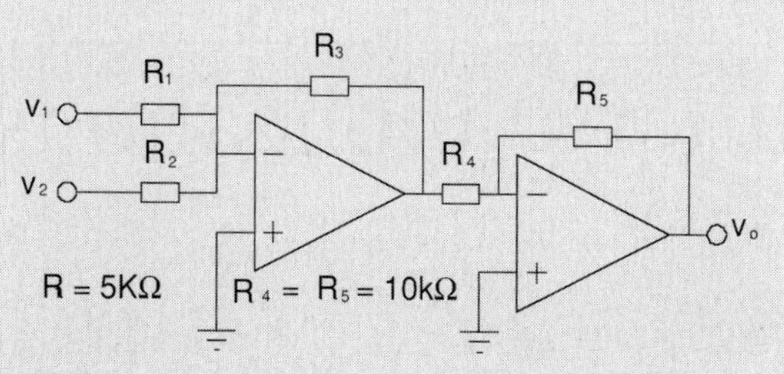

Figure 11.27

$R_1 = 5\ \text{k}\Omega$, $R_4 = R_5 = 10\ \text{k}\Omega$.

Find the values of R_2 and R_3 to give an output of $v_o = 2v_1 + 5v_2$

Answer: $R_2 = 2\ \text{k}\Omega$ and $R_3 = 10\ \text{k}\Omega$

The second Op-Amp is a 1 to 1 inverter. The output required from the first Op-Amp is

$$v_{o1} = 2v_1 - 5v_2$$

$$v_{o1} = -v_1 \frac{R_3}{R_1} - v_2 \frac{R_3}{R_2}$$

$$\frac{R_3}{R_1} = 2 \qquad \frac{R_3}{R_2} = 5$$

$$R_1 = 5\text{k} = \frac{R_3}{2}$$

$$R_3 = 2 \times 5\text{k} = 10\ \text{k}\Omega$$

$$R_2 = \frac{R_3}{5} = \frac{10\text{k}}{5} = 2\ \text{k}\Omega$$

PROBLEM 11.19 OP-AMP GENERAL

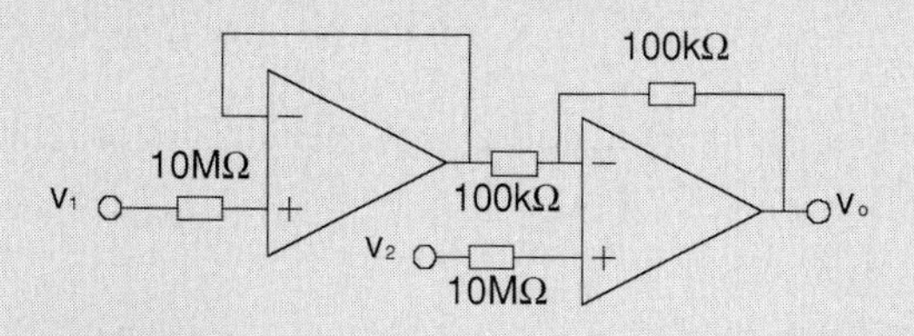

Figure 11.28

Find v_o.

Answer: $-v_1 + 2v_2$

a) $v_1 = 0$

$$v_+ = v_2$$

$$v_{oa} = v_2\left(1 + \frac{100}{100}\right) = 2v_2$$

b) $v_2 = 0$

$$v_+ = v_- = 0$$

$$v_{ob} = -v_1 \frac{100}{100} = -v_1$$

Total output

$$v_o = 2v_2 - v_1$$

PROBLEM 11.20 OP-AMP GENERAL

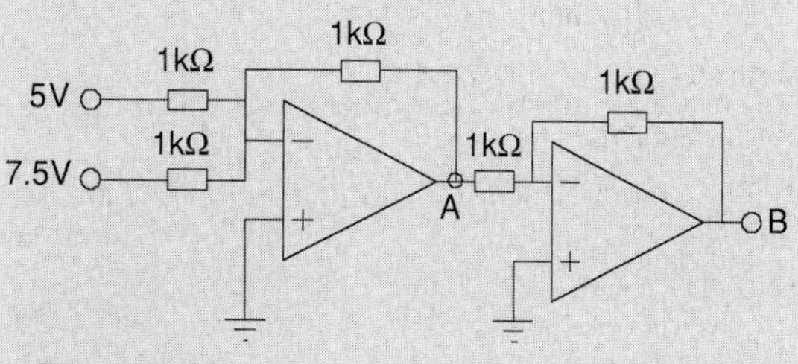

Figure 11.29

Find the voltages at A and B.

Answer: –12.5 V and 12.5 V

The first Op-Amp is summing and inverting with a ratio of 1 to 1.

$$V_A = -5 - 7.5 = -12.5\ \text{V}$$

The second Op-Amp is inverting also with a ratio of 1 to 1.

$$V_B = -V_A = 12.5\ \text{V}$$

12

Operational amplifiers: applications

Op-Amps nodal analysis

In some cases, nodal analysis can be an easy and interesting way of solving operational amplifier problems. It is based on the fact that for a large amplification, the input differential of an amplifier is very small and it can be approximated to zero.

Under this approximation, we can say that

$$V_+ = V_-$$

The nodes are usually, one at each input and one at the output, although you would place them wherever you need them. Sometimes one equation at one node is sufficient to solve the problem.

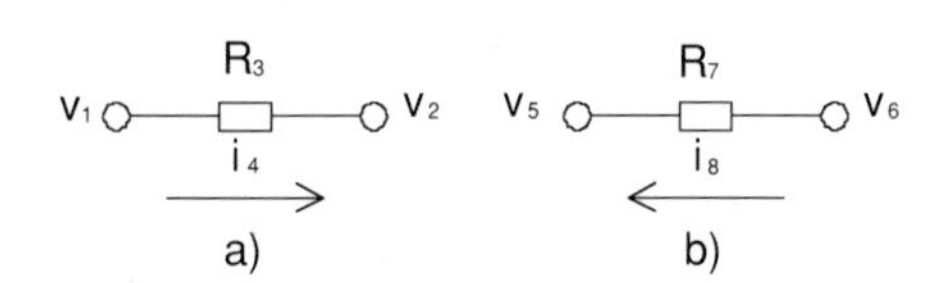

Figure 12.1

The important point here is to set the equation correctly according to the direction of the current. Let us look at Figure 12.1. In the first case, the current flows from v_1 to v_2. The equation for the current is

$$i_4 = \frac{v_1 - v_2}{R_3}$$

In the second case the current flows from v_6 to v_5. The equation for i_8 is

$$i_8 = \frac{v_6 - v_5}{R_7}$$

We assume that one voltage is larger than the other. The current will flow from the larger voltage to the smaller and the voltage difference will simply be the larger voltage minus the smaller voltage.

Nodal analysis is only a systematic application of KCL. It is therefore very important to select the current directions sensibly and to follow them with the corresponding equations.

Comparator

The Op-Amp can be used as a comparator in an open loop configuration. A linear voltage can be fed into one of the inputs. The other input can be held to ground. This arrangement is shown in Figure 12.2. The input and output waveforms are

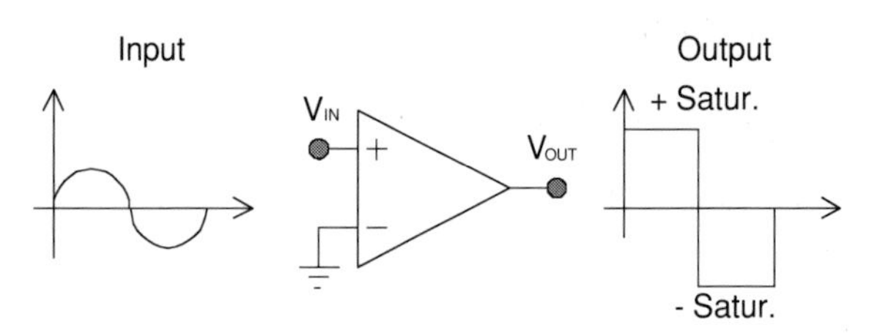

Figure 12.2

also shown. When the input is positive, the output is saturated positive. When the input is negative, the output is saturated negative. So, in a way, this circuit accepts an analogue input and will provide a digital output.

Using comparators, many different types of circuits and results can be achieved. What about a voltage level indicator, as we see in Figure 12.3?

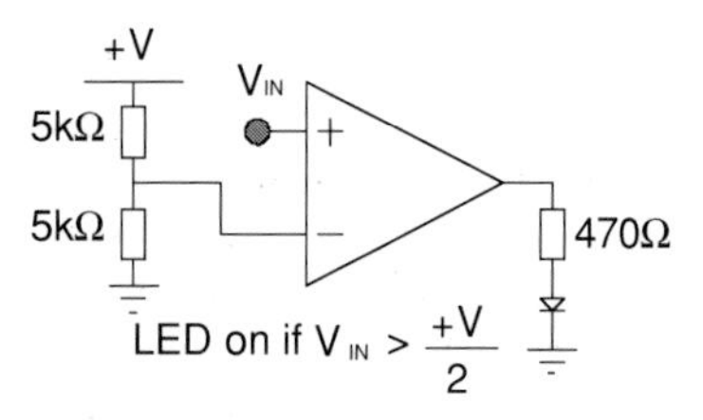

Figure 12.3

The potential divider selects half of the voltage at the negative input. For the LED to conduct and glow, the output has to be positive. For the ouput to be positive, the positive input has to be bigger than the negative input.

So

$$V_i > \frac{V}{2} \qquad \text{(will switch the LED on)}$$

If the opposite condition is required, this can be easily arranged as in Figure 12.4.

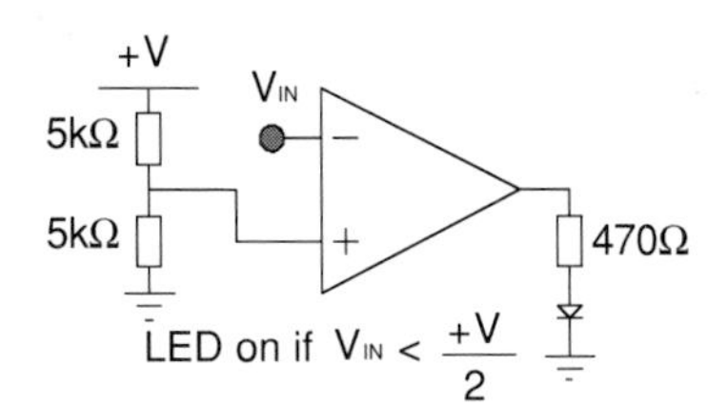

Figure 12.4

Window detector

If you only need to detect a band, within the range available, that is also possible and it is called a window detector. This can be seen in Figure 12.5. There are basically two comparators, one for each

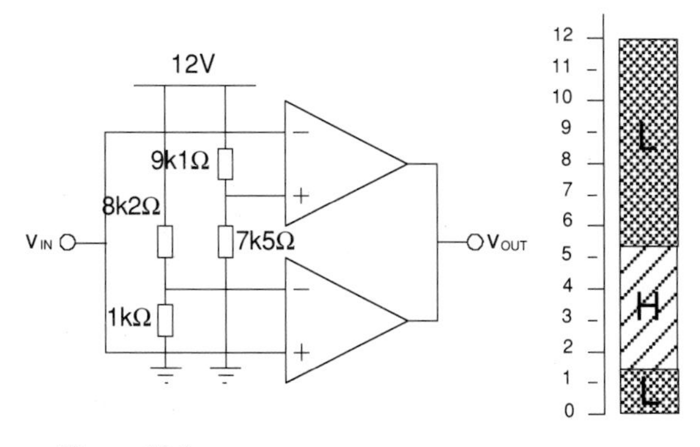

Figure 12.5

edge. The top one gives a low output at high voltage and the lower comparator gives a low for low voltage. In between, both give a positive output.

In order to get the window effect, the output will be high if both comparators give a high output.

Outside the high band we will have a situation where one comparator gives a low output whilst the other gives a high output. In this case we don't want a conflict between the two signals and we want the low signal to take preference.

These details are not shown in the circuit, but the problem would be resolved by using Op-Amps with open collector outputs and possibly a pull-up resistor in the circuit.

Comparator with hysteresis

A simple comparator has the problem that if the signal is around the switching point, the ouput will be jumping from one value to the other. We would prefer a clean jump from positive to negative or vice versa.

It is desirable in this case to introduce an extra condition in the circuit so as to achieve a differential band. This is called hysteresis. It is similar to the ordinary light switch which is not stable in the middle. It will take one position and it will stay there. But, it can also be forced in the opposite direction.

The way to achieve this is shown in Figure 12.6.

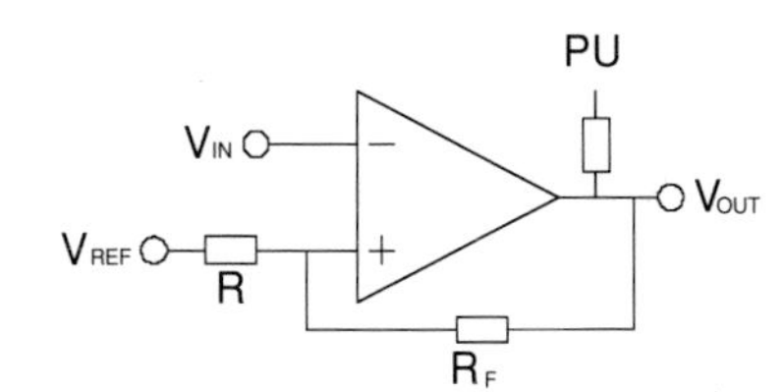

Figure 12.6

The hysteresis is:

$$\Delta V = V_{PP}\frac{R}{R + R_F}$$

The hysteresis is therefore a proportion of the V_{PP} voltage. This voltage is the total swing of the ouput. If the output varies from +12 V to –12 V, then V_{PP} is 24 V. The size of the differential is given by the amount of positive feedback allowed by the potential divider formed by R and R_F. The interesting point here is that the size of the hysteresis band does not depend on V_{REF} at all. See Figure 12.7. V_{REF} only has to do with the location of the hysteresis band, not its width.

In order to prove that this is the true, we will work out the size of the hysteresis (see Figure

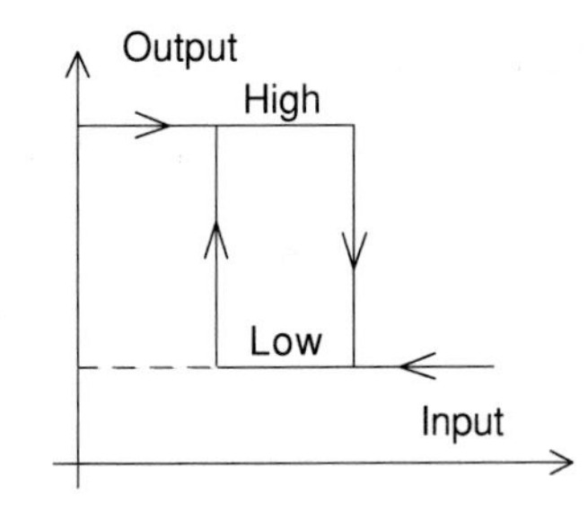

Figure 12.7

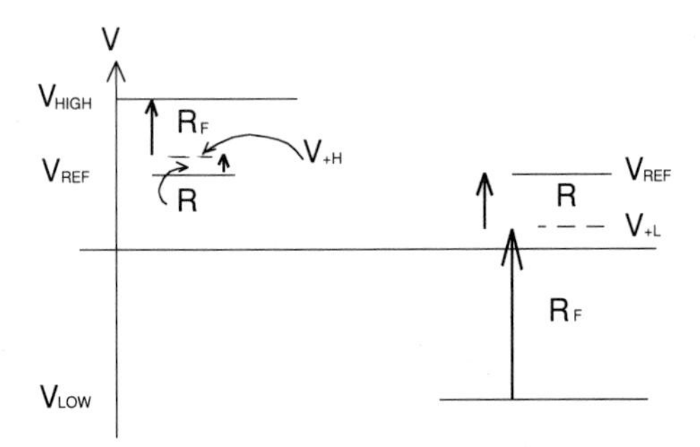

Figure 12.8

12.8). First of all, it must be clear that all the voltages are measured with respect to 0 V as a reference. The diagram on the left-hand side of Figure 12.8 shows the situation with the circuit of Figure 12.6, when the output is high (V_H).

We want to know the voltage at the positive terminal, when the output is high. We call the voltage at the positive terminal, V_{+H}.

$$V_{+H} = V_{REF} + (V_H - V_{REF})\frac{R}{R+R_F}$$

This can be checked against the diagram in the left part of Figure 12.8.

We refer to V_{+L} as the voltage at V_+ when the output is low:

$$V_{+L} = V_{REF} - (V_{REF} - (-V_L))\frac{R}{R+R_F}$$

We can also do the same equation, not this time from V_{REF}, but from V_L. We get:

$$V_{+L} = (V_{REF} - (-V_L))\frac{R_F}{R+R_F} + (-V_L)$$

The last two equations will reduce to

$$V_L = V_{REF}\frac{R_F}{R+R_F} - V_L\frac{R}{R+R_F}$$

The window is given by

$$W = V_{+H} - V_{+L}$$

$$W = V_{REF} + V_H\frac{R}{R+R_F} - V_{REF}\frac{R}{R+R_F}$$

$$-V_{REF}\frac{R_F}{R+R_F} + V_L\frac{R}{R+R_F}$$

$$\boxed{W = (V_H + V_L)\frac{R}{R+R_F}}$$

$V_H + V_L$ is the peak to peak voltage referred to before as V_{PP}.

Voltage to current converter

One interesting Op-Amp application is the voltage to current conversion. This conversion is widely used with a transducer to transmit the signals necessary for process control, usually in the range of 4 to 20 mA. The converter has to sink current into a number of different loads without changing the voltage current transfer characteristic. The simplest Op-Amp circuit for this is shown in Figure 12.9.

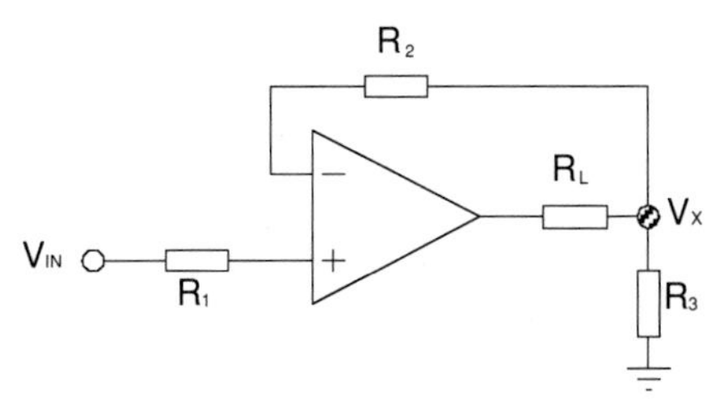

Figure 12.9

In this circuit:

$V_+ = V_i$ (no current through R_1)
$V_+ = V_-$
$V_X = V_-$ (no current through R_2)

I_L flows through R_L and R_3. Therefore:

$$V_x = I_L R_3$$

$$V_i = I_L R_3 \quad \text{and} \quad I_L = \frac{V_i}{R_3}$$

I_L is independant of R_L and it is proportional to V_i.

Although this circuit achieves the desired result, it has the complication that the load resistor is floating. That is to say that neither end can be grounded. Remember that the load resistor will be sited away from the control room, in the plant.

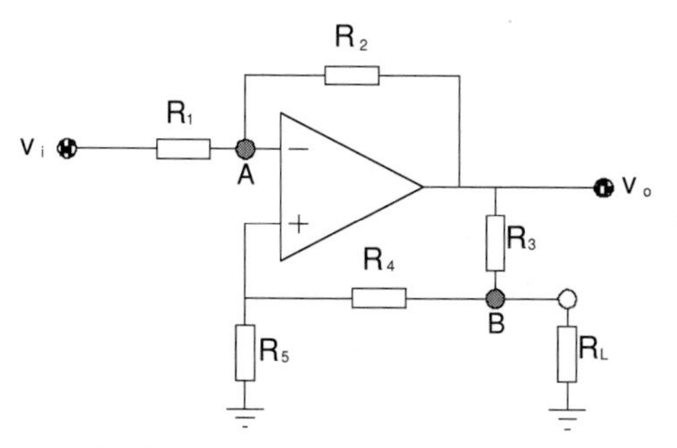

Figure 12.10

A variation of the circuit can be seen in Figure 12.10. The technique used here to solve this problem is to use node equations for A and B. This is the same as KCL or the first Kirchhoff's law. We later use the equation $V_+ = V_-$ at the input of the Op-Amp as an extra equation.

$$\frac{V_i - V_-}{R_1} = \frac{V_- - V_o}{R_2} \quad \text{(node A)} \tag{1}$$

$$\frac{V_o - V_+\dfrac{R_4+R_5}{R_5}}{R_3} = I_L + \frac{V_+\dfrac{R_4+R_5}{R_5}}{R_4+R_5} \quad \text{(node B)} \tag{2}$$

From the node A equation we can obtain an expression for V_-:

$$V_- = \frac{V_iR_2 + V_oR_1}{R_1+R_2} \tag{3}$$

From the node B equation we can get an expression for V_+:

$$V_+ = \frac{V_oR_5 - I_LR_3R_5}{R_3+R_4+R_5} \tag{4}$$

If the Op-Amp is working correctly and is not saturated, then we can equate $V_+ = V_-$

We obtain

$$V_iR_2(R_3+R_4+R_5) + V_oR_1(R_3+R_4+R_5)$$

$$= V_oR_5(R_1+R_2) - I_LR_3R_5(R_1+R_2) \tag{5}$$

As we are only interested in V_i and I_L, the equation can be greatly simplified if the V_o terms can be eliminated. The terms on V_o are made equal at both sides of the equation. In order to achieve this, we make:

$$R_1(R_3+R_4+R_5) = R_5(R_1+R_2) \tag{6}$$

This simplifies to

$$R_1(R_3+R_4) = R_2R_5$$

$$\frac{R_1}{R_2} = \frac{R_5}{R_4+R_3} \tag{7}$$

and by the laws of proportion:

$$\frac{R_1+R_2}{R_2} = \frac{R_3+R_4+R_5}{R_3+R_4} \tag{8}$$

From equation (5) we get:

$$V_i = -I_L\frac{R_3R_5(R_1+R_2)}{R_2(R_3+R_4+R_5)} \tag{9}$$

Using equation (8) on equation (9) we get:

$$V_i = -I_L\frac{R_3R_5R_2}{R_2(R_3+R_4)} \tag{10}$$

Using equation (7) on equation (10) we get:

$$\boxed{V_i = -I_L\frac{R_3R_1}{R_2}} \tag{11}$$

Again we obtain the desired result that V_i is proportional to I_L and independent of R_L.

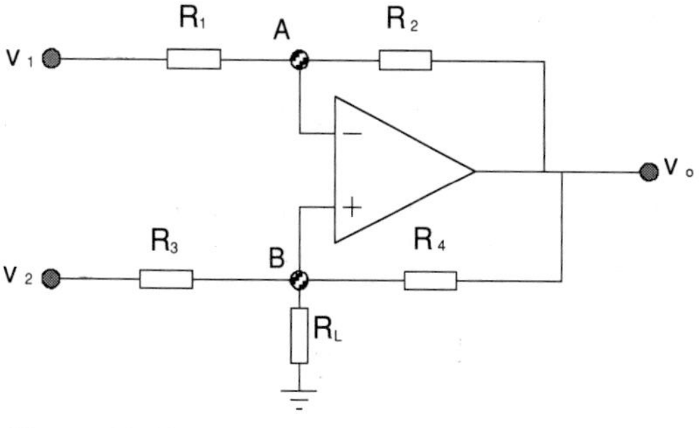

Figure 12.11

Another variation of the problem which helps to illustrate the technique for solving this type of problem is shown in Figure 12.11. We have nodes A and B again, where we can use Kirchhoff's current law (KCL) and the equation $V_+ = V_-$. At node A:

$$\frac{V_1 - V_-}{R_1} = \frac{V_- - V_o}{R_2}$$

At node B:

$$\frac{V_2 - V_+}{R_3} + \frac{V_o - V_+}{R_4} = I_L$$

We assume here that the first two currents are going into the node, whereas I_L is going out of the node. If this is not true, then we would get a minus sign in the result.

We find V_- from the first equation, V_+ from the second equation and we equate them, i.e.:

$$V_- = \frac{V_1R_2 + V_oR_1}{R_1 + R_2}$$

$$V_+ = \frac{V_2R_4 + V_oR_3 - I_LR_3R_4}{R_3 + R_4}$$

we obtain:

$$V_1R_2(R_3 + R_4) + V_oR_1(R_3 + R_4)$$
$$= V_2R_4(R_1 + R_2) + V_oR_3(R_1 + R_2)$$
$$- I_LR_3R_4(R_1 + R_2)$$

If $R_1 = R_3$ and $R_2 = R_4$ the term V_o disappears as they are equal and at both sides of the equation:

$$V_1R_2 - V_2R_4 = -I_LR_3R_4$$

$$\boxed{I_L = \frac{V_2}{R_1} - \frac{V_1}{R_1} = \frac{1}{R_1}(V_2 - V_1)}$$

The current I_L is independent of the load and it is a function of the voltage difference $V_2 - V_1$.

PROBLEM 12.1 OP-AMP APPLICATIONS

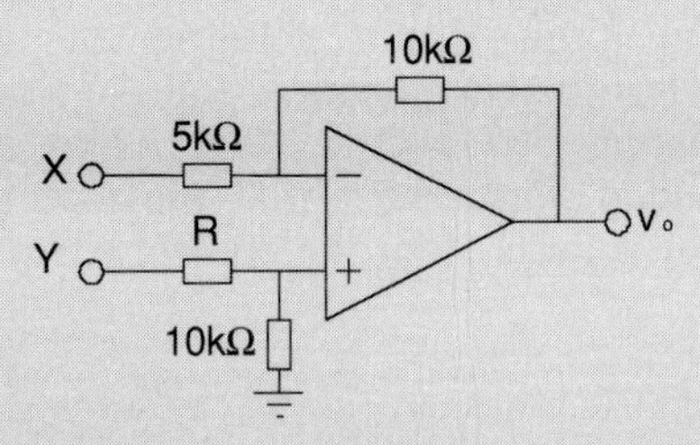

Figure 12.12

Using nodal analysis find R such that $v_o = \frac{Y}{3} - 2X$

Answer: 80 kΩ

At node v_-

$$\frac{X - v_-}{5k} = \frac{v_- - v_o}{10k} \qquad | \times 10k$$

$$2X - 2v_- = v_- - v_o$$

$$2X + v_o = 3v_-$$

At node v_+

$$\frac{Y - v_+}{R} = \frac{v_+ - 0}{10k}$$

$$Y10k - v_+10k = Rv_+$$

$$Y10k = v_+(R + 10k)$$

$$v_+ = Y\frac{10k}{R + 10k}$$

$$3v_+ = 3v_-$$

$$2X + v_o = 3Y\frac{10k}{R + 10k}$$

$$v_o = -2X + 3Y\frac{10k}{R + 10k}$$

Therefore

$$\frac{10k}{R + 10k} = \frac{1}{9}$$

$$90k = R + 10k$$

$$R = 80k$$

PROBLEM 12.2 OP-AMP APPLICATIONS

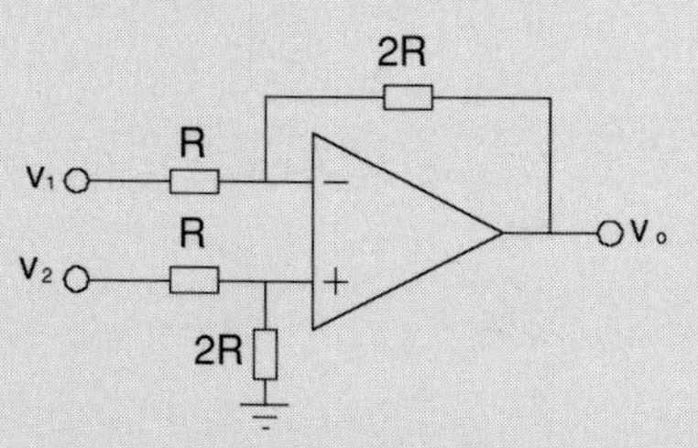

Figure 12.13

Show that $v_o = 2(v_2 - v_1)$

Answer: Yes

Using nodal analysis at node v_-

$$\frac{v_1 - v_-}{R} = \frac{v_- - v_o}{2R}$$

$$2Rv_1 - 2Rv_- = Rv_- - Rv_o$$

$$2Rv_1 + Rv_o = 3Rv_-$$

At node v_+

$$\frac{v_2 - v_+}{R} = \frac{v_+ - 0}{2R}$$

$$2Rv_2 - 2Rv_+ = Rv_+$$

$$2Rv_2 = 3Rv_+$$

$$v_+ = v_-$$

$$3Rv_+ = 3Rv_-$$

$$2Rv_2 = 2Rv_1 + Rv_o$$

$$Rv_o = 2Rv_2 - 2Rv_1$$

$$v_o = 2v_2 - 2v_1$$

$$v_o = 2(v_2 - v_1) \text{ q.e.d.}$$

PROBLEM 12.3 OP-AMP APPLICATIONS

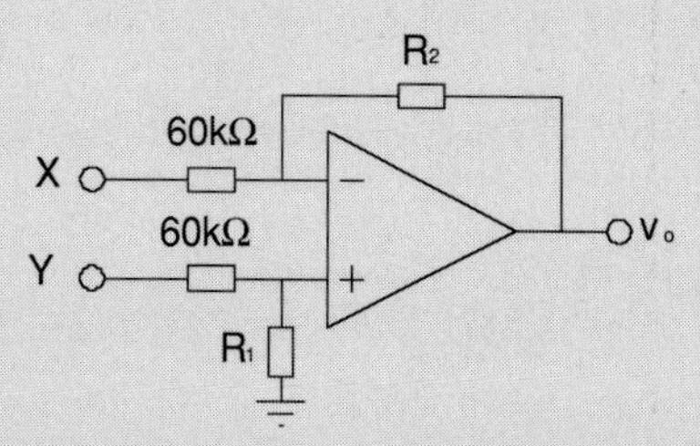

Figure 12.14

Find the values of R_1 and R_2 that will produce an output $v_o = \dfrac{Y}{4} - 3X$.

Answer: $R_1 = 4$ kΩ and $R_2 = 180$ kΩ

Use superposition. First we ground Y

$$v_{o1} = -X\frac{R_2}{60\text{k}}$$

This corresponds to the second term requested: R_2 must be 180 kΩ.

Now $X = 0$

$$v_+ = Y\frac{R_1}{60\text{k} + R_1}$$

$$v_{o2} = v_+\left(1 + \frac{180\text{k}}{60\text{k}}\right) = 4v_+$$

$$= 4Y\frac{R_1}{60\text{k} + R_1}$$

This must be equal to $\dfrac{Y}{4}$

$$\frac{1}{4} = 4\frac{R_1}{60\text{k} + R_1}$$

$$60\text{k} + R_1 = 16R_1$$

$$60\text{k} = 15R_1$$

$$R_1 = 4 \text{ k}\Omega$$

PROBLEM 12.4 OP-AMP APPLICATIONS

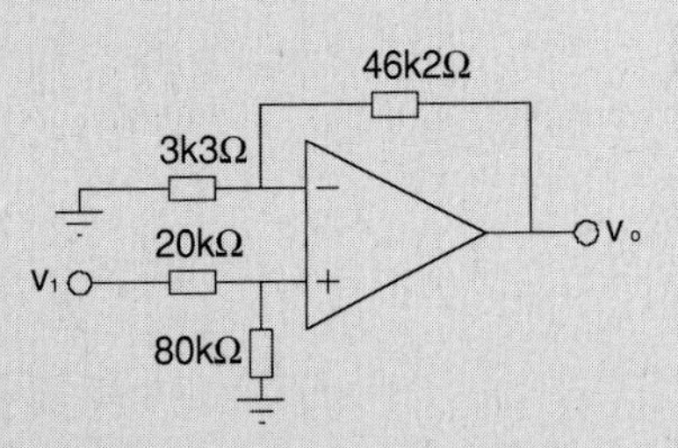

Figure 12.15

Using nodal analysis find v_o if v_i is equal to 750 mV.

Answer: 9 V

At node v_-

$$\frac{0-v_-}{3k3} = \frac{v_- - v_o}{46k2} \qquad | \times 46k2$$

$$-14v_- = v_- - v_o$$

$$v_o = 15v_-$$

At node v_+

$$\frac{0.75 - v_+}{20k} = \frac{v_+ - 0}{80k} \qquad | \times 80k$$

$$3 - 4v_+ = v_+$$

$$3 = 5v_+$$

$$v_+ = \frac{3}{5}$$

$$v_+ = v_-$$

$$v_o = 15v_+$$

$$= 15 \times \frac{3}{5} = 9 \text{ V}$$

PROBLEM 12.5 OP-AMP APPLICATIONS

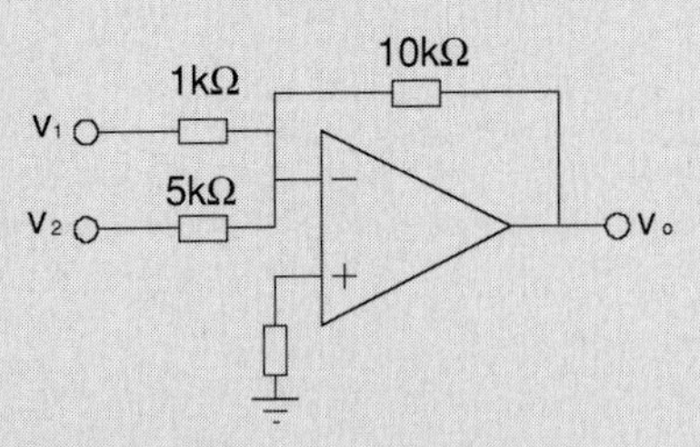

Figure 12.16

Using nodal analysis find v_o in terms of v_1 and v_2.

Answer: $-10v_1 - 2v_2$

At node v_- (KCL)

$$\frac{v_1 - v_-}{1k} + \frac{v_2 - v_-}{5k} = \frac{v_- - v_o}{10k} \qquad | \times 10k$$

$$10v_1 - 10v_- + 2v_2 - 2v_- = v_- - v_o$$

$$10v_1 - 13v_- + 2v_2 = -v_o$$

But $v_+ = v_- = 0$

$$10v_1 + 2v_2 = -v_o$$

$$v_o = -10v_1 - 2v_2$$

PROBLEM 12.6 OP-AMP APPLICATIONS

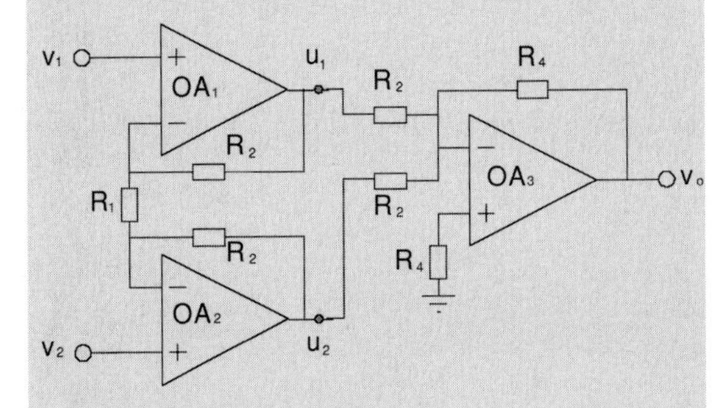

Figure 12.17

Find v_o as a function of $(u_2 - u_1)$ and as a function of $(v_2 - v_1)$.

Answer: $v_o = \frac{R_4}{R_3}(u_2 - u_1)$ and

$$v_o = \frac{R_4}{R_3}\left(1 + 2\frac{R_2}{R_1}\right)(v_2 - v_1)$$

At OA3

$$\frac{u_2 - v_+}{R_3} = \frac{v_+ - 0}{R_4}$$

$$R_4 u_2 - R_4 v_+ = R_3 v_+$$

$$\frac{R_4 u_2}{R_3 + R_4} = v_+$$

$$\frac{u_1 - v_-}{R_3} = \frac{v_- - v_o}{R_4}$$

$$R_4 u_1 - R_4 v_- = R_3 v_- - R_3 v_o$$

$$R_4 u_1 + R_3 v_o = (R_3 + R_4) v_-$$

Equating $v_+ = v_-$

$$\frac{R_4}{R_3 + R_4} u_2 = \frac{R_4 u_1 + R_3 v_o}{R_3 + R_4}$$

$$v_o = \frac{R_4}{R_3} \times (u_2 - u_1)$$

Now at 0A1 and 0A2. Current through R_1

$$i = \frac{v_1 - v_2}{R_1}$$

$$u_1 = v_1 + (v_1 - v_2)\frac{R_2}{R_1}$$

$$u_2 = v_2 - (v_1 - v_2)\frac{R_2}{R_1}$$

$$u_1 - u_2 = (v_1 - v_2) + (v_1 + v_2)\frac{2R_2}{R_1}$$

$$= (v_1 - v_2)\frac{R_1 + 2R_2}{R_1}$$

Replacing $u_1 - u_2$ we have

$$v_o = \frac{R_4}{R_3}(v_2 - v_1)\left(1 + \frac{2R_2}{R_1}\right)$$

PROBLEM 12.7 OP-AMP APPLICATIONS

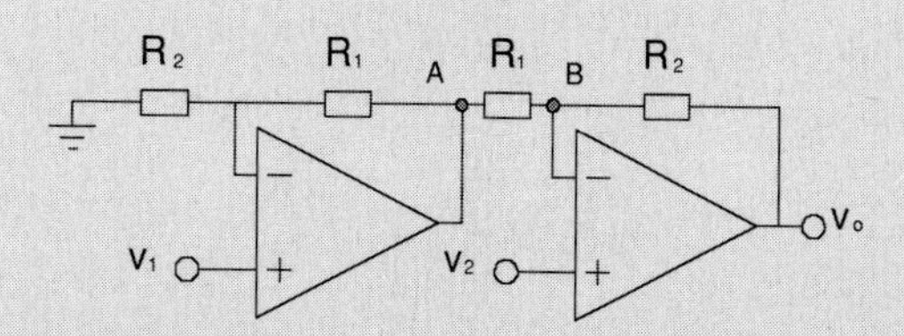

Figure 12.18

Find v_o as a function of v_1 and v_2.

Answer: $\frac{R_1 + R_2}{R_1}(v_2 - v_1)$

$$v_A = v_1\left(1 + \frac{R_1}{R_2}\right) = v_1\frac{R_1 + R_2}{R_2}$$

At the second Op - Amp

$$\frac{v_A - v_B}{R_1} = \frac{v_B - v_o}{R_2}$$

Replace $v_B = v_2$

$$(v_A - v_2)R_2 = R_1(v_2 - v_o)$$

$$R_2 v_A - R_2 v_2 = R_1 v_2 - R_1 v_o$$

Replace v_A

$$(R_1 + R_2)v_1 - R_2 v_2 = R_1 v_2 - R_1 v_o$$

$$(R_1 + R_2)v_1 - R_2 v_2 - R_1 v_2 = -R_1 v_o \quad | \times (-1)$$

$$-(R_1 + R_2)v_1 + R_1 v_2 + R_2 v_2 = R_1 v_o$$

$$(R_1 + R_2)(v_2 - v_1) = R_1 v_o$$

$$v_o = \frac{R_1 + R_2}{R_1}(v_2 - v_1)$$

PROBLEM 12.8 OP-AMP APPLICATIONS

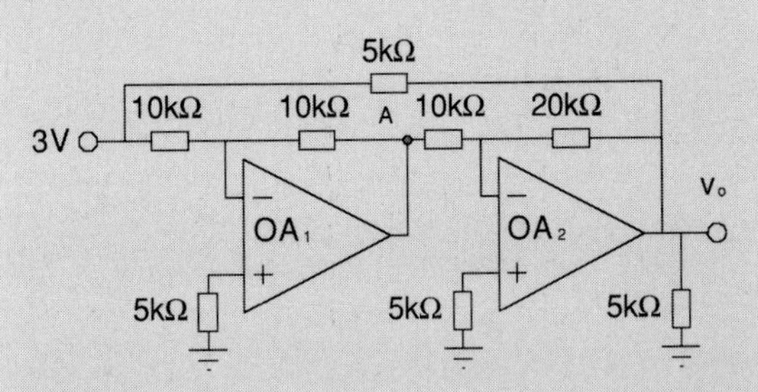

Figure 12.19

Find the current supplied by the output of OA1 and OA2.

> Answer: 0.6 mA and 2.1 mA

Current OA1

$$\frac{3}{10k} = 0.3 \text{ mA}$$

Output of OA1

$$v_A = -3\frac{10k}{10k} = -3 \text{ V}$$

The second Op -Amp has 0 V at v_-. The current also runs into A.

Current OA1

$$0.3 + 0.3 = 0.6 \text{ mA}$$

Current OA 2. This produces three currents

$$v_o = -v_A\frac{20k}{10k} = 3 \times 2 = 6 \text{ V}$$

$$I_1 = \frac{6}{5k} = 1.2 \text{ mA}$$

$$I_2 = \frac{6-3}{5k} = 0.6 \text{ mA}$$

$$I_3 = -\frac{v_A}{10k} = \frac{3}{10k} = 0.3 \text{ mA}$$

Current OA 2 = 1.2 + 0.6 + 0.3 = 2.1 mA

PROBLEM 12.9 OP-AMP APPLICATIONS

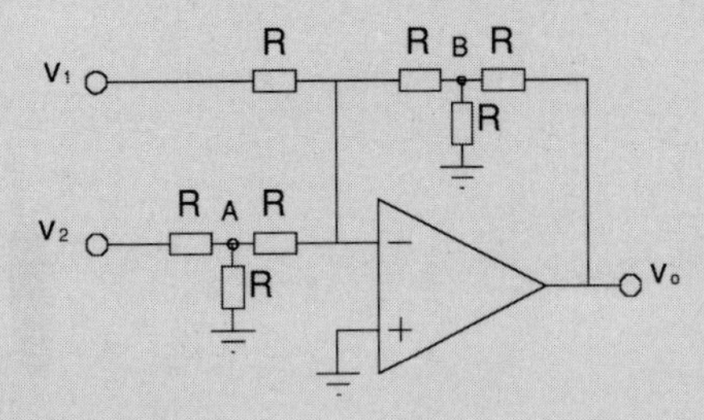

Figure 12.20

Find v_o as a function of v_1 and v_2.

> Answer: $v_o = -3v_1 - v_2$

Superposition $v_1 = 0$

Input to v_-. R in series with $R \| R$.

At A the input is

$$\frac{v_2}{3}$$

The branch to v_1 has 0 V at either end.

At point B we have

$$-\frac{v_2}{3}$$

The current from v_- to B is $v_2/3R$ and also from earth to B. So the current from B to v_o is

$$2\frac{v_2}{3R}$$

$$v_{o1} = -\frac{v_2}{3} - \frac{2v_2}{3R} \times R = -v_2$$

Superposition $v_2 = 0$

The current from v_1 is v_1/R and similarly from B to v_o the current is $2v_1/\text{R}$

$$v_{o2} = -v_1 - \frac{2v_1}{R} \times R = -3v_1$$

Total output

$$v_o = -3v_1 - v_2$$

PROBLEM 12.10 OP-AMP APPLICATIONS

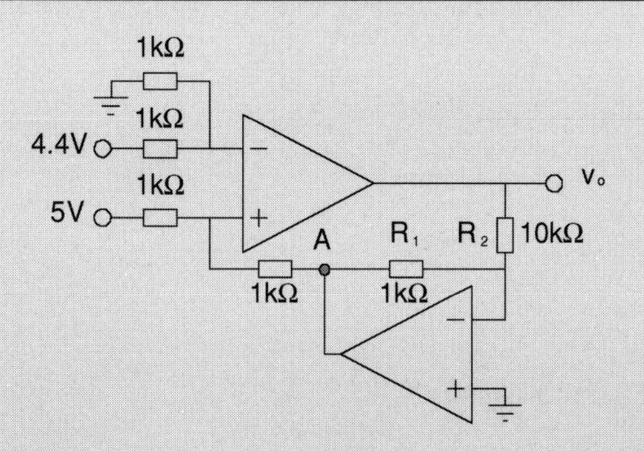

Figure 12.21

Find the value of the current through R_1.

Answer: 0.6 mA

$$v_- = \frac{4.4}{2} = 2.2 \text{ V}$$

The current from the 5.0 V source

$$i = \frac{5-2.2}{1\text{k}} = 2.8 \text{ mA}$$

$$v_+ = v_- = 2.2 \text{ V}$$

$$v_A = v_+ - iR$$

$$= 2.2 - 2.8 = -0.6 \text{ V}$$

R_1 current

$$\frac{0.6}{1\text{k}} = 0.6 \text{ mA}$$

PROBLEM 12.11 OP-AMP APPLICATIONS

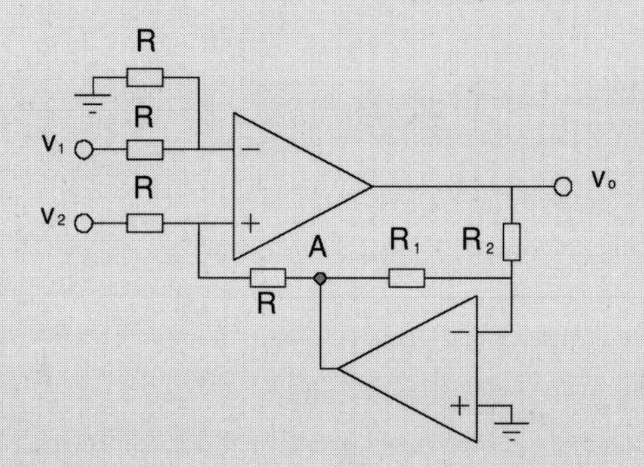

Figure 12.22

Find v_o in terms of v_1 and v_2 if $R_2 = 10R_1$.

Answer: $10(v_2 - v_1)$

The output at A from the lower Op -Amp is

$$v_A = -v_o \frac{R_1}{R_2} = -\frac{v_o}{10}$$

$$v_- = v_1 \frac{1}{2}$$

$$v_+ = v_-$$

Therefore the current from v_2 is

$$i = \frac{v_2 - v_+}{R}$$

$$v_A = v_+ - iR$$

$$= \frac{v_1}{2} - \frac{(v_2 - v_+)}{R} \times R$$

$$= \frac{v_1}{2} - v_2 + \frac{v_1}{2}$$

$$-\frac{v_o}{10} = v_1 - v_2$$

$$v_o = 10(v_2 - v_1)$$

PROBLEM 12.12 OP-AMP APPLICATIONS

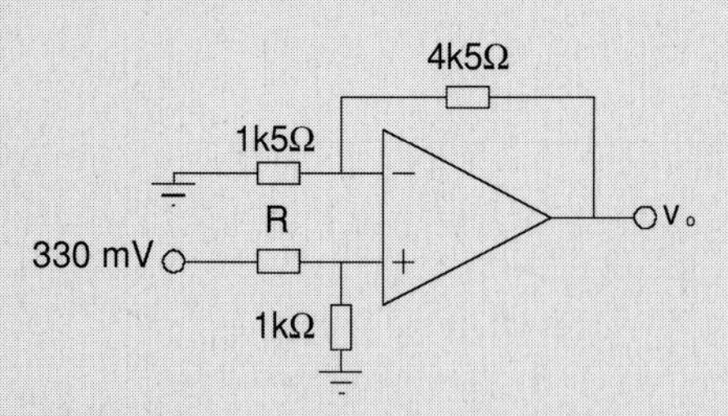

Figure 12.23

Calculate the value of R to produce an output of 880 mV.

> Answer: $R = 500\ \Omega$

Amplification

$$A = 1 + \frac{4k5}{1k5} = 4$$

$$v_- = \frac{880}{4} = 220\text{ mV}$$

$$v_+ = v_- = 220\text{ mV}$$

Proportion (or potential divider)

$$\frac{v_i}{R+1k} = \frac{v_+}{1k}$$

$$\frac{330}{R+1k} = \frac{220}{1k}$$

$$33\,000 = 22R + 22\,000$$

$$22R = 11\,000$$

$$R = 500\ \Omega$$

PROBLEM 12.13 OP-AMP APPLICATIONS

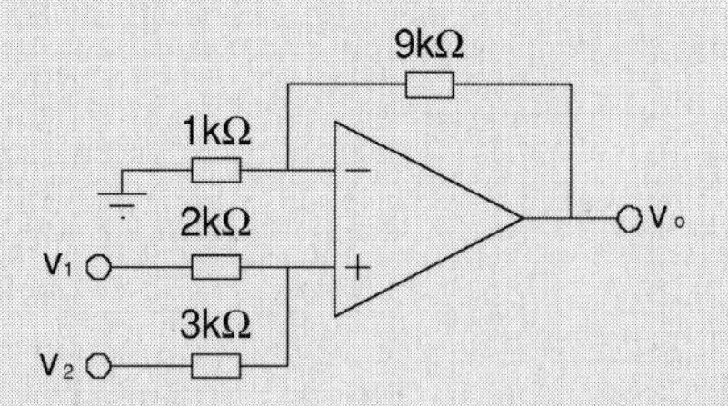

Figure 12.24

Find the value of v_o.

> Answer: $6v_1 + 4v_2$

We are going to find the value of v_+ and then, use the amplification to obtain the result.

Amplification

$$A = 1 + \frac{9k}{1k} = 10$$

$$v_+ = v_2 + \text{potential divider}$$

$$= v_2 + (v_1 - v_2) \times \frac{3k}{2k+3k}$$

$$= v_2 + (v_1 - v_2)\frac{3}{5}$$

$$= v_2 + \frac{3}{5}v_1 - \frac{3}{5}v_2$$

$$= \frac{3}{5}v_1 + \frac{2}{5}v_2$$

$$v_o = v_+ \times 10$$

$$= \left(\frac{3}{5}v_1 + \frac{2}{5}v_2\right)10$$

$$= 6v_1 + 4v_2$$

PROBLEM 12.14 OP-AMP APPLICATIONS

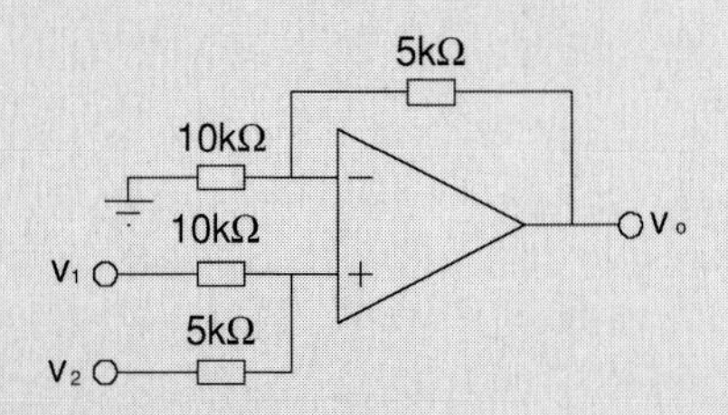

Figure 12.25

Using nodal analysis find v_o in terms of v_1 and v_2.

Answer: $v_o = \frac{v_1}{2} + v_2$

At node v_-

$$\frac{0 - v_-}{10k} + \frac{v_o - v_-}{5k} = 0 \qquad | \times 10k$$

$$0 - v_- + 2v_o - 2v_- = 0$$

$$2v_o = 3v_-$$

At node v_+

$$\frac{v_1 - v_+}{10k} + \frac{v_2 - v_+}{5k} = 0 \qquad | \times 10k$$

$$v_1 - v_+ + 2v_2 - 2v_+ = 0$$

$$v_1 + 2v_2 = 3v_+$$

Now we equate

$$v_+ = v_-$$

$$3v_+ = 3v_-$$

Therefore

$$2v_o = v_1 + 2v_2$$

$$v_o = \frac{v_1}{2} + v_2$$

PROBLEM 12.15 OP-AMP APPLICATIONS

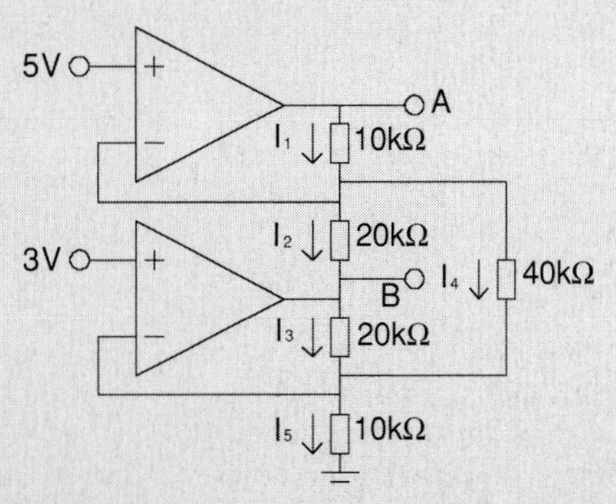

Figure 12.26

Find the voltages at A and B.

Answer: $V_A = 4$ V and $V_B = 8$ V

$$I_5 = \frac{3}{10k} = 0.3\,\text{mA}$$

$$I_4 = \frac{5-3}{40k} = 0.05\,\text{mA}$$

$$I_3 = I_5 - I_4 = 0.3 - 0.05 = 0.25\,\text{mA}$$

$$V_B = 3 + I_3R = 3 + 0.25 \times 20$$

$$= 3 + 5 = 8\text{ V}$$

$$I_2 = \frac{5-8}{20k} = -\frac{3}{20k} = -0.15\,\text{mA}$$

$$I_1 = I_4 + I_2 = 0.05 - 0.15 = -0.1\,\text{mA}$$

$$V_A = 5 + 10kI_1$$

$$= 5 - 0.1 \times 10 = 5 - 1 = 4\text{ V}$$

PROBLEM 12.16 OP-AMP APPLICATIONS

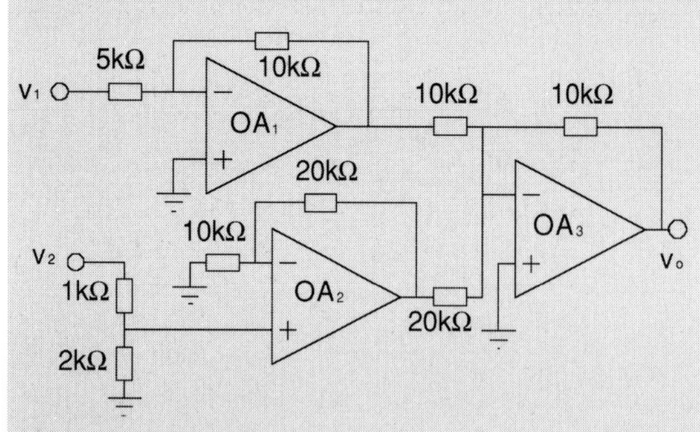

Figure 12.27

Find the value of v_o if $v_1 = 1$ V and $v_2 = 2$ V.

Answer: $v_o = 0$ V

Output OA1

$$v_{o1} = -v_1 \frac{10k}{5k} = -1 \times 2 = -2 \text{ V}$$

Output OA 2

$$v_{o2} = v_2 \times \frac{2}{3} \times \left(1 + \frac{20k}{10k}\right) = 2 \times 2 = 4 \text{ V}$$

$$v_o = -\left(-2 \times \frac{10k}{10k}\right) - 4 \times \frac{10k}{20k}$$

$$= 2 - 2 = 0 \text{ V}$$

Output at A

$$v_A = v_1\left(1 + \frac{190k}{10k}\right) = 20v_1$$

$v_B = v_2$ ($v_+ = v_-$ and no current on 10k resistor)

We now move to the final Op -Amp with inputs A and B.

Using superposition

$$v_B = 0$$

$$v_+ = v_A \times \frac{250k}{260k} = \frac{25}{26} v_A$$

$$v_{o1} = \frac{25}{26} \times v_A \times \left(1 + \frac{250k}{10k}\right)$$

$$= \frac{25}{26} \times v_A \times 26$$

$$= 25v_A = 25 \times 20v_1 = 500v_1$$

$$v_A = 0$$

$$v_+ = v_B \frac{10k}{260k} = v_B \frac{1}{26}$$

$$v_{o2} = v_B \frac{1}{26} \times 26 = v_B = v_2$$

Total output

$$v_o = v_2 + 500v_1$$

PROBLEM 12.17 OP-AMP APPLICATIONS

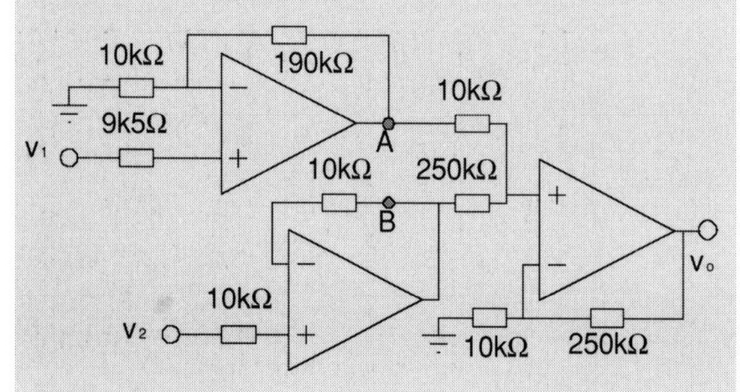

Figure 12.28

Demonstrate that the mathematical model for this Op-Amp is $V_o = V_2 + 500V_1$.

Answer: Yes

PROBLEM 12.18 OP-AMP APPLICATIONS

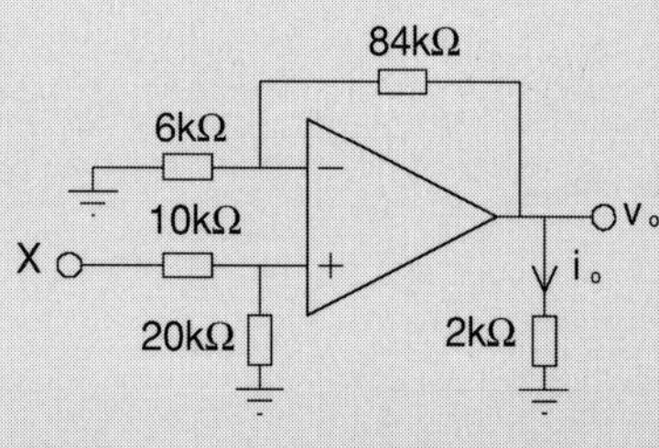

Figure 12.29

Find the gain of the Op-Amp and the output current i_o.

Answer: Gain = 15, $i_o = 5X$ mA

Gain

$$1+\frac{84\text{k}}{6\text{k}}=1+14=15$$

Output current

$$v_+ =\frac{20\text{k}X}{20\text{k}+10\text{k}}=\frac{2}{3}X$$

$$v_o =15v_+$$

$$15\frac{2}{3}X=10X$$

$$i_o =\frac{10X}{2\text{k}}=5X\text{ mA}$$

Using nodal analysis

$$\frac{X-v_+}{10\text{k}}+\frac{0-v_+}{20\text{k}}=0$$

$$2X-2v_+-v_+=0$$

$$2X=3v_+$$

$$v_+=\frac{2}{3}X$$

$$\frac{0-v_-}{6\text{k}}+\frac{v_o-v_-}{84\text{k}}=0$$

$$-14v_-+v_o-v_-=0$$

$$v_o =15v_-$$

$$v_+ =v_-$$

$$v_o =15v_-$$

$$=15\frac{2}{3}X$$

$$=10X$$

$$i_o =\frac{10X}{2\text{k}}=5X\text{ mA}$$

PROBLEM 12.19 OP-AMP APPLICATIONS

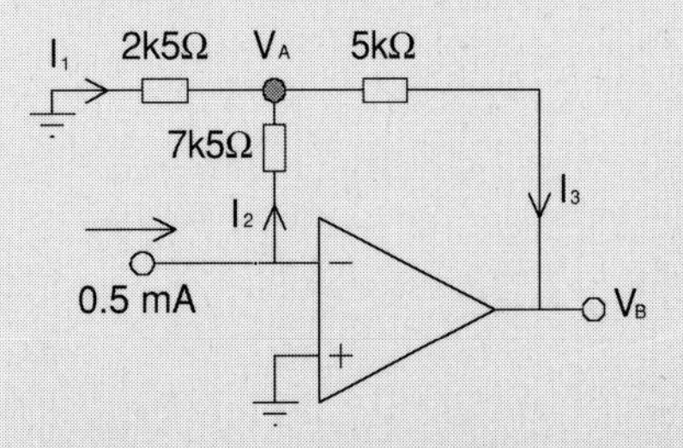

Figure 12.30

Find the value of I_1, I_2, I_3, V_A and V_B.

Answer: 1.5 mA, 0.5 mA, 2 mA, –3.75 V and –13.75 V

$$I_2 =0.5\text{ mA}$$

$$v_A =-0.5\times10^3\times7\text{k}5=-3.75\text{ V}$$

$$I_1 =\frac{3.75}{2\text{k}5}=1.5\text{ mA}$$

$$I_3 =I_1+I_2=2\text{ mA}$$

$$v_B =v_A-I_3 5\text{k}$$

$$=-3.75-2\times10^{-3}\times5\text{k}$$

$$=-3.75-10$$

$$=-13.75\text{ V}$$

PROBLEM 12.20 OP-AMP APPLICATIONS

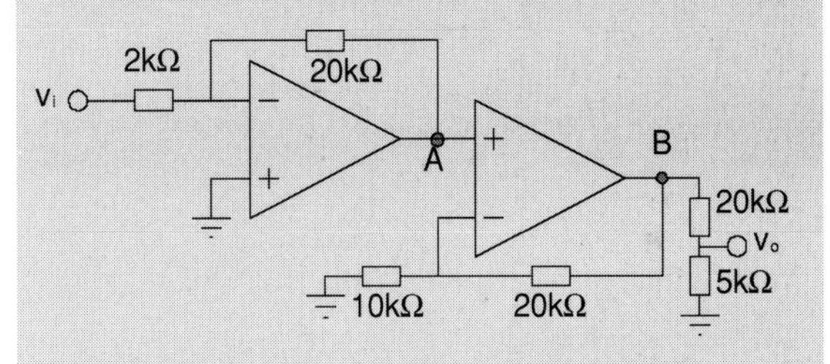

Figure 12.31

Find v_o if the input is 100 mV.

Answer: –600 mA

The first Op - Amp is inverting .

$$v_A = -v_i \frac{20k}{20k} = -10v_i$$

The second Op - Amp is non - inverting

$$v_B = v_A\left(1 + \frac{20k}{10k}\right)$$

$$= 3v_A$$

Potential divider output

$$v_o = v_B \frac{5k}{5k + 20k} = \frac{v_B}{5}$$

$$v_o = \frac{1}{5} \times 3v_A$$

$$= \frac{1}{5} \times 3 \times (-10v_1)$$

$$= -6v_1$$

$$v_o = -600 \text{ mV}$$

13

Operational amplifiers: oscillators

The circuit shown in Figure 13.1, represents a very user-friendly oscillator. It is a relaxation oscillator. This is the name given to an oscillator that produces a square wave.

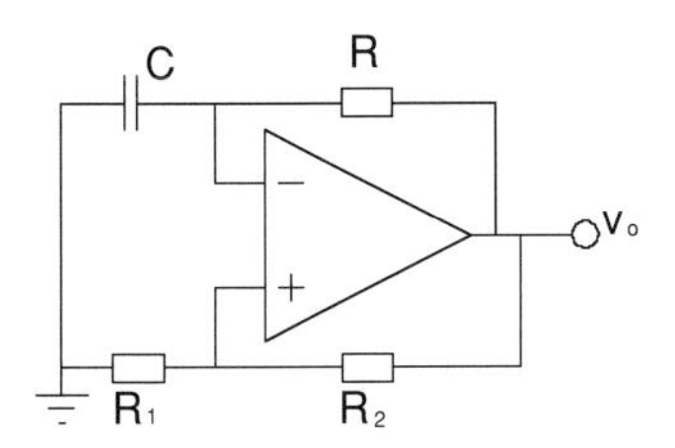

Figure 13.1

Due to the positive feedback loop, (R_1 and R_2), the output v_o will be unstable at switch on and the ouput will saturate at a positive or negative value. If the output is positive, the capacitor will charge through the R_C circuit towards the positive output voltage. The voltage v_+ at the Op-Amp is given by the potential divider network formed by R_1 and R_2 and this acts as a reference voltage point.

The capacitor would normally charge towards the positive saturation point, but before it gets to this value, it will match the value of v_+ and at this point everything will change. As the negative input v_-, will be larger than v_+, the output will saturate at the negative value. Now that we have a negative output value, the process can be repeated in the other direction, this time, the capacitor discharges. As the output changes from positive to negative and so on, the output is a square wave.

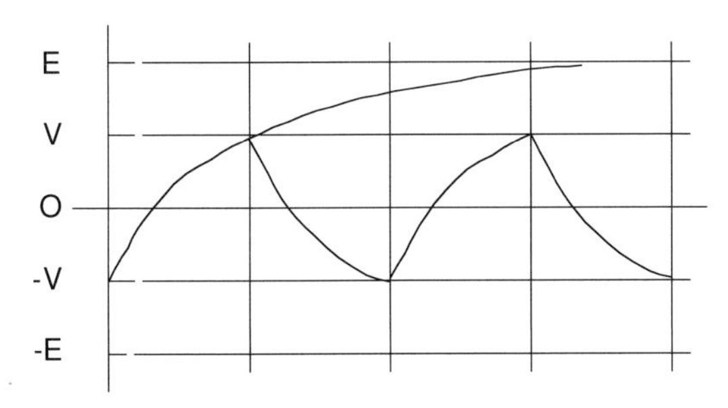

Figure 13.2

We now examine Figure 13.2, to find the frequency of oscillation. Clearly this has something to do with the charge and discharge curves.

E and $-E$ are the values of the saturated outputs, positive and negative, respectively. V and $-V$ are the maximum and minimum values at the positive feedback differential input V_+.
So

$$V = \frac{ER_1}{R_1 + R_2}$$

A simpler case is shown in Figure 13.3. If v is

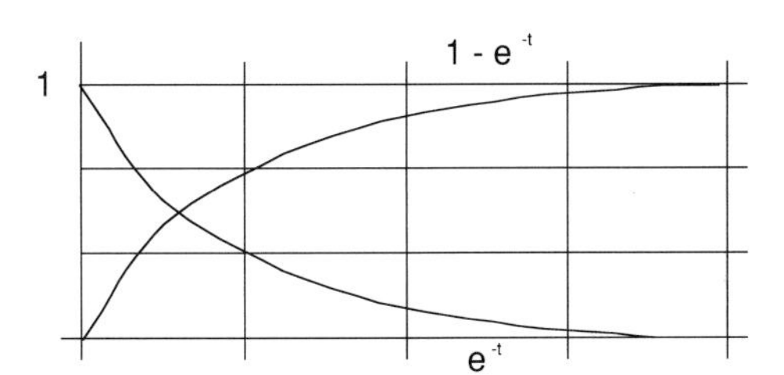

Figure 13.3

the instant voltage at the capacitor, or at v_- and if we are starting from zero the equation of the line would be

$$v = (E + V)(1 - e^{-\frac{t}{RC}})$$

But as we actually start from $-V$ we have to subtract $-V$. So the charging curve will be

$$v = (E + V)(1 - e^{-\frac{t}{RC}}) - V$$

$$v = E + V - V - (E + V)e^{-\frac{t}{RC}}$$

$$v = E - (E + V)e^{-\frac{t}{RC}}$$

Some people would have been able to write this equation by just looking at Figure 13.2.

Similarly, as they are symmetrical

$$v = -E + (E + V)e^{-\frac{t}{RC}}$$

This applies to the discharge curve, from V to $-V$.

Now that we have defined the charge and discharge curves, we can see that in order to calculate the frequency, or the period of this oscillation, we have two options. First, we can use the first equation and make $v = V$. That will give us half the time we are looking for, or secondly, we can use the second equation and use $v = -V$. Both alternatives should give the same result and you can verify this, as we are only going to carry out the second alternative here.

$$v = -V$$

$$-V = -E + (E + V)e^{-\frac{t}{RC}}$$

$$E - V = (E + V)e^{-\frac{t}{RC}}$$

$$\frac{E - V}{E + V} = e^{-\frac{t}{RC}}$$

Remember that V is the crossing point given by the potential divider effect of R_1 and R_2. In order to make it simpler, we will allow here $R_1 = R_2$.

$$\text{V} = \frac{R_1}{R_1 + R_2}E = \frac{E}{2}$$

$$e^{-\frac{t}{RC}} = \frac{E - \frac{E}{2}}{E + \frac{E}{2}} = \frac{1}{3}$$

Taking natural logarithms from both sides, we get

$$\frac{-t}{RC} = -1.0986 \text{ (say } -1.1)$$

therefore

$$\boxed{T = 2.2RC}$$

$$\boxed{f = \frac{0.455}{RC} \text{ Hz}}$$

Different resistors

More often than not, R_1 will be different from R_2 (see Figure 13.1). This can be taken into account by using the potential divider effect of R_1 and R_2 onto the crossing point defined above as V.

We start by rewriting the exponential equation we have already developed:

$$e^{-\frac{t}{RC}} = \frac{E - V}{E + V}$$

and with

$$V = \frac{R_1}{R_1 + R_2}E$$

As V is expressed as a function of E, the whole expression should simplify and appear as a function of R_1 and R_2 only.

$$\frac{E - V}{E + V} = \frac{E - \frac{R_1}{R_1 + R_2}E}{E + \frac{R_1}{R_1 + R_2}E} \qquad | \div E$$

$$= \frac{1 - \frac{R_1}{R_1 + R_2}}{1 + \frac{R_1}{R_1 + R_2}} \qquad | \times (R_1 + R_2)$$

$$= \frac{R_1 + R_2 - R_1}{R_1 + R_2 + R_1}$$

$$= \frac{R_2}{2R_1 + R_2}$$

We rewrite the expression as

$$e^{-\frac{t}{RC}} = \frac{R_2}{2R_1 + R_2}$$

Taking natural logarithms from both sides, we get:

$$-\frac{t}{RC} = \ln\frac{R_2}{2R_1 + R_2}$$

$$\boxed{t = -RC \ln\frac{R_2}{2R_1 + R_2}}$$

(ln is natural logarithm, base e)

This expression can be used to find any parameters when you know the other parameters. Note that if $R_1 = R_2$, we are back at the situation of the original formula.

555 Timer

The 555 timer circuit is seen in Figure 13.4. Basically, there are two Op-Amps that control the setting and resetting of a flip-flop register. The flip-flop operates an NPN transistor and the ouput stage. A capacitor, part of an external circuit, can be charged through a suitable circuit and it can be discharged by the NPN transistor.

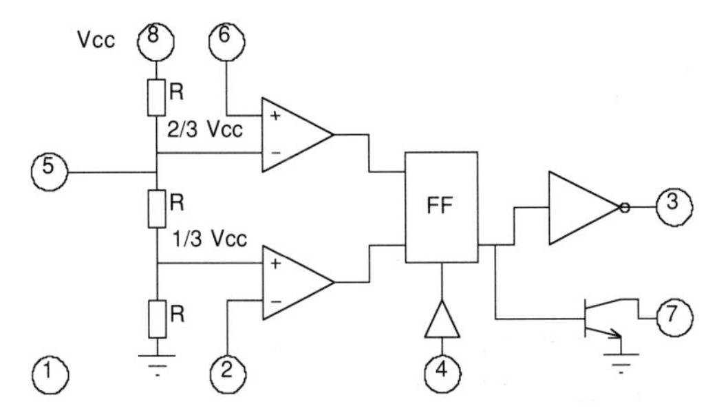

Figure 13.4

In Figure 13.5, we see such a configuration to operate the 555 as a multivibrator. There are many other possible uses of this chip as you will already know.

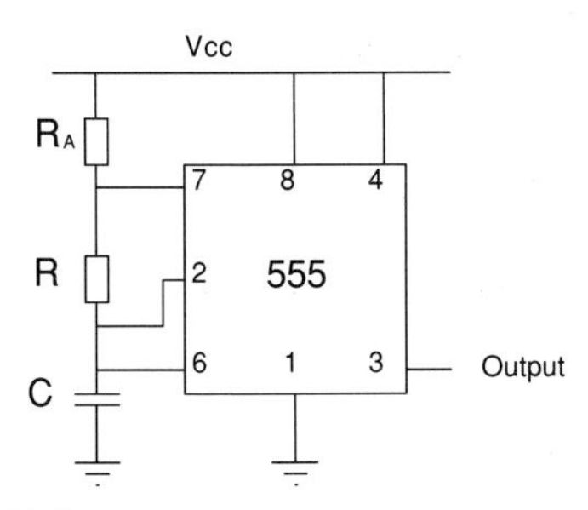

Figure 13.5

The capacitor C charges through R and R_A to V_{CC}, but before it reaches that value, the threshold on pin 6 (2/3 of V_{CC}) is reached and the flip flop is set. Output 3 goes high. The discharge transistor (on pin 7) goes on. The capacitor is forced to discharge by the low on pin 7. Whilst the capacitor is charged through R and R_A, the discharge takes place only through R and it is quicker. This feature can be useful in cases where you want to vary the duty cycle, or the mark to space ratio of the waveform.

The capacitor is now discharging and the value is monitored on pin 2 where it is compared with the 1/3 V_{CC} at the comparator. As soon as the capacitor voltage falls below 1/3 V_{CC}, the Op-Amp comparator will give a positive output and will reset the flip-flop, and we go back to square one.

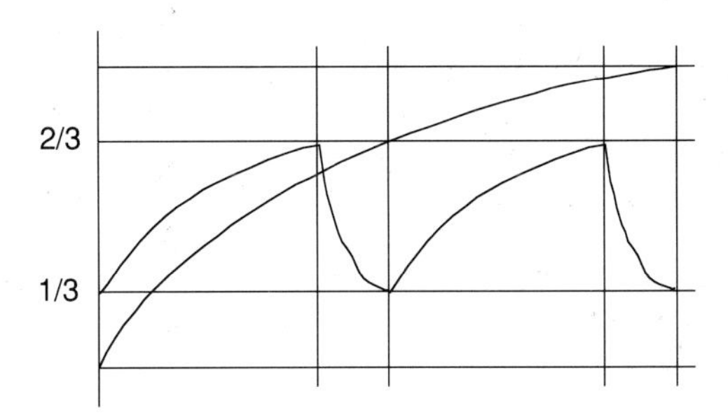

Figure 13.6

The waveform can be seen in Figure 13.6. We can identify three different curves. First, the one starting from 0. This is the initial charge, when we have just switched on. The capacitor is completely discharged and it will take longer to reach 2/3 V_{CC}. Once the multivibrator is working normally, the operation will be the charge curve from 1/3 V_{CC} to 2/3 V_{CC} and the discharge curve from 2/3 V_{CC} to 1/3 V_{CC}.

The calculations are lengthy, but quite simple and straightforward. They will be done as follows:

Charge curve
- Time to charge from 0 to 2/3 V_{CC}
- Time to charge from 0 to 1/3 V_{CC}
- The difference is the charge time.

Discharge curve
- Time to discharge from V_{CC} to 2/3 V_{CC}
- Time to discharge from V_{CC} to 1/3 V_{CC}
- The difference is the discharge time.

From 0 to $\frac{2}{3}V_{CC}$

$$\frac{2}{3} = 1(1 - e^{-\frac{t}{(R+R_A)C}})$$

$$e^{-\frac{t}{(R+R_A)C}} = \frac{1}{3}$$

Take natural logarithms and multiply by −1

$$\frac{t}{(R+R_A)C} = 1.0986 \quad t = 1.0986(R+R_A)C$$

From 0 to $\frac{1}{3}V_{CC}$

$$\frac{1}{3} = 1(1 - e^{-\frac{t}{(R+R_A)C}})$$

$$e^{-\frac{t}{(R+R_A)C}} = \frac{2}{3}$$

$$\frac{t}{(R+R_A)C} = 0.455 \quad t = 0.4055(R+R_A)C$$

Therefore

$$\Delta t = 0.693(R+R_A)C$$

$$t = 0.693(R+R_A)C$$

From V_{CC} to $\frac{2}{3}V_{CC}$

$$\frac{2}{3} = e^{-\frac{t}{RC}}$$

$$\frac{t}{RC} = 0.4055 \quad t = 0.4055RC$$

From V_{CC} to $\frac{1}{3}V_{CC}$

$$\frac{1}{3} = e^{-\frac{t}{RC}}$$

$$\frac{t}{RC} = 1.0986 \quad t = 1.0986RC$$

Therefore

$$\Delta t = 1.0986RC \quad 0.4055RC$$
$$= 0.693RC$$

The period of oscillation is therefore

$$T = 0.693(R + R_A)C + 0.693RC$$

$$T = 0.693(2R + R_A)C$$

$$f = \frac{1.443}{(2R+R_A)C}$$

Sinusoidal oscillators

In a sinusoidal oscillator, or linear oscillator, positive feedback is used to make it unstable and with the use of a frequency selective network, we can achieve oscillation at a given frequency.

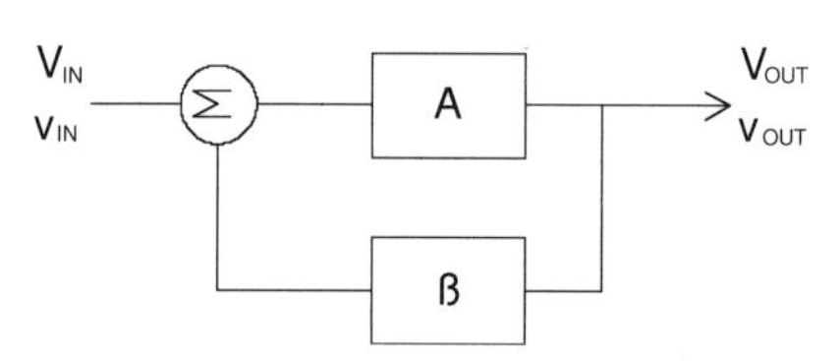

Figure 13.7

The general arrangement is shown in Figure 13.7.

$$v_{out} = A(v_{in} + \beta v_{out})$$

$$v_{out}(1 - \beta A) = Av_{in}$$

$$\frac{v_{out}}{v_{in}} = \frac{A}{1-\beta A}$$

If $\beta A = 1$ then, $\frac{v_{out}}{v_{in}} = \infty$

Although mathematically you are not allowed to divide by zero, in electronic terms this means that you can have an output, without having an input.

The conditions for oscillation are:

Magnitude of $\beta A = 1$

Overall phase shift = 0°

In some cases these conditions, although sufficient to maintain oscillations, are not sufficient to start the oscillations. It is therefore necessary to increase the gain at the start and decrease it to unity, once the oscillations have started. We will not be looking into these details. We will only look at two sinusoidal oscillators, the Wien bridge and the phase shift oscillator.

Wien bridge oscillator

The simplest oscillator is based on the Wien bridge. The Op-Amp is connected in the

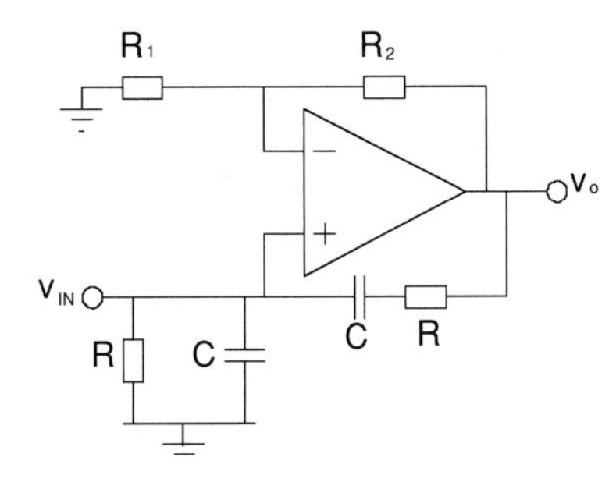

Figure 13.8

non–inverting configuration as in Figure 13.8. The Op-Amp gain is therefore $1 + R_2/R_1$. The loop gain is βA.

$$\beta A = \frac{\left(1+\frac{R_2}{R_1}\right)Z_P}{Z_P + Z_S}$$

Z_P is the parallel impedance and Z_S is the serial impedance.

The gain on the negative feedback determines the amplitude of the oscillation. The Wien bridge, on the positive feedback, determines the frequency of oscillation.

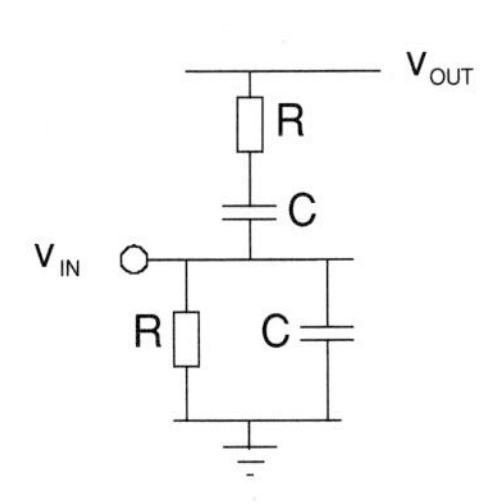

Figure 13.9

$$\frac{Z_P}{Z_P + Z_S} = \frac{\dfrac{R \times \dfrac{1}{j\omega C}}{R + \dfrac{1}{j\omega C}}}{\dfrac{R \times \dfrac{1}{j\omega C}}{R + \dfrac{1}{j\omega C}} + R + \dfrac{1}{j\omega C}} \mid \times \left(R + \frac{1}{j\omega C}\right)$$

$$= \frac{\dfrac{R}{j\omega C}}{\dfrac{R}{j\omega C} + \left(R + \dfrac{1}{j\omega C}\right)\left(R + \dfrac{1}{j\omega C}\right)} \mid \div R$$

$$\frac{Z_P}{Z_P + Z_S} = \frac{\dfrac{1}{j\omega C}}{\dfrac{1}{j\omega C} + \left(1 + \dfrac{1}{j\omega CR}\right)\left(R + \dfrac{1}{j\omega C}\right)} \mid \times j\omega C$$

$$= \frac{1}{1 + \left(j\omega C + \dfrac{1}{R}\right)\left(R + \dfrac{1}{j\omega C}\right)}$$

$$= \frac{1}{1 + j\omega CR + 1 + 1 + \dfrac{1}{j\omega CR}}$$

$$= \frac{1}{3 + j\omega CR + \dfrac{1}{j\omega CR}}$$

$$= \frac{1}{3 + j\left(\omega CR - \dfrac{1}{\omega CR}\right)}$$

We examine the term $Z_P/(Z_P + Z_S)$ in Figure 13.9.

The phase angle will be zero, if the imaginary part is zero. Therefore:

$$\omega CR = \frac{1}{\omega CR}$$

$$\boxed{\omega_o = \frac{1}{CR}}$$

At this frequency

$$\frac{Z_P}{Z_P + Z_S} = \frac{1}{3}$$

To obtain a loop gain of unity we need $R_2/R_1 = 2$

$$\beta A = (1+2)\frac{1}{3} = 1$$

Phase shift oscillator

The circuit is shown in Figure 13.10. In this case in order to obtain oscillations, the loop gain must have an overall phase shift of zero. As we are using the inverting input which provides 180° shift, the R_C network must produce another 180°, to make the overall phase shift 0° (see Figure 13.11).

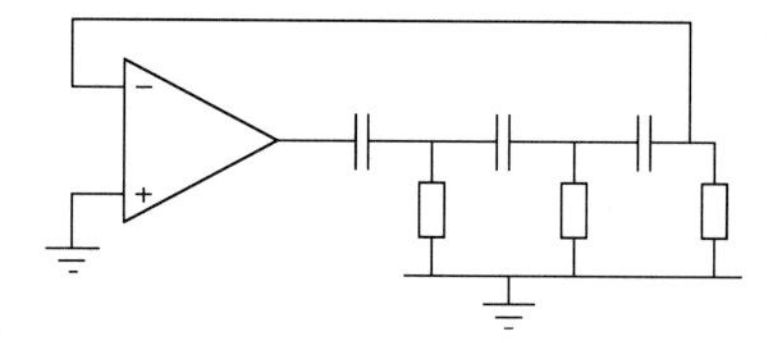

Figure 13.10

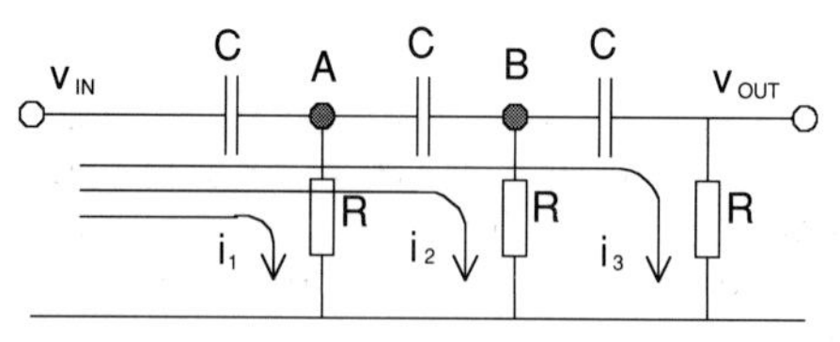

Figure 13.11

We want the value of v_o/v_i. We start from the output and work backwards to the input

$$v_B = v_o + i_3\frac{1}{j\omega C} \qquad i_3 = \frac{v_o}{R}$$

$$= v_o + \frac{v_o}{j\omega CR}$$

$$v_B = v_o\frac{j\omega CR + 1}{j\omega CR}$$

$$v_A = v_B + (i_2 + i_3)\frac{1}{j\omega C} \quad i_2 = \frac{v_B}{R}$$

$$= v_B + \frac{i_2}{j\omega C} + \frac{i_3}{j\omega C}$$

$$= v_B\left(1 + \frac{1}{j\omega CR}\right) + \frac{v_o}{j\omega CR}$$

$$= v_B\frac{j\omega CR + 1}{j\omega CR} + \frac{v_o}{j\omega CR}$$

We now replace v_B with its value obtained a few lines above.

$$v_A = v_o\left(\frac{j\omega CR + 1}{j\omega CR}\right)^2 + \frac{v_o}{j\omega CR}$$

$$= v_o\frac{(j\omega CR + 1)^2 + j\omega CR}{(j\omega CR)^2}$$

We now move to v_I

$$v_I = v_A + (i_1 + i_2 + i_3)\frac{1}{j\omega C} \qquad i_1 = \frac{v_A}{R}$$

$$v_A = v_o\left(\frac{j\omega CR + 1}{j\omega CR}\right)^2 + \frac{v_o}{j\omega CR}$$

$$= v_o\frac{(j\omega CR + 1)^2 + j\omega CR}{(j\omega CR)^2}$$

We now replace v_A by the value found a few lines above.

$$v_I = v_o\frac{(j\omega CR + 1)^2 + j\omega CR}{(j\omega CR)^2}\frac{j\omega CR + 1}{j\omega CR} +$$

$$+ \frac{v_B}{j\omega CR} + \frac{v_o}{j\omega CR}$$

$$v_I = v_o\frac{(j\omega CR + 1)^3 + (j\omega CR)^2 + j\omega CR}{(j\omega CR)^3} +$$

$$+ \frac{v_B}{j\omega CR} + \frac{v_o}{j\omega CR}$$

We now replace v_B and make the denominator $(j\omega CR)^3$

$$v_I = \frac{(j\omega CR + 1)^3 + (j\omega CR)^2 + j\omega CR}{(j\omega CR)^3} +$$

$$+ v_o\frac{(j\omega CR + 1)}{(j\omega CR)^2}\frac{j\omega CR}{j\omega CR} + v_o\frac{(j\omega CR)^2}{(j\omega CR)^3}$$

$$v_I = v_o\frac{(j\omega CR + 1)^3 + (j\omega CR)^2 + j\omega CR + \ldots}{(j\omega CR)^3}$$

$$\frac{\ldots + (j\omega CR)^2 + j\omega CR + (j\omega CR)^2}{(j\omega CR)^3}$$

$$v_I = v_o\frac{(j\omega CR + 1)^3 + 3(j\omega CR)^2 + 2j\omega CR}{(j\omega CR)^3}$$

The numerator can now be simplified separately. We will come back to this equation once we have simplified the numerator.

Numerator only:

$$(j\omega CR + 1)^3 + 3(j\omega CR)^2 + 2j\omega CR$$

$$= j^3\omega^3C^3R^3 + 3j^2\omega^2C^2R^2 + 3j\omega CR + 1$$

$$+3j^2\omega^2C^2R^2 + 2j\omega CR$$

$$= -j\omega^3C^3R^3 - 6\omega^2C^2R^2 + 5j\omega CR + 1$$

$$= 1 - 6\omega^2C^2R^2 + j(-\omega^3C^3R^3 + 5\omega CR)$$

Now that we have simplified the numerator we can go back to the general expression.

$$\frac{v_\text{I}}{v_\text{o}} = \frac{1 - 6\omega^2C^2R^2 + j(-\omega^3C^3R^3 + 5\omega CR)}{(j\omega CR)^3}$$

Inverting

$$\frac{v_\text{o}}{v_\text{I}} = \frac{-j\omega^3C^3R^3}{1 - 6\omega^2C^2R^2 + j(-\omega^3C^3R^3 + 5\omega CR)}$$

Multiplying by j up and down

$$\frac{v_\text{o}}{v_\text{I}} = \frac{\omega^3C^3R^3}{\omega^3C^3R^3 - 5\omega CR + j(1 - 6\omega^2C^2R^2)}$$

For oscillations the imaginary part must be zero

$$6\ \omega^2C^2R^2 = 1$$

$$\boxed{\omega_\text{o} = \frac{1}{\sqrt{6}CR}}$$

Remember that in this configuration we started with the inverting input, therefore the amplifier provides 180° shift. In order to achieve an overall phase shift of 0°, we need another 180° shift from the RC network.

Let us see what the effect of the network at the oscillation frequency is.

$$\frac{v_\text{o}}{v_\text{i}} = \frac{\omega^2C^2R^2}{\omega^2C^2R^2 - 5} = \frac{\frac{1}{6}}{\frac{1}{6} - 5} = \frac{1}{1 - 30}$$

$$\frac{v_\text{o}}{v_\text{i}} = -\frac{1}{29}$$

This minus sign signifies an 180° shift which we were looking for, but as the network produces an attenuation of 29, the amplifier must provide a gain of 29 to maintain the oscillations.

Complete phase shift oscillator

The complete amplifier can be seen in Figure 13.12. In this circuit $R_2 = R_3 = R_4 = R$. The numbers are only to facilitate our discussion. The situation here is that R_4 is in parallel with R_1. R_4 goes directly to earth, whereas R_1 goes to virtual earth. If R_4 value is R and the R_1 value is also R, $R_1 \parallel R_4$ will be $R/2$, which will upset the C_R circuitry.

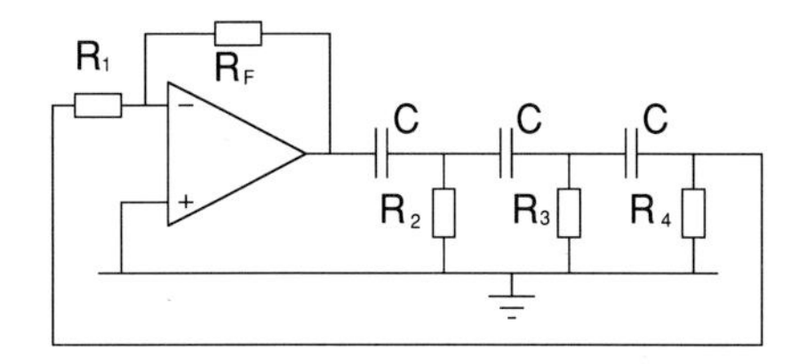

Figure 13.12

This problem can be alleviated by one of two possibilities. In the first place, we can make R_1 significantly larger than R_4 so that the loading effect of R_1 on R_4 will be reduced or even made negligible.

As a second alternative, we can make sure that:

$$R_2 = R_3 = R$$

and that:

$$R_1 \parallel R_4 = R$$

This can be achieved in a number of ways, for instance: $R_1 = 2R$ and $R_4 = 2R$. Additionally, this circuit needs a gain of 29 which can be achieved by making $R_\text{F} = 29R_1$.

Loading effect

A brief point to be mentioned here is the effect of loading on a circuit. The basic network is

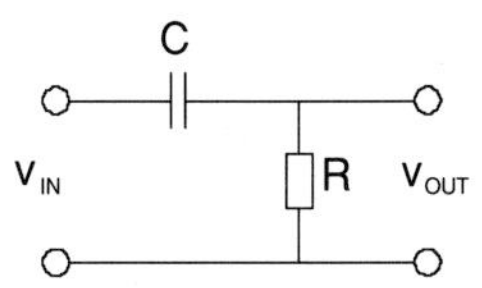

Figure 13.13

shown in Figure 13.13. It has the following transfer function:

$$\frac{v_\text{o}}{v_\text{i}} = \frac{R}{R + \frac{1}{j\omega C}} = \frac{j\omega CR}{j\omega CR + 1}$$

It would be easy to say that three of these networks would give a transfer function of

$$\frac{(j\omega CR)^3}{(j\omega CR+1)^3}$$

This would only be possible if between each network we have a buffer amplifier of amplification 1. Then the networks would be separated and the above formula would apply.

Due to the effect of loading the third network onto the second and the second network onto the first one, the correct formula is the one that we deduced earlier on page 226 and which gave a slight but very important difference.

$$\frac{v_o}{v_i} = \frac{(j\omega CR)^3}{(j\omega CR+1)^3 + 3(j\omega CR)^2 + 2j\omega CR}$$

It is also possible to exchange C and R in the phase shift oscillator. In this case we would arrive at a slightly different formula.

PROBLEM 13.1 OP-AMP OSCILLATORS

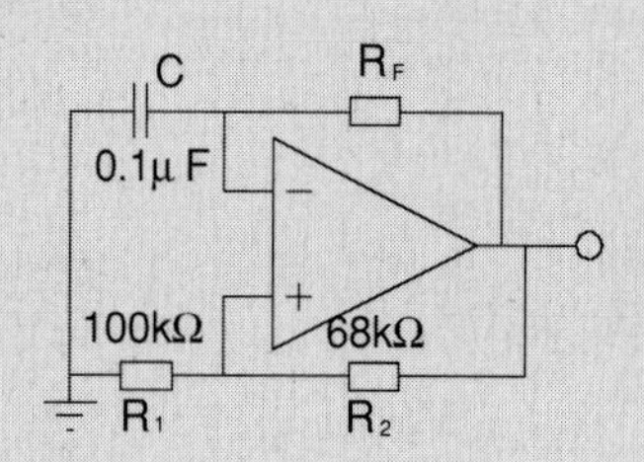

Figure 13.14

Find R_F to produce 100 Hz oscillations.

Answer: 36k45 Ω

$$2t = \frac{1}{f} = \frac{1}{100} = 10 \text{ ms}$$

$$t = 5 \text{ ms}$$

$$t = -RC \ln \frac{R_2}{2R_1 + R_2}$$

$$5 \times 10^{-3} = -R \times 0.1 \times 10^{-6} \times \ln \frac{68}{2 \times 100 + 68}$$

$$= -R \times 10^{-7} \times \ln \frac{68}{268}$$

$$= -R \times 10^{-7} \times (-1.371479)$$

$$R = \frac{5 \times 10^{-3}}{10^{-7} \times 1.371479} = 36\text{k}45\ \Omega$$

PROBLEM 13.2 OP-AMP OSCILLATORS

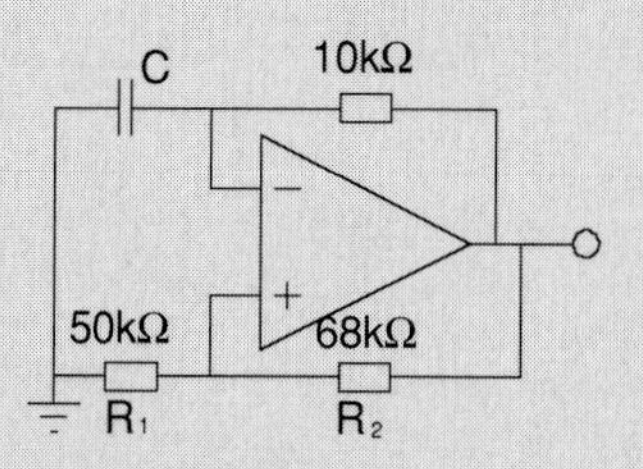

Figure 13.15

Find the value of C that will make the circuit oscillate at 125 Hz.

Answer: 0.4422 µF

$$2t = \frac{1}{f} = \frac{1}{125} = 0.008$$

$$t = 0.004 \text{ s}$$

$$t = -RC \ln \frac{R_2}{2R_1 + R_2}$$

$$0.004 = -10^4 \times C \times \ln \frac{68}{2 \times 50 + 68}$$

$$= -10^4 \times C \times (-0.904\,5)$$

$$C = \frac{0.004}{10^4 \times 0.904\,5} = 0.442\,2\ \mu\text{F}$$

PROBLEM 13.3 OP-AMP OSCILLATORS

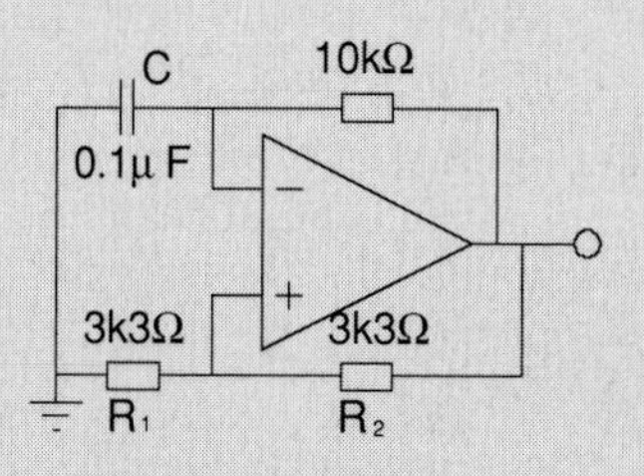

Figure 13.16

Find the frequency of oscillation.

Answer: 455 Hz

Charging time

$$t = -RC \ln \frac{R_2}{2R_1 + R_2}$$

as $R_1 = R_2$

$$t = -RC \ln \frac{1}{3}$$

$$t = RC \ln 3$$

$$= 10^4 \times 0.1 \times 10^{-6} \times 1.0986$$

$$t = 1.0986 \text{ ms}$$

$$f = \frac{1}{2t}$$

$2t$ is the time for charging and discharging

$$f = \frac{1}{2 \times 1.0986 \times 10^{-3}} = 455 \text{ Hz}$$

PROBLEM 13.4 OP-AMP OSCILLATORS

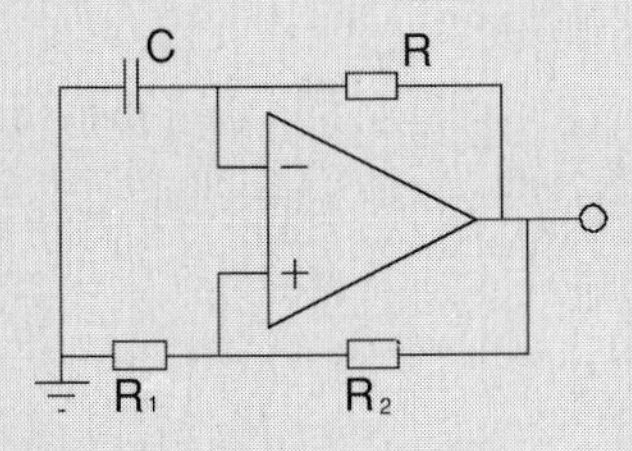

Figure 13.17

Design a relaxation oscillator where R_2 is three times R_1 and it oscillates at 1 kHz.

> Answer: R_1 = 10 kΩ, R_2 = 30 kΩ, R = 1 kΩ and C = 0.978 μF

Many other combinations of components will produce the same result. Here we see only one case.

Any value of R_1 and R_2 will do as long as they are in the requested proportion.

Say: R_1 = 10 kΩ R_2 = 30 kΩ

We now need to find the RC combination.

$$t = -RC \ln \frac{R_2}{2R_1 + R_2}$$

$$t = -RC \ln \frac{30}{50}$$

$$= RC \times 0.511$$

t is given by the frequency

$$2t = \frac{1}{f} \qquad t = \frac{1}{2f} = \frac{1}{2 \times 10^3}$$

$$t = 0.5 \text{ ms}$$

Therefore

$$RC = 0.5 \times 10^{-3} \times \frac{1}{0.511} = 0.978 \times 10^{-3}$$

If R = 1 kΩ, then, C = 0.978 μF.

PROBLEM 13.5 OP-AMP OSCILLATORS

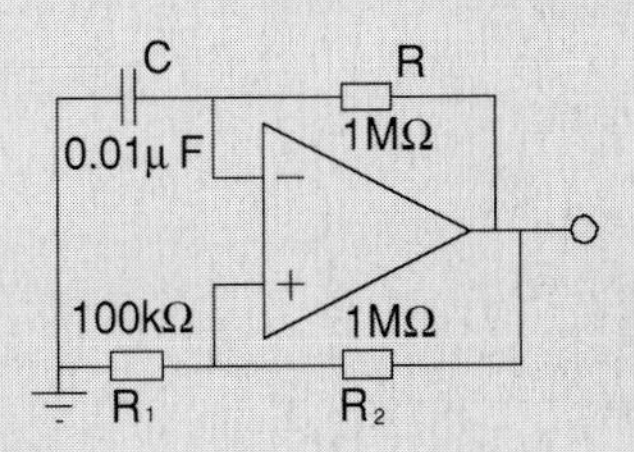

Figure 13.18

Find the frequency of oscillation.

> Answer: 274.7 Hz

$$t = -RC \ln \frac{R_2}{2R_1 + R_2}$$

$$= -RC \ln \frac{10^6}{2 \times 10^5 + 10^6}$$

$$= -RC \ln \frac{10}{12} = RC \times 0.182$$

$$t = 10^6 \times 0.01 \times 10^{-6} \times 0.182 = 1.82 \times 10^{-3}$$

$$f = \frac{1}{2t} = \frac{1}{2 \times 1.82 \times 10^{-3}} = 274.7 \text{ Hz}$$

PROBLEM 13.6 OP-AMP OSCILLATORS

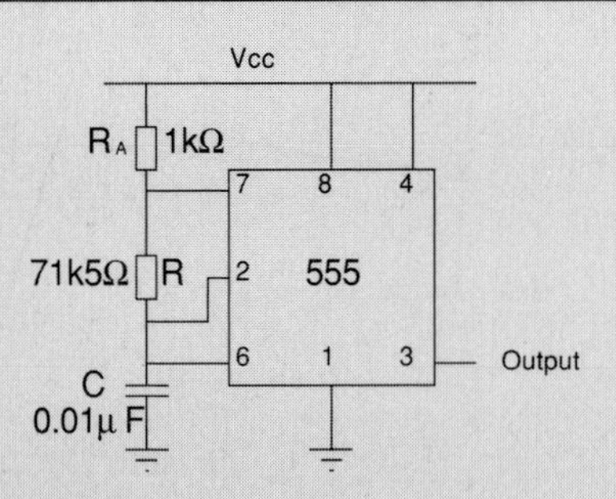

Figure 13.19

Find the frequency of oscillation.

Answer: 1002 Hz

$$f = \frac{1.443}{(2R + R_A)C}$$

$$= \frac{1.443}{(2 \times 71\,500 + 1000)0.01 \times 10^{-6}}$$

$$= 1002 \text{ Hz}$$

PROBLEM 13.7 OP-AMP OSCILLATORS

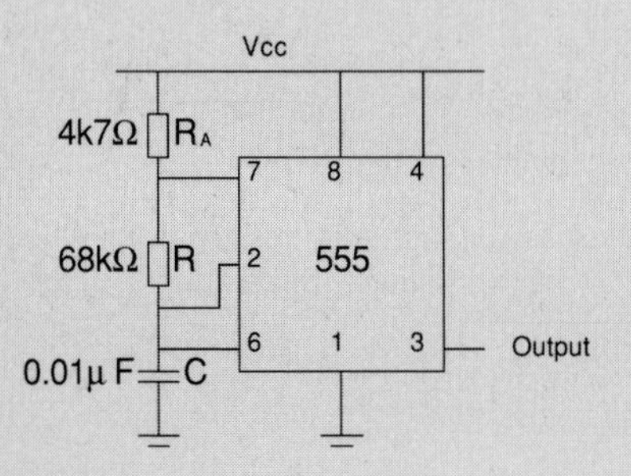

Figure 13.20

Find the frequency of oscillation as shown and then the oscillation when R is exchanged with R_A.

Answer: 1025.6 Hz and 1864.3 Hz

As shown

$$f_1 = \frac{1.443}{(2R + R_A)C}$$

$$= \frac{1.443}{(2 \times 6800 + 4700)0.01 \times 10^{-6}}$$

$$= \frac{1.443}{140\,700 \times 10^{-8}} = 1025.6 \text{ Hz}$$

Exchanged

$$f_2 = \frac{1443}{(2R + R_A)C}$$

$$= \frac{1.443}{(2 \times 4700 + 6800)0.01 \times 10^{-6}}$$

$$= \frac{1.443}{77\,400 \times 10^{-8}}$$

$$= 1864.3 \text{ Hz}$$

PROBLEM 13.8 OP-AMP OSCILLATORS

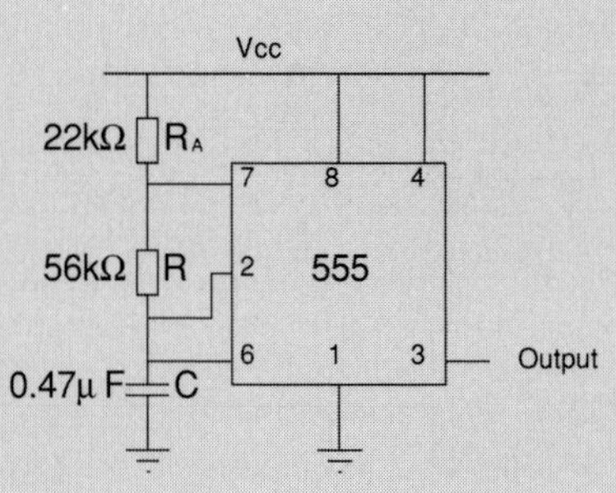

Figure 13.21

Find the duty cycle $\frac{t_c}{t_d}$.

Answer: 39 : 28

Charging constant, through both resistors

$$\tau_1 = (R_A + R)C = 78\,000 \times 47 \times 10^{-6}$$

$$= 0.036\,66$$

$$\tau_2 = RC = 56\,000 \times 0.47 \times 10^{-6}$$

$$= 0.0263\,2$$

Charging time

$$t_c = 0.693(R + R_A)C$$

Discharging time

$$t_d = 0.693RC$$

Ratio

$$\frac{t_c}{t_d} = \frac{R + R_A}{R} = \frac{78}{56} = \frac{39}{28}$$

PROBLEM 13.9 OP-AMP OSCILLATORS

Design a Wien bridge oscillator to work at 2.5 kHz.

> Answer: $C = 0.001\ \mu F$, $R = 63\,662\ \Omega$, $R_1 = 12\ k\Omega$, $R_2 = 24\ k\Omega$

Many components will achieve the same result. So we start by fixing an arbitrary value of capacitance.

$$C = 0.001\ \mu F$$

$$\omega = 2\pi f = \frac{1}{RC}$$

$$R = \frac{1}{2\pi fC}$$

$$= \frac{1}{2\pi \times 2.5 \times 10^3 \times 0.001 \times 10^{-6}}$$

$$R = 63\,662\ \Omega$$

In order to sustain the oscillations we need the condition

$$\frac{R_2}{R_1} = 2$$

Again we have several possibilities, so we 'fix' one and calculate the other. Say

$$R_1 = 12\ k\Omega$$

Therefore

$$R_2 = 24\ k\Omega$$

PROBLEM 13.10 OP-AMP OSCILLATORS

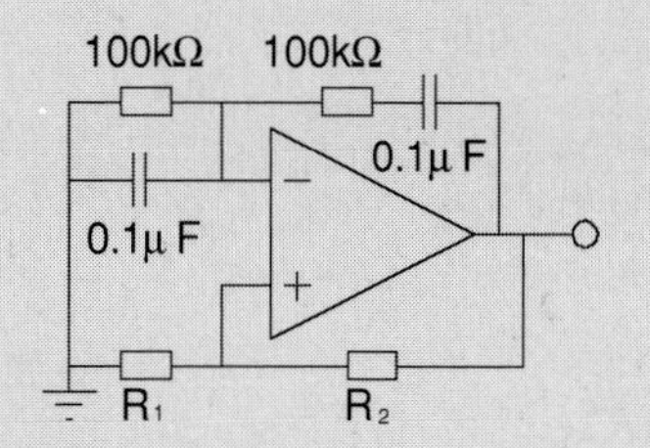

Figure 13.22

Find the frequency of oscillation.

> Answer: 15.92 Hz

$$\omega = \frac{1}{CR} = \frac{1}{0.1 \times 10^{-6} \times 10^5} = 100\ \text{rad/s}$$

$$f = \frac{100}{2\pi} = 15.92\ \text{Hz}$$

> NOTE: R_1 and R_2 will have to be adjusted to give the appropriate gain, but that is not part of the question.

PROBLEM 13.11 OP-AMP OSCILLATORS

Design a phase shift oscillator to work at 120 Hz.

> Answer: $C = 0.1\ \mu F$, $R = 5414.56\ \Omega$, $R_1 = 24\ k\Omega$, $R_2 = 696\ k\Omega$

$$\omega = \frac{1}{\sqrt{6}CR}$$

$$f = \frac{1}{2\pi\sqrt{6}CR}$$

We decide arbitrarily on a value of C. Say

$$C = 0.1\,\mu\text{F}$$

$$R = \frac{1}{2\pi\sqrt{6}Cf}$$

$$= \frac{1}{2\pi\sqrt{6} \times 0.1 \times 10^{-6} \times 120}$$

$$= 5414.56\ \Omega$$

In order to sustain oscillations, the amplifier must provide a gain of 29. A feedback loop is required of

$$\frac{R_2}{R_1} = 29$$

Say

$$R_1 = 24\ \text{k}\Omega$$

$$R_2 = 696\ \text{k}\Omega$$

PROBLEM 13.12 OP-AMP OSCILLATORS

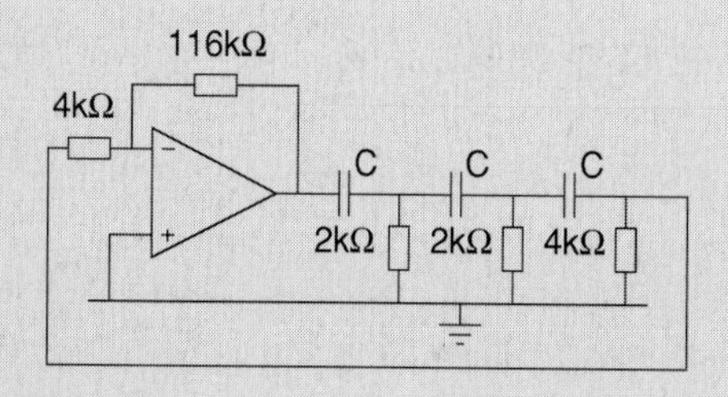

Figure 13.23

Find the value of C for the circuit to oscillate at 100 Hz.

Answer: 0.325 μF

$$f = 100 = \frac{1}{2\pi\sqrt{6}RC}$$

$$C = \frac{1}{2\pi\sqrt{6}fR}$$

$$= \frac{1}{2\pi\sqrt{6}100 \times 2000} = 0.325\,\mu\text{F}$$

Note that

$$4\text{k}\|4\text{k} = 2\ \text{k}\Omega$$

The last effective R from the CR network is also 2 kΩ. The gain required is also achieved.

$$\frac{116\text{k}}{4\text{k}} = 29$$

for the circuit to oscillate at 100 Hz.

PROBLEM 13.13 MISCELLANEOUS PROBLEMS

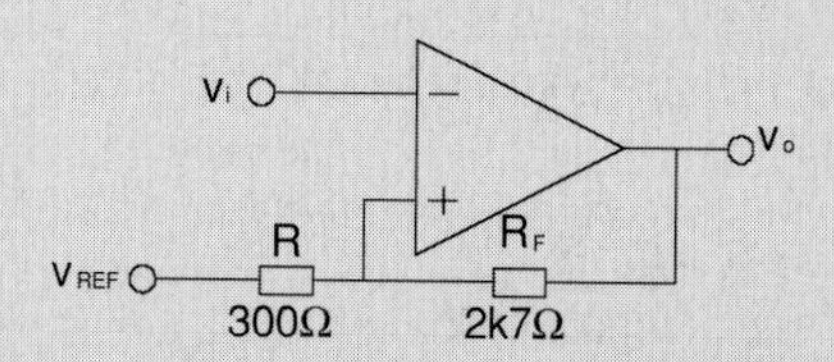

Figure 13.24

This Op-Amp with positive feedback, saturates at 10 V and 0 V. Calculate the hysteresis (i.e. the difference between high and low trigger points) for a reference voltage of 6 V.

Answer: 1 V

$$\text{Hys} = V_{\text{PP}}\frac{R}{R + R_{\text{F}}} = 10\frac{300}{2\text{k}7 + 300} = 1\ \text{V}$$

This is regardless of the value of v_{REF}.

Case a)

$v_{\text{REF}} = 6$ V output 10 V

$$v_+ = 6 + (10 - 6) \times \frac{300}{2\text{k}7 + 300} = 6.4\ \text{V}$$

Case b)

$v_{\text{REF}} = 6$ V output 0 V

$$v_+ = 6\frac{2\text{k}7}{300 + 2\text{k}7} = 5.4\ \text{V}$$

$$\text{Hys} = 6.4 - 5.4 = 1\ \text{V}$$

PROBLEM 13.14 MISCELLANEOUS PROBLEMS

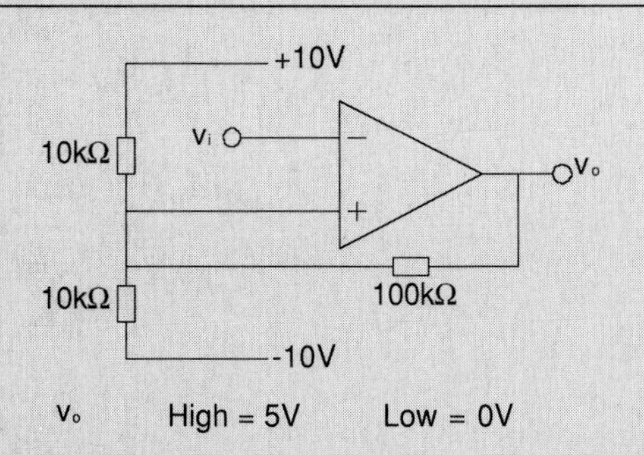

Figure 13.25

Determine the upper and lower switching points of the comparator (with positive feedback).

Answer: High = 0.238 V, low = 0 V

$$\frac{10-v_+}{10k}+\frac{5-v_+}{100k}+\frac{-10-v_+}{10k}=0 \quad | \times 100k$$

$$100-10v_+ + 5 - v_+ - 100 - 10v_+ = 0$$

$$21v_+ = 5$$

$$v_+ = \frac{5}{21} = 0.238 \text{ V}$$

Case b) v_o at 0 V

The circuit is now completely symmetrical and we should expect v_+ to be 0 V. We will see.

$$\frac{10-v_+}{10k}+\frac{0-v_+}{100k}+\frac{-10-v_+}{10k}=0 \quad | \times 100k$$

$$100-10v_+ - v_+ - 100 - 10v_+ = 0$$

$$21v_+ = 0$$

$$v_+ = 0 \text{ V}$$

PROBLEM 13.15 MISCELLANEOUS PROBLEMS

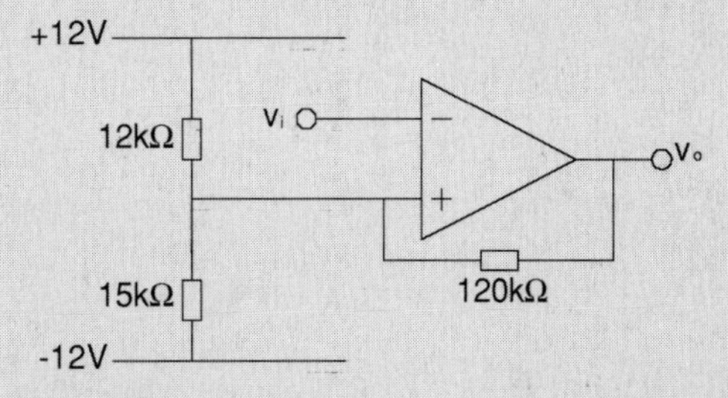

Figure 13.26

The output of the comparator varies from 0 V to 5 V. Find the upper and lower switching levels.

Answer: 1.53 V and 1.26 V

This is a case of a small amount of positive feedback to provide hysteresis for the comparator. There are many ways of solving this problem. We will use Thevenin–Norton conversions. At the 5 V output we have:

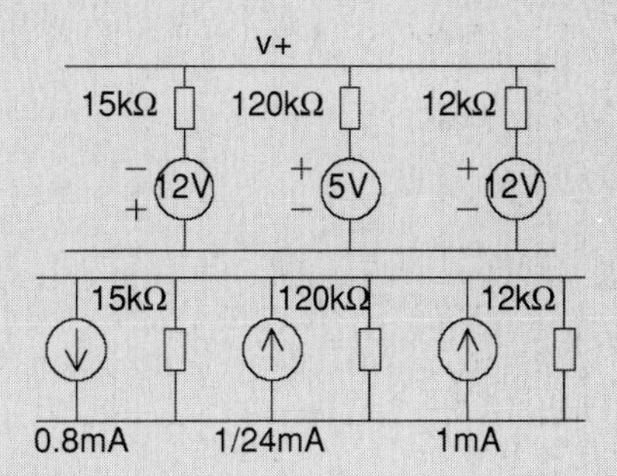

Figure 13.27

All the resistors can be joined in one

$$R = 15\text{k}||120\text{k}||12\text{k} = 6315.789\ \Omega$$

All the currents are joined into one

$$\frac{1}{24}+1-0.8=\frac{5.8}{24}$$

$$v_+ = \frac{5.8}{24}\times 10^3 \times 6315.789 = 1.526\,3 \text{ V}$$

For the second level the output is 0 V.

The combined resistor is the same.

$$R = 6315.789$$

The two currents are

$$1-0.8 = 0.2 \text{ mA}$$

$$v_+ = 0.2\times 10^{-3}\times 6315.789 = 1.263 \text{ V}$$

PROBLEM 13.16 MISCELLANEOUS PROBLEMS

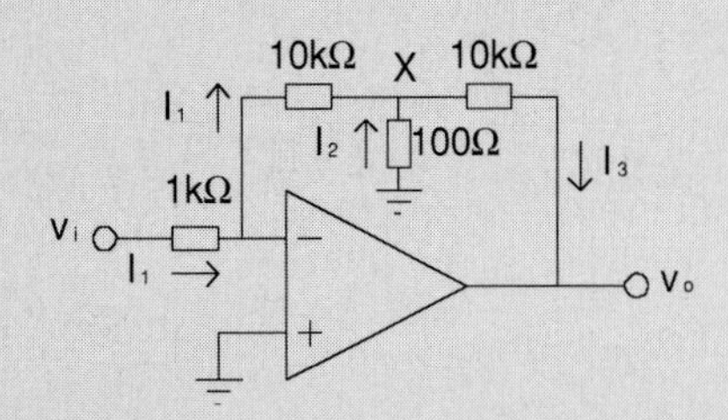

Figure 13.28

a) Find $\frac{v_o}{v_i}$

b) If $v_i = 5$ mV calculate the voltage at X and at v_o.

Answer: a) –1020, b) –50 mV and 5.1 V

$$v_x = -v_i \frac{10\text{k}}{1\text{k}} = -10v_i$$

$$I_1 = \frac{v_i}{1\text{k}}$$

$$I_2 = \frac{10v_i}{100} = \frac{v_i}{10}$$

$$I_3 = I_1 + I_2 = \frac{v_i}{1000} + \frac{v_i}{10} = \frac{101v_i}{1000}$$

$$v_o = v_x - I_3 \times 10\text{k}$$

$$= -10v_i - \frac{101v_i}{1000} \times 10\,000$$

$$= -10v_i - 1010v_i = -1020v_i$$

$$\frac{v_o}{v_i} = -1020$$

For $v_i = 5$ mV

$$v_x = -10v_i = -50 \text{ mV}$$

$$v_o = -1020v_i = -5.1 \text{ V}$$

PROBLEM 13.17 MISCELLANEOUS PROBLEMS

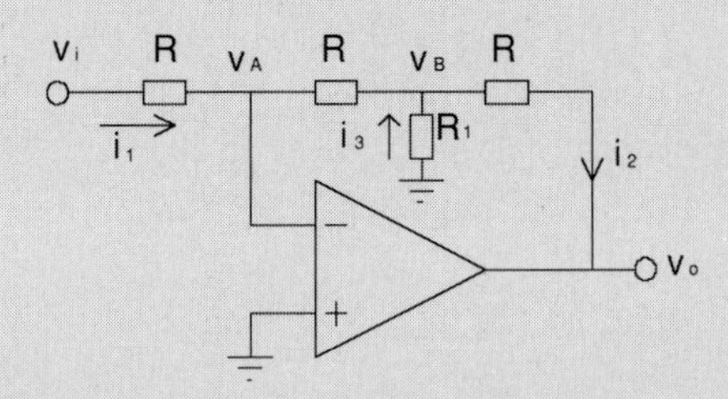

Figure 13.29

R is 10 kΩ and $v_o = -100v_i$. Calculate R_1.

Answer: 102 Ω

$v_A = 0$ V (virtual earth)

$v_B = -v_i$ (equal to $i_1 \times 10$k)

$$v_o = -100v_i$$

$$i_2 = \frac{v_B - v_o}{10\text{k}} = \frac{-v_i - (-100v_i)}{10\text{k}}$$

$$= \frac{99v_i}{10\text{k}}$$

$$v_B = -i_3 R_1$$

$$i_3 = \frac{v_i}{R_1}$$

KCL $i_1 + i_3 = i_2$

$$\frac{v_i}{10\text{k}} + \frac{v_i}{R_1} = \frac{99v_i}{10\text{k}}$$

$$\frac{1}{10\text{k}} + \frac{1}{R_1} = \frac{99}{10\text{k}} \qquad | \times 10\text{k}R_1$$

$$R_1 + 10\text{k} = 99R_1$$

$$10\text{k} = 98R_1$$

$$R_1 = \frac{10\text{k}}{98} = 102.040\,8\ \Omega$$

PROBLEM 13.18 MISCELLANEOUS PROBLEMS

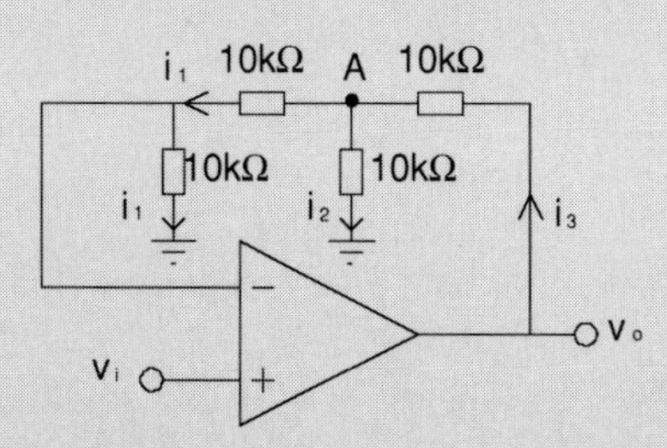

Figure 13.30

Find the value of the transfer function $\frac{v_o}{v_i}$.

Answer: $\frac{v_o}{v_i} = 5$

$v_+ = v_- = v_i$

$i_1 = \frac{v_i}{10k}$

There is no current going to v_-, i_1 appears in two branches. At point A we have $2v_i$

$$i_2 = \frac{2v_i}{10k} = 2i_1$$

$$i_3 = i_2 + i_1 = 3i_1$$

$$v_o = v_A + i_3 10k$$

$$= 2v_i + 3i_1 10k$$

$$= 2v_i + 3v_i$$

$$= 5v_i$$

$$\frac{v_o}{v_i} = 5$$

PROBLEM 13.19 MISCELLANEOUS PROBLEMS

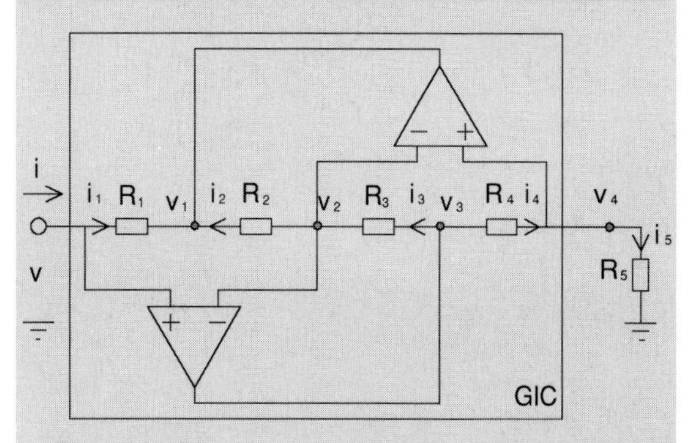

Figure 13.31

Find the value of R_{IN}.

Answer: $R_{IN} = \frac{R_1}{R_2}\frac{R_3}{R_4}R_5$

$$v = v_2 = v_4$$

$$i = i_1 \qquad i_2 = i_3 \qquad i_4 = i_5$$

$$i_5 = i_4 = \frac{v_4}{R_5} = \frac{v}{R_5}$$

$$v_3 = v_4 + i_4 R_4 = v + v\frac{R_4}{R_5}$$

$$= v\left(1 + \frac{R_4}{R_5}\right)$$

$$i_3 = \frac{v_3 - v_2}{R_3} = \frac{v\left(1 + \frac{R_4}{R_5}\right) - v}{R_3} = v\frac{1}{R_3}\frac{R_4}{R_5}$$

$$v_1 = v_2 - i_2 R_2 = v - \frac{R_2}{R_3}v\frac{R_4}{R_5}$$

$$= v\left(1 - \frac{R_2}{R_3}\frac{R_4}{R_5}\right)$$

$$i_1 = \frac{v - v_1}{R_1} = \frac{1}{R_1}\left(v - v\left(1 - \frac{R_2}{R_3}\frac{R_4}{R_5}\right)\right)$$

$$i_1 = \frac{v}{R_1}\frac{R_2}{R_3}\frac{R_4}{R_5}$$

$$R_{IN} = \frac{v}{i} = v\frac{R_1}{v}\frac{R_3}{R_2}\frac{R_5}{R_4}$$

$$R_{IN} = \frac{R_1}{R_2}\frac{R_3}{R_4}R_5$$

The general impedance converter will change the value of R_5 by a factor made out of the four internal resistances, as shown.

PROBLEM 13.20 MISCELLANEOUS PROBLEMS

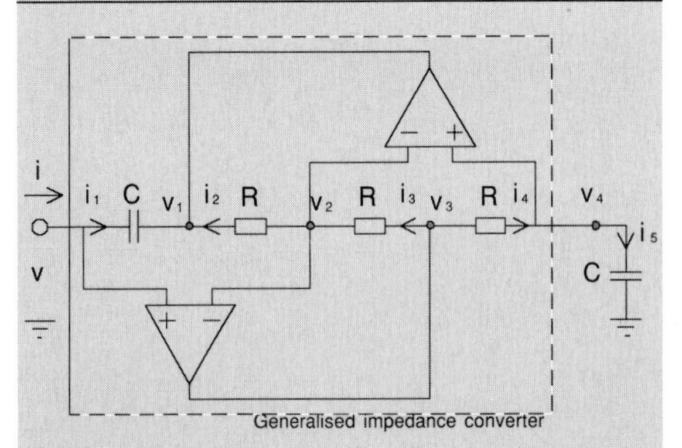

Figure 13.32

Find the value of Z_{IN}.

Answer: $\dfrac{1}{\omega^2 C^2 R}$

Due to the properties of the Op-Amps, high input impedance and the virtual short circuit input, we have

$$v = v_2 = v_4$$

$$i = i_1 \qquad i_2 = i_3 \qquad i_4 = i_5$$

$$i_5 = v_4 j\omega C = vj\omega C$$

$$v_3 = v_4 + i_4 R = v + vj\omega CR$$

$$= v(1 + j\omega CR)$$

$$i_3 = \frac{v_3 - v_2}{R} = v(1 + j\omega CR) - v$$

$$= \frac{vj\omega CR}{R} = vj\omega C$$

$$v_1 = v_2 - i_2 R = v(1 - j\omega CR)$$

$$i_1 = (v - v_1)j\omega C = (v - v(1 - j\omega CR))j\omega C$$

$$Z_{IN} = \frac{v}{i} = \frac{1}{(1 - 1 + j\omega CR)j\omega C}$$

$$= -\frac{1}{\omega^2 C^2 R}$$

In this case there is a frequency -dependent negative resistance.

14

Bandwidth gain considerations

At midband frequency, the capacitors of a bipolar junction transistor (BJT) amplifier, C_1, C_2 and C_E are all conducting. They are designed to be a short circuit at the frequencies of interest which correspond to the midband. This can be seen in Figure 14.1. We now look more closely at the left corner frequencies in Figure 14.1. There

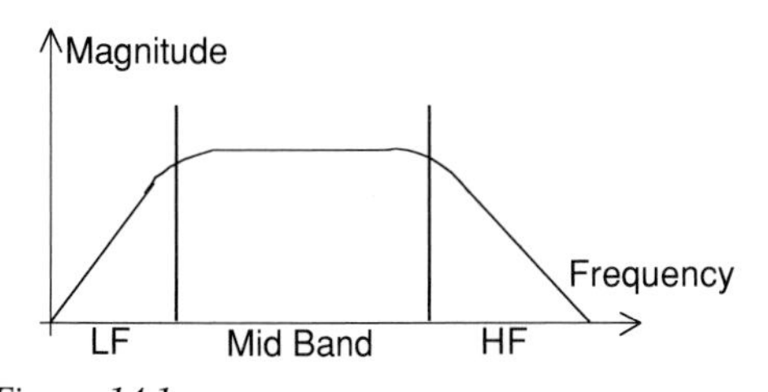

Figure 14.1

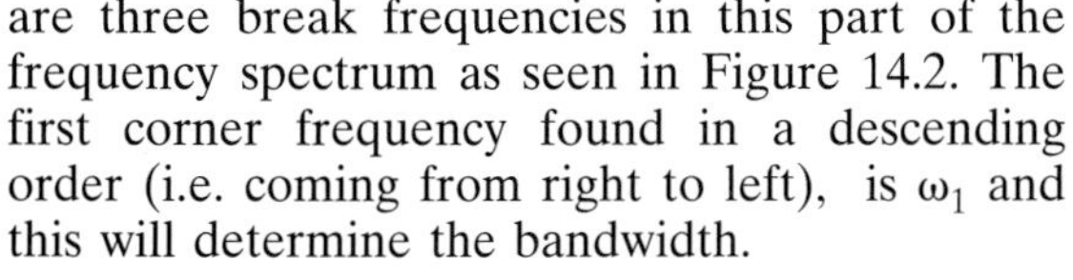

are three break frequencies in this part of the frequency spectrum as seen in Figure 14.2. The first corner frequency found in a descending order (i.e. coming from right to left), is ω_1 and this will determine the bandwidth.

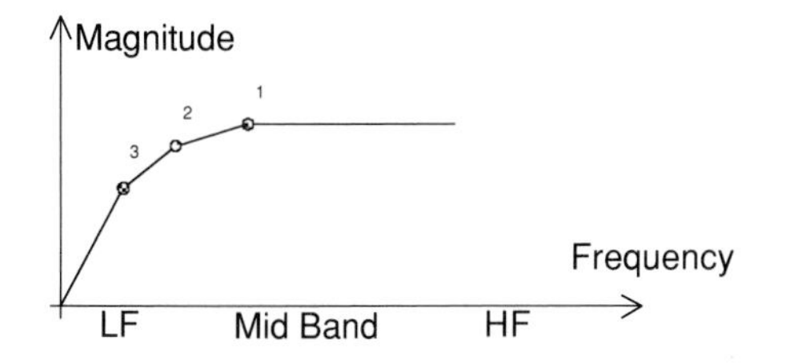

Figure 14.2

In most applications, it is required that the gain is constant down to a given frequency. The gain is considered constant if it stays within

$$\frac{1}{\sqrt{2}} = 0.707$$

This is:

$$20 \log 0.707 = -3.010\,3 \text{ dB}$$

Therefore, the definition of bandwidth is the range of frequencies where the gain, expressed in decibels stays within a 3 dB gap.

Frequency response of an *RC* network

Using our knowledge of Bode plots, we can examine the network shown in Figure 14.3.

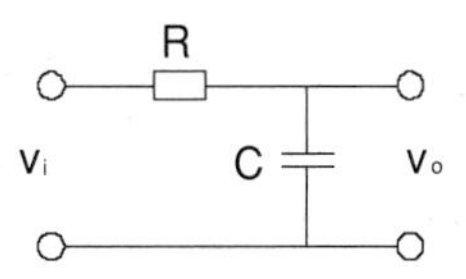

Figure 14.3

The transfer function is:

$$\frac{v_o}{v_i} = \frac{\frac{1}{j\omega C}}{R + \frac{1}{j\omega C}} \qquad | \times j\omega C$$

$$= \frac{1}{j\omega CR + 1}$$

We identify this as a type D Bode function.

There is a corner frequency at:

$$\omega CR = 1$$

$$\omega = \frac{1}{CR}$$

As can be seen in Figure 14.4.

This Bode plot shows the error, which is greatest at the corner frequency. This error can be calculated at the corner frequency.

The magnitude or gain is:

$$G = 20 \log \frac{1}{\sqrt{2}}$$

$$G = 20 \log 0.707 \approx -3 \text{ dB}$$

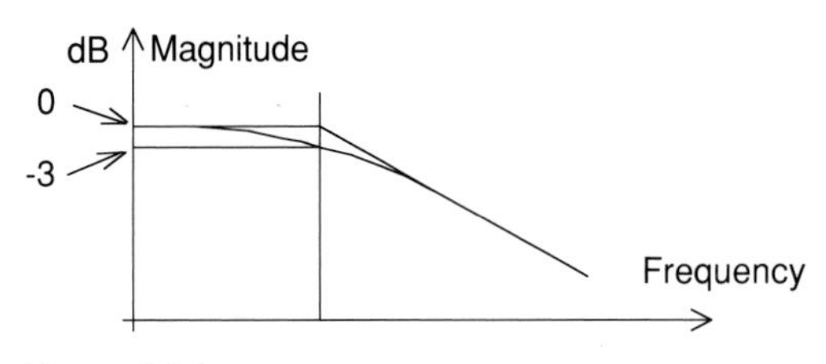

Figure 14.4

In Figure 14.4, we can see the straight line approximation and, the rounded part which corresponds to the actual curve. The approximation can be seen to be quite good, as the curve adjusts asymptotically to the straight line.

This particular response, is exactly the response of the gain of the transistor at high frequency. We see this in Figure 14.5. Notice that in this representation we have omitted the low frequency slope, this is due to the scale we are using. The high frequency takes place at MHz whereas the low frequency is just a few Hz.

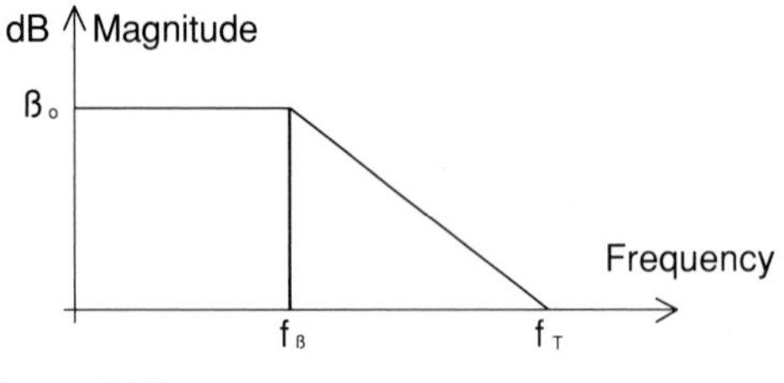

Figure 14.5

In this picture we have introduced three new terms: β_o, f_β and f_T. They are very important parameters for high frequency operation. β_o is the midband gain of the transistor, which at midband is considered to be constant. f_β is the bandwidth. It is taken from 0 frequency, until the gain starts to drop. Remember that in a straight line approximation this is the corner frequency, or the point where the line bends. f_T is the transition frequency. The gain is progressively reduced until a value of 1 is reached. There is no more amplification beyond this point and the amplifier doesn't work any more.

$$\beta = 1$$

$$20 \log 1 = 0 \text{ dB}$$

If we are plotting in decibels, the crossing is the 0 dB level. For this reason we have the following accepted names for f_T:

Unity gain frequency
Cross over frequency
Gain bandwidth product

The latter will be explained later.

Compensation

We know that the conditions for oscillation for an amplifier with positive feedback is that the loop gain be 1.

$$\beta A = 1$$

On an amplifier, however, the instability of oscillations is to be avoided at all cost. We must therefore avoid the condition where βA approaches 1.

Practical amplifiers usually have three poles and they will tend to oscillate if they are in a feedback configuration. This is why compensation has to be employed.

The 741 Op-Amp has internal compensation provided by the manufacturer and this results in a very narrow band response in the open loop. The single pole is located at 10Hz.

With a single pole roll-off crossing at 1MHz, the other poles that exist are located above the 1 MHz frequency. This guarantees a phase margin of 45° and the amplifier is stable for all values of loop gain.

We mentioned that the conditions for oscillation were $\beta A = 1$. There are two conditions here. The magnitude has to be one and the phase angle must be zero. As these are important parameters, we define the gain margin and the phase margin to allow us to manage the instability problem.

If the phase shift is 180° on an inverting amplifier, the amount by which the gain may be increased to make $\beta A = 1$ is the gain margin. If the loop gain has a value of 1 when the phase angle is ϕ, then the difference $180° - \phi$ is the phase margin. Whenever there is a possibility of instability, the problem can be cured by the addition of a compensation network.

In the case of the single pole compensation, we add an extra corner frequency to modify the total response. By placing the new corner frequency at a lower frequency than that of the first existing corner frequency we narrow the bandwidth and modify the magnitude and phase angle of the response. The position of this extra pole is very important and visualising the response with the help of Bode plots is very useful. In a way we achieve stability by a reduction of the bandwidth of the amplifier.

Next, we look at three networks that can be used as equalising networks.

Phase lead network

The phase lead network is shown in Figure 14.6.

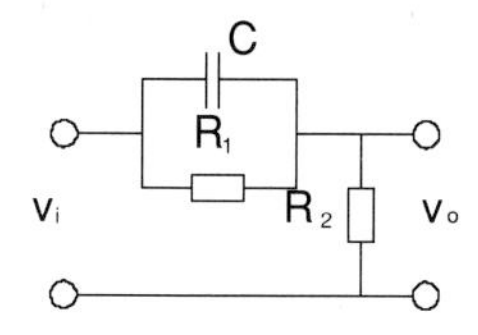

Figure 14.6

The transfer function is given by:

$$\frac{v_o}{v_i} = \frac{R_2}{Z(s)}$$

First we calculate $Z(s)$

$$Z(s) = \frac{R_1 \times \frac{1}{sC}}{R_1 + \frac{1}{sC}} + R_2 \quad | \times sC$$

$$= \frac{R_1}{sR_1C + 1} + R_2$$

$$= \frac{R_1 + sR_1R_2C + R_2}{sR_1C + 1}$$

Factorising $(R_1 + R_2)$:

$$Z(s) = \frac{(R_1 + R_2)\left(s\frac{R_1R_2}{R_1 + R_2}C + 1\right)}{sR_1C + 1}$$

$$\frac{v_o}{v_i} = \frac{R_2(sR_1C + 1)}{(R_1 + R_2)\left(s\frac{R_1R_2}{R_1 + R_2}C + 1\right)}$$

In this case:

$$R_1 > R_1 \parallel R_2$$

which means that we have a zero first and then the pole. This, in Bode form, can be seen in Figure 14.7. The zero and the pole have been placed over a decade apart to make the drawing easier to read. This particular network is widely used in feedback systems as a stabilising circuit.

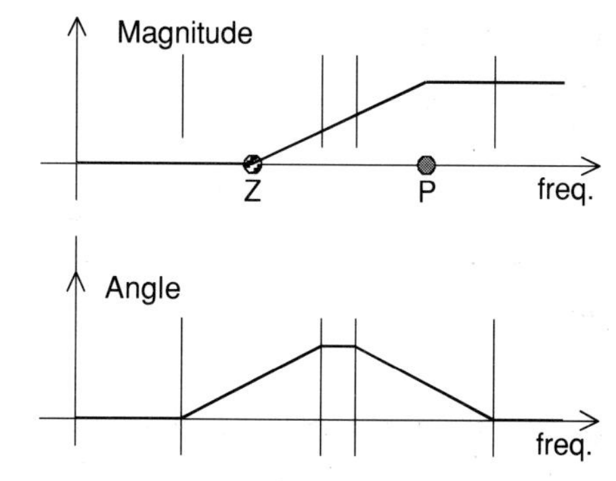

Figure 14.7

The level of the magnitude of this compensation network is attenuated to start with and later, in frequency terms, it settles at 0 dB level. In order to confirm this we divide by s top and bottom and then make s tend to infinity.

$$\frac{v_o}{v_i} = \frac{R_2}{R_1 + R_2} \frac{\left(R_1C + \frac{1}{s}\right)}{\left(\frac{R_1R_2}{R_1 + R_2}C + \frac{1}{s}\right)}$$

as $s \to \infty$

$$\frac{v_o}{v_i} = \frac{R_2}{R_1 + R_2} \frac{R_1C}{\frac{R_1R_2}{R_1 + R_2}C} = 1$$

in decibels : $20 \log 1 = 0$ dB

See Figure 14.8. We have put the response of a three-pole amplifier as a standard amplifier. We have in the same figure included the effect of the equalising network (or compensating network). In order to make it easier to see what is compensated and what is the original this has been mentioned in the graph. In addition we have shaded the gap between the original and the compensated response.

Phase lag network

The next equalising network that we examine is shown in Figure 14.9.

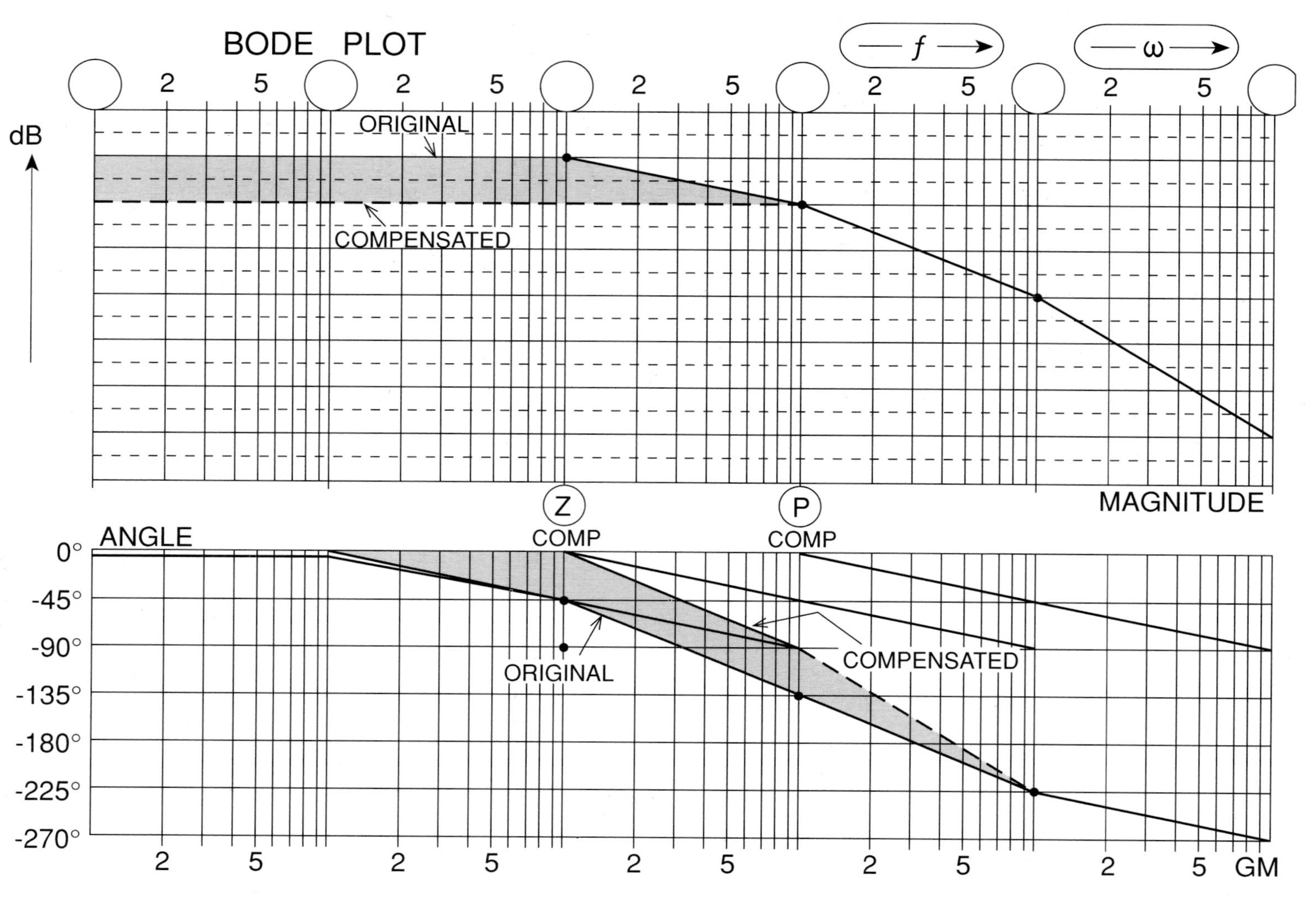
BODE PLOT
f
ω
dB
ORIGINAL
COMPENSATED
MAGNITUDE
Z
COMP
P
COMP
ANGLE
0°
-45°
-90°
-135°
-180°
-225°
-270°
ORIGINAL
COMPENSATED
GM

Figure 14.8

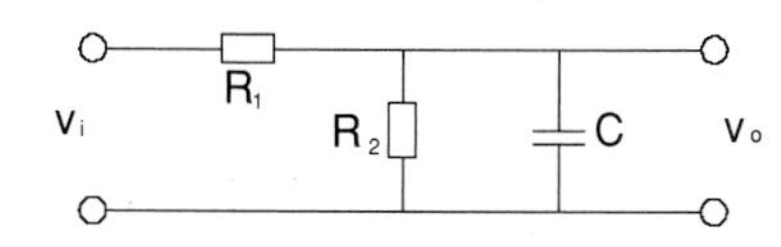

Figure 14.9

The transfer function is given by:

$$\frac{v_o}{v_i} = \frac{\dfrac{R_2 \dfrac{1}{sC}}{R_2 + \dfrac{1}{sC}}}{R_1 + \dfrac{R_2 \dfrac{1}{sC}}{R_2 + \dfrac{1}{sC}}} \quad | \times \left(R_2 + \frac{1}{sC} \right)$$

$$= \frac{\dfrac{R_2}{sC}}{R_1 R_2 + \dfrac{R_1}{sC} + \dfrac{R_2}{sC}} \quad | \times sC$$

$$= \frac{R_2}{R_1 R_2 sC + R_1 + R_2}$$

$$= \frac{R_2}{R_1 + R_2} \frac{1}{\dfrac{R_1 R_2}{R_1 + R_2} sC + 1}$$

The Bode plot of this can be seen in Figure 14.10. With this network a break is introduced at a particular frequency. The gain rolls off at 20 dB per decade.

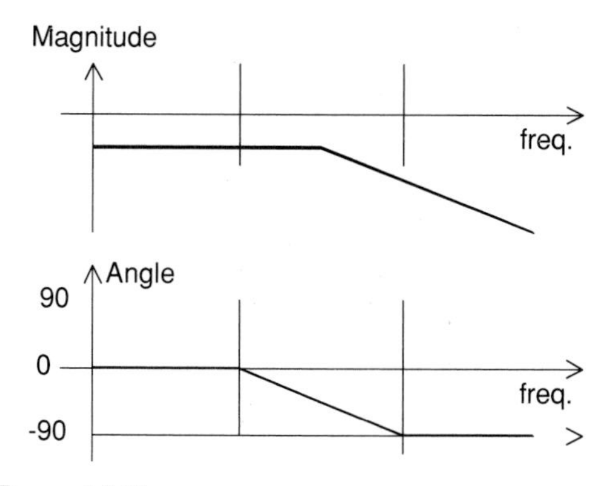

Figure 14.10

The network itself has an attenuation given by the factor

$$\frac{R_2}{R_1 + R_2}$$

Expressed in decibels this is

$$20 \log \frac{R_2}{R_1 + R_2} \text{ dB}$$

As can be seen from Figure 14.11.

We have shown the same three-pole amplifier as we used before to facilitate comparison. The effect is quite pronounced in the magnitude and phase angle responses. The location of the pole is indicated and it is at 100 Hz.

Pole zero equalising network

We will now see a compensating network, as shown in Figure 14.12. The transfer function is:

$$\frac{v_o}{v_i} = \frac{R_2 + \dfrac{1}{j\omega C}}{R_1 + R_2 + \dfrac{1}{j\omega C}} \quad | \times j\omega C$$

$$= \frac{j\omega C R_2 + 1}{j\omega C (R_1 + R_2) + 1}$$

There are a pole and a zero and the Bode functions are type C and type D.

This transfer function starts at 0 dB. This is because when ω tends to zero the transfer function takes the value of one. This in decibels is 0 dB.

As far as the angle is concerned, the angle of the denominator is larger than the angle of the numerator as $C(R_1 + R_2)$ is larger than CR_2. The

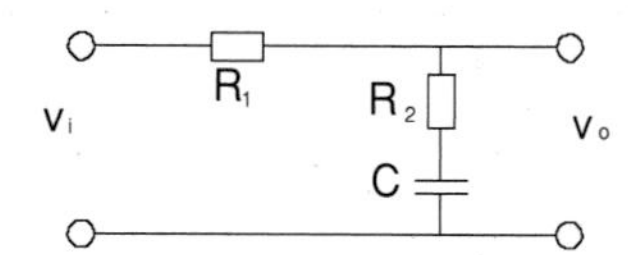

Figure 14.12

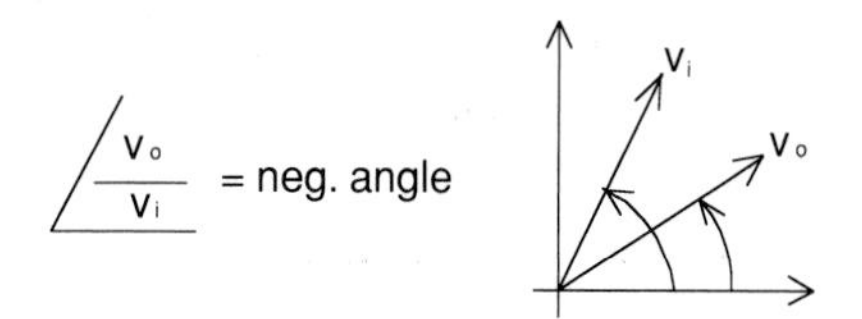

Figure 14.13

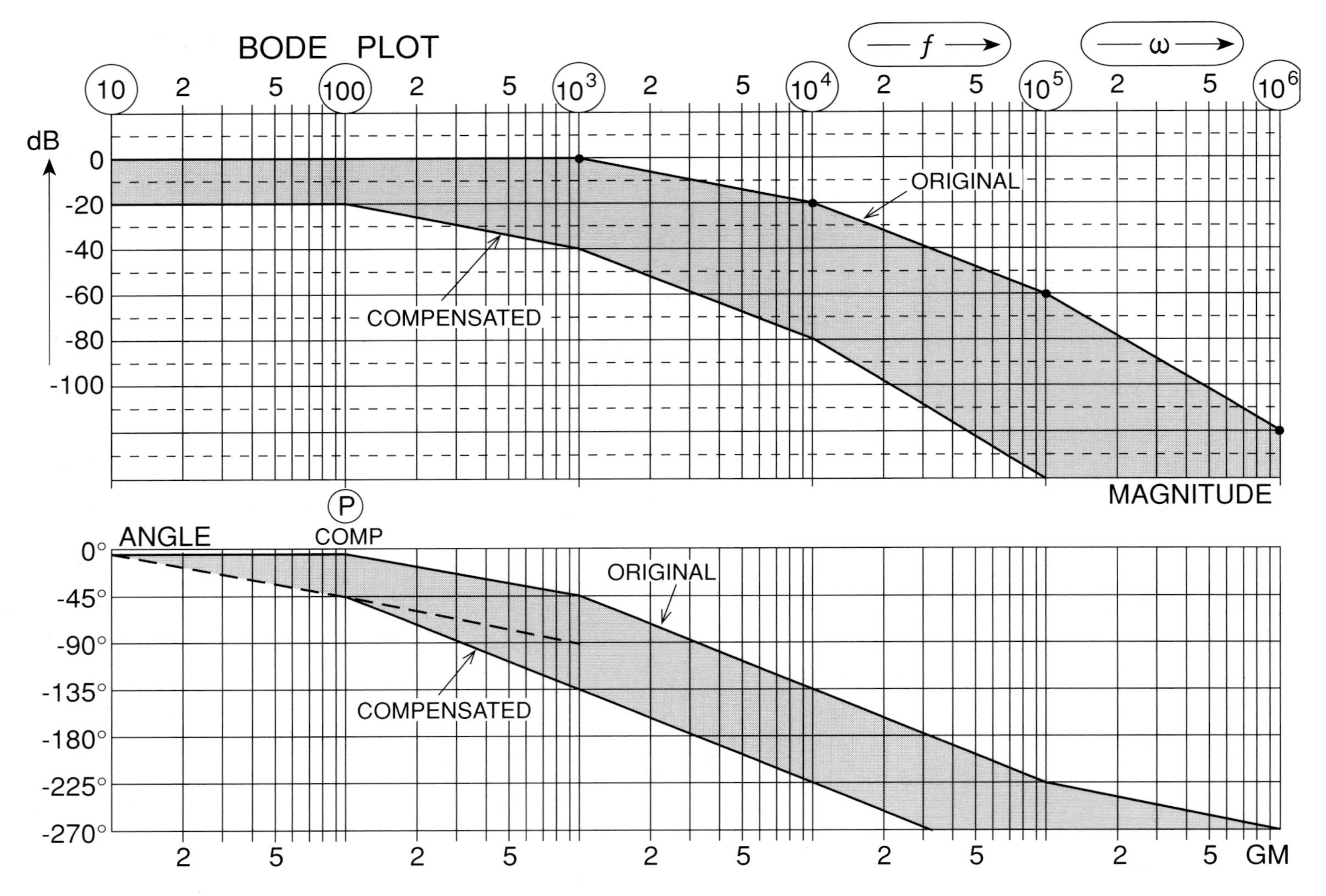
BODE PLOT
f
ω
10
2
5
100
10³
10⁴
10⁵
10⁶
dB
0
-20
-40
-60
-80
-100
ORIGINAL
COMPENSATED
MAGNITUDE
P
COMP
ANGLE
0°
-45°
-90°
-135°
-180°
-225°
-270°
ORIGINAL
COMPENSATED
GM

Figure 14.11

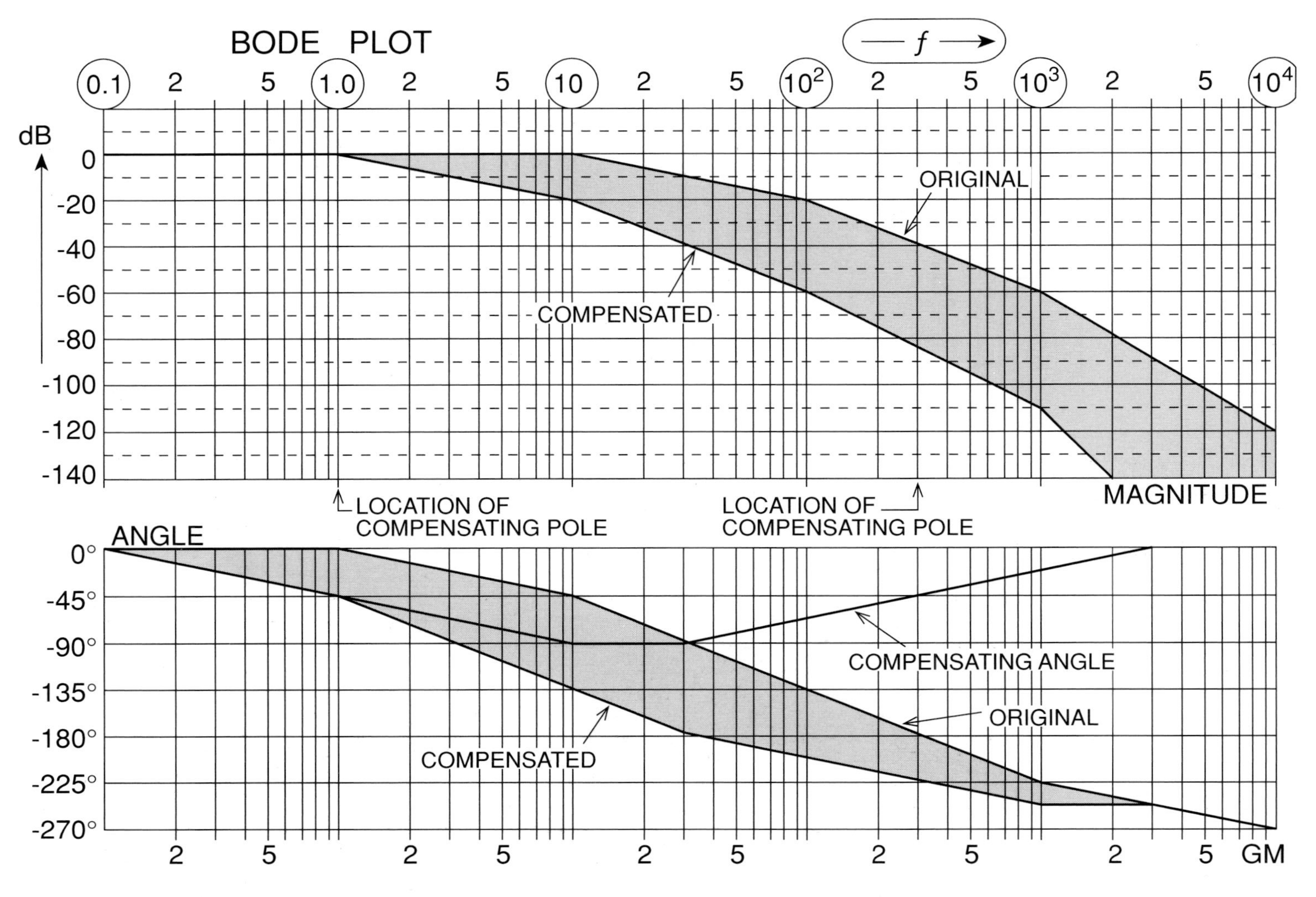
BODE PLOT
f
0.1
1.0
10
10^2
10^3
10^4
dB
0
-20
-40
-60
-80
-100
-120
-140
ORIGINAL
COMPENSATED
MAGNITUDE
LOCATION OF COMPENSATING POLE
LOCATION OF COMPENSATING POLE
ANGLE
0°
-45°
-90°
-135°
-180°
-225°
-270°
COMPENSATING ANGLE
ORIGINAL
COMPENSATED
GM

Figure 14.14

total phase angle (numerator minus denominator) is therefore negative (see Figure 14.13). The senses of the vector are taken in an anticlockwise direction. In this case the phase shift of the output v_o lags the phase shift of the input v_i. With lead compensation the angle is positive, with lag compensation the angle is negative.

Again we see the effect of this compensation in the standard three-poles amplifier used before (see Figure 14.14). In the normal circuit, at the second pole, the phase angle is –135° giving a phase margin of 45°. If we want full compensation then we would have to arrange a gain of 1 at the frequency where the phase margin is 45°.

In this case the gain of the uncompensated circuit for a phase angle of –180° is approximately –41 dB, at a frequency of approximately 300 Hz. The gain margin is therefore 41 dB.

With compensation we see that the frequency for a phase angle of –180° is approximately 30 Hz and the gain is –45 dB. The gain margin in this case is 45 dB.

It is clear that the location of poles and zeros of the compensating network, in relation to the poles and zeros of the circuit can drastically alter the shape of the response.

Measurement of f_β and f_T

f_T transition frequency
f_β bandwidth
These parameters are measured in the common emitter configuration, with the transistor output in short circuit.

The model that we use is seen in Figure 14.15. As we short circuit the ouput, $C_{B'C}$ and $C_{B'E}$ are in parallel. $R_{B'E}$ and $R_{B'C}$ are also in parallel, but as $R_{B'C}$ is much larger, it can be neglected. The result appears in Figure 14.16.

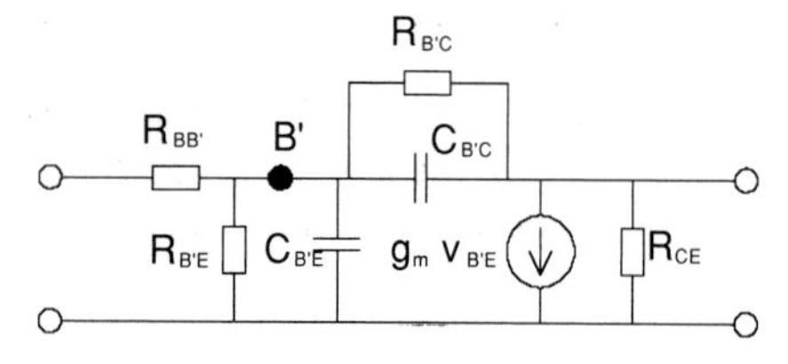

Figure 14.15

The short circuit output current is:

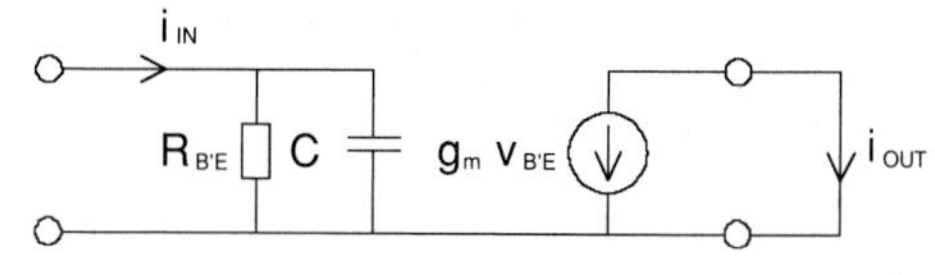

Figure 14.16

$$i_{out} = -g_m v_{B'E}$$

At the input circuit $v_{B'E}$ is given by Ohm's law

$$v_{B'E} = i_{IN} \frac{R_{B'E}\dfrac{1}{j\omega C}}{R_{B'E} + \dfrac{1}{j\omega C}}$$

therefore

$$i_{IN} = v_{B'E}\frac{1 + j\omega C R_{B'E}}{R_{B'E}}$$

The current amplification A_i is

$$A_i = \frac{i_{OUT}}{i_{IN}} = \frac{-g_m v_{B'E}}{v_{B'E}\dfrac{1 + j\omega C R_{B'E}}{R_{B'E}}}$$

$v_{B'E}$ can be simplified

$$A_i = \frac{-g_m R_{B'E}}{1 + j\omega C R_{B'E}}$$

We are dealing with the transistor under short circuit. The incoming current is i_B and the output current is i_C, the amplification is therefore β (or h_{FE}).

> NOTE: In certain problems, the transistor gain is stated as different for AC and DC. In these cases the norm is to use β_{DC} and β_{AC}, or f_{FE} for DC and h_{fe} for AC.

From the Bode theory we know that if ω << 1, we can neglect the ω term. Under the straight line approximation we are at the constant β range, or midband and we can call this β_o. So:

$$\beta_o = g_m R_{B'E}$$

and we can rewrite A_i as follows:

$$A_i = \frac{-\beta_o}{1 + j\omega C\dfrac{\beta_o}{g_m}}$$

The –20 dB per decade slope can be obtained using the Bode theory. If $\omega >> 1$, the term 1 disappears and we have:

$$A_i = \frac{g_m}{\omega C} = \frac{1}{\omega r_e C}$$

C is still $C_{B'E} + C_{B'C}$.

The corner frequency which is the bandwidth of the transistor under short circuit output, is given by:

$$1 = \omega C \frac{\beta_o}{g_m}$$

$$\omega = \frac{g_m}{\beta_o C}$$

so

$$f_B = \frac{g_m}{\beta_o 2\pi (C_{B'E} + C_{B'C})}$$

This can be seen in Figure 14.17. What is difficult to understand about this diagram is that: $f_T = \beta_o f_\beta$. We will demonstrate this twice. First, by looking at the mathematical relationship and later by looking at the electronic relationship.

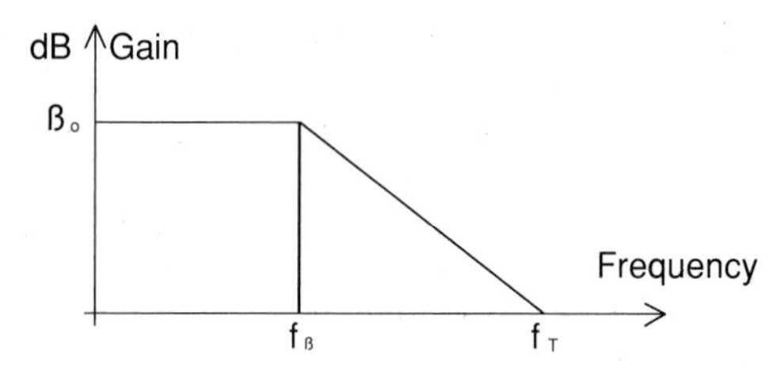

Figure 14.17

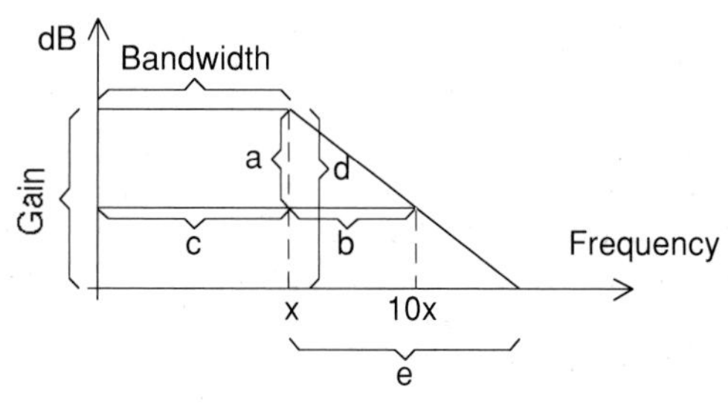

Figure 14.18

Mathematical relationship

We refer to Figure 14.18. The slope of the incline is –20 dB per decade, as we saw from the Bode theory. Due to this, the portion b is equal to a. The distance b is that of one decade from the value X to a value of $10X$. So the distance a should correspond to 20 dB. (20 dB per decade).

$$20 \log a = 20 \text{ dB}$$

$$\log a = 1$$

$$a = 10$$

So

$$a = b$$

Due to the same proportion, d is equal to e. Therefore in the above picture:

$$c \times a = c \times b$$

In log form:

$$\log ca = \log c + \log a$$

$$\log ce = \log c + \log e = \log c + \log d$$

Therefore at the limit:

$$\log f_T = \log ce = \log c + \log e$$

$$\log f_T = \log cd = \log c + \log d$$

Therefore: $f_T = cd$

$$f_T = \text{gain} \times \text{bandwidth}$$

Furthermore, any point along the slope is the corner of a rectangle of equivalent area, that is to say, the base times the height is constant. We see this in the second of the two illustrations that follow (see Figure 14.19). The scales have been arranged in such a way that the slope is 45°. It is easy to accept that the distance a is the same as the distance b.

In this case $\beta_o = 20$ dB

$$20 \log \beta_o = 20$$

$$\log \beta_o = 1$$

$$\beta_o = 10$$

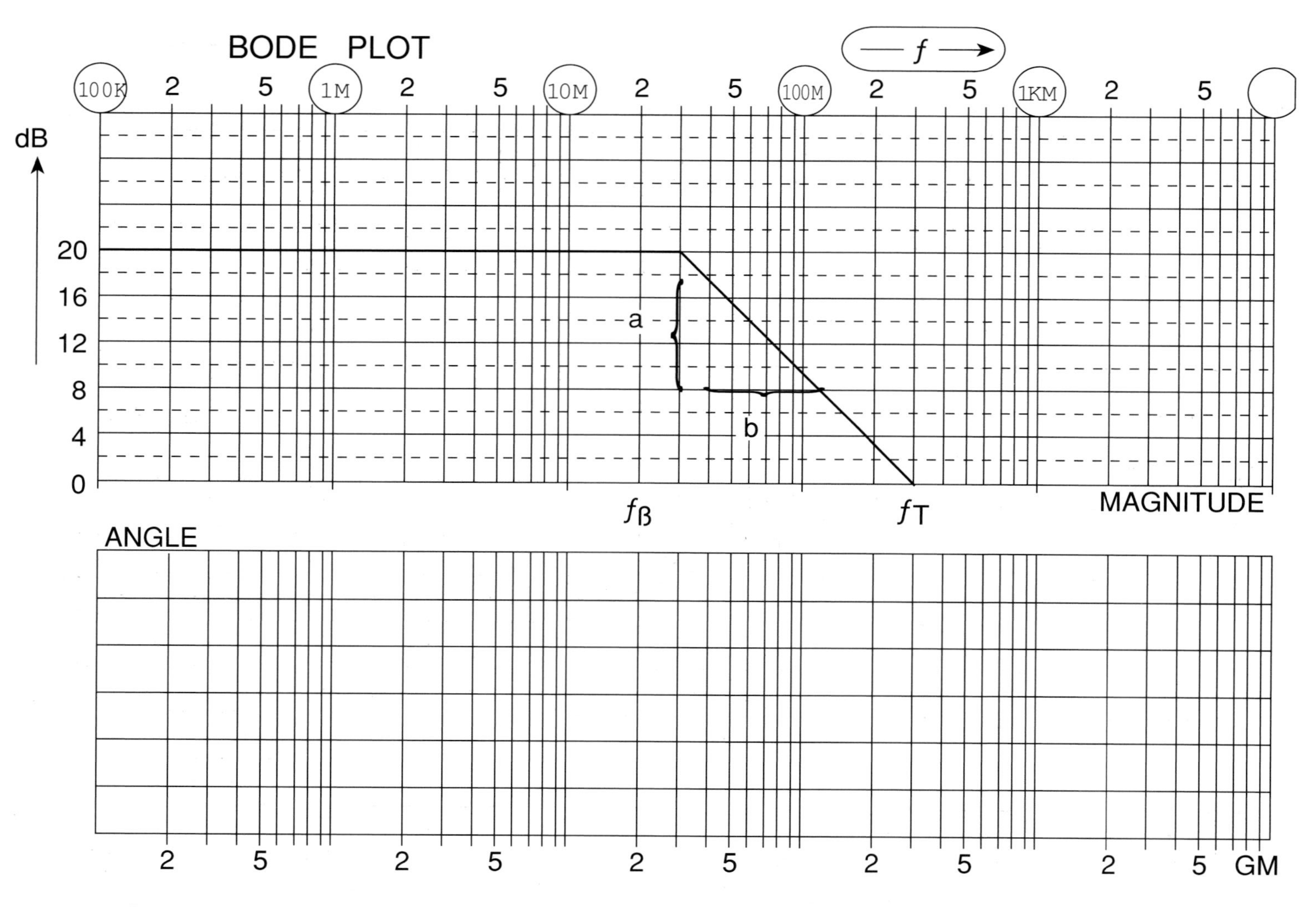
BODE PLOT
f
100K
1M
10M
100M
1KM
2
5
GM
dB
20
16
12
8
4
0
a
b
f_β
f_T
MAGNITUDE
ANGLE

Figure 14.19

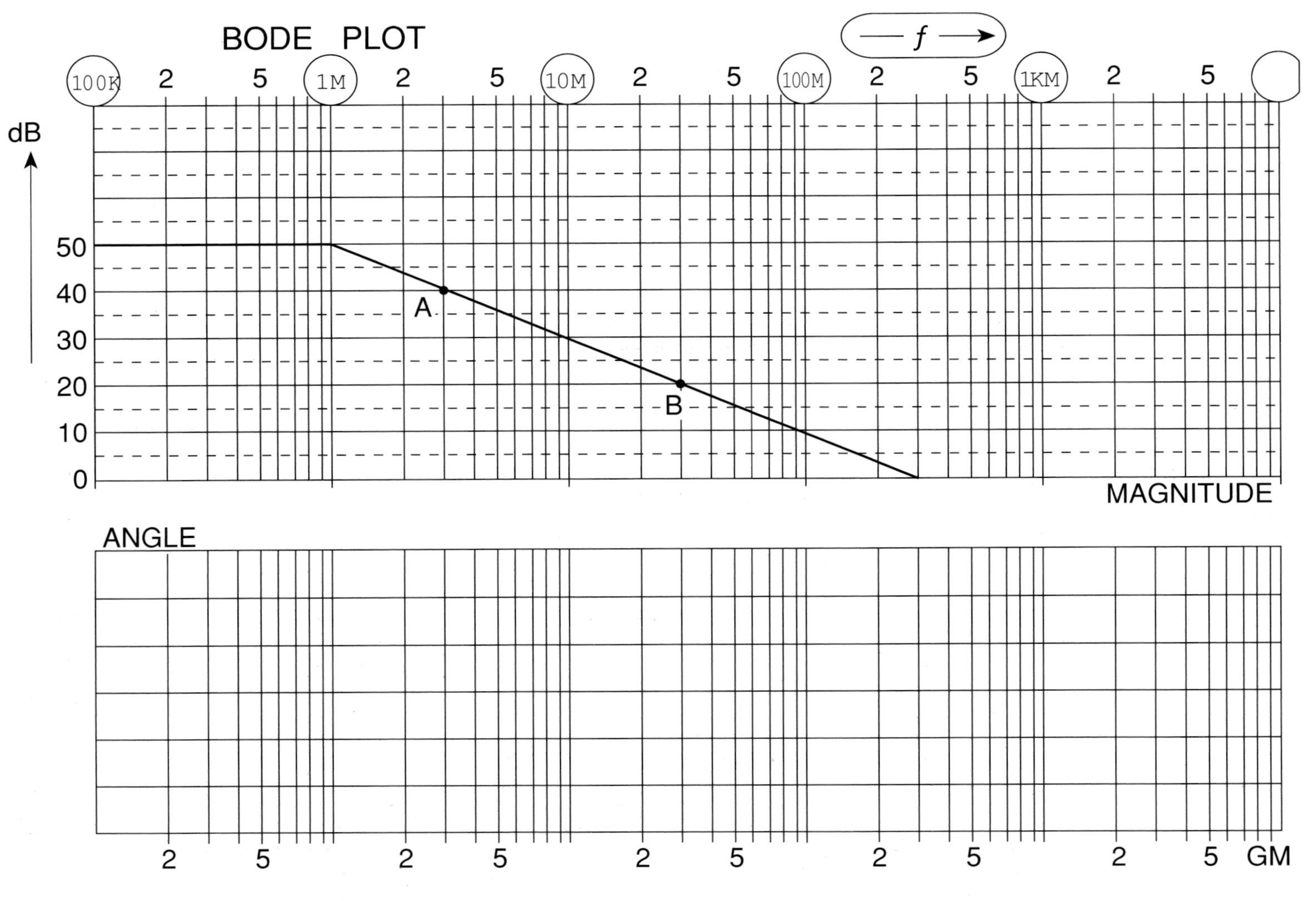
BODE PLOT
f
100K
1M
10M
100M
1KM
2
5
dB
50
40
30
20
10
0
A
B
MAGNITUDE
ANGLE
GM

Figure 14.20

$$\beta_o = \frac{f_T}{f_\beta} = \frac{300 \text{ MHz}}{30 \text{ MHz}} = 10$$

We now change the scale as we see in Figure 14.20.

In this case, $\beta_o = 300$

$\beta_o = 20 \log 300 = 49.54$ dB

$$f_B = \frac{f_T}{\beta_o} = \frac{300 \text{ MHz}}{300} = 1 \text{ MHz}$$

Now the case of the magic rectangles.

At point A:

$20 \log \beta = 40$ dB

$\log \beta = 2$

$\beta = 100$

$f_T = \beta_o \times f_\beta = 100 \times 3 \text{ MHz} = 300$ MHz

At point B:

$20 \log \beta = 20$

$\log \beta = 1$

$\beta = 10$

$f_T = \beta_o \times f_\beta = 10 \times 30 \text{ MHz} = 300$ MHz

This is valid at any point in the slope. The product β times bandwidth is constant and equal to the transition frequency.

Electronic relationship

We look at Figure 14.21. The transition frequency f_T comes from the necessity of knowing the maximum useable high frequency of the transistor. The frequency at which the absolute value of the short circuit common emitter current reduces to one (0 value in dB), is termed the transition frequency.

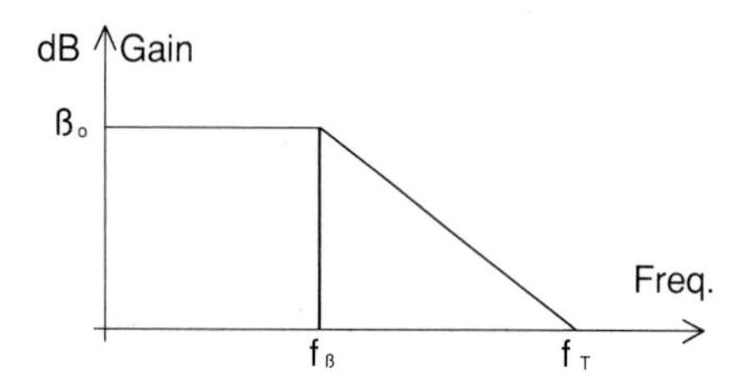

Figure 14.21

We go back to the general equation and the Bode approximation:

$$A_i = 1 = \frac{\beta_o}{1 + j\omega C \frac{\beta_o}{g_m}}$$

$$1 = \frac{1}{\frac{1}{\beta_o} + j\omega C \frac{1}{g_m}}$$

If $\beta_o >> 1$, then we can ignore the term 1, leaving

$$1 = \frac{1}{jC \frac{1}{g_m}}$$

$$\omega_T = \frac{1}{jC \frac{1}{g_m}}$$

$$f_T = \frac{g_m}{2\pi C} \qquad (C = C_{B'E} + C_{B'C})$$

But f_B was

$$f_B = \frac{g_m}{\beta_o} \frac{1}{2\pi(C_{B'E} + C_{B'C})}$$

Therefore

$$\boxed{f_T = \beta_o \times f_\beta}$$

Ideal versus practical Op-Amp

The ideal Op-Amp has the following characteristics

1 Open loop voltage gain	infinite	(10^6)
2 Bandwidth	infinite	(1 MHz)
3 Input resistance	infinite	(2 MΩ)
4 Output resistance	zero	(50 Ω)

The main difference between an ideal Op-Amp and a practical one lies in the voltage gain. In the ideal Op-Amp the gain is considered to be infinite whereas in practice it is very high. Practical figures are shown in brackets in the list above. But the main difference in the practical Op-Amp is that the voltage gain decreases as the frequency increases.

Some Op-Amps allow a capacitor to be added externally to provide compensation, but most Op-Amps are internally compensated. That is to say they have *RC* networks which introduce a pole at a given point providing a single pole compensation, and therefore giving a predictable response. When there is internal compensation, the manufacturers produce curves with details of the bandwidth – gain and phase against frequency for the open loop response.

Open loop voltage gain

Open loop gain is the ratio of the output voltage to the input voltage of an Op-Amp without feedback. This is illustrated in Figure 14.22. The gain decreases with increase in frequency.

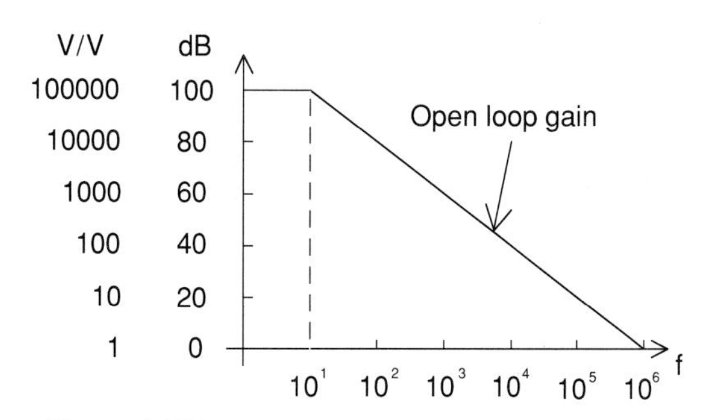

Figure 14.22

This characteristic correspond to the Bode straight line approximation of a one-pole response. If we call the frequency of the corner frequency ω_c then the type D factor will be given by:

$$1+\frac{s}{\omega_c}$$

So, *G(s)*, the open loop response will be given by

$$G(s)=G_o\frac{1}{1+\dfrac{s}{\omega_c}}$$

G_o is the open loop gain at DC and very low frequency. This is shown to be 100 dB. The corner frequency is at 10Hz. The curve will reach 0 dB at the transition frequency f_T. The 0 dB value correspond to a ratio of 1 in volt per volt. This is why this point is also termed the unity gain frequency (UGF). This value also corresponds to the gain bandwidth product (GBP).

Closed loop bandwidth

We are now interested in the effect of feedback on the bandwidth. We use negative feedback as seen in Figure 14.23. *A* is the amplification seen

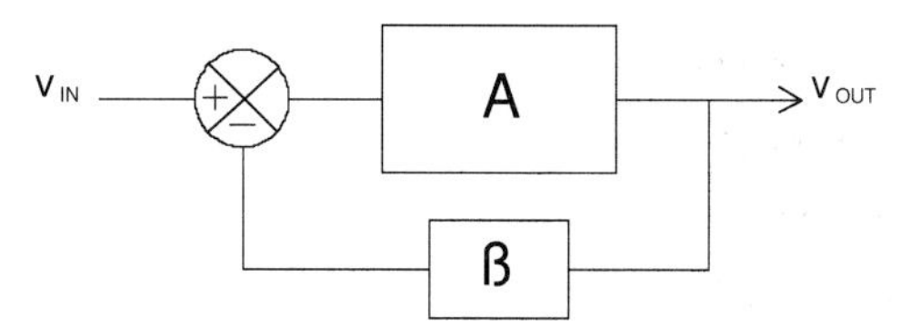

Figure 14.23

before in the case of open loop gain. β is the fraction of the output which is fed back.

$$(v_{IN}-\beta v_{OUT})A=v_{OUT}$$

$$Av_{IN}=v_{OUT}+\beta A v_{OUT}$$

$$A_{CL}=\frac{v_{OUT}}{v_{IN}}=\frac{A}{1+\beta A}$$

In a closed loop situation with negative feedback the gain of the open loop amplifier is divided by 1 + β*A*.

If β*A* is much greater than 1, then the closed loop gain can be approximated to 1/β. This lower gain, corresponding to the closed loop gain can be plotted on the previous response. We see this in Figure 14.24. We see that the closed loop gain intercepts the open loop gain at f_{CLC} (closed loop

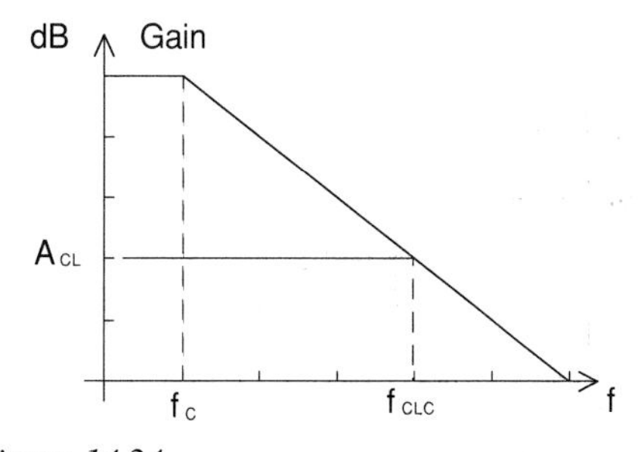

Figure 14.24

corner frequency). This means that at a lower gain we have a larger bandwidth and that the bandwidth increases as the closed loop gain decreases.

These curves are the same type of curve as the transistor under short circuit high frequency test. If represented by the frequency in a logarithmic scale and the gain in decibels, then we have the 20 dB per decade incline and we should be able to verify that the gain bandwidth product is constant. We see this in Figure 14.25.

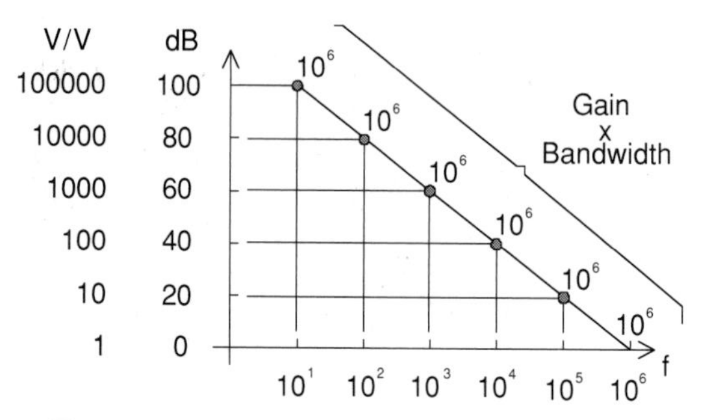

Figure 14.25

We can also verify the values of the closed loop gain for given values of β.

For $\beta = 0.001$ and $A = 10^5$

$$A_{CL} = \frac{A}{1+\beta A} = \frac{10^5}{1+0.001\times 10^5} = 10^3$$

A is the open loop gain. We ignored the 1 in the above equation to arrive to the result.

For $\beta = 0.01$ and $A = 10^5$

$$A_{CL} = \frac{A}{1+\beta A} = \frac{10^5}{1+0.01\times 10^5} = 10^2$$

We have shown this in Figure 14.26. Negative feedback gives greater stability, less distortion and an increased bandwidth.

By using internal compensation, the voltage gain of an Op-Amp becomes entirely predictable, independent of the internal details or even the particular Op–Amp used.

Op-Amps have a very high gain and this is normally in excess of what is required for a particular application. It is reasonable to sacrifice part of this gain in favour of stability, no distortion and bandwidth improvements.

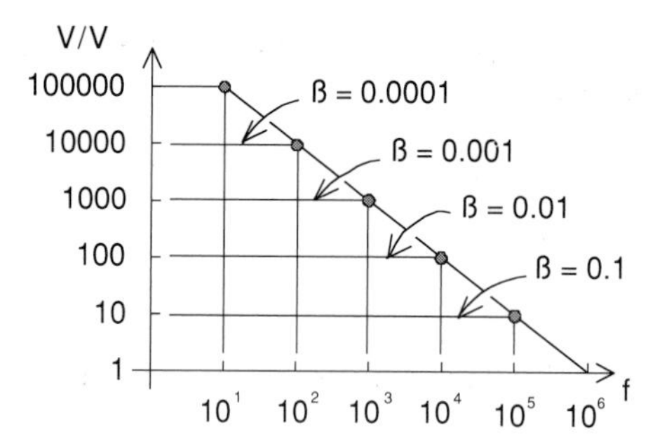

Figure 14.26

Miller effect on Op-Amps

A degradation of the frequency response, especially at high frequencies, results from a capacitance between the input and the output of an Op-Amp, as seen in Figure 14.27.

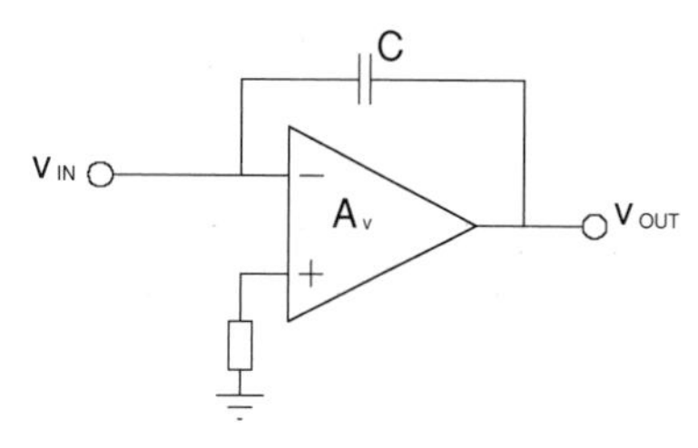

Figure 14.27

As we saw from the Miller's theorem, we need an inverting amplifier. There is no meaning to the theorem if the amplification is non-inverting as we would end up with an unwanted minus sign.

The transformation after applying Miller's theorem appears on Figure 14.28.

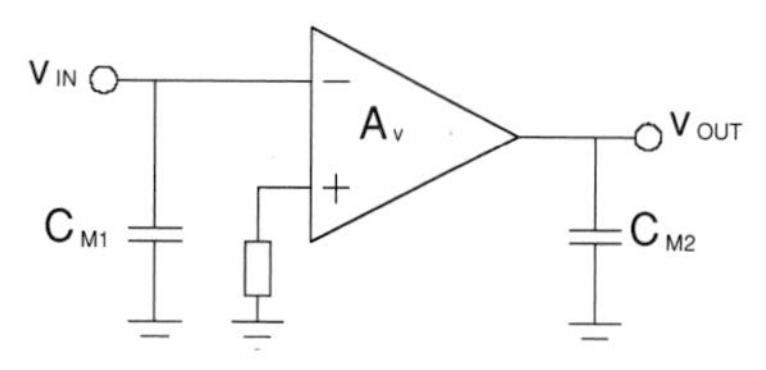

Figure 14.28

$$C_{1M} = C(1 - A_v)$$

$$C_{2M} = C\left(1 - \frac{1}{A_v}\right) \approx C$$

If C was 28 pF and we have an amplification of –180, then

$$C_{1M} = 28(1 - (-180))$$

$$C_{1M} = 5040 \text{ pF}$$

This capacitor C_{1M}, combined with the input resistance R_i would form a corner frequency given by

$$f_c = \frac{1}{2\pi R_i C_{1M}}$$

The capacitance between input and output is seen magnified at the input. This is the Miller's effect. Because the Miller's effect reduces the gain, we can accept a reasonable approximation, the use of the midband gain to calculate Miller's equivalent capacitance. Electronics is the art of approximating. You should be proficient enough to know when to approximate and when not to approximate.

$$i = \frac{\left(v_i - \left(-\frac{v_o}{A}\right)\right)}{R_1}$$

$$v_o = -\frac{v_o}{A} - iR_2$$

$$v_o = -\frac{v_o}{A} - \frac{R_2}{R_1}\left(v_i + \frac{v_o}{A}\right)$$

$$v_o = -\frac{v_o}{A} - \frac{R_2}{R_1}v_i - \frac{R_2}{R_1}\frac{v_o}{A}$$

$$v_o + \frac{v_o}{A} + \frac{R_2}{R_1}\frac{v_o}{A} = -\frac{R_2}{R_1}v_i$$

$$v_o\left(1 + \frac{1}{A} + \frac{R_2}{R_1}\frac{1}{A}\right) = -\frac{R_2}{R_1}v_i$$

$$\frac{v_o}{v_i} = -\frac{R_2}{R_1}\frac{1}{1 + \frac{1}{A}\left(1 + \frac{R_2}{R_1}\right)}$$

PROBLEM 14.1 GAIN BANDWIDTH

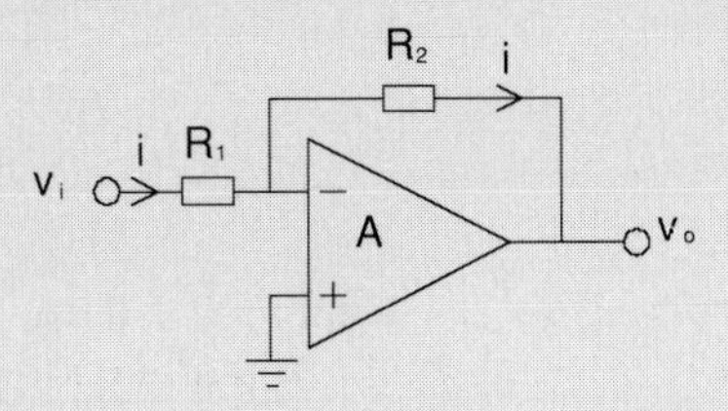

Figure 14.29

Non-inverting Op-Amp. Find the transfer function, assuming finite gain.

Answer: $\frac{v_o}{v_i} = -\frac{R_2}{R_1}\frac{1}{1 + \frac{1}{A}\left(1 + \frac{R_2}{R_1}\right)}$

Input

$$\frac{v_o}{A}$$

$$v_- = -\frac{v_o}{A}$$

PROBLEM 14.2 GAIN BANDWIDTH

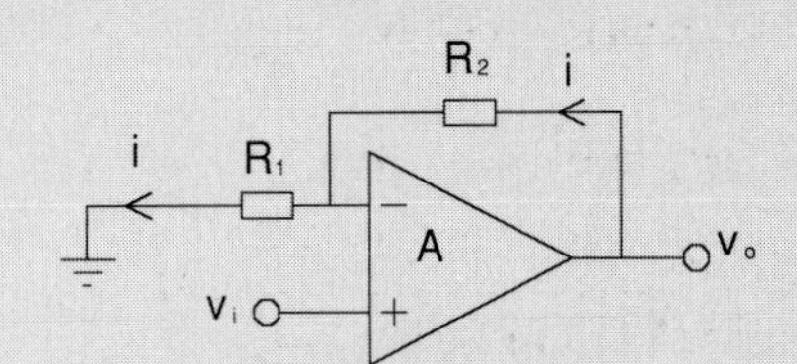

Figure 14.30

Find the transfer function assuming finite gain.

Answer: $\frac{v_o}{v_i} = \frac{\left(1 + \frac{R_2}{R_1}\right)}{1 + \left(1 + \frac{R_2}{R_1}\right)A}$

The input voltage is

$$\frac{v_o}{A}$$

$$v_- = v_i - \frac{v_o}{A}$$

Therefore

$$i = \frac{v_i - \frac{v_o}{A}}{R_1}$$

$$v_o = v_i - \frac{v_o}{A} + iR_2$$

$$v_o = v_i - \frac{v_o}{A} + \frac{R_2}{R_1}\left(v_i - \frac{v_o}{A}\right)$$

$$v_o + \frac{v_o}{A} + \frac{R_2}{R_1}\frac{v_o}{A} = v_i + \frac{R_2}{R_1}v_i$$

$$v_o\left(1 + \frac{1}{A} + \frac{R_2}{R_1}\frac{1}{A}\right) = v_i\left(1 + \frac{R_2}{R_1}\right)$$

$$\frac{v_o}{v_i} = \frac{1 + \frac{R_2}{R_1}}{1 + \left(\frac{1 + \frac{R_2}{R_1}}{A}\right)}$$

Ideal value

$$-\frac{9}{1} = -9$$

For $A = 10^3$

$$\frac{v_o}{v_i} = -\frac{R_2}{R_1}\frac{1}{1 + \frac{1}{A}\left(1 + \frac{R_2}{R_1}\right)}$$

$$= -\frac{9}{1}\frac{1}{1 + \frac{10}{1000}}$$

$$= -9 \times 0.99$$

$$= -8.91$$

Difference

$$9 - 8.91 = 0.09$$

A difference of 1% (with reference to 9)

For $A = 10^5$

$$\frac{v_o}{v_i} = -\frac{9}{1}\frac{1}{1 + \frac{10}{100\,000}}$$

$$= -9 \times 0.9999 = 8.999$$

Difference

$$9 - 8.999 = 0.001$$

A difference of 0.01%

PROBLEM 14.3 GAIN BANDWIDTH

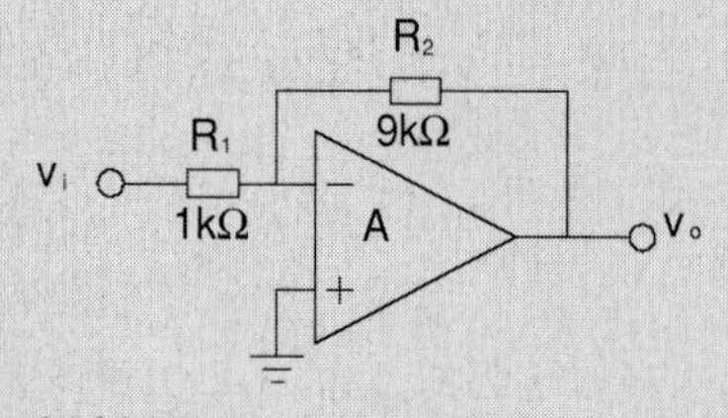

Figure 14.31

Inverting Op-Amp. Find the deviation of the closed loop gain from the ideal value for $A = 10^3$ and $A = 10^5$.

Answer: 1% and 0.01%

PROBLEM 14.4 GAIN BANDWIDTH

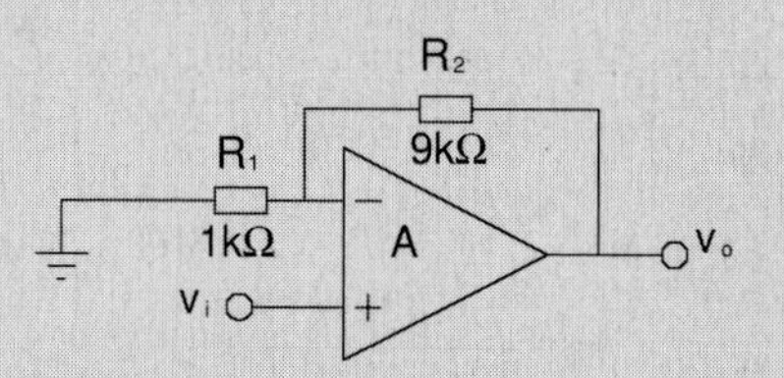

Figure 14.32

Find the deviation of the closed loop gain from the ideal value for $A = 10^3$ and $A = 10^5$.

Answer: 1% and 0.01%

Ideal value

$$1 + \frac{9k}{1k} = 10$$

For $A = 10^3$

$$\frac{v_o}{v_i} = \frac{1 + \frac{R_2}{R_1}}{1 + \frac{1 + \frac{R_2}{R_1}}{A}}$$

$$= \frac{10}{\left(1 + \frac{10}{1000}\right)} = \frac{10}{1 + 0.01}$$

$$= \frac{10}{1.01} = 9.9$$

A difference of 1%

For A = 10^5

$$\frac{v_o}{v_i} = \frac{10}{1 + \frac{10}{100\,000}} = \frac{10}{1.0001}$$

$$= 9.999$$

A difference of 0.001 which is 0.01%

PROBLEM 14.5 GAIN BANDWIDTH

An Op-Amp has f_T = 1 MHz.

Find the corner frequency for gains of 1000, 100 and 1.

Answer: 1 kHz, 10 kHz and 1 MHz

Transition frequency

$$f_T = \text{gain} \times \text{bandwidth}$$

$$1\,\text{MHz} = \left(1 + \frac{R_2}{R_1}\right) B$$

For a gain of 1000

$$\text{BW} = \frac{10^6}{1000} = 1\ \text{kHz}$$

For a gain of 100

$$\text{BW} = \frac{10^6}{100} = 10\ \text{kHz}$$

For a gain of 1

$$\text{BW} = \frac{10^6}{1} = 1\,\text{MHz}$$

PROBLEM 14.6 GAIN BANDWIDTH

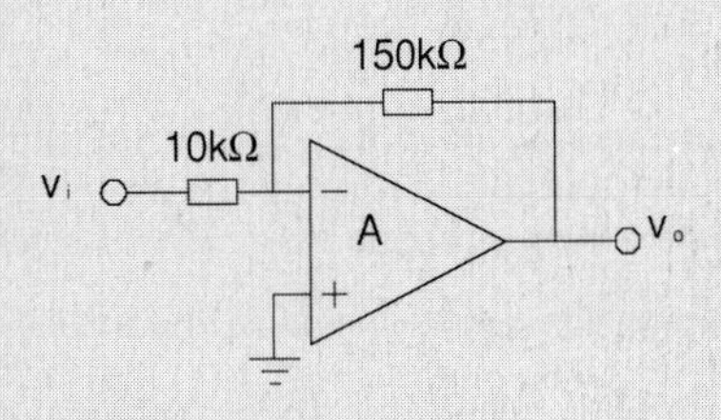

Figure 14.33

Find the cut-off frequency in the closed loop configuration shown if the open loop gain–bandwidth product is equal to 1 MHz.

Answer: 62.5 kHz

Amplification

$$1 + \frac{150k}{10k} = 16$$

Bandwidth

$$\text{BW} = \frac{f_T}{\text{gain}} = \frac{1\,000\,000}{16} = 62.5\ \text{kHz}$$

PROBLEM 14.7 GAIN BANDWIDTH

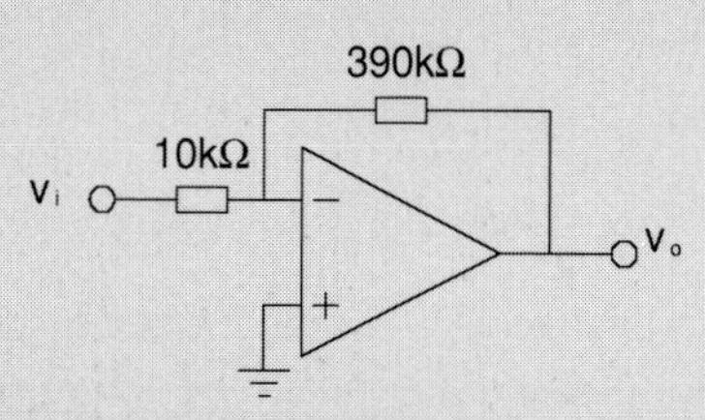

Figure 14.34

f_T = 1 MHz and the slew rate is 1 V/μs. Find the maximum frequency of a 0.1 V peak sine wave, without distortion of the output.

Answer: 25 kHz

There are two factors contributing to the limit of the maximum frequency. The first is the bandwidth due to the gain. The second is the slew rate.

Amplification

$$1+\frac{390k}{10k}=40$$

Bandwidth

$$BW=\frac{1\,000\,000}{40}=25\text{ kHz}$$

$$v_o=0.1\times 40=4\text{ V peak}$$

Slew rate

For a sine wave

$$v(t)=k\sin\omega t$$

The maximum slew rate is given by

$$S_{MAX}=k\omega$$

k is the peak value of sine wave (in volts) and ω is the angular frequency.

$$S_{MAX}=k2\pi f$$

$$f=\frac{S}{k2\pi}=\frac{10^6}{4\times 2\times \pi}=39.79\text{ kHz}$$

In this case the limiting factor is the gain giving a limit of 25 kHz.

PROBLEM 14.8 GAIN BANDWIDTH

An Op-Amp has a slew rate of 0.5 V/μs. The input signal is a ramp which increases at a rate of 0.8 V in 20 μs. Find the maximum gain of the closed loop Op-Amp without exceeding the slew rate.

Answer: 12.5

With the increase of gain, the rate of change increases as we have a bigger signal in the same amount of time.

Rate of change

$$\frac{0.8}{20\times 10^6}=40\text{ kV/s}$$

A slew rate of $0.5\text{ V}/\mu\text{s}$ correspond to

$$500\,000\text{ V/s}\quad\text{or } 500\text{ kV/s}$$

The maximum gain is

$$\frac{500}{40}=12.5$$

PROBLEM 14.9 GAIN BANDWIDTH

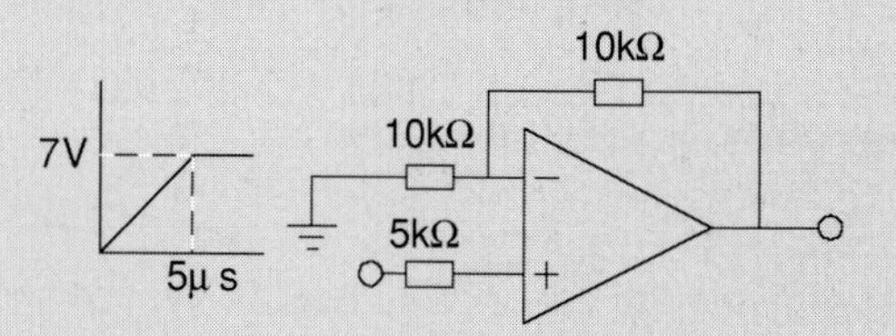

Figure 14.35

The Op-Amp has a slew rate of 4 V/μs and a transition frequency of 2 MHz. Will it cope with the signal shown?

Answer: Yes

$$\text{Gain}=1+\frac{10k}{10k}=2$$

The output will vary 14 V in 5 μs

The rate of 4 V/μs, corresponds to 3.5 μs for 14 V. So we are alright with the slew rate.

Bandwidth

$$BW=\frac{f_T}{\text{gain}}=\frac{2}{2}=1\text{ MHz}$$

Rise time (for single pole response) is

$$t = \frac{0.35}{\text{BW}}$$

$$= \frac{0.35}{1} = 0.35\ \mu s$$

Both conditions are met. The Op-Amp will cope.

We need to modify R_F to allow 50 kHz.

$$\text{For } K = \frac{S}{(2\pi f)} = \frac{0.5 \times 10^6}{2\pi 5000} = 1.59$$

$$\text{gain} = \frac{K}{\text{input signal}} = \frac{1.59}{0.2828} = 5.62$$

This is achieved with $R_F = 56\text{k}2\ \Omega$

PROBLEM 14.10 GAIN BANDWIDTH

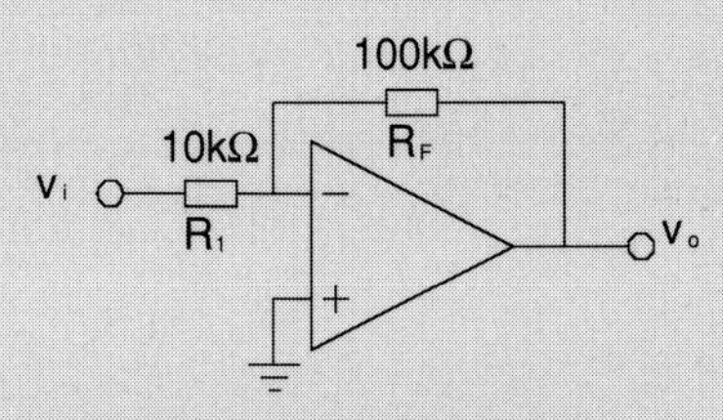

Figure 14.36

The Op-Amp has a slew rate of 0.5 V/μs and f_T = 1 MHz. The input signal is a 0.2 V rms sine wave. Can it be used up to 50 kHz?
What new value of R_F would work up to 50 kHz?

Answer: No, $R_F = 56\text{k}2\ \Omega$

$$\text{Gain} = -\frac{100\text{k}}{10\text{k}} - 10$$

Bandwidth

$$\text{BW} = \frac{f_T}{\text{gain}} = \frac{1\ \text{MHz}}{10} = 100\text{k Hz}$$

No problem here.

Now the slew rate

Input signal 0.2 V rms sine wave

Peak value

$$0.2 \times \sqrt{2} = 0.2828$$

$$\text{K} = \text{gain} \times \text{peak value}$$

$$= 10 \times 0.2828 = 2.828$$

Maximum frequency allowed by slew rate

$$f = \frac{S}{(2\pi K)} = \frac{0.5 \times 10^6}{(2\pi \times 2.828)} = 28139\ \text{Hz}$$

Answer: No. The slew rate will not allow 50 kHz.

PROBLEM 14.11 GAIN BANDWIDTH

An Op-Amp has a DC gain of 106 dB and a transition frequency of 2 MHz. Find the gain at 1 kHz and the frequency for a gain of 100.

Answer: 2000 and 20 kHz

$$20 \log X = 106$$

$$\log X = 5.3$$

$$X = 200\,000$$

$$f_T = \beta B$$

$$200\,000 = \beta \times 1000$$

$$\beta = 2000$$

Gain at 1 kHz

$$f_T = \beta B$$

$$2\,000\,000 = 100 B$$

$$B = 20\,000\ \text{Hz}$$

PROBLEM 14.12 GAIN BANDWIDTH

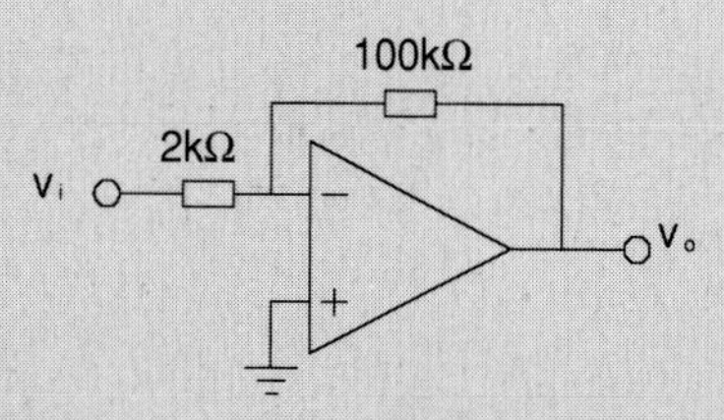

Figure 14.37

If the closed loop gain is to be –49.5, find the value of the open loop gain and the value of the loop gain.

Answer: A = 5049, loop gain = 99

For a gain less than ideal, we have

$$\frac{v_o}{v_i} = -\frac{R_2}{R_1}\left(\frac{1}{1+\frac{1}{A}\left(1+\frac{R_2}{R_1}\right)}\right)$$

$$49.5 = 50\frac{1}{1+\frac{1}{A}(1+50)}$$

$$49.5 + \frac{1}{A} \times 49.5 \times 51 = 50$$

$$\frac{2524.5}{A} = 0.5$$

$$A = 5049$$

Loop gain

The loop gain is the product of the amplification multiplied by the feedback factor.

The feedback factor is

$$\frac{2k}{2k} + 100k = \frac{1}{51}$$

$$A\frac{1}{51} = \text{LG}$$

$$\text{LG} = \frac{5049}{51} = 99$$

PROBLEM 14.13 GAIN BANDWIDTH

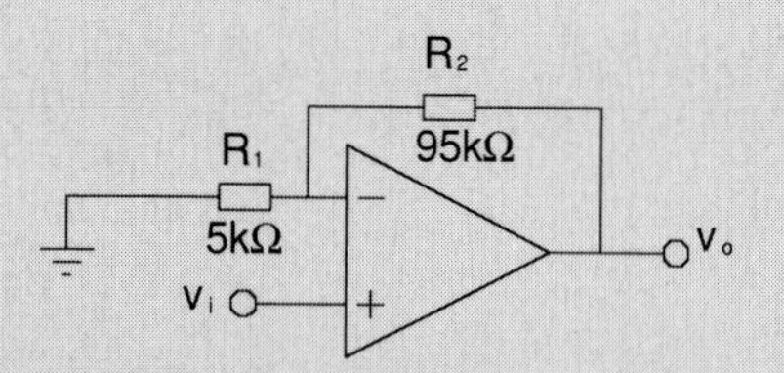

Figure 14.38

f_T = 1.2 MHz. Find the closed loop bandwidth and the closed loop gain at 600 kHz.

Answer: 60 kHz and 2

Gain

$$1 + \frac{95}{5} = 20$$

$$\text{BW} = \frac{f_T}{\text{gain}} = \frac{1\,200\,000}{20} = 60 \text{ kHz}$$

At 600 kHz

$$\text{gain} = \frac{f_T}{\text{BW}} = \frac{1\,200\,000}{600\,000} = 2$$

PROBLEM 14.14 GAIN BANDWIDTH

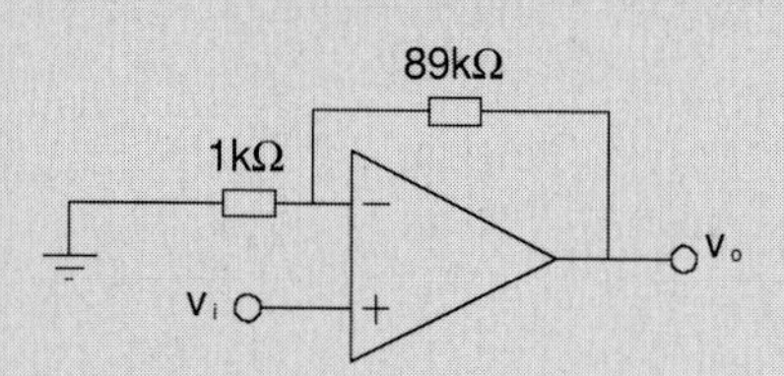

Figure 14.39

The Op-Amp has an open loop gain of 200 000 with a corner frequency of 5 Hz. Find the gain bandwidth product and the closed loop 3 dB frequency.

Answer: 1 MHz and 11 111 Hz

$f_T = \text{gain} \times \text{BW}$

$= 200\,000 \times 5 = 1\text{ MHz}$

Corner frequency

Closed loop gain

$$\text{gain} = 1 + \frac{89\text{k}}{1\text{k}} = 90$$

$$\text{BW} = \frac{f_T}{\text{gain}} = \frac{1\,000\,000}{90} = 11\,111\text{ Hz}$$

PROBLEM 14.15 GAIN BANDWIDTH

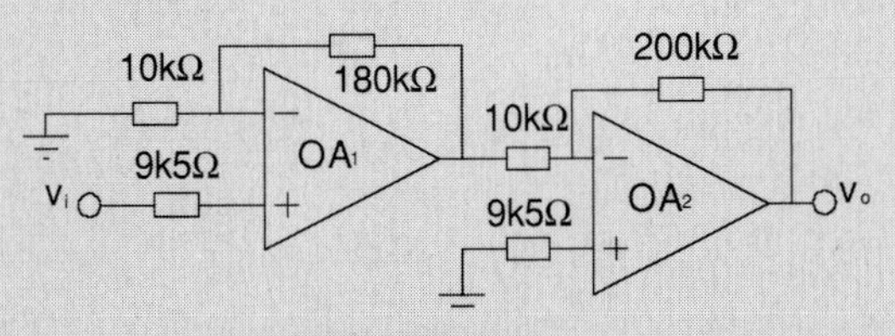

Figure 14.40

The transition frequency is 10^6 Hz for these amplifiers. Find v_o in terms of v_i and the maximum useful bandwidth.

Answer: 399 and 47.62 kHz

Both amplifiers are non-inverting.

OA1

$$A_1 = 1 + \frac{180}{10} = 19$$

OA 2

$$A_2 = 1 + \frac{200}{10} = 21$$

Total $A = 19 \times 21 = 399$

Bandwidth OA1

$$BW = \frac{f_T}{\text{gain}} = \frac{10^6}{19} = 52.63\text{ kHz}$$

Bandwidth OA 2

$$\text{BW} = \frac{f_T}{\text{gain}} = \frac{10^6}{21} = 47.62\text{ kHz}$$

The limit is OA 2.

PROBLEM 14.16 GAIN BANDWIDTH

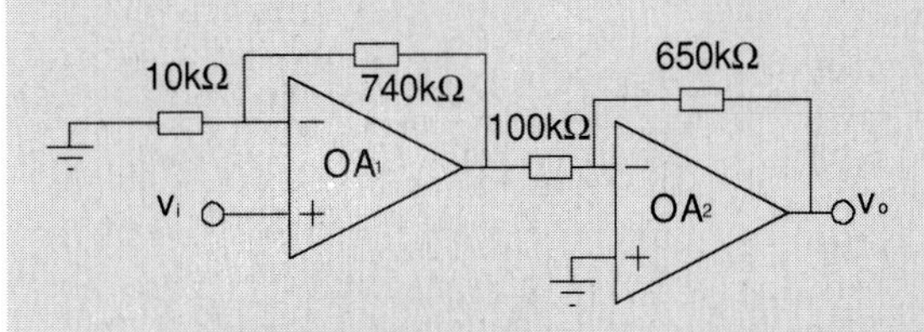

Figure 14.41

The transition frequency for both amplifiers is 750 kHz. Find the upper corner frequency of the system.

Answer: 10 kHz

Gain OA1

$$1 + \frac{740}{10} = 75$$

Gain OA 2

$$-\frac{650\text{k}}{100\text{k}} = -6.5$$

$f_T = \text{gain} \times \text{BW}$

$$\text{BW} = \frac{f_T}{\text{gain}}$$

$$\text{BW}_1 = \frac{750\,000}{75} = 10\text{ kHz}$$

$$\text{BW}_2 = \frac{750\,000}{6.5} = 115.38\text{ kHz}$$

The system is limited by the first Op-Amp. The upper corner frequency of the system is 10 kHz.

PROBLEM 14.17 GAIN BANDWIDTH

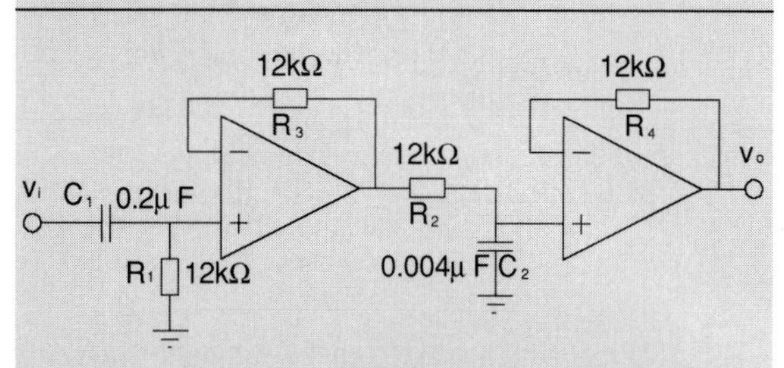

Figure 14.42

Find the bandwidth of this filter.

Answer: 3244 Hz

High pass

$$f_1 = \frac{1}{2\pi R_1 C_1} = \frac{1}{2\pi \times 12 \times 10^3 \times 0.2 \times 10^{-6}}$$

$$= 66.31 \text{ Hz}$$

$$f_2 = \frac{1}{2\pi R_2 C_2} = \frac{1}{2\pi \times 12 \times 10^3 \times 0.004 \times 10^{-6}}$$

$$= 3.31 \text{ kHz}$$

$$\text{Bandwidth} = 3310 - 66$$

$$= 3244 \text{ Hz}$$

PROBLEM 14.18 GAIN BANDWIDTH

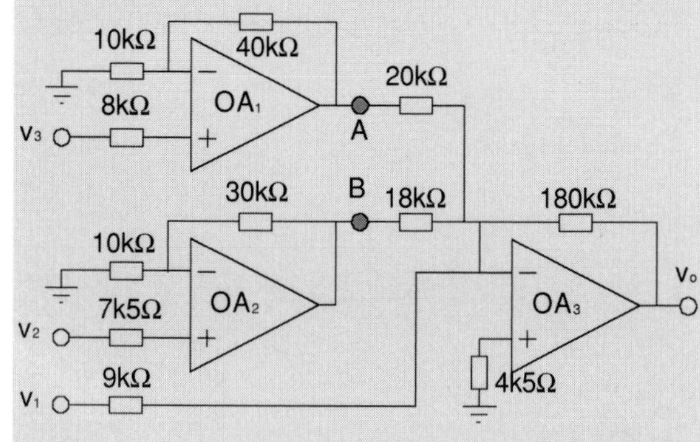

Figure 14.43

Find v_o in terms of v_1, v_2 and v_3. The transition frequency is 1 MHz.

Will this amplifier work at 55 kHz?

Answer: No

OA 2

$$v_B = v_2\left(1 + \frac{30\text{k}}{10\text{k}}\right) = 4v_2$$

OA 3

$$v_A = v_3\left(1 + \frac{40\text{k}}{10\text{k}}\right) = 5v_3$$

OA 1

$$v_o = -9v_A - 10v_B - 20v_1$$

$$v_o = -20v_1 - 9 \times 5v_3 - 10 \times 4v_2$$

$$= -20v_1 - 40v_2 - 45v_3$$

$$\text{gain} = \frac{f_T}{\text{BW}} = \frac{1\,000\,000}{55\,000} = 18.18$$

The amplifier will not work at 55 kHz. OA2 and OA3 are alright, but OA1 has the signal v_1 with a gain of 20 which is more than 18.18. This amplifier will work only up to 50 kHz.

Index